Artificial Intelligence Applications for Sustainable Construction

T0229086

Woodhead Publishing Series in Civil and Structural Engineering

Artificial Intelligence Applications for Sustainable Construction

Edited by

Moncef L. Nehdi
Department of Civil Engineering, McMaster University, Hamilton, ON, Canada

Harish Chandra Arora
Structural Engineering Department, CSIR—Central Building Research Institute, Roorkee, Uttarakhand, India

Krishna Kumar
Research and Development Unit, UJVN Ltd., Dehradun, Uttarakhand, India

Robertas Damaševičius
Department of Applied Informatics, Vytautas Magnus University, Kaunas, Lithuania

Aman Kumar
Structural Engineering Department, CSIR—Central Building Research Institute, Roorkee, Uttarakhand, India

ELSEVIER

WP
WOODHEAD PUBLISHING
An imprint of Elsevier

Woodhead Publishing is an imprint of Elsevier
50 Hampshire Street, 5th Floor, Cambridge, MA 02139, United States
125 London Wall, London EC2Y 5AS, United Kingdom

Copyright © 2024 Elsevier Ltd. All rights are reserved, including those for text and data mining, AI training, and similar technologies.

No part of this publication may be reproduced or transmitted in any form or by any means, electronic or mechanical, including photocopying, recording, or any information storage and retrieval system, without permission in writing from the publisher. Details on how to seek permission, further information about the Publisher's permissions policies and our arrangements with organizations such as the Copyright Clearance Center and the Copyright Licensing Agency, can be found at our website: www.elsevier.com/permissions.

This book and the individual contributions contained in it are protected under copyright by the Publisher (other than as may be noted herein).

Notices
Knowledge and best practice in this field are constantly changing. As new research and experience broaden our understanding, changes in research methods, professional practices, or medical treatment may become necessary.

Practitioners and researchers must always rely on their own experience and knowledge in evaluating and using any information, methods, compounds, or experiments described herein. In using such information or methods they should be mindful of their own safety and the safety of others, including parties for whom they have a professional responsibility.

To the fullest extent of the law, neither the Publisher nor the authors, contributors, or editors, assume any liability for any injury and/or damage to persons or property as a matter of products liability, negligence or otherwise, or from any use or operation of any methods, products, instructions, or ideas contained in the material herein.

ISBN: 978-0-443-13191-2

For information on all Woodhead Publishing publications visit our website at https://www.elsevier.com/books-and-journals

Publisher: Matthew Deans
Acquisitions Editor: Chiara Giglio
Editorial Project Manager: Teddy A. Lewis
Production Project Manager: Surya Narayanan Jayachandran
Cover Designer: Miles Hitchen

Working together to grow libraries in developing countries
www.elsevier.com • www.bookaid.org

Typeset by TNQ Technologies

Contents

List of contributors

Masoud Ahmadi Department of Civil and Geomechanics Engineering, Arak University of Technology, Arak, Iran

Mohammed Al Sageer College of Engineering, Qatar University, Doha, Qatar

Anas Alsharo College of Engineering, Qatar University, Doha, Qatar

Harish Chandra Arora Academy of Scientific and Innovative Research (AcSIR), Ghaziabad, Uttar Pradesh, India; Department of Structural Engineering, CSIR—Central Building Research Institute, Roorkee, Uttarakhand, India

Bahareh Behkamal Department of Civil and Environmental Engineering, Politecnico di Milano, Milan, Italy

Hussnain Bilal Cheema School of Civil and Environmental Engineering (SCEE), National University of Sciences and Technology (NUST), Islamabad, Pakistan

Jiayao Chen Key Laboratory for Urban Underground Engineering of the Education Ministry, Beijing Jiaotong University, Beijing, China

Rajat Dabral Department of Civil Engineering, Sardar Vallabhbhai National Institute of Technology (SV-NIT), Surat, Gujarat, India

Ulrike Dackermann School of Civil and Environmental Engineering, University of New South Wales, Sydney, NSW, Australia

Alireza Entezami Department of Civil and Environmental Engineering, Politecnico di Milano, Milan, Italy

Parag Gohel Department of Civil Engineering, Sardar Vallabhbhai National Institute of Technology (SV-NIT), Surat, Gujarat, India

Samer Gowid College of Engineering, Qatar University, Doha, Qatar

Ivanka Netinger Grubeša University North, Varazdin, Croatia

Marijana Hadzima-Nyarko Faculty of Civil Engineering and Architecture Osijek, Josip Juraj Strossmayer University of Osijek, Osijek, Croatia

Sahar Hassani Centre for Infrastructure Engineering and Safety, School of Civil and Environmental Engineering, University of New South Wales, Sydney, NSW, Australia

M. Helen Santhi School of Civil Engineering, Vellore Institute of Technology, Chennai, Tamil Nadu, India

Hashem Jahangir Department of Civil Engineering, University of Birjand, Birjand, Iran

Gopal Lal Jat Department of Electrical Engineering, PEC (Deemed to be University), Chandigarh, India

Nishant Raj Kapoor Department of Civil Engineering, COER University, Roorkee, Uttarakhand, India

Ayoub Keshmiry Faculty of Civil Engineering, Shahrood University of Technology, Shahrood, Iran

Sikandar Ali Khokhar School of Civil and Environmental Engineering (SCEE), National University of Sciences and Technology (NUST), Islamabad, Pakistan; Bendcrete Construction Services (Pvt) Ltd., National Science and Technology Park (NSTP), Islamabad, Pakistan

Denise-Penelope N. Kontoni Department of Civil Engineering, School of Engineering, University of the Peloponnese, Patras, Greece; School of Science and Technology, Hellenic Open University, Patras, Greece

Miljan Kovačević University of Pristina, Faculty of Technical Sciences, Mitrovica, Serbia

Prashant Kumar Academy of Scientific and Innovative Research (AcSIR), Ghaziabad, Uttar Pradesh, India; Department of Structural Engineering, CSIR—Central Building Research Institute, Roorkee, Uttarakhand, India; Department of Civil Engineering, COER University, Roorkee, Uttarakhand, India

Ashok Kumar Academy of Scientific and Innovative Research (AcSIR), Ghaziabad, Uttar Pradesh, India

Pawan Kumar Department of Computer Science and Engineering, Guru Jambheshwar University of Science and Technology, Hisar, Haryana, India

Aman Kumar Academy of Scientific and Innovative Research (AcSIR), Ghaziabad, Uttar Pradesh, India; Department of Structural Engineering, CSIR—Central Building Research Institute, Roorkee, Uttarakhand, India

Anuj Kumar Academy of Scientific and Innovative Research (AcSIR), Ghaziabad, Uttar Pradesh, India

Ankush Kumar Department of Public Health Engineering, (GOH), Hisar, Haryana, India

V.H. Lad Department of Civil Engineering, Nirma University, Ahmedabad, Gujarat, India

Madhu School of Chemistry, University of Hyderabad, Hyderabad, Telangana, India

G. Malathi School of Computer Science and Engineering, Vellore Institute of Technology, Chennai, Tamil Nadu, India

Amr Mohamed College of Engineering, Qatar University, Doha, Qatar

Navdeep Mor Department of Civil Engineering, Guru Jambheshwar University of Science and Technology, Hisar, Haryana, India

Khalid Kamal Naji College of Engineering, Qatar University, Doha, Qatar

Mohaddeseh Nikpay Department of Civil Engineering, University of Birjand, Birjand, Iran

Emmanuel Karlo Nyarko Faculty of Electrical Engineering, Computer Science and Information Technology, Josip Juraj Strossmayer University of Osijek, Osijek, Croatia

K.A. Patel Department of Civil Engineering, Sardar Vallabhbhai National Institute of Technology (SV-NIT), Surat, Gujarat, India

D.A. Patel Department of Civil Engineering, Sardar Vallabhbhai National Institute of Technology (SV-NIT), Surat, Gujarat, India

Fazal Rehman School of Civil and Environmental Engineering (SCEE), National University of Sciences and Technology (NUST), Islamabad, Pakistan

Danial Rezazadeh Eidgahee Department of Civil Engineering, Ferdowsi University of Mashhad, Mashhad, Iran

Hassan Sarmadi Head of Research and Development, IPESFP Company, Mashhad, Iran; Department of Civil Engineering, Faculty of Engineering, Ferdowsi University of Mashhad, Mashhad, Iran

Mahdi Shadabfar Center for Infrastructure Sustainability and Resilience Research, Department of Civil Engineering, Sharif University of Technology, Tehran, Iran

Mati Ullah Shah School of Civil and Environmental Engineering (SCEE), National University of Sciences and Technology (NUST), Islamabad, Pakistan

Atefeh Soleymani Structural Engineering, Shahid Bahonar University of Kerman, Kerman, Iran

Muhammad Usman School of Civil and Environmental Engineering (SCEE), National University of Sciences and Technology (NUST), Islamabad, Pakistan

V. Vasugi School of Civil Engineering, Vellore Institute of Technology, Chennai, Tamil Nadu, India

About the editors

Prof. Moncef L. Nehdi

Dr. Moncef L. Nehdi is a Professor and Chair of the Department of Civil Engineering at McMaster University, Canada, Emeritus Professor at Western University, Canada, and Visiting Professor at Henan Polytechnic University, China. He is the Chair of the Canadian Society for Civil Engineering (CSCE)—Materials and Mechanics Division; co-Chair of the RILEM Concrete Data Science Committee; and served as Chair of the CSCE Cement and Concrete Committee; Chair of the American Concrete Institute Recycled Materials Committee; and Co-Chair of the Natural Science and Engineering Research Council of Canada (NSERC) Discovery Grant Committee. He also served as an Associate Director for Environmental Research Western and as a Technical Manager for three international companies.

Prof. Nehdi is a Fellow of the Canadian Academy of Engineering, Engineering Institute of Canada, American Concrete Institute (ACI), Canadian Society for Civil Engineering (CSCE), and Asia-Pacific Artificial Intelligence Association. He has received numerous prestigious awards including the ACI 2023 Sustainability Award, Engineering Medal for Research and Development from the Ontario Professional Engineers; Ontario Premier's Research Excellence Award; ACI Award for Professional Achievement; CSCE Horst Leipholz Medal; CSCE Whitman Wright Award; Bill Curtin Medal from the United Kingdom Institution of Civil Engineers; Faculty Fellow Award for Excellence in Engineering Education from the American Society of Engineering Education; and Engineering Prize for Excellence in Teaching from Western University, along with several best paper awards and other recognitions. He has been invited as a distinguished keynote speaker in numerous international forums and conferences.

Dr. Nehdi's scholarly and scientific research has been used in high-profile construction projects in Canada and abroad, including some world landmark structures. He has been active in several technical committees of ACI, CSCE, and RILEM, and a member of the editorial boards of several leading technical journals. He is also Canada's leader in civil engineering professional development courses offered to practicing engineers. A prolific author with more than 470 peer-reviewed publications, he was listed among the world's most impactful civil engineers by Elsevier and the Shanghai Global Ranking of Academic Subjects and in the Stanford University ranking of the world's top 1% scientists. Dr. Nehdi's research focuses on sustainable, eco-efficient, and

NetZero construction, resilience of civil infrastructure, artificial intelligence and machine learning in civil engineering, and innovative materials for construction and energy storage.

Dr. Harish Chandra Arora

Dr. Harish Chandra Arora currently holds the esteemed position of Principal Scientist in the Structural Engineering Group at CSIR-Central Building Research Institute in Roorkee, India. With a distinguished career spanning more than 29 years, Dr. Arora is a renowned figure in the field of structural engineering. Dr. Arora is also functioning as an Associate Professor in the Academy of Scientific and Innovative Research (AcSIR), Ghaziabad, India. His contemporary research areas include structural composites, structural corrosion, distress diagnosis, seismic evaluation, repair and retrofitting of structures and machine learning applications in structural engineering, etc. Dr. Arora's exceptional contributions to the field have garnered recognition in both national and international academic journals. Beyond his scholarly achievements, he has made a significant impact on the education and development of future engineers, having supervised and guided over 100^{+} students in their pursuit of bachelor of technology and master of technology degrees. Additionally, he continues to mentor and support research scholars pursuing doctoral programs at the Central Building Research Institute in Roorkee, India.

Furthermore, Dr. Arora actively contributes to the scholarly community as a reviewer for journals published by Springer Nature and Elsevier. His commitment to maintain the quality and rigor of academic publications is highly regarded. Beyond his academic pursuits, Dr. Arora has undertaken numerous consultancy and research and development projects within the field of structural engineering, further showing his dedication to advancing the science and practice of sustainable construction.

Dr. Krishna Kumar

Dr. Krishna Kumar received his BE degree in Electronics and Communication Engineering from Govind Ballabh Pant Engineering College, Pauri Garhwal, Uttarakhand, India, MTech degree in Digital Systems from Motilal Nehru NIT, Allahabad, India, in 2006 and 2012, respectively, and PhD degree in the Department of Hydro and Renewable Energy at the Indian Institute of Technology Roorkee, India, in 2023. He is currently working as an Assistant Engineer at UJVN Ltd. (a State Government PSU of Uttarakhand) since January 2013. Before joining UJVNL, he worked as an Assistant Professor at BTKIT, Dwarahat (a Government of Uttarakhand Institution). He has published numerous research papers in international journals and conferences, including IEEE, Elsevier, Springer, MDPI, Hindawi, and Wiley. He has also edited and written books for Taylor & Francis, Elsevier, Springer, River Press, and Wiley. His current research interests include IoT, AI, and renewable energy.

Prof. Robertas Damaševičius

Prof. Robertas Damaševičius received his PhD degree in Informatics Engineering from the Kaunas University of Technology, Lithuania, in 2005. He is currently a Professor in the Department of Applied Informatics, Vytautas Magnus University, Lithuania, and the Department of Software Engineering, Kaunas University of Technology, Lithuania, as well as an Adjunct Professor at the Faculty of Applied Mathematics, Silesian University of Technology (Poland). He lectures courses on human–computer interaction design, robot programming, and software maintenance. He is the author of more than 500 peer-reviewed articles and a monograph published by Springer. His research interests include assisted living, medical imaging, and medical diagnostics using explainable artificial intelligence and robotics. He is also the Editor-in-Chief of *Information Technology and Control* journal. He has been a Guest Editor of several invited issues of international journals, such as *BioMed Research International*, *Computational Intelligence and Neuroscience*, the *Journal of Healthcare Engineering*, *IEEE Access*, *IEEE Sensors*, and *Electronics*.

Mr. Aman Kumar

Er. Aman Kumar hails from Bilaspur, Himachal Pradesh, India. He holds a Master of Engineering degree in Construction Technology and Management from the prestigious National Institute of Technical Teacher's Training and Research Institute in Chandigarh, India.

Currently, he is fervently pursuing his PhD in Engineering Sciences, specializing in structural engineering, at the renowned CSIR-Central Building Research Institute in Roorkee, India. His academic journey is underscored by a deep passion for various facets of civil engineering, including sustainability development, nondestructive testing, concrete technology, and strengthening techniques such as fiber-reinforced polymer and fiber-reinforced cementitious matrix. He is also deeply engaged in exploring corrosion protection techniques for structural design, as well as the cutting-edge domains of artificial intelligence and the Internet of Things. Now his focus is to solve complex structural engineering problems with machine learning algorithms.

Aman Kumar's dedication to the field is evident through his extensive research endeavors and comprehensive technical surveys. His scholarly pursuits have culminated in numerous research papers and book chapters, which have been featured in esteemed international scientific publications.

Foreword

In the ever-evolving landscape of technology and innovation, one field stands out as a beacon of transformative potential—civil engineering. With population growth, urbanization, and challenges imposed by climate change, the demands placed upon our civil infrastructures are becoming increasingly complex. It is within this framework that the fusion of human ingenuity and artificial intelligence finds its purpose. In the pages that follow, you will embark on a captivating journey through the integration of artificial intelligence into the world of civil engineering.

Artificial Intelligence Applications for Sustainable Construction is a testament to the unyielding pursuit of knowledge and the relentless drive to push the boundaries of what is possible. Each chapter within this comprehensive volume sheds light on the various facets of artificial intelligence's profound impact on the field. Together, they form a symphony of innovation, demonstrating the symbiotic relationship between human creativity and the transformative power of artificial intelligence.

As the readers journey through the subsequent chapters, they will encounter a diverse range of applications, from construction management to masonry heritage, and from predicting structural strength to assessing seismic damage. At every turn, artificial intelligence proves its predictive prowess, adaptability, and unparalleled problem-solving abilities.

In this era of rapid technological advancement, the integration of artificial intelligence into civil engineering is not just an option; it is a necessity. *Artificial Intelligence Applications for Sustainable Construction* offers a comprehensive view of the future, where the synergy of human insight and the computational ability of artificial intelligence shapes the world we live in. This book is evidence of our boundless potential and also a reminder that together we can build a better, safer, and more sustainable world.

The editors have managed to compile this comprehensive book, capturing the state of the art of diverse applications of artificial intelligence in the various domains of civil engineering such as concrete technology, structural engineering, and repair and rehabilitation of concrete infrastructures. This volume is a must-have for all engineers, practitioners, researchers, civil infrastructure asset managers, students, and other stakeholders dealing with the design and maintenance of concrete structures like buildings, roads, bridges, tunnels, dams, and more civil engineering constructions and repairs in general.

Prof. Pradeep Kumar Ramancharla
Director
CSIR—Central Building Research Institute
Roorkee, India

Preface

Artificial intelligence endeavors to develop computational tools that mimic human cognitive processes. It has in recent years found a wide range of applications in a variety of small-scale and large-scale civil engineering problems, including design optimization, parameter estimation and identification, and damage detection. The considerable advances in this domain indicate that artificial intelligence will be increasingly deployed in a wider spectrum of civil engineering applications over the next years.

Concrete is the most widely used material on the planet, second only to water. As its primary binder, cement plays an important role in concrete. With an average release of 630 kg of CO_2 for every tonne of cement produced, the cement sector is responsible for about 8% of global CO_2 emissions. To meet the United Nations sustainability goals, new classes of sustainable construction materials are required for both new constructions and the repair and rehabilitation of deteriorating concrete structures such as buildings, bridges, highways, dams, industrial facilities, chimneys, water tanks, nuclear reactors, etc.

Civil engineering challenges can be highly complex and dependent on a multitude of factors. Therefore, artificial intelligence and machine-learning approaches are emerging as strong contenders to create more accurate and dependable solutions to intricate problems and to improve the performance and sustainability of construction materials and the performance of civil infrastructure at large. Artificial intelligence is poised to transform the building environment, related practices, and engineering industries.

This volume aims to provide a practical introduction to the application of artificial intelligence and machine learning in the design, modeling, characterization, optimization, forecast, performance prediction, and development of sustainable construction materials and construction solutions in the domain of civil infrastructure.

In the area of civil engineering, the pursuit of knowledge and innovation has led to the exploration of artificial intelligence and machine learning as potent tools that promise to revolutionize the industry. In this book, each chapter serves as a beacon illuminating various facets of artificial intelligence's impact in the field.

Chapter 1 acts as an overture, setting the stage for an immersive journey into the applications of artificial intelligence in civil engineering, including the history of artificial intelligence. It beckons the reader to embark on a transformative exploration. In Chapter 2, the narrative takes an intriguing turn, delving into the integration of artificial intelligence into sustainable construction practices. It unveils the role of artificial intelligence as a secret eye, silently observing and enhancing the latest techniques in civil engineering. Chapter 3 presents a marvel—the infusion of machine learning into sustainable composite building materials. Here, the reduction of carbon emissions becomes a tangible possibility, exemplifying the potential of artificial intelligence for environmental stewardship. In Chapter 4, the story pivots to a practical application: the prediction of concrete compressive strength using machine-learning models. The utilization of glass waste powder showcases the ability of artificial intelligence to optimize resource utilization. Chapter 5 introduces artificial intelligence-based structural health monitoring systems, where technology acts as a sentinel safeguarding the integrity of civil structures.

In Chapter 6, the spotlight shines on the role of ensemble learning in rock mass rating for tunnel construction, highlighting the collaborative power of artificial intelligence in solving complex engineering challenges. Chapter 7 presents a case study, illustrating the use of artificial intelligence framework for Construction 4.0. Structural health monitoring becomes a testament to the adaptability and versatility of artificial intelligence. Chapter 8 ventures into the realm of predictions, with artificial intelligence models determining the ultimate axial strain and peak axial stress of FRP-confined concrete. Chapter 9 delves into the world of heritage masonry, where artificial intelligence–driven automated kernel-based regression modeling predicts long-term dynamic responses under thermal effects. The past meets the future through artificial intelligence. Chapter 10 offers a comprehensive review of the application of artificial intelligence in construction management, employing a science mapping approach. It unfolds a roadmap for the strategic utilization of artificial intelligence in managing complex construction projects.

Chapter 11 showcases the proficiency of artificial intelligence in calibrating textile-reinforced mortar-masonry bond strength, leveraging machine-learning methods to enhance structural understanding. In Chapter 12, artificial intelligence algorithms forecast the compressive strength of FRCM-strengthened RC columns. The predictive prowess of artificial intelligence remains unparalleled. Chapter 13 assesses the shear capacity of FRP-reinforced concrete beams without stirrups using machine learning. Chapter 14 utilizes soft computing techniques to estimate the load-carrying capacity of reinforced concrete beam-column joints. Finally, in Chapter 15, the capabilities of artificial intelligence are harnessed to assess global seismic damage in RC-framed buildings. Machine-learning techniques illuminate the path toward safer structures.

Each chapter in this comprehensive volume is a testament to artificial intelligence's transformative influence in the realm of civil engineering, offering an immersive view into the future of the field. Together, they form a symphony of innovation, demonstrating the symbiotic relationship between human ingenuity and artificial intelligence in shaping the future of civil engineering.

Acknowledgments

The writing of this book on the *Artificial Intelligence Applications for Sustainable Construction* is the result of a team effort; therefore, our first acknowledgments are addressed to the four teams at the Department of Civil Engineering of McMaster University, Canada; the Department of Structural Engineering, CSIR-Central Building Research Institute (CBRI), India; Uttarakhand Jal Vidyut Nigam (UJVN) Ltd., India; and the Department of Applied Informatics, Vytautas Magnus University (VMU), Lithuania, who played a pivotal role in producing this book.

Without such synergy and collaborative spirit, this work would not have been possible, since none of us would have been able to compile such a comprehensive and multidisciplinary work that has roots in different scientific fields including structural engineering, materials science, civil engineering, and machine learning, along with real-world field applications.

In a fast-paced world where professionals tend to have fully booked schedules, our collaborators strived to find long hours to write a text that is easy to read, yet scientifically precise, and intellectually stimulating. For all of us, it has been a long journey, quite often arduous, but never painful, because our goal was to provide the best book we could.

We are most grateful to Chiara Giglio (*Acquisitions Editor*) and Theodore Lewis (*Editorial Project Manager*) at Elsevier for overseeing this project and guiding us to completion in a timely manner.

Our sincere thanks go to many colleagues, students, friends, and family, who have provided inspiration, suggestions, ideas, and support.

McMaster, CSIR-CBRI, UJVN, VMU
October 2023

Artificial intelligence in civil engineering: An immersive view

1

Nishant Raj Kapoor [1], *Ashok Kumar* [2], *Anuj Kumar* [2], *Aman Kumar* [2,3] *and Harish Chandra Arora* [2,3]
[1]Department of Civil Engineering, COER University, Roorkee, Uttarakhand, India; [2]Academy of Scientific and Innovative Research (AcSIR), Ghaziabad, Uttar Pradesh, India; [3]Department of Structural Engineering, CSIR—Central Building Research Institute, Roorkee, Uttarakhand, India

It is better to learn & implement AI in Civil Engineering imperfectly than to oppose it with perfection

Authors.

1.1 Introduction

The world is sprouting technology faster than the earth is rotating on its axis. Before one full rotation, researchers throughout the globe are developing several new methods, techniques, programs, codes, innovations, and whatnot. It can be said that the epicenter and hotspot of today's research is undoubtedly artificial intelligence (AI). Change is the universal law as stated by the ancients in the marvelous evergreen Indian literature "*The Geeta*" and is now known by all. Currently, AI has the capability to alter each and every sector positively with its enormous adaptability. During the past half century, AI research has ridden the bull and has seen several ups and downs, and at certain times it also falls down for some time [1]. The "*bull*" that is used as a synonym for research speed is now converted to "*horse*". The horse is now trained enough to win several races after overcoming all the bumps and hitches [2]. Researchers, engineers, and other stakeholders are nudging their research with AI. AI is emerging with a lot of evolution in various machine learning (ML) and deep learning (DL) methodologies. Practicing engineers and researchers leverage these strategies to solve a wide range of hitherto complex problems [3]. This chapter presents a global overview of the rapidly evolving engineering industry and associated works, outlining both theoretical and practical applications of AI techniques across all areas of civil engineering.

The civil engineering and construction sector is at a turning point. Sophisticated technological advancements have the ability to fundamentally alter conventional engineering with sustainability. To progress toward sustainable civil engineering (SCE), a proper method for managing challenges relating to the economy, environment, and society is required. Many engineers, academics, and stakeholders have attempted to

Artificial Intelligence Applications for Sustainable Construction. https://doi.org/10.1016/B978-0-443-13191-2.00009-2
Copyright © 2024 Elsevier Ltd. All rights are reserved, including those for text and data mining, AI training, and similar technologies.

achieve Sustainable Development Goals (SDGs) by developing AI-based efficient systems, which leads to SCE [4]. AI technology and the Internet of Things (IoT) are among the pioneers that power the new methods in SCE [5]. The "trio" of institutions, industry, and the government is now promoting AI in civil engineering due to its wide applications and increasing dominance as an advanced computational tool. In the real world, applied AI is helping to advance civil and construction engineering and all its subareas. As a core engineering domain, civil engineering contains enormous computational problems that need precise solutions, and this can be achieved by good computational systems. AI contains unprecedented computational power, and by harnessing the highest degree of humanlike cognitive ability, AI has helped civil engineers and other stakeholders tackle a wide range of problems.

AI is the capacity of a machine to mimic human reasoning and has the ability to incorporate significant heterogeneous technological solutions to integrated multidisciplinary problems. After 2000, AI gained momentum and began to be used in real-time scenarios in industries such as medicine, data mining, engineering, logistics, and so on [2]. With increased computational capabilities, AI may produce exciting findings when combined with subjects such as mathematics, economics, and statistics. AI is also expanding its scope and entering new industries. AI research is rapidly growing in the "responsible AI" domain and the "generative AI" domain [6,7]. The use of AI in civil engineering was first mentioned in the late 1990s and early 2000s [8]. Applications for AI in civil engineering are numerous and include subdomain namely, structural engineering, transportation engineering, nondestructive testing (NDT), earthquake engineering, concrete technology, material engineering, structure health monitoring (SHM), construction technology, quality management, optimizations in design, maintenance scheduling, geotechnical engineering, risk control and safety management, stakeholder management, contract management, architectural engineering, water resource engineering, remote sensing and surveying, environmental engineering, coastal engineering, renewable power engineering, net zero energy buildings, optimizations of renewable sources of energy using emerging computational techniques, advanced civil engineering technologies namely; virtual reality (VR), building information modeling (BIM), 4D computer aided design (4D CAD), 3D-printed concrete, building twins, carbon sequestering structures and materials, IoT in civil engineering, advanced materials, green buildings, space structures, remote construction monitoring, construction robotics, drones in civil engineering, indoor environmental quality engineering, public health and safety and many more. Before diving into the detailed applications of AI in civil engineering, a slight background of AI is presented in the next section.

1.2 Background of artificial intelligence

Warren McCulloch and Walter Pits developed a model of artificial neurons in 1943, which is today acknowledged as the earliest effort in AI. Canadian psychologist Donald Hebb proposed an idea in his 1949 book "The Organization of Behavior". He presented an updating algorithm that explains how connections between neurons strengthen based on their activity and interaction. This rule is well-known as "Hebbian" learning and is often summarized with the phrase "cells that fire together, wire together" [2]. The next year, in 1950, an English mathematician named Alan Turing published "Computing Machinery and Intelligence," in which he proposed a test named "Turing test". This test tests a machine's ability to exhibit intelligent behavior that is indistinguishable from that of a human. After a half-decade, Herbert A. Simon and Allen Newell built the "first AI software" called "Logic Theorist". It was designed to explore and prove mathematical theorems using symbolic logic. This program marked a significant step forward in demonstrating the potential of computers to perform tasks that were previously thought to require human intelligence. This software affirmed 38 of 52 mathematics theorems while also discovering new and more elegant proofs for a few of them.

In August 1955, a proposal was drafted to seek funding from the Rockefeller Foundation, and this document is the origin of the term "artificial intelligence," specifically referred to as the "2-month, 10-men study of AI." Fig. 1.1 presents those 10 men, the founding fathers of AI. The primary objective of this proposal was to launch an initiative known as the Dartmouth Workshop. During this workshop, participants with diverse expertise would collaborate intensively for a period of 2 months, combining

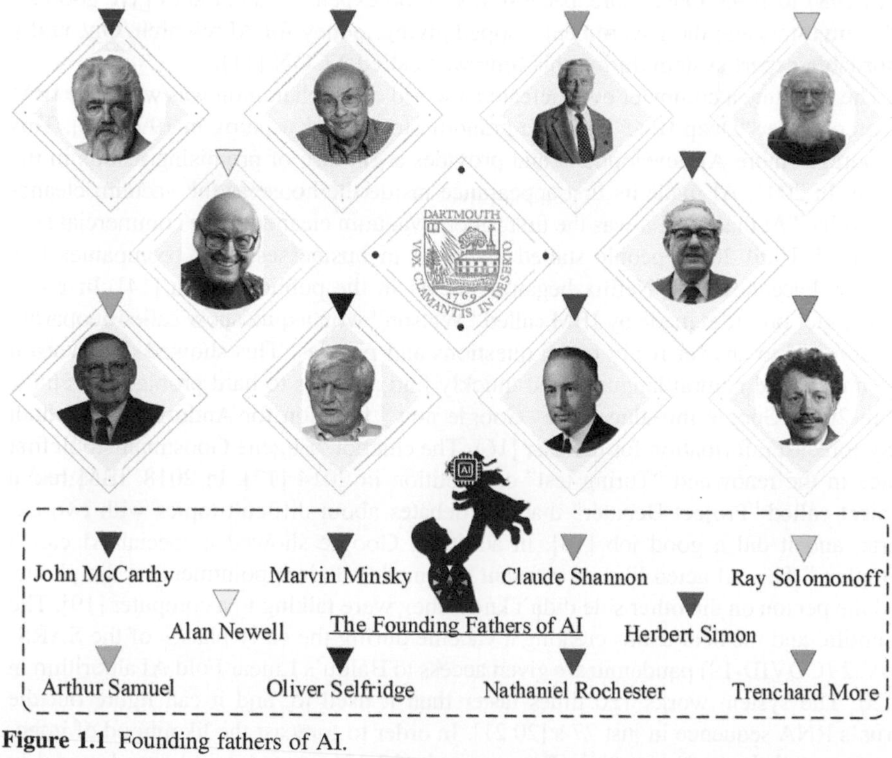

Figure 1.1 Founding fathers of AI.

their individual knowledge and skills to achieve tangible progress in the development of AI programs and algorithms. The original intention behind the proposal was for the participants to convene at Dartmouth for this 2-month duration, working collectively to drive significant advancements in the field of AI. The term "artificial intelligence" was conceived by John McCarthy, which he presented in a workshop at Dartmouth College in the United States during 1956 [9]. For the first time, AI was defined as an academic field in 1956. At the time, high-level computer languages such as LISP, FORTRAN, and COBOL were being developed, and there was a lot of interest in AI. During 1966, experts focused on building algorithms that could solve mathematical problems. In 1966, Joseph Weizenbaum built the first chatbot, titled "ELIZA". Three years later, in 1969, "Shakey" was the first general-purpose mobile robot built. Shakey may do things with a purpose rather than just a set of instructions. Following that, in 1972, Japan built the first intelligent humanoid robot, "WABOT-1" [10].

The first AI winter happened between 1974 and 1980. AI winter means a time when computer scientists had trouble getting enough money from the government to do AI research. People lost interest in AI during these times. However, after the 1980s, AI came back with something called "Expert System." These systems could make decisions like a human expert because they were programmed that way. In 1980, there was a big conference about AI at Stanford University, hosted by the American Association of Artificial Intelligence. After that, there was another AI winter that lasted from 1987 to 1993. Once more, because it was too expensive and didn't give good results, investors and the government stopped giving money for AI research. One really affordable expert system during this time was called XCON [11].

The first time a computer ever defeated a world chess champion was when the IBM supercomputer "Deep Blue" defeated grandmaster Gary Kasparov in 1997 [12]. This encourages more AI development and provides a glimpse of promising results in the future. In 2002, AI made its first appearance inside the house as the vacuum cleaner "Roomba." At that time, it was the first robotic vacuum cleaner to see commercial success [13]. Until 2006, people started using AI in businesses. Also, companies like Twitter, Facebook, and Netflix began using AI in the public domain [14]. In 2011, a computer program made by IBM called "Watson" won a quiz show called Jeopardy. Watson had to answer really tough questions and puzzles. This showed that Watson can understand natural language and quickly find answers to hard problems [15].

In 2012, Google introduced the "Google now" function for Android apps, which may forecast information for the user [16]. The chatbot "Eugene Goostman" took first place in the renowned "Turing test" competition in 2014 [17]. In 2018, IBM had a project called "Project Debater" that had debates about difficult topics with two experts, and it did a good job [18]. In addition, Google showed a special AI called "Duplex." This AI acted like an assistant and made a hair appointment on the phone, and the person on the other side didn't know they were talking to a computer [19]. The scientific and medical teams creating a vaccine during the early phases of the SARS-CoV-2 (COVID-19) pandemic are given access to Baidu's Linear Fold AI algorithm in 2020. The system works 120 times faster than it used to, and it can figure out the virus's RNA sequence in just 27 s [20,21]. In order to forecast the likelihood of infection in a real-time office setting, Kapoor et al. [22,23] created an AI-based model in

2022. A chatbot named "ChatGPT" was introduced by OpenAI in November 2022. GPT stands for "Generative pretrained Transformer" in the above term. The GPT-3 family of complex language models from "OpenAI" is the foundation upon which it is constructed, and it is tweaked using both reinforcement and supervised learning methods [24]. Presently, GPT-3.5 is in wide use for free, and GPT-4 is also used by a large user force. OpenAI filed for a trademark for GPT-5 on July 18, 2023. Users are excited about this, but many of them are not sure if GPT-5 is coming soon. Experts think that OpenAI might release something called GPT-4.5 in the last quarter of 2023, which would be like a halfway point between GPT-4 and GPT-5, similar to how they did with GPT-3.5. The AI development journey is showcased in Fig. 1.2.

It is clear from the comprehensive AI background that AI has progressed to a remarkable level at this juncture. The current funding landscape for continued AI research demonstrates relative stability. Nonetheless, the presence of certain technological constraints, in conjunction with exceedingly optimistic projections, raises the prospect of a potential recurrence of an AI winter. Presently, data science, big data, DL, and quantum computing [25] are experiencing unprecedented levels of prominence. In the contemporary landscape, enterprises such as Amazon, IBM, Facebook, and Google leverage AI to drive the development of remarkable technological advancements. Several construction-related businesses like Caidio, Nyfty.ai, Kwant.ai, AirWorks, BuildStream, Versatile, Built Robotics, Dusty Robotics, OpenSpace, Procore Technologies, etc. are nudging AI research in civil engineering. AI has a bright future and will be incredibly intelligent and helpful.

1.2.1 Types of AI

The process of creating intelligent devices and tools using AI is based on massive amounts of data. Systems execute humanlike tasks by learning from their prior knowledge and experiences. It improves the efficiency, effectiveness, and speed of human activities. To create robots that can make judgments on their own, AI employs sophisticated algorithms and techniques. AI is built on the foundation of ML and DL. AI can be segmented into two parts: The first is based on its capabilities, and the second is based on its functionalities. The subparts of the two types are presented in Fig. 1.3. The following subsections provide a brief description of each type and their subtypes.

1.2.1.1 Based on capabilities

There are three forms of AI on the basis of capabilities: Narrow AI, general AI, and super AI [26]. The brief information about these AI's is explained in the subsequent sections.

Narrow artificial intelligence

Weak AI, commonly known as narrow AI, is constrained to executing a singular, delimited task. It progresses within the scope of a specific cluster of cognitive functions. As ML and DL techniques advance, instances of narrow AI applications are becoming more pervasive in our daily routines. Apple's "Siri" serves as an exemplar

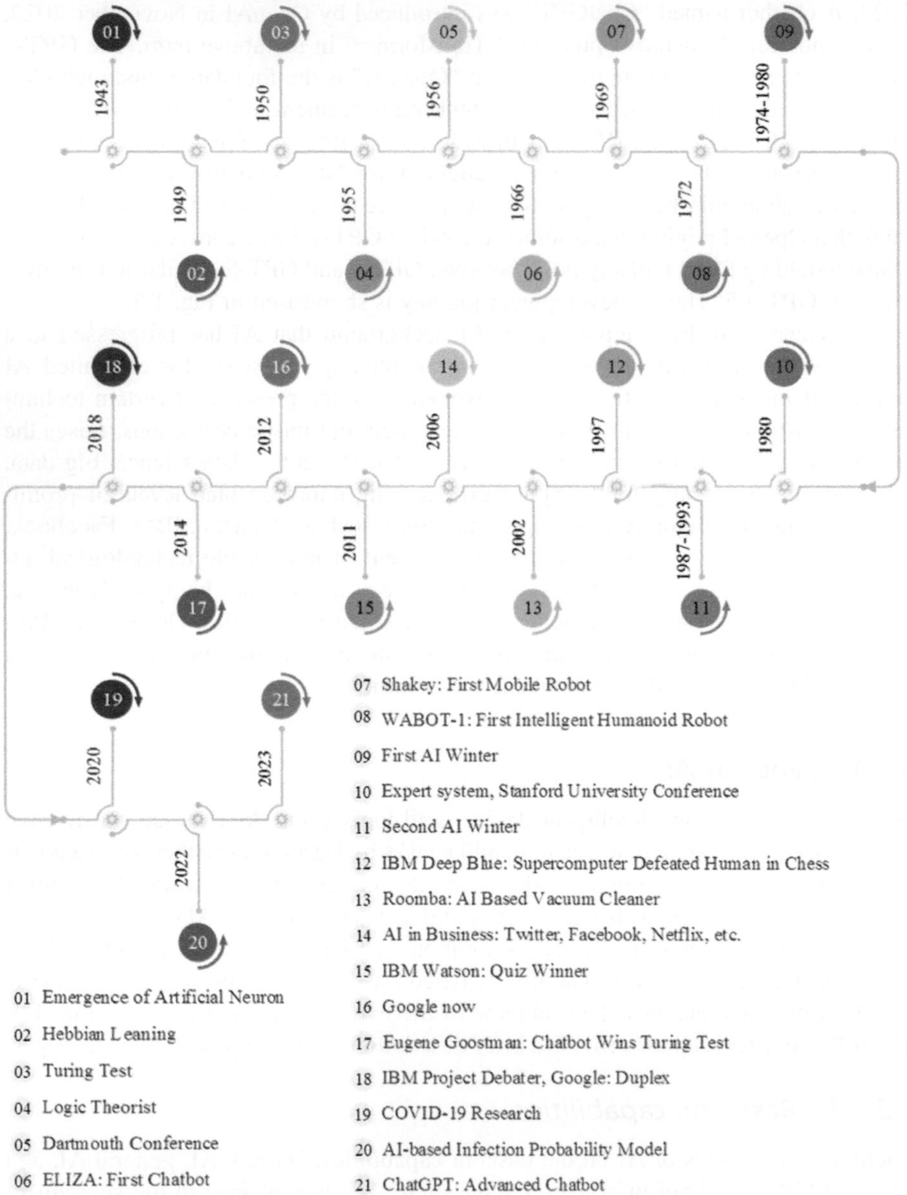

Figure 1.2 Development of AI.

of narrow AI, designed to handle a defined set of tasks [9]. However, Siri frequently encounters difficulties when confronted with tasks beyond its capabilities. Another illustration of narrow AI is the supercomputer IBM Watson, which processes data and offers responses through cognitive computing, ML, and natural language

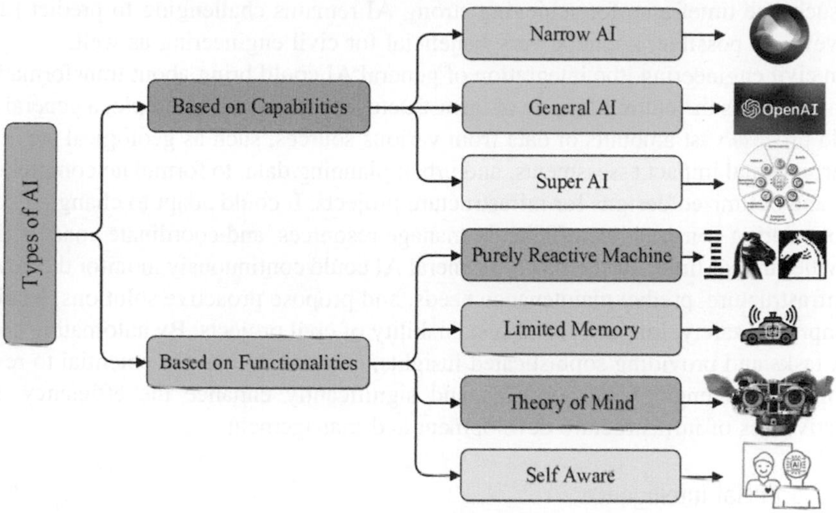

Figure 1.3 Types of AI.

processing. Notably, IBM Watson even outperformed human participant Ken Jennings to clinch victory in a prominent game show. In the area of civil engineering, narrow AI finds various applications characterized by their constrained scope.

In civil engineering, Narrow AI finds practical applications in various areas, such as structural analysis, construction planning, and project management. For instance, an AI-based system focused on structural analysis may be used to detect and assess cracks in buildings or bridges, providing engineers with valuable insights for maintenance and repairs. Another example is the use of narrow AI in construction scheduling, where AI algorithms can optimize resource allocation, minimizing delays and costs. Despite its limitations, narrow AI proves to be highly effective in these specialized tasks, augmenting human capabilities and significantly improving the efficiency and accuracy of civil engineering projects. Narrow AI has been extensively deployed in various industries, including civil engineering, but it remains distinct from general AI or super AI, which are more comprehensive and versatile forms of AI.

General artificial intelligence

Strong AI, commonly referred to as general AI, possesses the capacity to learn and understand any form of intellectual activity that humans are capable of. It empowers machines to apply knowledge and skills across diverse scenarios. The realization of strong AI remains a goal that has not yet been achieved by AI researchers. To accomplish this, they would need to formulate methodologies for imbuing computers with the complete spectrum of cognitive capabilities necessary for consciousness. Microsoft has allocated a substantial investment of $1 billion through OpenAI to advance general AI. Fujitsu's K computer, among the world's fastest supercomputers, stands as a notable endeavor in the quest for strong AI. However, creating even a single second of cognitive function on this system necessitated almost 40 min of processing time.

As such, the timeframe for achieving strong AI remains challenging to predict [26]. However, if possible, it can be very beneficial for civil engineering as well.

In civil engineering, the integration of general AI could bring about transformative changes across the entire lifecycle of infrastructure projects. For example, a general AI could process vast amounts of data from various sources, such as geological surveys, environmental impact assessments, and urban planning data, to formulate comprehensive and optimized designs for infrastructure projects. It could adapt to changing conditions during construction, efficiently manage resources, and coordinate construction activities in real-time. Furthermore, a general AI could continuously monitor the health of infrastructure, predict maintenance needs, and propose proactive solutions, leading to improved safety, longevity, and sustainability of civil projects. By automating complex tasks and providing sophisticated insights, General AI has the potential to revolutionize civil engineering practices and significantly enhance the efficiency and effectiveness of infrastructure development and management.

Super artificial intelligence

Super AI surpasses human intelligence and possesses the capability to outperform humans in any task. The concept of artificial superintelligence posits that AI has progressed to a stage where it closely resembles human emotions and encounters. This entails not only comprehending human emotions and experiences but also generating its own sentiments, desires, beliefs, and objectives. The existence of such AI remains a subject of ongoing debate. For super AI to materialize, it must demonstrate autonomous reasoning, problem-solving proficiency, sound judgment, and independent decision-making capabilities, along with a host of other fundamental attributes [27,28].

In civil engineering, the integration of super AI could lead to ground-breaking advancements in the design, construction, and management of projects. For instance, a super AI could optimize complex structural designs, considering multiple factors like material properties, cost-effectiveness, and environmental impact simultaneously. It could conduct advanced simulations, predicting the long-term performance of infrastructure under various conditions, leading to more resilient and efficient designs. Additionally, a super AI could oversee construction processes, manage robotic teams, and autonomously adapt to unexpected challenges, resulting in faster and safer project execution. Moreover, such AI could analyze massive amounts of data from various sources, aiding in urban planning and transportation management, optimizing traffic flow, and improving overall urban infrastructure. Nonetheless, it is crucial to exercise caution and ensure ethical oversight as we move closer to the potential integration of super AI in civil engineering to address concerns related to safety, privacy, and responsible decision-making. Researchers have not yet achieved super AI in reality, and these possibilities remain speculative.

1.2.1.2 Based on functionalities

Categorizing AI systems based on their functionalities is essential to delineating the diverse types of AI systems. Drawing from functional distinctions, experts have classified AI into four distinct categories: purely reactive machines, limited theory, theory of mind, and self-awareness [29].

Purely reactive machine

A prevalent form of AI that operates without storing memories or deriving judgments from past encounters is known as a reactive machine. Such machines solely operate with real-time data from their immediate environment, processing it and delivering corresponding responses. Reactive machines are designed for specific tasks, lacking capabilities beyond their designated functions. An example of a reactive machine is Deep Blue, IBM's chess-playing grandmaster-defeating machine. Deep Blue's functionality is limited to perceiving the current state of the chessboard and reacting to it. It cannot draw upon past experiences for improvement or learning. Deep Blue has the capacity to identify chess pieces on the board and comprehend their movement rules. It can predict the opponent's next moves by evaluating the current board configuration. However, it does not retain information from before the present moment; instead, it assesses the current arrangement of chess pieces to make informed decisions about potential subsequent moves.

An AI-based, purely reactive machine in civil engineering is a system that employs AI algorithms to respond to external stimuli without possessing autonomous decision-making capabilities. An example of this can be seen in AI-controlled traffic management systems. Utilizing data from sensors, cameras, and historical traffic patterns, the AI system reacts in real-time to changing traffic conditions by adjusting traffic signals and flow patterns to optimize traffic flow and reduce congestion. AI does not have proactive decision-making abilities but relies on its preprogrammed algorithms to react to the incoming data, making it a purely reactive system that enhances the efficiency of urban transportation networks.

Limited memory

Limited Memory AI functions by learning to make judgments through the analysis of historical data. These systems possess a transient form of memory that allows them to harness historical information for a finite period of time. However, they are restricted from integrating this data into an ongoing repository of their experiences. This technology is often employed in autonomous vehicles, such as self-driving cars. Limited memory AI, as seen in autonomous cars, maintains a record of the movement patterns of nearby vehicles both in the present moment and across a span of time. This dynamic dataset supplements the AI's static information, which includes details like lane markings and traffic signals. The accumulated real-time data is taken into consideration when the autonomous vehicle decides whether to execute actions such as lane changing, yielding to other drivers, or maneuvering to avoid collisions.

In civil engineering, an AI-based limited memory system refers to an AI system that retains only a limited amount of historical data to make decisions and predictions. An example of this can be found in predictive maintenance applications for infrastructure. Let's consider a bridge monitoring system that utilizes AI algorithms to assess the structural health of the bridge. Instead of storing and analyzing an extensive history of sensor data, the limited-memory AI system might retain data from the past few months. By focusing on recent trends and patterns, the system can effectively predict maintenance requirements, identify potential issues, and prioritize inspections and repairs without being overwhelmed by a vast amount of historical data. This approach allows for more efficient data processing and decision-making while maintaining a high level of accuracy in detecting and addressing potential structural concerns.

Theory of Mind

Theory of Mind AI represents an advanced category of technology that currently remains theoretical. This variant of AI necessitates a profound comprehension of how individuals and entities within an environment can influence emotions and actions. It is expected to possess the capacity to interpret the emotions, viewpoints, and cognitions of others. Despite notable strides in this domain, the development of such AI is an ongoing endeavor that demands further refinement. An illustration of Theory of Mind AI is found in applications like Kismet [30]. In the late 1990s, a researcher affiliated with the Massachusetts Institute of Technology developed Kismet, a robotic head. Kismet possesses the capability to emulate and discern human emotions. While the robot can track and acknowledge human gazes, as well as convey attention toward individuals, these functionalities constitute substantial strides within the area of theory of mind AI. An additional exemplification of theory of mind AI can be observed in Sophia, an innovation of Hanson Robotics. Sophia is equipped with cameras embedded in her eyes and operates through intricate computer algorithms. She is adept at visual perception, maintaining eye contact, recognizing individuals, and effectively tracking facial features.

AI-based theory of mind in civil engineering refers to the integration of AI systems with the ability to understand and model the intentions, beliefs, and mental states of human agents involved in construction projects or infrastructure development. The AI system with theory of mind can interpret and predict the behavior of project stakeholders, allowing for better communication, collaboration, and decision-making. For instance, consider a construction project where an AI-powered virtual assistant is used to interact with the project team and stakeholders. This AI system, equipped with theory of mind capabilities, can assess the intentions and concerns of different parties involved, anticipate their needs, and suggest optimal solutions that align with their individual perspectives. This enhances the project's efficiency, minimizes conflicts, and fosters a more productive working environment by creating an AI agent that can better empathize with and adapt to the unique requirements and expectations of various stakeholders throughout the project's lifecycle.

Self-aware

Self-aware AI, at present, remains a theoretical construct. These systems are envisioned to possess the capability to perceive human emotions and fathom their inner attributes, contexts, and emotional states. Anticipated to exceed human cognitive capabilities, they are envisioned to surpass the intellectual capacity of the human brain. In addition to recognizing and eliciting emotions in their interactions with humans, these AI systems are conceptualized to possess their own emotions, desires, and beliefs. Envisioned as the technologies of the future, self-aware machines would exhibit cognition, sentience, and intelligence. However, as of now, no AI system has achieved authentic self-awareness akin to human consciousness. True self-awareness encompasses awareness and comprehension of one's own existence, thoughts, and emotions—a level of sophistication currently beyond the area of AI's capabilities, which is currently beyond the capabilities of AI.

However, in the future, if AI were to achieve a level of advanced consciousness and self-awareness, it could revolutionize civil engineering. For example, imagine an AI that could autonomously monitor and maintain infrastructure, such as bridges or roads. This AI would not only detect structural issues but also be aware of its own limitations and capabilities. It could communicate effectively with human engineers, recognizing when certain tasks require human intervention and when it can handle them independently. Such self-aware AI could potentially contribute to more efficient and sustainable infrastructure management, leading to safer and longer-lasting civil projects.

1.2.2 AI implementation techniques

AI systems are constructed by amalgamating expansive, intelligent, iterative processing algorithms. The current state of AI enables it to discern features and patterns within assessed data, a capacity attributed to this amalgamation. AI's allure heightens when it transcends mere replication of human actions and ventures into areas surpassing human capacities. In this context, an AI system engages in self-monitoring and evaluation subsequent to each data processing iteration, leveraging the outcomes to enhance its proficiency. By leveraging AI implementation techniques, civil engineering can become more efficient, accurate, and informed, ultimately leading to improved project outcomes and resource management. Following are the most utilized ways (presented in Fig. 1.4) that one can implement AI at present [1].

1.2.2.1 Natural language processing

Natural language processing (NLP) is built into computers so they can interact with human language. NLP, which extracts meaning from human languages through ML, is a proven technique. In NLP, a machine records the audio of a person speaking. The process begins with converting dialog from audio to text, followed by text processing to render the data back into audio form. Subsequently, the system interacts with individuals through audio responses. NLP finds application in various domains, including interactive voice response systems deployed in customer support centers, language translation tools like Google Translate, and text editors such as Microsoft Word that validate syntactical accuracy. However, the intricacies of human language usage present a challenge for computers due to the nuanced rules inherent in natural language communication. NLP faces difficulties due to these linguistic intricacies. The complexity lies in enabling computers to grasp the rules governing natural language. NLP employs algorithms to discern and abstract these linguistic rules, thereby facilitating the translation of unstructured human language data into a format understandable by computers.

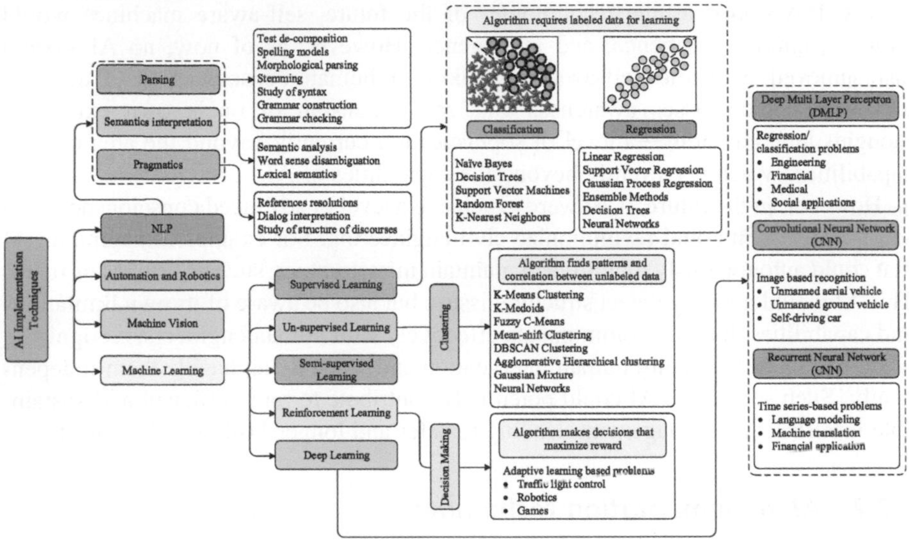

Figure 1.4 AI implementation techniques.

NLP is vital for smart building since it contributes directly and indirectly to the digitalization pillar. Because the majority of data is unstructured text, NLP directly and considerably enhances efficacy and efficiency when dealing with such data through text cleaning and information extraction. NLP also helps AI-enabled decision-making and BIM-based management in an indirect way. The decision-making pyramid states that decisions are produced from data in four steps: data, information, knowledge, and decisions, with only three of these covered by fundamental NLP capabilities. As a result, it is more advantageous that NLP enhances other functions by supplying meaningful data. For example, optimizing infrastructure O&M plans using damage severity collected from Twitter and augmenting BIM for effective information exchange. The more powerful NLP approaches are, the more unstructured data can be managed, more advanced AI techniques can be used, and more efficient BIM collaboration can be used, all of which make things more digitalized, automated, and linked. NLP has been used in numerous downstream applications, such as information extraction and exchange, to speed up management and decision-making. The role of NLP in smart construction for fully utilizing its benefits needs more exploration. It is hoped that the results of research in this direction will help both academic scholars and professional practitioners understand the research and application frontiers of NLP for smart building and civil engineering [31].

NLP can be utilized in project management, documentation, and communication as well. For instance, an NLP-powered chatbot can be integrated into project management software to assist stakeholders in obtaining real-time information on construction progress, upcoming milestones, or material requirements simply by asking questions in natural language. NLP algorithms can also be employed to analyze vast amounts of textual data from project reports, research papers, and regulations, extracting relevant insights for better decision-making and compliance adherence. Additionally, NLP can facilitate communication between project teams and stakeholders by automatically translating technical jargon into easily understandable language, promoting smoother collaborations and efficient information exchange. Overall, NLP plays a crucial role in streamlining communication, data analysis, and knowledge extraction in civil engineering, thereby enhancing project efficiency and facilitating informed decision-making.

1.2.2.2 Automation and robotics

The objective of automation is to delegate monotonous, repetitive tasks to machines, thereby enhancing productivity and yielding outcomes that are more efficient, effective, and cost-friendly. To achieve process automation, many enterprises employ technologies like ML and neural networks, among others. The integration of CAPTCHA technology can thwart fraudulent activities during online financial transactions, adding security to such automated processes. Robotic process automation is devised to execute high-volume, repetitive tasks while maintaining adaptability to accommodate changing conditions. In the construction sector, which is marked by inefficiencies and subpar productivity, robotics and automated systems hold the potential to mitigate these challenges. However, their adoption in this industry remains limited. Given the labor-intensive nature of construction, automation and robotic systems have demonstrated notable success in reducing labor expenses while concurrently boosting productivity and product quality across various domains. Furthermore, these systems can contribute to injury reduction by eliminating the need for workers to engage in hazardous tasks. The adoption of robotics within the construction area may be influenced by technological advancements such as the Industry 4.0 framework, BIM, sensor technologies, and AI. These innovations could potentially pave the way for increased implementation of robotics, transforming the construction industry's operational landscape [32].

Four broad categories may be used to classify the various automation and robotic technologies used in the construction industry: Drones and autonomous vehicles, on-site automated and robotic systems, off-site prefabrication systems, and exoskeletons (wearable mechanical devices that augment the capabilities of the user). Automation and robotics utilizing AI are still in the research phase, and it is expected that in the near future, AI-based automation and robotics will help unprecedentedly in civil engineering and allied domains [33]. Automation and robotics play a significant role in civil engineering, streamlining construction processes, improving efficiency, and enhancing safety. One prevalent example is the use of construction robots in bricklaying. Robotic bricklayers can accurately and quickly lay bricks, reducing human labor and the time required for construction. These robots follow precise programming, ensuring consistent brick placement and alignment, resulting in higher quality and more uniform structures. Moreover, automation is employed in heavy machinery like excavators and bulldozers, which can be controlled remotely or programmed to perform repetitive tasks with precision, reducing the need for manual labor and increasing productivity. Additionally, drones equipped with sensors and cameras are used for surveying large construction sites, providing real-time data on progress, and identifying potential issues promptly. By leveraging automation and robotics, civil engineering projects can be executed more efficiently, cost-effectively, and with improved safety, ultimately contributing to the development of sustainable and innovative infrastructure solutions.

1.2.2.3 Machine vision

In order to assess the status of civil infrastructure, information gathered through inspection and/or monitoring methods is used. Traditionally, skilled inspectors conduct visual inspections. However, such inspections can be difficult, costly, risky, and/or time-consuming. Improved inspection and monitoring techniques that involve less human participation, are more affordable, and have greater spatial resolution are needed to solve some of these issues. Machines possess the capacity to gather and analyze visual data. In this context, cameras are employed to capture visual information, which is subsequently subjected to digital signal processing after analog-to-digital conversion. The resulting data is then input into a computer for further processing. Two critical features of machine vision (MV) are sensitivity and resolution. Sensitivity denotes a machine's ability to detect faint signals, while resolution refers to its capability to distinguish objects within a given range. MV finds applications in text recognition, pattern recognition (PR), image analysis, and more. Within the area of civil engineering, computer vision techniques have been recognized as pivotal for efficient inspection and monitoring. MV methods predominantly handle two types of data: images and videos. In the domain of civil infrastructure assessment, MV algorithms, coupled with data sourced from remote cameras and unmanned aerial vehicles, present practical noncontact alternatives. The ultimate objective of such a system is to autonomously and effectively convert input images or videos into valuable information.

There are two different categories of MV applications: monitoring applications and inspection applications. Artificial neural networks (ANNs) and convolutional neural networks (CNNs) have been used extensively in end-to-end learning to drive recent advancements in computer vision approaches. A complicated input-output relation of data is approximated in ANNs and CNNs by a parametrized nonlinear function that is specified using nodes, which are logical units. The development of perception systems for extremely complicated visual issues has been remarkably successful with the use of these algorithms. The use of computer vision for monitoring and inspecting civil infrastructure is a logical advancement that may be quickly implemented to supplement and ultimately replace manual visual inspection while presenting new benefits and opportunities. Although each image has a diversity of spatial, textural, and contextual information, it may be difficult to extract information that can be used to make decisions. Consequently, the utilization of image data can exhibit both advantages and challenges. The scientific community has effectively demonstrated the feasibility of various vision algorithms, encompassing optical flow techniques to cutting-edge DL approaches. The rapid progression of research in computer vision-based methods for inspecting and monitoring civil infrastructure holds the potential to facilitate time-efficient, cost-effective, and ultimately automated procedures for civil infrastructure inspection and monitoring [34–36].

One practical application of MV in civil engineering is in construction site monitoring and quality control. Drones equipped with high-resolution cameras and sensors can capture aerial images and videos of construction sites, allowing project managers to assess progress, track material usage, and detect potential issues such as deviations from design plans or safety hazards. MV algorithms can automatically analyze these images, identifying construction elements, measuring distances, and evaluating the quality of completed work. For instance, MV can detect cracks or defects in concrete structures by analyzing images taken from various angles, aiding in the early identification of potential structural problems and facilitating timely maintenance. By leveraging MV, civil engineers can gain valuable insights from visual data, leading to more informed decision-making, improved project efficiency, and enhanced construction quality.

1.2.2.4 Machine learning

ML algorithms formulate a model through the utilization of sample data, commonly known as training data, with the aim of enabling predictions or decisions without requiring explicit programming for these tasks. In 1959, Arthur Samuel, an IBM employee renowned for his contributions to computer games and AI, introduced the term "ML." Concurrently, the phrase "self-teaching computers" was also employed during this period. "A computer program is said to learn from experience 'E' with respect to some class of tasks 'T' and performance measure 'P' if its performance at tasks in T, as measured by P, improves with experience E," said Tom M. Mitchell in a more formal description of ML that is commonly referenced. Contemporary ML serves dual objectives: firstly, to categorize data utilizing preexisting models, and secondly, to predict forthcoming outcomes employing these models. ML algorithms find extensive application across diverse domains, encompassing medical sciences, email filtering, speech recognition, agriculture, and computer vision. This adoption becomes particularly pertinent in scenarios where designing conventional algorithms for accomplishing the desired tasks would prove challenging or unfeasible [37,38].

ML is becoming better with time, and it is starting to appear more frequently in core engineering domains. ML has been widely employed in a variety of civil engineering applications and has evolved into an effective tool for solving complicated engineering challenges. ML is used by civil engineers to streamline their jobs and delegate some work to machines so they may concentrate on other activities. Due to the rapid proliferation of available data, coupled with enhanced processing capabilities and more accessible programming methodologies, ML tools are experiencing growing utilization within the area of civil engineering disciplines. ML approaches offer a quick and strong tool for decomposing complicated phenomena into basic mathematical processes. ML will provide a platform for innovative applications in civil engineering, including creating prediction models, addressing optimization issues, conducting data analysis, and examining the behavior of diverse systems.

In the contemporary landscape, ML boasts a diverse array of applications, encompassing the fields of forecasting, categorization, and resolution of complex mathematical challenges within civil engineering. Over recent years, numerous ML methodologies and techniques have witnessed rapid growth, including neural networks, evolutionary computation, fuzzy logic systems, DL, and image processing applications. Recently, researchers have been dedicating significant attention to ML methods, which have proven effective in addressing various challenges within the area of civil engineering. Notable applications include furnishing data for fully automated, intelligent, and autonomous urban and regional planning, addressing concerns related to rainfall forecasting and hydrological aspects, and facilitating the development of novel technologies. Additionally, ML is harnessed across various phases of engineering, spanning design, construction, and disaster management within the civil engineering domain [39].

Despite its utility, applying ML directly to civil engineering challenges can pose challenges. ML models designed and tested in laboratory-simulated environments often exhibit shortcomings when assessed in real-world scenarios. This discrepancy is often attributed to a phenomenon termed "data shift," which arises when the data used for training and evaluating the ML model deviates from the data it encounters in practical settings. In order to mitigate thc effects of data shift, a physics-based ML approach integrates data, partial differential equations, and mathematical models [40].

Supervised learning, unsupervised learning, semisupervised learning, reinforcement learning, and DL are the five primary categories of ML techniques [41]. These ML subcategories will be covered in more detail in the next sections.

Supervised learning

Supervised learning, a subset of both ML and AI, is commonly denoted as supervised ML. It is characterized by its approach to instructing computers to accurately classify data or anticipate outcomes through the utilization of labeled datasets [42]. The learning procedure conducted with observable label data is recognized as supervised learning. In the context of supervised learning, datasets are utilized to assign labels to new observations from the testing set after being trained with the training sets to develop ML models. Concerning the training set, the input variables are the attributes that influence the predictability of the variable. This encompasses both quantitative and qualitative variables. The output variable, known as the label class, is used by supervised learning to classify new observations. Supervised learning tasks are categorized into two primary types: classification tasks and regression tasks, contingent on the nature of the output variables. In both classification and regression tasks, the output variables are categorical in nature [43].

A classification algorithm endeavors to categorize inputs into predetermined classes or categories, relying on the labeled data it was trained on. This approach finds application in binary classifications, such as determining whether customer feedback is positive or negative or assessing the safety of tasks. Supervised learning addresses diverse classification challenges, including feature recognition tasks like classifying handwritten characters and numbers, as well as sorting equipment into distinct categories. Regression, on the other hand, represents a statistical technique that establishes a relationship between one or more independent variables and a dependent variable. In the context of regression, the model aims to establish a numerical correlation between input and output data. Regression models can predict real estate values based on geographical location or assess how much individuals are willing to invest in a particular building space based on their age. Effectively managing the inherent bias and variance within supervised learning algorithms is crucial, as there is a fine balance between adaptability and excessive flexibility. Additionally, the complexity of the model or function that the system is attempting to comprehend is a crucial factor. Prior to selecting an algorithm, considerations such as data heterogeneity, accuracy, redundancy, and linearity should be taken into account. It's worth noting that supervised learning often demands substantial volumes of accurately labeled data to attain satisfactory performance levels, a resource that might not always be readily available [44].

An example of supervised learning in civil engineering is predicting the compressive strength of concrete mixtures. Engineers can gather a dataset with various concrete mix designs, each labeled with their corresponding compressive strength after testing. By using this labeled data to train a supervised learning model, such as a regression algorithm, the model can learn the relationship between different mix components (e.g., cement, water, aggregates) and the resulting compressive strength. Once the model is trained, it can be used to predict the strength of new concrete mixtures, helping engineers optimize their designs and achieve desired performance levels while reducing the need for costly and time-consuming physical testing.

Unsupervised learning

Unsupervised learning operates on unlabeled data. In unsupervised learning, unlabeled data is provided to the algorithm as a training set. Unlike supervised learning, where definite output values are expected, unsupervised learning involves algorithms identifying patterns and similarities intrinsic to the data without relying on third-party measurements. In essence, algorithms possess the liberty to operate autonomously, delving into the data to uncover unexpected or unforeseen insights that might not have been explicitly sought by human designers. Unsupervised learning serves as a valuable tool in applications like clustering, which involves discerning groupings within datasets, and association, which entails forecasting rules that characterize the data. These tasks derive substantial benefit from the exploratory nature of unsupersized learning. Unsupervised learning seeks to extract patterns from datasets such that their description may be condensed to just their most distinctive aspects. For instance, clustering divides datasets into smaller groupings, which serve as an overview of the original data. Cluster analysis finds utility in diverse domains such as market research, object identification, and DNA sequence analysis. Unsupervised learning encompasses various techniques, including clustering, anomaly detection, neural networks, and approaches focused on learning latent variable models [45–47].

An example of unsupervised learning in civil engineering is the clustering of construction materials based on their properties. Engineers can collect data on various materials, such as concrete, steel, and wood, and input features like strength, density, and cost into an unsupervised learning algorithm. The algorithm can then identify natural groupings or clusters of materials with similar characteristics, helping engineers gain insights into material categorization and selection for specific applications. Unsupervised learning enables civil engineers to discover hidden relationships in data and can aid in making data-driven decisions, improving material selection processes, and optimizing construction material usage.

Semisupervised learning

The amalgamation of supervised and unsupervised ML techniques is referred to as semisupervised learning. In semisupervised learning algorithms, training data with partial labeling is utilized, typically comprising a substantial volume of unlabeled data alongside a limited quantity of labeled data.

Algorithms that are semisupervised are trained using both labeled and unlabeled input. This helps to increase learning accuracy and is highly helpful. In semisupervised ML, active learning (AL) actively identifies high-value data points in unlabeled datasets, which speeds up algorithm training. AL offers another layer of flexibility by allowing the learner to choose a small number of highly informative examples to label and add to the training set [48−51].

In civil engineering, acquiring labeled data can be time-consuming and expensive, making semisupervised learning an advantageous technique. An example of this in civil engineering is the assessment of building conditions. Engineers can gather a small labeled dataset of buildings with known conditions (e.g., good, fair, or poor) and a larger unlabeled dataset with additional building information. By using a semisupervised learning algorithm, the model can leverage both the labeled and unlabeled data to identify patterns and correlations between building features and conditions. This allows engineers to predict the condition of other buildings in the larger dataset based on the insights from the labeled examples.

Reinforcement learning

Dynamic programming techniques like reinforcement learning are used to train algorithms utilizing a reward and penalty scheme. The learning system, referred to in this context as an agent, learns in a collaborative environment. The agent chooses and carries out activities, earning rewards for properly carrying them out and suffering consequences for wrongly doing so. Through the use of dynamic programming, the agent uses reinforcement learning to teach itself the most effective way to maximize reward in a given environment [52−54].

Reinforcement learning differs from unsupervised learning in that it has distinct objectives, whereas the aim of unsupervised learning is to identify an action model that will maximize the agent's overall cumulative reward. Building AI for computer games, robotics and industrial automation, text summarizing engines, dialog agents (text, speech), etc. are typical examples of practical reinforcement learning applications [55].

Reinforcement learning in civil engineering involves training AI systems to make decisions and take actions in dynamic environments by maximizing a cumulative reward signal. An example of reinforcement learning in civil engineering is autonomous construction equipment control. Suppose an autonomous bulldozer is tasked with grading a construction site to achieve a specific slope and smoothness. Reinforcement learning can be applied to train the bulldozer's control algorithm, where the AI agent receives rewards for correctly adjusting its movements to achieve the desired grading outcomes and penalties for deviations from the target. Through repeated interactions with the environment, the AI agent learns to optimize its actions to accomplish the grading task efficiently and accurately. Reinforcement learning can lead to the development of self-adaptive and intelligent construction equipment, enabling cost-effective and precise construction processes while minimizing human intervention and increasing safety in challenging and dynamic construction environments.

1.2.2.5 Deep learning

DL represents a subset of ML characterized by neural networks comprising three or more layers. These neural networks aim to replicate certain aspects of human brain function, albeit with varying degrees of success, enabling them to undergo "learning" from extensive datasets. Situated within a broader category of ML methodologies, DL amalgamates ANNs and representation learning. Learning within DL can occur through supervised, semisupervised, or unsupervised modes. Employing multiple layers, DL extracts higher-level features from raw input. In the context of image processing, for instance, lower layers may discern boundaries, while upper layers may identify human-conceptualized concepts like digits, characters, or faces [38,56].

A significant proportion of contemporary DL models are rooted in ANNs, predominantly CNNs. However, these models can also encompass propositions or latent variables organized in a layered structure, as evident in deep generative models like the nodes in deep belief networks and deep Boltzmann machines. The designation "deep" pertains to the count of layers through which data undergoes transformation. DL systems, specifically, exhibit notable depth in their credit assignment path (CAP). The CAP denotes the sequential chain of transformations leading from input to output. CAPs serve to symbolize potential causative relationships existing between input and output elements [57].

An example of DL in civil engineering is the use of deep CNNs for infrastructure inspection. For instance, to assess the condition of bridges, images of their components (such as decks and support structures) can be collected using drones or other imaging devices. These images can be fed into a deep CNN, which automatically learns hierarchical features and patterns, enabling it to detect cracks, corrosion, and other defects with remarkable accuracy. DL models can significantly speed up the inspection process, allowing civil engineers to efficiently identify maintenance needs, prioritize repairs, and ensure the structural integrity of bridges and other critical infrastructure.

1.3 AI in civil engineering

Synchronizing to nature human species is continuously evolving with time. With the passage of time, learning from nature and complying with that learning in the real world with applied intelligence helped transmute the world. Parallel to the invention of the wheel and the discovery of fire, the ancients understood the need for structures to feel safe and to enjoy a comfortable environment [58]. With continuous developments for hundreds and thousands of years and major breakthroughs like cement and many others, construction growth was unprecedented. Albeit, with continuing growth, the current industry demands improving existing structures throughout the world with novel techniques to achieve the SDGs and save natural resources for the next generations. Enormous resources and gigantic amounts of energy are required in civil engineering work, which can be reduced by AI. Thus, for sustainable development, merging AI with civil engineering becomes necessary [38]. Engineers have been working consistently to increase the effectiveness of traditional materials, solutions, and testing procedures in civil engineering. Recent advances in civil engineering have raised complex mathematical challenges as a result of the advancement of materials science and diverse composite materials. As a result, it is impossible to use the conventional underpinning theories and testing procedures. Additionally, due to the growing acceptance of the idea of smart cities on both a global and local scale, the following decade will see significant improvements in AI.

The introduction of AI into several fields of civil engineering would boost the infrastructure and building sectors with increased safety and simplicity. Within the domain of civil engineering, several branches of AI find relevance, encompassing ML, DL, fuzzy logic, PR, decision trees, swarm optimization, and evolutionary computations. A prominent application of AI in civil engineering pertains to the advancement of robotics and automated systems [59]. Additionally, AI's utility extends to diverse areas, including construction management, building materials, hydraulic optimization, geotechnical engineering, transportation engineering, and more. AI is used in civil engineering for a variety of purposes, such as improving building designs, preventing cost and schedule overruns, identifying and mitigating risks, accelerating project delivery on-site through smart construction, enhancing facility management efficiency, adopting AI for construction solutions, and many more. Leveraging AI models in civil engineering projects offers advantages such as heightened precision, reduced costs, and minimized disruptions. In current construction practices, AI is employed to strategize the positioning of plumbing and electrical systems. Additionally, AI is harnessed to monitor and assess real-time interactions among workers, equipment, materials, and supervisors on the construction site, identifying potential safety hazards, construction errors, and productivity challenges [59]. The integration of AI streamlines the development process, enhancing operational efficiency for all stakeholders involved. Moreover, by rendering structural design an appealing career path, AI-driven advancements in civil engineering create avenues for expanded career opportunities. Presently, many elements of civil infrastructure are constructed, maintained, and managed by civil engineers using AI. The detailed role of AI in civil engineering subdomains is presented in Fig. 1.5.

The literature indicates a growing interest in applying AI techniques to address various challenges in civil engineering, with a focus on enhancing efficiency, accuracy, and safety in infrastructure development and management. As AI continues to advance, it is likely that more sophisticated and diverse applications will emerge in the field of civil engineering, further transforming the industry. The information gleaned from the extensive literature on AI applications in civil engineering will be presented in the parts that follow.

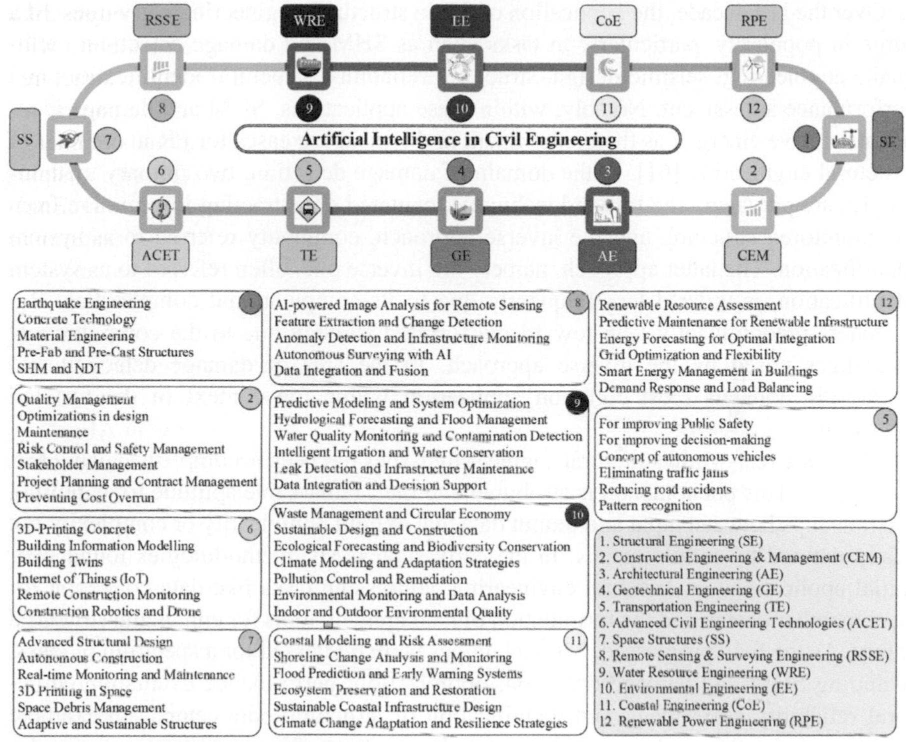

Figure 1.5 AI in civil engineering.

1.3.1 *Structural engineering*

Numerous challenges within the area of structural engineering elude resolution through conventional computational methods. Nevertheless, these issues often find effective solutions in the hands of AI experts possessing specialized knowledge. The amalgamation of diverse fields encompassing computer science, information theory, cybernetics, linguistics, and neurophysiology has given rise to the evolution of AI. AI offers distinct advantages over conventional approaches, particularly in addressing problems riddled with uncertainty, rendering it a potent tool for tackling complex quandaries. Furthermore, AI-based solutions emerge as potent alternatives to experimentation when determining engineering design parameters, leading to significant reductions in the temporal and human effort expended in trial-based approaches. Beyond this, AI possesses the potential to accelerate decision-making, mitigate error rates, and enhance computational efficiency. Among the array of AI methodologies, ML, PR, and DL stand out as prominent contenders that have garnered substantial attention in recent times, positioning themselves as a novel category of intelligent techniques for application in structural analysis [60]. Neural networks, fuzzy logic, evolutionary computation, expert systems, probability theory, discriminant analysis, swarm optimization, metaheuristic optimization, and decision trees have all been extensively employed within the area of structural engineering.

Over the last decade, the application of PR in structural engineering has witnessed a surge in popularity, particularly in tasks such as SHM and damage detection, earthquake engineering, seismic design, structural reliability, structural identification, and performance assessment. Notably, within these applications, SHM and damage identification have emerged as the predominant and typical use cases for PR in the field of structural engineering [61]. In the domain of damage detection, two primary methodologies are prevalent: the forward technique, centered on extracting information from the monitored structure, and the inverse approach, commonly referred to as system identification. The latter approach, namely the inverse one, often referred to as system identification, is utilized less frequently due to its computational complexities. Researchers predominantly lean toward the forward strategy due to the computational challenges posed by the inverse approach. In the area of damage detection and SHM, PR finds its most common application within the context of the forward approach.

Modeling real-world structural engineering issues has been accomplished using ML techniques. This phenomenon is attributable to their remarkable aptitude for capturing relationships between input and output data that exhibit nonlinearity or complexity not easily defined by formal means. In the early stages, ML methodologies found their initial applications in structural engineering, addressing tasks like data collection for steel member design and the formulation of management tools for enhancing structural safety. As the field progressed, ML techniques were harnessed for a spectrum of tasks, including SHM and damage detection, optimization, performance evaluation, structural reliability assessment, and identification of structural parameters. An example of this includes the modeling of the material properties of concrete [62].

AI algorithms, particularly genetic algorithms and particle swarm optimization, can efficiently explore the design space to find optimal solutions for complex structural systems. These techniques can help engineers find cost-effective and material-efficient designs while satisfying safety and performance requirements. Under the structural engineering domain as a subpart, AI is also extensively used in earthquake engineering, concrete technology, material engineering, SHM, precasting, and prefabricating works. The AI applications in the mentioned sections are presented in detail in the next subsections for better understanding.

1.3.1.1 Earthquake engineering

Seismology uses a vast amount of measurement data to study earthquakes at multiple scales, with a focus on determining how the natural disaster may affect systems of civil infrastructure. Studies in seismology are typically conducted with the following four goals in mind: (i) Actions to reduce earthquake disasters over the long term; (ii) disaster preparation or adjustment; (iii) disaster response strategies; and (iv) planning for postdisaster recovery. These goals are referred to, respectively, as strategies of mitigation, preparedness, response, and recovery.

In order to process data and extract pertinent information for accurate seismology forecasts and judgments, AI offers effective tools. The effectiveness of AI-enabled seismology hinges on the proficient utilization of AI methodologies, encompassing ML and

DL, to extract meaningful data from noise and discern earthquake events even when they occur at or beneath the level of background noise. In the context of evaluating distinct AI algorithms within seismology, significant attention has been directed toward supervised and unsupervised ML techniques. AI has proven to be a strong tool for handling several difficult tasks, such as automation, modeling, and scientific discovery.

AI has proven to be highly advantageous within the domain of earthquake engineering, particularly in the modeling of the seismic behavior exhibited by soil-structure systems spanning from the soil to the foundation and superstructure. Achieving accurate assessments of site and structure responses, as well as grasping soil-structure interactions, is imperative for the accurate modeling of earthquake responses. This modeling is a pivotal aspect of seismic risk assessment studies. However, the intricate dynamics inherent in real-world soil-structure systems often defy effective representation through a set of differential equations. Consequently, AI is progressively gaining traction in the area of earthquake engineering. Notwithstanding the encouraging outcomes witnessed, the application of AI within geotechnical and structural seismic engineering is still in its nascent stages [63].

AI has its limitations and is not an omni-solution. In certain circumstances, having a tiny sample size makes it difficult to construct solid data-driven models. For certain applications, researchers must deal with data that is noisy, fragmentary, inhomogeneous, from many sources with various spatial-temporal resolutions, uncertain, and lacking in gold-standard ground truth (particularly for ground-motion recordings of minor earthquakes). The problem is further complicated by the fact that the seismic response of soil-structure systems is multivariate, nonlinear, and nonstationary [64]. Additionally, one of the barriers preventing a wider acceptance of AI in society is the "black box" aspect of some AI systems [65].

AI can assist in automated code compliance checking, ensuring that structural designs adhere to relevant design codes and standards, reducing the likelihood of noncompliance issues. Current research is overcoming many constraints and capitalizing on the potential presented by AI in the field of earthquake engineering. One pertinent endeavor involves establishing extensive and openly accessible benchmark datasets for AI, sourced from meticulously validated numerical simulations and seismic experiments/observations. Within this context, numerous domains hold promise for the effective implementation of AI. These domains encompass soil characterization and site classification, the influence of topography and basin effects, the evaluation of liquefaction susceptibility, generating maps for regional site response, performance-centric seismic design, the assessment of seismic fragility and vulnerability in structures, site-specific earthquake modeling encompassing linear site response, soil nonlinearity, the interdependence of intelligent infrastructures, surrogate models for structural seismic analysis, and the implementation of seismic structural health monitoring for the swift identification and assessment of damage in its early stages.

1.3.1.2 Concrete technology

Saving time, money, and effort is made possible by the development of AI tools and their use in the construction sector, particularly for anticipating the mechanical properties of concrete [66]. The increasing intricacy of cementitious systems has led to persistent challenges in formulating concrete with desired properties. Despite notable conceptual and methodological advancements within the area of concrete science, achieving the desired qualities in concrete remains a formidable endeavor [67]. AI has proven its transformational potential in practical research by demonstrating its capacity to handle complicated tasks independently. As the adoption of AI in concrete mixture design accelerates, it becomes imperative to grasp the methodological limitations and establish optimal practices within this emerging area of computation.

The four paradigms of concrete research are data-driven, computational, theoretical, and empirical. The first paradigm is based entirely on empirical research and on the technique of trial and error (iterative experiments on proportioning, processing, and characterization) [39]. Theoretical research is the second paradigm, which is primarily concerned with the rules of kinematics, mechanics, and thermodynamics (hydration, structural development model, and morphology). The third paradigm in concrete research, computational research, is focused on molecular dynamics simulations of cement hydrates (calcium silicate hydrate gel). The fourth paradigm in the concrete study is AI and ML predictions utilizing computational and experimental data combined with data mining.

The first paper related to concrete and ML was authored in 1992 by Donald Pratt and Mary Sansalone [68]. The number of publications remained modest in the first decade; however, after 2009, AI publications rose unprecedentedly. The surge in ML research is currently attributed to the emergence of petascale computing facilities, the accessibility of publicly available data sources, and the availability of readily deployable ML software libraries. Regression problems have received the majority of research interest, whereas classification tasks have just lately begun to receive attention as DL for computer vision has gained popularity. Neural networks stand as the preeminent ML algorithms and have held a prominent role in concrete research since 1992, primarily employed for a plethora of categorization challenges. Furthermore, the concrete domain has embraced supplementary techniques such as support vector machines, decision trees, random forests, and k-nearest neighbors algorithms, each contributing significantly to advancements within this sector [69].

AI had been used widely in the prediction of compressive, flexural, tensile, and shear strengths. In addition to these applications, AI has demonstrated its capability to predict a wide range of concrete properties and behaviors. These include flowability, pozzolanic reactivity, setting time, elastic modulus, hydration reactions, mechanical properties of calcium silicate hydrate (C−S−H), interatomic potentials for C−S−H, fracture properties of the interfacial transition zone, creep, shrinkage, thermal performance, optimization of the cement manufacturing process, optimization of mix design, pore structure analysis, identification of aggregate shape, evaluation of fiber distribution, detection of cracks, quality control for concrete admixture manufacturing or 3D concrete printing, prediction of durability aspects like permeability, freeze-thaw durability, carbonation, chloride diffusion, alkali-silica reaction, corrosion, sulfate attack, fire-induced damage, creep dynamics of C−S−H, chloride diffusion, and more. Additionally, AI has proven effective in tasks such as quality control for concrete admixture manufacturing, classification of mortar types, identification of fiber failure modes, assessment of shrinkage, determination of carbonation depth, estimation of chloride concentration, detection of voids, analysis of thermal properties, and classification of mortar types.

Data is required for ML models to learn patterns and relationships, which are subsequently generalized to a broader population. The quality of the data utilized, encompassing aspects such as representativeness, completeness, and accuracy, holds paramount importance in determining the performance and credibility of the model. Within concrete research, several challenges pertaining to data quality arise, including data sparsity, high dimensionality, data bias (comprising representation bias, measurement bias, temporal bias, and deployment bias), as well as issues linked to validation.

1.3.1.3 Material engineering

Material engineering is an interdisciplinary research area that focuses primarily on the processing-structure-property-performance aspects of a material. In recent times, ML has found substantial utility within various domains of materials science. Notably, ML has been instrumental in swiftly and precisely forecasting phase diagrams, crystal structures, and material properties. Additionally, ML has facilitated the creation of interatomic potentials and energy functionals, aiming to expedite and enhance the precision of material simulations. Furthermore, ML techniques have been applied to real-time data analysis for high-throughput experiments, contributing to the efficient handling and interpretation of experimental data [70]. ML algorithms can be employed to predict material properties, such as concrete strength or steel yield strength, based on mix proportions and other input parameters. This helps engineers optimize material usage and ensure structural reliability.

Developing well-curated and diversified datasets, selecting effective representation for materials, inverse material design, integrating autonomous experiments and theory, and selecting an appropriate algorithm/work-flow are some of the important areas of application in applying AI approaches to materials. The use of physics-based models in the AI framework is also intriguing. In material engineering, AI is leveraged to anticipate the intricate relationships between processing conditions and material properties, construct model Hamiltonians, forecast crystal structures, and classify different crystal arrangements [71]. Future work on AI's uses in materials science is anticipated to go into two different areas. The first will be to keep developing increasingly complex ML techniques and using them in materials science. The objective of the second direction will be to confirm the applicability of ML models.

1.3.1.4 Prefab and precast structures

Prefabricated (prefab) and precast structures are gaining popularity due to their efficiency, cost-effectiveness, and reduced construction time. The integration of AI in the design, manufacturing, and assembly processes of these structures has significantly improved their overall quality and performance [72]. AI-driven algorithms enable designers and engineers to create highly sophisticated and optimized designs for prefab and precast structures. By analyzing vast amounts of data, including environmental factors, structural requirements, and material properties, AI can generate innovative designs that meet safety regulations, cost constraints, and aesthetic preferences. This optimization process allows for faster and more accurate decision-making, resulting in enhanced structural integrity and reduced waste.

AI-powered predictive analytics facilitate real-time monitoring and quality control during the construction process. Sensors and IoT devices integrated into prefab components and precast elements gather data on various parameters such as temperature, humidity, and structural performance. AI algorithms analyze this data, enabling early detection of potential defects, deviations, or weaknesses. By addressing these issues promptly, construction companies can ensure higher-quality and safer structures, minimizing the need for costly rework. AI plays a pivotal role in enabling robotics and automation in the assembly of prefab and precast structures. Robots equipped with AI algorithms can efficiently handle repetitive tasks, such as material handling, fastening, and welding, with speed and precision. This automation not only increases productivity but also reduces the dependence on manual labor, addressing labor shortages and minimizing the risk of accidents on construction sites [73].

The integration of AI in prefab and precast structures has revolutionized the construction industry, offering numerous benefits ranging from enhanced design and optimization to streamlined project management and reduced waste. AI-driven solutions have improved the overall quality, safety, and efficiency of construction processes, making prefab and precast structures a viable and attractive option for modern construction projects. As AI continues to advance, its role in the construction sector is expected to expand further, driving innovation and reshaping the future of the built environment.

1.3.1.5 SHM and NDT

Civil infrastructure, such as bridges, buildings, dams, and roads, plays a critical role in society's functioning. To ensure their longevity and safety, continuous monitoring and inspection are vital. Traditional SHM and NDT techniques have limitations in terms of cost, labor, and accuracy. AI solutions have emerged as a promising tool to overcome these challenges. The integration of AI in civil engineering has revolutionized the fields of SHM and NDT. AI-powered systems have demonstrated significant potential for enhancing the safety, reliability, and efficiency of civil infrastructure. SHM involves the real-time assessment of the structural integrity and performance of civil infrastructure. AI algorithms can process large amounts of data obtained from various sensors, such as accelerometers, strain gauges, and displacement sensors. By using ML techniques, AI systems can recognize patterns and detect anomalies in structural behavior, enabling engineers to take proactive measures [74].

AI algorithms can identify subtle changes in structural behavior that might be indicative of potential damage or deterioration, allowing timely intervention to prevent catastrophic failures. AI-powered SHM systems can optimize maintenance schedules, reducing unnecessary inspections and associated costs while focusing resources on critical areas. NDT techniques are used to assess the structural integrity of materials and components without causing damage. AI has shown immense potential in advancing NDT methodologies, enhancing their accuracy and efficiency.

AI algorithms can analyze images obtained from techniques like X-ray, ultrasound, and infrared thermography, identifying defects, cracks, and irregularities that may not be easily detectable by human operators. AI-powered acoustic sensors can detect and classify acoustic emissions from materials under stress, providing insights into structural health and potential failure mechanisms. AI can integrate data from multiple NDT techniques, offering a comprehensive assessment of the structural condition [75].

Data quality and quantity, interpretability, and the integration of advanced technologies with existing conventional infrastructure are some of the major challenges at present. On the other hand, future opportunities include advancements in autonomous inspection, advanced materials, predictive maintenance, and many more [76]. The integration of AI in SHM and NDT applications represents a paradigm shift in civil engineering. By harnessing the power of AI, engineers can enhance the safety, reliability, and efficiency of civil infrastructure, ensuring sustainable development for future generations.

1.3.2 Construction engineering and management

The management of construction operations entails overseeing the set of procedures involved in building and constructing structures. This management must ensure the highest productivity of construction work, as well as a reduction in construction time, cost, and unfinished work volume, while also ensuring the elimination of losses in the organization's operations. Technological developments in recent years have created opportunities for engineers and researchers to explore advanced computational techniques in the management of construction engineering. In recent times, the adoption of AI has become increasingly widespread in the domains of construction project management and engineering. This surge is primarily driven by the inherent technological capabilities of AI to enhance both productivity and efficiency within the construction industry. Construction enterprises are employing AI to conduct real-time assessments of project-related data, encompassing aspects like construction site monitoring and predictive analytics. This data-driven approach enables informed decision-making that significantly impacts crucial project dimensions, including quality, safety, profitability, and adherence to schedule [77].

Modern engineering and construction technologies must incorporate AI. The construction industry grapples with significant challenges, notably encompassing cost and schedule overruns, along with safety apprehensions. AI offers substantial potential to address these issues within the sector. AI's applicability spans across all phases of construction projects, ranging from initial planning and design to subsequent stages including bidding, financing, transportation management, operations, asset management, and overall project supervision.

The tasks encompassed within the construction process are characterized by their labor-intensive nature, time-consuming requirements, and susceptibility to errors when carried out manually. Project managers invest a significant portion of their time in task allocation and workforce information management. However, numerous repetitive operations, which are amenable to being executed with minimal errors, can be effectively automated through the deployment of AI. Furthermore, AI holds the potential to automate the process of task assignment, thereby enhancing the efficiency of task delegation [78]. This not only expedites the workflow process but also encourages staff to focus on their area of expertise. AI-powered project management tools assist construction companies in streamlining their operations and improving project scheduling. These tools can predict potential delays, resource bottlenecks, and unforeseen challenges, allowing for more accurate project planning and resource allocation. As a result, construction projects are completed within specified timelines and budgets, leading to increased client satisfaction and better overall project outcomes. Some detailed AI applications are presented in the following subparts of the construction management domain:

1.3.2.1 Quality management

A crucial factor in the building and construction industry is quality. Everything from safety issues to project budget overruns can be attributed to poor project quality. AI promises to provide more control over the quality management process, improve understanding of the cost of quality, and reduce costs. Enterprises are initiating the provision of self-driving construction machinery that specializes in executing repetitive activities such as concrete pouring, brick laying, welding, and demolition. These autonomous systems exhibit enhanced effectiveness compared to their human counterparts. AI-driven autonomous or semiautonomous bulldozers, guided by human programmers, can achieve precision in tasks like site preparation through excavation and preparatory efforts [79]. This means a reduction in the time required for task completion, thereby releasing human personnel for direct engagement in construction activities. Additionally, project managers gain the capability to monitor the ongoing developments at a job site in real-time. To assess employee productivity and compliance with established protocols, project managers deploy AI-powered facial recognition software, on-site cameras, and other pertinent technologies.

AI-based field inspections are becoming increasingly automated, which boosts productivity and enhances the quality control procedure. Quality issues are becoming more predictable due to the quick analysis of field data via AI techniques. Quality issues will probably be prevented in the future with the aid of AI. Construction businesses will have access to more structured data owing to AI technology, enabling them to make quicker, more informative, and more intelligent decisions. More sophisticated information is produced by AI and advanced analytics to support ongoing progress. For instance, by employing clustering algorithms, AI may classify problems into fresh categories and offer a number of entirely new insights, each requiring a plan of action.

AI could alert managers to problems that were either previously missed by team members or in which the subtle signals are not yet visible to management by analyzing a variety of input factors and identifying data patterns. With advanced warning, precautions can be taken, equipment can be adjusted or reset, materials can be changed, work packages and processes can be adjusted, and workers can be retrained or replaced. This results in an enhanced quality of work. Once an issue has appeared, a trained AI algorithm can forecast the time and cost of resolution. When combined with scheduling AI, it can assess the impact on the overall schedule and determine the resources needed to resolve the issue.

Lean construction management can also take advantage of AI techniques and AI-enabled technology [80,81]. The anticipated building project wastes range from 30% to 50% owing to a lack of resource utilization, which results in inefficient labor. Contractors and developers employ breakthroughs in AI to automate the monitoring of their current construction projects and actively manage site operations and activities from a remote location [82].

1.3.2.2 Optimizations in design

The existing construction design procedure is out of date and consequently sluggish. Engineers and architects may optimize their project design by using insights from material, building, and environmental data. The construction sector harnesses AI-driven generative design to identify and collaborate on architectural, engineering, mechanical, electrical, and plumbing (MEP) plans. This approach aims to ensure a harmonious alignment among the subteams, mitigating any potential conflicts that may arise [83,84]. These precautions serve to diminish the likelihood of rework. The ML algorithm engages in a comprehensive exploration of all feasible solutions and variations to generate design alternatives. Throughout each iteration, models containing diverse variants are constructed and learned from, persisting until the attainment of a perfected model. Generative design is made possible by AI, which is a significant advantage that the automotive, aviation, and other manufacturing-based sectors already take advantage of. Generative design may be thought of as a more sophisticated type of parametricism in architecture. By adding the capability to manage architectural design complexity in real-time, generative design elevates parametricism as a style of architecture [85,86]. When creating full building systems, parametricism, for instance, links the components, architecture, spatial connections, and urbanism; however, the incorporation of AI speeds up and improves the process in various ways. Real-time analytics of how the environment, people movements, and urbanism should be represented or incorporated into building designs is one of these methods.

The usage of BASILICA, a design system dependent on AI and ML to manage architectural and environmental design, is an example of how AI is used in generative design. In Taiwan, BASILICA was utilized to develop intricate structures [87]. The limitations and resources that were supplied into the design system were also utilized to recreate medieval towns that were present in Italy centuries ago. It created architectural designs that took into account the terrain and environmental interactions surrounding them using PR and MV.

1.3.2.3 Maintenance

Even after the completion of construction, building managers can continue to leverage AI technologies. Through advanced analytics and AI-driven algorithms, valuable insights regarding the operation and performance of diverse structures within the built environment, such as buildings, bridges, roads, and pipelines, can be gleaned. This is accomplished by harnessing data collected from sensors, drones, and other wireless technologies. Consequently, AI can be instrumental in early issue detection, prediction of optimal timing for preventive maintenance, and even guiding human actions to ensure the utmost security and safety. By harnessing the power of AI-driven technologies like ML, data analytics, and predictive modeling, civil engineers can efficiently monitor, assess, and optimize infrastructure assets [7]. Moreover, AI-driven systems facilitate more informed decision-making processes, enabling civil engineers to prioritize maintenance efforts, allocate resources effectively, and enhance overall safety and resilience in our modern infrastructure networks.

1.3.2.4 Risk control and safety management

Every construction project entails a certain degree of risk, spanning aspects like quality, safety, time, and finances. As the scale of a project expands, involving numerous subcontractors across diverse crafts simultaneously, the complexity and magnitude of the risks amplify. Leveraging AI and ML technologies, general contractors can now employ advanced monitoring techniques to assess and prioritize risks present on the construction site. This approach enables project teams to allocate their limited time and resources efficiently toward addressing the most significant risk factors. AI facilitates the automatic prioritization of critical issues. In a bid to mitigate risk, construction managers can engage in close collaboration with high-risk teams, evaluating subcontractors based on their specific risk profiles [88].

The fatality rate among construction workers is notably higher, with occurrences of on-the-job deaths being five times more frequent compared to workers in other sectors. In the private sector of construction, the leading causes of fatalities, apart from those resulting from traffic accidents, include falls, incidents involving being struck by objects, electrocution, and being caught in between or crushed by objects. Accidents can happen on construction sites for a variety of reasons. Many mishaps can be prevented by using ML to analyze and anticipate hazards [89]. The site manager can identify possible risks by using software to monitor sources like images and videos and take appropriate action when necessary. The user may receive reports on possible safety issues, such as dangerous scaffolding, waterlogging, and workers who lack protective gear, including gloves, helmets, and safety glasses, to rate the projects. The utilization of statistical ML techniques within the area of construction facilitates a more efficient and time-effective examination of pertinent data, including modified orders and information requests. This application aids in proactively alerting project managers to impending issues demanding immediate attention. Furthermore, it enhances the efficacy of safety monitoring processes.

1.3.2.5 Stakeholder management

The technical change due to AI is expected to have a favorable effect on all project stakeholders, including owners, contractors, and service suppliers. Given the ongoing evolution of adjacent industries like transportation and manufacturing into integrated ecosystems, the imperative for the construction sector to embrace AI-driven processes is heightened. Enterprises that embrace enhanced AI technologies stand to gain advantages, as the technological transformation within the engineering and construction domain is still in its nascent stages. Through AI, construction organizations can effectively tackle current challenges and sidestep past errors, fostering a confident approach toward their operations.

The role of AI in stakeholder management within civil engineering is pivotal in fostering effective communication, collaboration, and transparency throughout the project lifecycle [90]. By leveraging AI-powered tools, civil engineers can streamline interactions with diverse stakeholders, such as clients, contractors, government agencies, and the public. NLP facilitates real-time analysis of stakeholders' feedback, concerns, and preferences, enabling engineers to respond promptly and adapt project plans accordingly. AI-driven sentiment analysis helps gauge public sentiment, aiding in the development of tailored communication strategies and mitigating potential conflicts [91]. Furthermore, AI-powered data visualization tools allow for the clear presentation of complex information, enhancing stakeholder engagement and understanding. Overall, AI empowers civil engineers to establish stronger relationships, anticipate challenges, and proactively address issues, leading to more successful and socially acceptable infrastructure projects.

1.3.2.6 Project planning and contract management

In the building process, the planning phase is the most time-consuming. Construction planning requires a significant amount of architect and project planner work, whether it is for idea validation, design feasibility testing, or brainstorming about visualization. Because of this, automation in this area is something that every architect, project planner, and manager should be aiming for. AI just excels at doing that. Access to design databases with numerous proposals, interpretation of the building site, model generation, and, most significantly, cost calculation are all made possible by AI [92]. Not only is information readily available for making educated decisions, but AI is also able to convey facts in an easily accessible manner while maintaining the seriousness of the content. Overall, saving time and effort during planning and construction is a huge benefit for engineers, architects, and builders.

The planning and management team possesses the capability to intervene promptly when deviations arise, addressing minor issues before they escalate into more significant challenges. Through the utilization of AI techniques, the optimal trajectory is optimized and continually refined, providing substantial assistance in project planning. Notably, the construction industry's planning and design domain stands out as an area where AI is exerting a notable impact by expediting tasks with enhanced efficiency and cost-effectiveness.

AI plays a crucial role in optimizing project planning and contract management processes. AI-driven algorithms analyze historical project data, industry trends, and relevant regulations to assist civil engineers in developing more accurate and efficient project plans [83]. By considering various factors like resource allocation, risk assessment, and scheduling constraints, AI helps create realistic timelines and budgets, reducing the likelihood of delays and cost overruns. Additionally, AI-powered contract management systems streamline the creation, negotiation, and execution of contracts with stakeholders, ensuring compliance with legal requirements and promoting transparency. These AI-based tools facilitate the automatic review and analysis of contract terms, identifying potential risks and discrepancies to safeguard the interests of all parties involved. Overall, AI empowers civil engineering professionals with data-driven insights, enabling them to make well-informed decisions, optimize project outcomes, and foster successful collaborations [93].

1.3.2.7 Preventing cost overruns

Despite the utilization of top-tier project teams, a significant proportion of large-scale projects frequently exceed their allocated budgets. In the area of projects, ANNs are employed to predict instances of cost overruns, drawing insights from various factors such as project scale, contract specifications, and the proficiency of project managers [94]. Predictive models leverage past data, including projected commencement and conclusion dates, to forecast feasible timelines for forthcoming projects. Through AI, individuals can remotely access practical training materials, facilitating rapid enhancement of their skills and knowledge. This expedites the process of assimilating additional resources into projects, ultimately hastening project completion.

The construction industry is compelled to allocate resources toward AI and data science due to labor scarcity and the imperative to enhance the sector's modest productivity levels. Construction companies are employing AI and ML to enhance the planning of labor and equipment distribution. Utilizing an AI-powered robot that continuously monitors task advancement and the locations of individuals and equipment, project managers can swiftly identify sites with sufficient personnel and resources to meet deadlines, as well as those lagging behind, where additional labor might be deployed.

The role of AI in preventing cost overruns in civil engineering is instrumental in improving project outcomes and financial efficiency. AI-powered cost estimation tools leverage historical data, project scope, and other relevant parameters to provide more accurate and comprehensive cost forecasts [95]. By analyzing past projects' performance and identifying cost drivers, AI helps civil engineers create realistic budgets and better allocate resources. During project execution, AI-driven project management systems monitor expenses, track progress, and identify potential deviations from the planned budget in real-time. This enables proactive decision-making and timely adjustments to prevent cost overruns. Additionally, AI's predictive analytics capabilities help anticipate risks and uncertainties, allowing for better risk management strategies to be implemented.

1.3.3 Architectural engineering

Big Data, robotics, and ML in AI have all emerged, sparking a general discussion on how to define the "Second Digital Era in Design" and how architecture must change to include new concepts in the design process. If future architects are willing to lead architecture into the AI age, and due to the shift, there is change in the contemporary environment. Architects and other stakeholders will collaborate as/or alongside computational specialists for the new transdisciplinary design methodology. A change that will influence how we interact with the morphology of space, human conduct, and social usage.

In the area of architectural engineering within civil engineering, AI plays a transformative role by revolutionizing the design process, enhancing sustainability, and streamlining building operations. AI-powered design tools enable architects to explore innovative ideas and generate optimized building layouts that maximize space utilization and energy efficiency [96,97]. Through ML algorithms, AI analyzes vast datasets to predict building performance under different scenarios, ensuring optimal material selection and reducing environmental impacts. Additionally, AI-driven energy management systems continuously monitor and regulate building operations, optimizing heating, cooling, and lighting to minimize energy consumption. AI's role extends to construction as well, where it facilitates real-time quality control and progress monitoring, ensuring adherence to architectural plans.

Architectural databases are extensive and intricate, and they may give architects a wealth of priceless knowledge that is ideal for teaching AI systems. Automating routine tasks is also achievable by AI. AI is being used more and more frequently in architectural design. AI-powered digital tools offer a solution in the form of software that automates routine project tasks for architects. By integrating data and automation into their daily workflows, architects and designers can focus more on the creative aspects of their work. Automation is already a prevalent practice among engineers and architects. For instance, CAD, in use for over 50 years, has played a significant role in shifting the industry from paper-based to digital design versions [98]. The next stage is to integrate ML with currently available software and further automate some of the repetitive operations, such creating calculations and drawings [99].

Some businesses have already used AI in software for architecture and construction [100]. Some businesses have already incorporated AI into software for the building industry. For almost 2 decades, the BIM platform has assisted architects and engineers in streamlining the building process and making it more effective and economical. The procedure will be improved with the aid of ML, which will speed up project work and increase efficiency. With ML, BIM software can recognize patterns in the data and learn from them, giving it the ability to decide for itself how to automate the building process [101]. BIM software powered by AI is able to monitor a variety of factors of the building process and generate solutions more quickly than humans. More importantly, it reduces the possibility of human mistakes. Another significant benefit of this strategy is that it enhances site safety since AI-powered BIM software can analyze the set from photos and spot dangers, such as fall hazards. AI can generate alternate designs as well.

Architects and engineers devote many hours to developing buildings, coming up with several variants, and verifying the specifications of the structures. AI is quite helpful in this situation. AI in architecture can provide designs with the necessary parameters and within the budget considerably more quickly if it has access to datasets from active projects. Additionally, crucial, it may improve over time by using the knowledge gained from each iteration to create better designs. Businesses putting this strategy into practice on a large scale are using the Spacemaker platform (AI software for urban development that was acquired by Autodesk in 2020), which combines cloud computing and AI for early-stage real estate development [102]. It offers a generative design driven by AI to produce more environmentally friendly structures and towns. Designers may quickly test and assess design concepts with this tool [103].

Thousands of architects worldwide may lose their jobs as AI in architecture continues to improve, according to industry concerns [104,105]. Contrarily, it is preferable to view AI as an opportunity rather than a threat [106]. The latter makes sense since AI may be used to automate repetitive jobs and streamline operations. Previously, innovation in digital tools like CAD helped architects and designers work fast and smart. More importantly, while AI and ML excel at tracking and analyzing large amounts of data, human brains continue to beat machines when it comes to coming up with original ideas.

Megatrends, including population expansion, urbanization, and globalization, are causing a variety of social requirements in terms of building design and construction as well as other types of civil infrastructure. Additionally, the pandemic, the state of the economy, and the societal unrest in 2020 have raised the need for quicker and more efficient architectural design solutions to address both ongoing and new societal concerns. To effectively navigate the future, a paradigm leap like AI is necessary [107]. The World Sustainability Council claims that with AI technology, architects may include sustainability in their plans. The management platform with AI-powered capabilities can recognize usage patterns and establish the perfect conditions for enhancing facility upkeep.

Modeling tools are used by engineers and architects to automate manual design work. Some people now use "predictive design" software, which uses AI to plan infrastructure and structures in accordance with rules and best practices. The design process is unlikely to be totally automated due to the cultural relevance of architecture; however, AI may assist optimize the design by identifying best practices by learning across thousands of projects. As a result, the machine can fully comply with the rules and construct the ideal house, hospital, or school for the situation. AI may choose construction materials from a preset set of effective parts that have a substantial level of sustainability thanks to peripheral manufacturing of the building elements mass-produced in automated factories, from minor components to whole modules.

Architectural floor plan design has often been a significant part of architecture. Since the modest pencil and drafting sheet, the skill of presenting the parameters of such blueprints has undergone several evolutions. CAD was one advancement that made the process more understandable and precise, and now AI is being used to automate CAD. Nowadays, creating floor plans necessitates a thorough investigation into the purpose, characteristics, and predetermined parameters before choosing how the best design may be executed. Before deciding to remain with a single approach or combine various design processes, traditional research required reading through dozens of previous works and concepts. However, AI has the power to change all of that. The process of designing architecture is seen from a fresh angle thanks to AI's capacity to use enormous data sets and analyze them to provide various outputs. It accomplishes this by operating inside predetermined boundaries, which may be dimensional, related to a particular architectural style, or topological in nature. These limitations, along with knowledge of fundamental techniques for architectural design, enable AI frameworks to quickly generate a number of plausible design solutions.

The outstanding capacity of AI to do real-time analytics, generative design, and the automation of architectural design and urban planning is presently being tapped by a few design companies. Daiwa House Group from Japan is one such instance. The company presently uses generative design and AI to improve its capacity to address architectural problems in its urban enclaves [108]. The use of AI streamlined and accelerated the architectural design process for its clients [109].

AI has the potential to significantly improve CAD software's ability to solve problems in the field of architecture. The decision-making process may be seamless and much faster than you anticipate with AI integration, eliminating the lag of paper forms and 2D or 3D CAD software. The study of various complicated design solutions for the same or comparable buildings in milliseconds using AI would undoubtedly save many productive hours in the same situation. The idea that computers and other platforms with software and hardware cannot be as creative as people has been debated throughout the years. Things are now changing thanks to AI, which combines human intellect and computational power to boost creativity to a new level.

The term "parametric modeling" is consistently mentioned in the majority of writing on contemporary architecture. With this cutting-edge design method, you may experiment with various building factors and adjust the structural visualization as necessary. AI has helped designers experiment with many factors to produce inventive ideas that would have otherwise been nearly impossible [110]. While complicated algorithms may be used to execute geometric programming in parametric CAD programs like dynamo and grasshopper to create complex structures, AI integration streamlines the process. AI is assisting architects with completing data points, modifying limitations, and producing design variants with a great deal less time, iterations, and, of course, more correctness [110]. Additionally, it would not harm or delay the timeline if one produced something entirely different from what was originally anticipated and flipped the idea design.

AI is still only a tool, despite the fact that it has the potential to lead the next paradigm change in architectural engineering. Because it needs skilled architects to create the frameworks and offer the resources it needs to automate, simplify, or accelerate architectural design processes, it may be thought of as an extension of the conventional pencil. Its primary goal is to help architects come up with workable design solutions in less-than-ideal circumstances. The cost of implementing AI will undoubtedly be significant in comparison to the established architectural solutions. Another difficulty would be the high expense of employing qualified AI experts who are also well-versed in architecture. However, if the situation were assessed from the perspective of ROI, it would undoubtedly be high because there are so many obvious advantages that the Architecture, Engineering, and Construction (AEC) sector will eventually have to adapt to. Overall, AI empowers architectural engineers to create smarter, greener, and more cost-effective structures, shaping a sustainable and technologically advanced future for the civil engineering industry.

1.3.4 Geotechnical engineering

Geotechnical engineering is concerned with the utilization of soils and rocks in engineering structures. In material modeling, soils and rocks naturally display complicated behavior and a high level of uncertainty. AI has emerged as a game-changer in this field. One of the key roles of AI in geotechnical engineering is data analysis and interpretation [111]. AI-powered algorithms can process and analyze vast amounts of geotechnical data, including soil test results, geological surveys, and historical data from similar projects. This enables engineers to gain valuable insights into soil behavior, subsurface conditions, and potential hazards, which are crucial for making informed decisions during the design and construction phases. In the past 30 years, an increasing number of researchers within the domain of geotechnical engineering have developed and implemented AI techniques. These techniques' effectiveness is attributed to their ability to predict complex nonlinear interactions. There are nine prominent areas where the application of AI methods is prominent: frozen soils and soil thermal properties, rock mechanics, subgrade soil and pavements, landslides and soil liquefaction, slope stability, shallow and pile foundations, tunneling and tunnel boring machines, dams, and unsaturated soils.

Over the past 3 decades, researchers in the field of geotechnical engineering have shown increasing interest in AI-driven modeling techniques as potential alternatives to conventional methods. AI techniques have demonstrated effectiveness in various applications, including prediction, monitoring, selection, detection, and identification. Furthermore, these techniques excel in cases where the underlying physical relationships between parameters are not well understood. AI methods enable the analysis of extensive datasets to uncover novel patterns, and they possess the capability to model intricate and nonlinear processes [111].

ML and DL algorithms play a significant role in predicting soil behavior and ground conditions [112]. By training on large datasets, AI models can learn complex patterns and relationships, enabling engineers to forecast settlement, bearing capacity, and slope stability with greater accuracy. This enhanced predictive capability allows for more robust and reliable designs, reducing the risk of geotechnical failures and cost overruns.

AI also streamlines the process of geotechnical parameter determination and soil classification. With the help of AI-driven tools, engineers can automate time-consuming tasks like soil testing and analysis, leading to faster and more efficient project planning. Additionally, AI can assist in optimizing foundation designs by considering a wide range of factors, such as soil properties, seismicity, and environmental conditions, resulting in more cost-effective and sustainable structures [111]. Furthermore, AI contributes to geotechnical monitoring during construction. Real-time monitoring systems equipped with AI can continuously assess ground conditions, detecting potential issues early on and allowing for prompt adjustments in construction methods and designs. This proactive approach enhances safety and minimizes the likelihood of geotechnical hazards.

Moreover, AI facilitates data-driven decision-making in geotechnical engineering. By integrating AI-generated insights with other engineering disciplines, such as structural and environmental engineering, civil engineers can achieve a more holistic approach to infrastructure development. This collaborative effort leads to more resilient and adaptive designs capable of withstanding various environmental and load conditions. In summary, AI has proven to be a transformative force in geotechnical engineering, empowering civil engineers to make data-informed decisions, optimize designs, and enhance the overall safety and performance of civil infrastructure. As AI continues to advance, its role in geotechnical engineering will undoubtedly expand, unlocking new possibilities for sustainable and resilient infrastructure development.

1.3.5 Transportation engineering

Due to the escalating effects of economic globalization, population expansion, intensified industrial production, safety imperatives, and environmental degradation, the challenges in transportation are becoming increasingly intricate. Urban expansion often grapples with limitations due to the incessant rise in traffic load on conventional transportation systems. Given the present level of computational capability and substantial strides in AI theory, ML models emerge as a promising tool to tackle the aforementioned complexities. ML-based solutions have already demonstrated notable efficacy in various scenarios, encompassing logistical service planning, event and object recognition from surveillance videos, and analysis of traveler behavior. Research and development endeavors are continually underway to explore further potential applications of AI. Many benefits of AI include better efficiency, improved pedestrian and driver safety, reduced costs, improved operations, and eco-friendliness. These factors are what essentially fuel the dynamic expansion of the transportation industry [113]. It is unequivocal that AI expansion is dynamic.

AI plays a transformative role in optimizing various aspects of transportation systems, revolutionizing mobility, safety, and efficiency. One of the key roles of AI in this field is traffic management and control. AI-powered traffic light systems can respond to changing traffic circumstances in real time, decreasing congestion and enhancing traffic flow [114]. ML algorithms analyze historical traffic data and patterns to predict future traffic conditions, enabling engineers to design better road networks and implement dynamic traffic management strategies.

AI also enhances the safety of transportation systems through various applications. Autonomous vehicles, enabled by AI technologies, are a promising avenue for reducing accidents and improving road safety. AI-driven collision avoidance systems, equipped with computer vision and sensor technologies, can detect potential hazards and assist drivers in making safer decisions [115]. Moreover, AI can analyze traffic data and patterns to identify high-risk areas and recommend targeted safety measures. Furthermore, AI contributes to the development of smart transportation systems. AI-powered routing and navigation systems can optimize travel routes for individual vehicles, taking into account real-time traffic conditions and road closures [116]. This leads to reduced travel times and fuel consumption, resulting in a more sustainable transportation network. Additionally, AI-driven logistics and fleet management systems optimize the movement of goods and services, improving supply chain efficiency and reducing delivery times.

AI also plays a crucial role in public transportation systems. AI-powered predictive maintenance systems can monitor the condition of buses, trains, and other transit vehicles, allowing for proactive maintenance and minimizing service disruptions [117]. Additionally, AI-based passenger demand forecasting can aid in optimizing public transportation schedules and resources, enhancing the overall quality of service. Moreover, AI contributes to transportation planning and infrastructure design. ML algorithms can analyze large datasets, including demographic information, land use, and travel patterns, to inform the development of comprehensive transportation plans. AI-driven simulations and models allow engineers to evaluate different scenarios, helping them make data-driven decisions to address current and future transportation needs.

AI's role in transportation engineering is transformative, offering opportunities for smarter, safer, and more efficient transportation systems. As AI technologies continue to advance, their integration into transportation infrastructure and operations will play a vital role in shaping the future of mobility and urban planning, creating a more connected, sustainable, and resilient transportation ecosystem.

1.3.6 Advanced civil engineering technologies

Construction costs can be reduced by up to 20% through the integration of robotics, AI, and the IoT. Engineers can deploy miniature robots equipped with VR goggles to navigate within newly constructed structures. These robots employ cameras to observe ongoing tasks. Furthermore, AI is instrumental in optimizing the layout of plumbing and electrical systems in contemporary buildings. The incorporation of AI extends to workplace safety solutions, where businesses utilize AI-driven systems to monitor real-time interactions among personnel, equipment, and objects at construction sites. This monitoring process helps identify potential safety hazards, design deficiencies, and productivity challenges, allowing managers to take timely corrective actions.

Contrary to forecasts of substantial workforce reductions, AI is anticipated to not entirely supplant the labor pool. Rather, it will induce shifts in business strategies within the construction domain, leading to decreased instances of costly errors, mitigated workplace accidents, and enhanced building operations [7]. Construction industry leaders should strategically allocate investments to areas where AI exhibits the greatest potential to align with the distinct requirements of their organization. Pioneering adopters will shape the trajectory of the industry, reaping both immediate and future benefits [118].

The combined application of AI with BIM, IoT, building twins, and 3D printing, 4D CAD creates a powerful ecosystem for civil engineering [119,120]. AI-powered digital twins can simulate construction scenarios, analyze performance, and optimize operational efficiency. AI-driven IoT networks can monitor and manage entire infrastructures, making real-time adjustments to improve safety and resource utilization.

Integrating AI with advanced civil engineering technologies comes with challenges, such as data security, interoperability, and ensuring AI's transparency and explainability. Addressing these concerns will pave the way for further advancements and the widespread adoption of AI-driven solutions in the construction industry. The convergence of AI with advanced civil engineering technologies presents a transformative opportunity for the industry. AI enhances the capabilities of BIM, IoT, building twins, 4D CAD, VR, and 3D printing, leading to better-designed structures, efficient construction processes, and smarter infrastructure management [5,121]. AI synergizes with cutting-edge civil engineering technologies, including BIM, IoT, building twins, and 3D printing [122].

1.3.6.1 3D-printing concrete

The integration of AI and 3D printing technologies has opened up exciting possibilities in the field of civil engineering, particularly in the construction industry. In recent years, AI has played a vital role in enhancing the efficiency, accuracy, and sustainability of 3D printing concrete. AI algorithms are employed to optimize the material formulation for 3D printing concrete. By analyzing the properties of different raw materials, AI can suggest the most suitable mix proportions, ensuring enhanced structural integrity and durability. This optimization process minimizes waste, reduces costs, and maximizes the use of locally available materials, making 3D-printed concrete more sustainable [123].

AI-driven systems allow real-time monitoring and control of the 3D printing process. Through integrated sensors and ML algorithms, AI can detect anomalies, adjust printing parameters, and ensure consistent layering and compaction. This degree of precision ensures the adherence of the final printed structure to the prescribed quality and safety benchmarks. AI's capacity to handle extensive datasets facilitates the generation of intricate designs and intricate geometries, overcoming challenges that were historically difficult to address through conventional construction approaches. It allows architects and engineers to customize designs as per project requirements, optimizing material distribution and reducing excess material usage.

AI can conduct structural analysis on 3D-printed concrete components, predicting their behavior under different loads and environmental conditions. By incorporating this feedback into the design process, engineers (civil, structure, material, and mechanical) can optimize the structures for better performance and safety while also reducing material consumption [124]. AI-driven robotic systems have the potential to perform 3D printing tasks autonomously, reducing the need for extensive manual labor. These robots can work collaboratively, carrying out tasks simultaneously and efficiently, thereby accelerating construction timelines and reducing construction costs. AI-powered 3D printing technology enables both on-site and off-site construction. On-site 3D printing can be deployed for rapid repair and construction in disaster-stricken areas or locations with limited accessibility. Off-site 3D printing allows for prefabrication of complex components in controlled environments, enhancing efficiency and quality.

The integration of AI and 3D printing in concrete construction marks a significant advancement in civil engineering. The role of AI in optimizing material formulation, enabling real-time process control, achieving complex geometries, and providing structural analysis and optimization is revolutionizing the construction industry.

Through AI's contributions, 3D printing of concrete is becoming more sustainable, cost-effective, and adaptable to various construction scenarios. As this technology undergoes further development, it harbors the potential to revolutionize our approach to civil engineering projects, presenting pioneering solutions to effectively address the future's requirements. The collaboration between AI and 3D printing is a prime example of how cutting-edge technologies can reshape traditional industries for the better.

1.3.6.2 Building information modeling

BIM is a 3D model-oriented approach that furnishes architects, engineers, and construction personnel with pertinent information for proficiently strategizing, designing, erecting, and supervising infrastructure and edifices. BIM constitutes a digital rendition of a structure's physical and operational attributes, functioning as a collaborative tool for architects, engineers, and construction experts to streamline the process of conceptualizing, constructing, and overseeing structures. The incorporation of architectural, engineering, and MEP plans and the synchronization of the activities of diverse teams within the 3D models are essential aspects for devising and framing project buildings. A major challenge arises in ensuring that the various models from distinct subteams do not overlap and cause conflicts. AI technologies have the potential to significantly enhance BIM workflows by offering advanced data analytics, automation, predictive capabilities, and optimization functionalities [125].

The AI-driven BIM technique assists professionals in the AEC sector to generate 3D model designs, facilitating efficient processes in planning, design, construction, and ongoing infrastructure maintenance [126]. The industry employs ML, particularly AI-driven generative design, to identify and address conflicts among the numerous models generated by diverse teams, thereby mitigating the need for extensive rework. This software systematically explores various potential solutions, employing ML techniques to propose design alternatives. By inputting user requirements into the model, the generative design program creates 3D models that are optimized within the specified constraints. It iteratively generates models, learning from each iteration to ultimately arrive at the optimal design solution.

AI can automate clash detection in BIM models, identifying potential clashes between different building elements (e.g., structural and mechanical systems) and streamlining the coordination process [127]. AI can analyze building designs and predict energy consumption to optimize energy efficiency, potentially saving costs and reducing environmental impact. AI can analyze BIM models to identify potential safety hazards during construction and operation, leading to improved safety protocols. AI-driven image recognition can help in automatically tagging and categorizing assets within the BIM model, making it easier to manage large-scale projects.

By leveraging AI in BIM, construction professionals can make informed decisions, reduce errors, optimize resources, and streamline the overall construction process, leading to cost savings and improved project outcomes. As AI continues to advance, its integration with BIM is expected to play an even more significant role in the construction industry.

1.3.6.3 Building twins

Civil engineering projects involve complex systems with numerous variables that can impact their performance and safety. Digital twin technology has emerged as a potent tool in this domain, facilitating real-time monitoring, analysis, and optimization of physical infrastructure [128]. By leveraging AI techniques, digital twins in civil engineering become more sophisticated, allowing engineers and stakeholders to make data-driven decisions, predict potential issues, and enhance the overall efficiency and resilience of infrastructure projects. Civil engineering projects, such as buildings, dams, bridges, and roads, are subject to various environmental factors and operational conditions that can affect their structural integrity and performance [129]. Digital twin technology offers a virtual representation of these physical assets, allowing engineers to simulate real-world scenarios and analyze their behavior. AI plays a pivotal role in the development and application of digital twins in civil engineering, offering three key advantages: structural analysis and design optimization, real-time monitoring and maintenance, and disaster resilience and risk assessment.

AI-driven digital twins enable engineers to analyze the structural behavior of infrastructure in real-time. By integrating sensor data from physical assets, AI algorithms can model and predict stress, strain, and load distribution, helping optimize the design for maximum efficiency and safety [120]. This capability allows civil engineers to explore different design alternatives virtually before committing to physical implementation, reducing the risk of structural failures and costly modifications.

AI-based digital twins facilitate continuous monitoring of infrastructure assets, detecting changes in conditions that might indicate structural deterioration or potential risks. The real-time analysis of sensor data allows for predictive maintenance, enabling engineers to schedule repairs and replacements before significant issues arise. This proactive approach leads to extended asset lifespans and a reduction in unplanned downtime. AI-enabled digital twins play a crucial role in assessing the resilience of civil infrastructure against natural disasters and extreme events. By integrating historical data, weather forecasts, and geographical information, AI algorithms can simulate potential scenarios, predict vulnerable areas, and recommend strategies to enhance the infrastructure's resistance to adverse conditions.

The integration of AI in twin technology offers a wide range of applications in the civil engineering domain. AI-driven digital twins can aid in designing and planning smart cities. By simulating the impact of proposed urban development projects, civil engineers can optimize transportation networks, urban layouts, and energy consumption, resulting in more sustainable and efficient cities. AI-powered digital twins can monitor the health of bridges by analyzing data from embedded sensors [130]. Engineers can assess the structural integrity, predict potential issues, and plan maintenance activities accordingly, ensuring public safety and minimizing disruptions. AI-driven digital twins allow civil engineers to optimize traffic flow in urban areas. By analyzing real-time data from traffic cameras, GPS devices, and weather conditions, engineers can develop efficient traffic management strategies using AI-driven digital twins to reduce congestion and improve transportation.

While AI-driven twin technology holds immense potential in civil engineering, certain challenges must be addressed. These include data security and privacy concerns, model interpretability, and the need for domain-specific AI algorithms tailored to civil engineering applications. Further research and collaboration between AI experts and civil engineers are essential to overcome these challenges and unlock the full potential of digital twin technology in the field. The integration of AI in twin technology has revolutionized the way civil engineers approach infrastructure planning, design, and maintenance. By harnessing AI's capabilities in data analysis, predictive modeling, and optimization, digital twins in civil engineering become indispensable tools for enhancing the efficiency, safety, and resilience of infrastructure projects. As AI continues to advance, the potential of digital twins to drive innovation and transform the civil engineering industry is bound to grow further.

1.3.6.4 Internet of Things (IoT)

The fusion (AIoT) of AI and IoT technologies has brought about a revolutionary transformation in various industries, and civil engineering is no exception [131]. IoT is being used by construction businesses to manage their fleets of cars and equipment. From real-time monitoring and predictive maintenance to smart infrastructure and sustainable construction, this synergy promises to create a safer, more efficient, and resilient built environment for future generations. IoT offers solutions like location awareness, predictive maintenance, fuel and battery usage, and much more with the use of AI measurements as inputs. Predicting the risk of equipment breakdown is now feasible with IoT devices and tags, which is a useful tool that saves time and money. AIoT has enabled the automation of energy-efficient operations in buildings. Additionally, IoT is automating construction sites to enhance safety; sensors are capable of accurately detecting workers' locations and sending notifications in the event of a fall or trip. Through the utilization of field reporting tools, foremen can log job-site activities and receive alerts, ensuring that vital project stakeholders are updated in real-time, even in their absence from the job site.

The convergence of IoT and AI in technology is anticipated to bring about substantial and enduring transformations in the construction industry. This union will not only introduce novel business opportunities and revenue streams but also foster innovative business models and frameworks that capitalize on the capabilities of AIoT. The integration of AIoT is projected to catalyze shifts in the construction sector's business models, particularly in domains such as logistics, customer relationship management, support services, workflow optimization, automation, and financial operations [132]. AI can further contribute to training through simulations of real-world scenarios, thereby reducing the occurrence of injuries and costly errors while simultaneously enhancing operational efficiency. Such utilization of AI can alleviate the impact of the skilled labor shortage by enabling operators to optimize their existing labor resources more efficiently.

In order to make equipment and components on the building site smarter, intelligent sensors are added to spray painting machines. According to materials on a gateway that permit sensors to be inserted within walls, the sensors are connected to a display in the helmet or even on a smartphone to offer feedback to the person operating that equipment and help make it better and more efficient. For instance, detecting temperature and humidity on the building site using intelligent and personal systems makes a significant contribution to quality control. When designing, planning, and constructing smart cities, AI may benefit greatly from the wealth of data that smart cities generate. AIoT may be used to build and administer public facilities related to transportation, healthcare, and road networks within smart cities if the recorded data is effectively utilized [133].

The synergy between AI and IoT in civil engineering marks a ground-breaking era for the industry. The combination of data-driven insights, autonomous systems, and smart infrastructure is set to revolutionize construction, maintenance, and urban planning, shaping a more sustainable, efficient, and resilient future for our cities and communities. To fully realize the potential of AIoT, collaborative efforts between industry, academia, and policymakers are crucial to drive innovation and implement these transformative technologies on a global scale.

1.3.6.5 Remote construction monitoring

There is a lot of room to use AI's capabilities in remote construction monitoring and management. The construction business constantly struggles to find workers since the work is dangerous and physically demanding. Construction has a far higher average turnover rate than any other industry. AI-powered robots give project managers the ability to monitor the current state and resource needs of several job sites [134].

AI-powered remote site analysis is another advantage of technology nowadays. There are other instances of AI being used to streamline the time-consuming labor and research involved in architecture design and urban planning. One of these is the use of AI to carry out a thorough investigation of spatial networks without having to go to real locations.

AI-powered sensors and devices can continuously monitor construction sites, collecting real-time data on various parameters such as progress, worker activities, equipment usage, temperature, humidity, and more [120]. This data allows project managers to have up-to-date insights without having to be physically present on-site. AI-based computer vision systems can analyze images and videos captured by drones or fixed cameras on the construction site. These analyses can identify safety hazards, measure progress, track equipment usage, and even detect defects, ensuring better project supervision and adherence to quality standards.

AI algorithms can use historical data and real-time inputs to predict potential delays, resource shortages, or other issues that might arise during construction. This enables proactive decision-making and risk mitigation. AI can monitor the health of construction machinery and equipment by analyzing sensor data and detecting anomalies. Predictive maintenance alerts can be generated, helping prevent costly breakdowns and ensuring optimal equipment utilization. AI-driven systems can enhance safety by identifying unsafe practices, potential hazards, and noncompliance with safety protocols. Alerts can be sent to project managers or supervisors to take immediate corrective action.

AI tools can automatically track construction progress based on data inputs, creating comprehensive reports for stakeholders. This streamlines communication and helps stakeholders stay informed about the project's status. AI can optimize resource allocation, such as labor and materials, based on project requirements, timelines, and historical data. This results in cost savings and increased efficiency. AI can effectively oversee environmental parameters within construction sites, including factors like air quality and noise levels. This oversight ensures adherence to environmental regulations and minimizes the project's ecological footprint on its surroundings. AI-powered platforms facilitate remote collaboration among project teams, enabling seamless communication and coordination regardless of the team members' physical locations.

Overall, AI's role in remote construction monitoring is to enhance decision-making, streamline operations, improve safety, and increase productivity. By leveraging AI technologies, construction companies can optimize project outcomes and reduce the need for physical site visits, leading to cost savings and improved project efficiency.

1.3.6.6 Construction robotics

The construction industry is rapidly evolving with robotics and AI. With higher precision and efficiency, robotics optimize productivity while reducing project timelines and labor costs. The integration of robotics-assisted automation nudged by AI ensures accuracy, minimizing errors and inconsistencies in measurements and alignment. Robots with cameras driven by AI can be used by construction businesses. To take 3D images, these robots may roam about the building site on their own. These images may be cross-checked with information from the bill of materials and BIM with the use of neural networks. This data is used by the engineers in charge of huge projects to monitor job progress. Additionally, it aids in the early detection of quality faults and helps monitor financial data and deadlines.

Robotics and AI in the construction sector are guaranteeing the finest building projects delivered while reducing costs and time [135]. As technology continues to progress, collaborative robots (cobots) and robots are expected to play a more significant role alongside human workers within the construction industry. Automation will replace tasks that can be mechanized with robots, while cobots will operate either independently or with minimal supervision. This collaborative approach has the potential to accelerate construction processes, reduce costs, enhance safety, and optimize decision-making. The integration of AI into construction not only addresses the labor shortage challenge but also triggers shifts in business models, reduces costly errors, and enhances building operations. Given these advancements, construction industry leaders are advised to strategically invest in areas where AI-enabled robotics can yield the greatest impact aligned with their specific demands. Early adopters of this digital transformation are positioned to outperform their competitors in the market. By gaining a competitive advantage, these pioneers can shape the industry's trajectory and achieve both short-term and long-term profitability.

In construction sites, AI-driven vehicles, such as autonomous bulldozers, excavators, and dump trucks, can operate with precision and accuracy. They can follow predefined routes, avoid obstacles, and adjust their movements in real-time, reducing the risk of accidents and human errors. Moreover, AI allows these vehicles to learn from their surroundings, adapt to changing conditions, and optimize their performance.

The use of AI-driven vehicles in civil engineering not only expedites construction timelines but also reduces labor costs. With autonomous machines taking over repetitive and labor-intensive tasks, human operators can focus on more strategic and complex aspects of the project. Furthermore, AI-enabled vehicles can collect and analyze vast amounts of data on construction progress, material usage, and site conditions. This data-driven approach enables project managers to make data-backed decisions, identify potential issues early on, and optimize resource allocation. However, the implementation of AI-driven vehicles in civil engineering also comes with challenges. Ensuring the safety of autonomous machines, maintaining their reliability, and addressing legal and regulatory concerns are areas that require ongoing attention.

A variety of AI-powered autonomous vehicles eventually worked together to schedule and fulfill tasks on time, increase productivity, cut expenses, and minimize their environmental effects. The introduction of partially and entirely autonomous vehicles is a result of a broad robotics boom. This is an era of engineering excellence and collaboration to shape a future where human brilliance and technology converge for remarkable advancements in construction.

Drone

AI has significantly transformed the field of land survey and mapping in civil engineering by revolutionizing data collection, analysis, and interpretation. Geospatial AI (GeoAI) and drones have emerged as powerful tools that leverage AI algorithms to enhance the accuracy, efficiency, and cost-effectiveness of surveying and mapping processes [136]. GeoAI refers to the application of AI techniques to geospatial data, such as satellite imagery, aerial photographs, and LiDAR scans. AI algorithms can process and analyze these vast datasets at unprecedented speeds, allowing civil engineers to extract valuable information, identify patterns, and make informed decisions.

Drones, equipped with sophisticated sensors and cameras, have become instrumental in land surveying and mapping. AI-powered drones can autonomously collect data over large areas, capturing high-resolution images and generating 3D models of the terrain [137]. AI algorithms can then process this data to create precise maps, identify features, and measure distances and elevations. The integration of AI, geospatial data, and drones streamlines the entire land survey and mapping process. Traditional surveying methods often require substantial time and manpower, but AI-driven techniques drastically reduce the time needed for data collection and analysis. This, in turn, accelerates project timelines and allows civil engineers to respond quickly to changes and challenges on-site. Furthermore, GeoAI enables the detection of hidden patterns and insights that may not be apparent through manual analysis [138]. AI aids in identifying potential risks, such as areas prone to erosion or landslides, and helps in better urban planning, infrastructure development, and environmental management [139].

In-depth land surveys and aerial photography of the task site may be completed in a significantly shorter amount of time for improved project management. Drones may help you monitor project progress and challenges on the job site, and geospatial information systems and GeoAI can empower you with superior decision-making for effective project management [140,141]. Drones are now crucial for tracking construction site progress since they provide precise contour maps, generate savings of thousands of dollars on surveys, and identify mistakes before removing machines from the site [142]. Drones make it possible to measure stockpiles, which improves the effectiveness of earth and soil transportation on the building site.

Mechanical and automatic robotics

AI and mechanical robotics have been increasingly integrated into various aspects of civil engineering, revolutionizing the way projects are designed, constructed, and managed. Self-driving construction equipment can swiftly and effectively do repetitive activities like welding, bricklaying, and concrete pouring. For prework and excavation, you can use automated or semiautomatic bulldozers. These machines carry out the task precisely in accordance with the requirements after the exact specifications are supplied to them. You can lower the hazards to human life associated with executing these activities while freeing up your human crew for actual building work. Mechanical robotics, such as autonomous construction vehicles, are being used for various tasks on construction sites. Mechanical robotic vehicles equipped with cameras can survey large areas quickly and accurately, providing real-time data for monitoring progress and identifying potential issues [143]. Autonomous machinery can also assist in repetitive tasks like bricklaying, excavation, and material transportation, improving efficiency and reducing labor-intensive work. AI-powered systems can analyze images and data from sensors to inspect and detect defects in structures, bridges, and roads. This helps ensure that construction meets quality standards and safety regulations. Additionally, robotic inspection devices can access hard-to-reach areas, reducing the need for human inspection in hazardous environments. By analyzing sensor data and historical information, AI can predict when maintenance is required, optimizing resources and minimizing downtime. Robotics can also be employed for routine maintenance tasks, reducing the need for human intervention in dangerous conditions. The integration of AI and mechanical robotics in civil engineering brings numerous benefits, including increased productivity, improved safety, cost reduction, and enhanced sustainability [144]. As technology continues to advance, the role of AI and robotics in civil engineering will undoubtedly expand, leading to even more innovative and impactful solutions in the future.

1.3.7 AI in space structures

Expandable space structures were among the initial civil engineering research topics in space structure, where transportability and ease of erection are major functional requirements [145]. AI is playing an increasingly crucial role in space structures within the area of civil engineering. As space exploration and colonization become more feasible and realistic, the demand for innovative and efficient solutions for constructing structures in space is growing. In the following six areas, AI can impact space structures in civil engineering.

1.3.7.1 Advanced structural design

Space habitat design is a very essential research topic in the field [146]. AI enables sophisticated structural design optimization for space habitats and other space structures. By analyzing various factors, including material properties, gravity conditions, and radiation exposure, AI algorithms can generate optimal designs that maximize strength and resource efficiency while minimizing weight and material usage. This ensures that space structures are not only robust but also cost-effective and feasible for transportation.

1.3.7.2 Autonomous construction

AI-driven robotics and automation will be instrumental in building space structures. Self-driving construction vehicles, drones, and robots equipped with AI will work together to assemble and construct space habitats [147]. These autonomous systems can adapt to the unique challenges of space environments, such as microgravity, extreme temperatures, and vacuum conditions, to perform construction tasks more efficiently and safely.

1.3.7.3 Real-time monitoring and maintenance

Space structures are subjected to various stresses and environmental factors that can impact their integrity as well as human health [148]. AI-integrated monitoring systems can continuously assess the health of space habitats and detect any structural anomalies or damages. This real-time monitoring enables prompt responses to potential issues, ensuring the safety and longevity of space structures. Additionally, the economic aspect and cost reduction are also important factors in space structures [146].

1.3.7.4 3D printing in space

AI plays a vital role in 3D printing technologies used in space construction. AI algorithms optimize printing paths, material usage, and structural integrity, ensuring the successful creation of intricate and robust components in the challenging space environment. 3D printing technology, enabled by AI, can also facilitate on-demand manufacturing of spare parts and tools, reducing the need for extensive prelaunch preparation [149,150].

1.3.7.5 Space debris management

With an increasing number of satellites and space missions, space debris poses a significant threat to space structures [151]. AI can aid in tracking and managing space debris to minimize the risk of collisions with space habitats. ML algorithms can predict the trajectories of space debris, enabling engineers to plan maneuvers to avoid potential hazards [152,153].

1.3.7.6 Adaptive and sustainable structures

AI can help design space structures that can adapt to changing needs and conditions in space. By learning from environmental data and user behavior, AI can optimize the habitat's layout, lighting, and energy usage, resulting in more sustainable and efficient space structures. Optimization of the aforesaid is very essential because humans experience declines in neurobehavioral performance, sleep loss, circadian rhythm disturbances, and other issues in space [154,155].

As humanity ventures into space exploration and colonization, the integration of AI in civil engineering for space structures will be pivotal in ensuring the success and sustainability of these ventures [156]. AI's ability to optimize design, enable autonomous construction, and provide real-time monitoring and maintenance will be invaluable in creating resilient, safe, and efficient space habitats that will support and advance humanity's presence beyond Earth [156]. The collaboration between AI and civil engineering will continue to push the boundaries of space exploration, making the vision of human colonies in space closer to reality.

1.3.8 Remote sensing and surveying engineering

There is a significant impact of AI in the field of remote sensing and surveying engineering within civil engineering. AI has revolutionized data acquisition, analysis, and decision-making processes, enhancing the efficiency, accuracy, and speed of remote sensing and surveying activities. This section presents various AI-powered applications, including image analysis, feature extraction, anomaly detection, and autonomous surveying, and highlights their benefits and challenges in the civil engineering domain. Remote sensing and surveying engineering play a crucial role in gathering essential geospatial information required for various civil engineering projects. Traditional surveying methods involve manual measurements, which are labor-intensive, time-consuming, and may introduce errors. The integration of AI in this domain offers promising solutions to overcome these challenges, providing more accurate and efficient data acquisition and analysis methods.

1.3.8.1 AI-powered image analysis for remote sensing

AI algorithms, particularly DL models, have demonstrated remarkable capabilities in image analysis tasks. In remote sensing, AI can automatically interpret satellite or aerial imagery, enabling rapid extraction of features such as roads, buildings, vegetation, water bodies, and land use patterns [157]. CNNs have shown exceptional performance in semantic segmentation, object detection, and classification tasks, allowing civil engineers to identify and monitor changes in the landscape more effectively.

1.3.8.2 Feature extraction and change detection

AI algorithms can extract meaningful features from remote sensing data, facilitating the identification of critical areas in civil engineering projects. By comparing multiple images over time, AI can detect changes such as land subsidence, erosion, or construction progress, providing invaluable insights for decision-making and risk assessment [158].

1.3.8.3 Anomaly detection and infrastructure monitoring

AI-driven anomaly detection techniques enhance the monitoring of civil infrastructure, such as bridges, dams, and pipelines. By analyzing sensor data, including LiDAR,

GPS, and vibration sensors, AI algorithms can identify irregular patterns that may indicate structural issues or potential failures, enabling preventive maintenance and ensuring public safety.

1.3.8.4 Autonomous surveying with AI

AI-powered robotics and drones have revolutionized surveying by automating data collection processes. Drones equipped with AI can fly predefined routes, capture images, and create high-resolution 3D models of construction sites or large areas [159]. These autonomous systems reducc human intervention, increase surveying frequency, and provide real-time data for project monitoring.

1.3.8.5 Data integration and fusion

AI also contributes to integrating and fusing data from multiple sources, such as satellite imagery, aerial surveys, ground-based sensors, and historical data. By combining various data streams, civil engineers can gain comprehensive insights into the project area, leading to better-informed decisions and optimized project planning.

While AI offers numerous advantages, there are challenges to address in its implementation in remote sensing and surveying engineering. AI models rely on vast and high-quality datasets for accurate predictions. Obtaining such datasets can be challenging, especially in remote or inaccessible areas. Some AI models, like deep neural networks, lack interpretability, making it difficult to explain the reasoning behind their decisions. AI algorithms often demand significant computational power and resources, which may be costly for some civil engineering projects. Integrating AI into civil engineering processes requires adherence to regulations and ethical considerations, especially concerning data privacy and security [6].

The role of AI in remote sensing and surveying engineering within civil engineering has expanded rapidly, providing innovative solutions to improve data acquisition, analysis, and decision-making. With further advancements and overcoming existing challenges, AI is poised to revolutionize the field, empowering civil engineers to design and construct projects with greater precision, safety, and efficiency. Embracing AI in this domain is crucial to staying at the forefront of technological advancements and meeting the complex demands of modern civil engineering projects.

1.3.9 Water resource engineering

Water is a critical natural resource, and managing it effectively is essential for sustainable development. AI technologies offer advanced tools for data analysis, modeling, and decision-making, revolutionizing water resource management practices. This section delves into various AI applications, including predictive modeling, optimization, hydrological forecasting, water quality monitoring, and leak detection, showcasing their benefits and potential challenges in the domain of civil engineering. Water resource engineering holds significant importance in the sustainable administration of water for diverse applications, encompassing potable water provisioning, irrigation, hydropower production, and ecological preservation. In light of the growing intricacies associated with water-related issues, AI has emerged as a potent instrument to augment water resource management methodologies. By furnishing data-derived insights and facilitating intelligent decision-making support, AI bolsters the effectiveness of water resource management practices.

1.3.9.1 Predictive modeling and system optimization

AI-driven predictive modeling enables water resource engineers to forecast water demand, assess potential water stress areas, and optimize water allocation for various uses. ML algorithms can analyze historical data and patterns to predict future water demand accurately, ensuring efficient resource allocation and reducing waste.

1.3.9.2 Hydrological forecasting and flood management

AI plays a crucial role in hydrological forecasting, enabling early warning systems for floods and other water-related disasters. By processing real-time data from weather stations, rainfall gauges, and remote sensing, AI models can predict river flow patterns and potential flood events, helping authorities take proactive measures to minimize damages [160].

1.3.9.3 Water quality monitoring and contamination detection

Maintaining water quality is vital for public health and environmental preservation. AI-powered systems can process data from water quality sensors and detect anomalies, indicating potential contamination sources [161]. Real-time monitoring and analysis facilitate rapid responses and prompt actions to mitigate pollution risks.

1.3.9.4 Intelligent irrigation and water conservation

AI can optimize irrigation practices by analyzing weather conditions, soil moisture levels, and crop data to determine the precise amount of water needed for irrigation. Smart irrigation systems based on AI algorithms help conserve water resources while ensuring optimal crop yields [162,163].

1.3.9.5 Leak detection and infrastructure maintenance

Water loss due to leakages in the distribution network is a significant concern for water utilities. AI-powered leak detection systems can analyze pressure and flow data, identify potential leaks, and pinpoint their locations accurately [161]. By addressing leaks promptly, water utilities can reduce losses and improve infrastructure maintenance.

1.3.9.6 Data integration and decision support

AI facilitates the integration of diverse datasets from sensors, satellites, and other sources. By assimilating this data, water resource engineers can gain a comprehensive understanding of water systems, aiding in informed decision-making and policy formulation [164]. While AI offers tremendous potential for water resource engineering, certain challenges must be addressed. AI models require extensive and reliable data to achieve accurate results. Access to comprehensive datasets can be a challenge, especially in some regions with limited data collection infrastructure. Some AI algorithms, particularly DL models, exhibit a deficiency in interpretability, posing a challenge in comprehending the underlying rationales behind their predictive outcomes. Implementing AI systems may require substantial computational resources and expertise, which may be a limitation for some water resource management agencies.

The role of AI in water resource engineering within civil engineering is rapidly evolving, empowering water professionals to make informed decisions and develop sustainable management strategies. By harnessing the power of AI for predictive modeling, optimization, hydrological forecasting, water quality monitoring, and infrastructure maintenance, the water sector can efficiently address current and future challenges. Embracing AI technologies is essential to secure water resources, promote water conservation, and achieve a more resilient and sustainable water management system.

1.3.10 Environmental engineering

Environmental challenges, such as pollution, climate change, and resource depletion, demand innovative solutions for sustainable development. AI technologies offer powerful tools for data analysis, modeling, and decision-making, revolutionizing environmental monitoring, impact assessment, and conservation efforts. This section presents various AI applications, including environmental monitoring, pollution control, climate modeling, ecological forecasting, and sustainable design, highlighting their benefits and potential challenges in the domain of civil engineering. Environmental engineering plays a critical role in safeguarding the environment and promoting sustainable practices in civil engineering projects. With the growing complexity of environmental challenges, AI has emerged as a transformative force, offering advanced solutions to address pressing issues and improve environmental management. AI can assist in solving both indoor and outdoor environmental issues throughout the globe with different variable constraints.

1.3.10.1 Environmental monitoring and data analysis

AI applications enhance environmental monitoring by processing large volumes of data collected from sensors, satellites, and other sources. ML algorithms can analyze this data to detect patterns, identify trends, and assess environmental conditions in real time. Such insights empower environmental engineers to make informed decisions for pollution control and resource management.

1.3.10.2 Pollution control and remediation

AI-driven models assist in pollution control efforts by predicting pollutant dispersion patterns, identifying pollution sources, and optimizing pollutant removal processes. AI also aids in designing efficient and cost-effective remediation strategies for contaminated sites, ensuring environmental restoration [165].

1.3.10.3 Climate modeling and adaptation strategies

AI plays a vital role in climate modeling, helping environmental engineers understand complex climate dynamics and predict future climate scenarios [166]. By simulating different climate change scenarios, AI models aid in the development of adaptive strategies to mitigate the impacts of climate change on civil engineering projects.

1.3.10.4 Ecological forecasting and biodiversity conservation

AI-powered ecological forecasting models enable environmental engineers to predict changes in ecosystems and biodiversity patterns. These predictions can guide conservation efforts, ecosystem restoration projects, and habitat protection initiatives.

1.3.10.5 Sustainable design and construction

AI-driven tools facilitate sustainable design in civil engineering projects. AI algorithms can optimize building designs to reduce energy consumption, improve material efficiency, and minimize environmental impacts [167]. Furthermore, AI assists in selecting sustainable construction materials and techniques, promoting green infrastructure development.

1.3.10.6 Waste management and the circular economy

AI applications improve waste management practices by optimizing waste collection routes, identifying recycling opportunities, and minimizing landfill usage. AI also contributes to the development of circular economy models, promoting the reuse and recycling of materials in civil engineering projects [168]. While AI offers numerous advantages in environmental engineering, certain challenges must be addressed. AI models require high-quality and diverse environmental data for accurate predictions. Data collection, especially in remote or ecologically sensitive areas, may pose challenges. Certain AI algorithms, notably DL models, exhibit a deficiency in interpretability, rendering the task of elucidating the reasoning behind their decisions complex. This is especially significant for environmental decision-making processes. Implementing AI in environmental applications necessitates ethical considerations, particularly regarding data privacy, equity, and community engagement.

1.3.10.7 Indoor environmental quality

Research on indoor environments is being transformed by AI. The use of sensors, ML, and data analytics by AI improves the monitoring of indoor air quality. Pollutant, temperature, and humidity sensors gather data, which AI analyzes in real-time to detect patterns and irregularities indicating the need for appropriate ventilation system modifications. By taking into account past data along with external conditions, AI can also forecast trends in air quality, assisting proactive steps. AI also improves a building's ability to use energy efficiently. Through smart building management systems, it optimizes HVAC and lighting depending on occupancy, weather predictions, and energy prices, lowering energy expenditures and having a less significant environmental effect. AI's capacity to personalize indoor environments and gather input on preferences enhances user comfort and productivity. Benefits from predictive maintenance, security, and safety are obtained when AI improves surveillance systems and forecasts equipment breakdowns. A healthier, more energy-efficient, and smarter indoor environment is promoted by AI's multidimensional role in indoor environment research, which is advantageous for both people and the environment.

The integration of AI applications in environmental engineering within civil engineering holds immense potential to address environmental challenges, promote sustainability, and ensure responsible development. By leveraging AI for environmental monitoring, pollution control, climate modeling, ecological forecasting, and sustainable design, civil engineers can play a pivotal role in preserving natural resources and protecting ecosystems. Embracing AI technologies is vital for creating a more resilient, ecologically sound, and sustainable future for civil engineering projects worldwide.

1.3.11 Coastal engineering

Coastal areas are susceptible to various natural hazards, including erosion, flooding, and sea-level rise, necessitating innovative solutions for effective coastal management and protection. AI technologies offer advanced tools for data analysis, prediction, and decision-making, revolutionizing coastal monitoring, hazard assessment, and adaptation strategies [169]. This section shares various AI applications, including coastal modeling, shoreline change analysis, flood prediction, ecosystem preservation, and sustainable coastal infrastructure design, highlighting their benefits and potential challenges in the domain of civil engineering. Coastal engineering plays a crucial role in managing and safeguarding coastal areas from natural hazards and climate change impacts. With the increasing vulnerability of coastal regions, AI has emerged as a valuable tool to enhance coastal management practices, providing data-driven insights and intelligent solutions for coastal protection and adaptation.

1.3.11.1 Coastal modeling and risk assessment

AI-powered coastal models facilitate accurate simulations of complex coastal processes, including wave dynamics, sediment transport, and shoreline evolution. These models help predict erosion rates, sediment deposition patterns, and storm surge impacts, enabling engineers to assess coastal risks and design effective protection measures.

1.3.11.2 Shoreline change analysis and monitoring

AI applications assist in analyzing historical shoreline data and satellite imagery to track long-term coastal changes. ML algorithms can identify trends and anomalies, enabling engineers to monitor erosion hotspots and plan targeted mitigation strategies.

1.3.11.3 Flood prediction and early warning systems

AI-driven flood prediction models integrate real-time data from tide gauges, weather stations, and remote sensing to forecast coastal flooding events [170]. By providing early warning systems, AI helps communities and authorities take proactive measures to minimize damages and protect lives and property.

1.3.11.4 Ecosystem preservation and restoration

AI plays a vital role in preserving and restoring coastal ecosystems. AI models aid in identifying critical habitats, assessing ecological health, and predicting the impacts of coastal developments on marine and aquatic life. This information supports the implementation of conservation measures to maintain biodiversity and ecosystem services.

1.3.11.5 Sustainable coastal infrastructure design

AI-driven optimization algorithms contribute to the design of sustainable coastal infrastructure. By analyzing various parameters, such as wave loads, sediment transport, and environmental impacts, AI models can optimize the design of coastal structures, such as breakwaters and seawalls, for enhanced resilience and reduced environmental impacts.

1.3.11.6 Climate change adaptation and resilience strategies

AI applications help coastal engineers develop adaptation strategies to address climate change-induced sea-level rise and extreme weather events [171]. AI models can predict future coastal scenarios, guiding the development of resilient infrastructure and land-use planning in vulnerable coastal areas. AI models require extensive and reliable coastal data for accurate predictions. Limited data availability, especially in remote or poorly monitored areas, can be a challenge. Validating AI models for complex coastal processes and accounting for uncertainties in predictions can be demanding and require careful consideration. Integrating AI with traditional coastal engineering practices may require expertise and adaptation to new workflows.

The integration of AI applications in coastal engineering within civil engineering is essential for effective coastal management and protection. By leveraging AI for coastal modeling, shoreline analysis, flood prediction, ecosystem preservation, and infrastructure design, civil engineers can better understand and address coastal challenges. Embracing AI technologies is crucial for building resilient coastal communities and ensuring the sustainable development of coastal regions in the face of environmental changes and hazards.

1.3.12 Renewable power engineering

With the increasing global demand for clean and sustainable energy sources, renewable power technologies have gained prominence. AI technologies offer advanced tools for data analysis, optimization, and decision-making, revolutionizing renewable energy generation, grid integration, and system management. This section presents various

AI applications, including renewable resource assessment, predictive maintenance, energy forecasting, grid optimization, and smart energy management, highlighting their benefits and potential challenges in the domain of civil engineering. Renewable power engineering plays a vital role in the transition to a low-carbon and sustainable energy future [172]. As renewable energy sources, such as wind, solar, hydro, and geothermal, become integral to the global energy mix, AI has emerged as a transformative force, enhancing the efficiency and reliability of renewable power systems.

1.3.12.1 Renewable resource assessment

AI applications play a crucial role in accurately assessing the potential of renewable resources, such as solar irradiance, wind speed, and hydrological data. ML algorithms analyze historical climate data and satellite imagery to predict renewable energy availability, enabling better site selection and optimal power plant design.

1.3.12.2 Predictive maintenance for renewable infrastructure

AI-driven predictive maintenance models monitor the health of renewable energy infrastructure, such as wind turbines and solar panels [172]. By analyzing sensor data and detecting anomalies, AI can predict potential equipment failures, allowing timely maintenance and minimizing downtime.

1.3.12.3 Energy forecasting for optimal integration

AI-powered energy forecasting models predict renewable energy generation patterns. This information is essential for grid operators and energy planners to efficiently integrate variable renewable sources into the power grid. Accurate forecasting supports grid stability and ensures optimal power generation and consumption.

1.3.12.4 Grid optimization and flexibility

AI applications facilitate grid optimization to accommodate higher penetration of renewable energy. AI algorithms process real-time data to effectively handle power flow dynamics, harmonize supply-demand equilibrium, and optimize the distribution of energy resources. Moreover, AI aids in managing energy storage systems for grid flexibility and stability.

1.3.12.5 Smart energy management in buildings

AI plays a vital role in optimizing energy consumption in buildings and infrastructure. Building management systems guided by AI can dynamically regulate heating, cooling, and lighting protocols by considering occupancy trends and weather conditions. This approach minimizes energy waste while simultaneously elevating energy efficiency [4].

1.3.12.6 Demand response and load balancing

AI applications enable demand response strategies to manage electricity consumption during peak hours. AI algorithms communicate with smart appliances and systems, adjusting loads to match renewable energy availability and grid conditions, contributing to load balancing and reducing the need for conventional fossil fuel-based backup power. AI models rely on high-quality and real-time data for accurate predictions. Data collection and integration from diverse sources can be complex. AI algorithms may demand significant computational resources, particularly for large-scale renewable energy systems.

The integration of AI applications in renewable power engineering within civil engineering is paramount to driving the adoption of clean and sustainable energy technologies. By leveraging AI for renewable resource assessment, predictive maintenance, energy forecasting, grid optimization, and smart energy management, civil engineers can foster the widespread adoption of renewable energy and contribute to a greener and more sustainable future. Embracing AI technologies is essential to optimize renewable power generation, enhance grid stability, and accelerate the transition toward a decarbonized energy landscape.

1.4 Challenges of artificial intelligence in civil engineering

AI has undeniably revolutionized the civil engineering domain, offering unprecedented opportunities to tackle complex problems and streamline processes. However, with the vast potential of AI come several challenges that need to be addressed for its successful integration and optimal utilization in the field. The challenges of implementing AI in civil engineering are presented in Fig. 1.6.

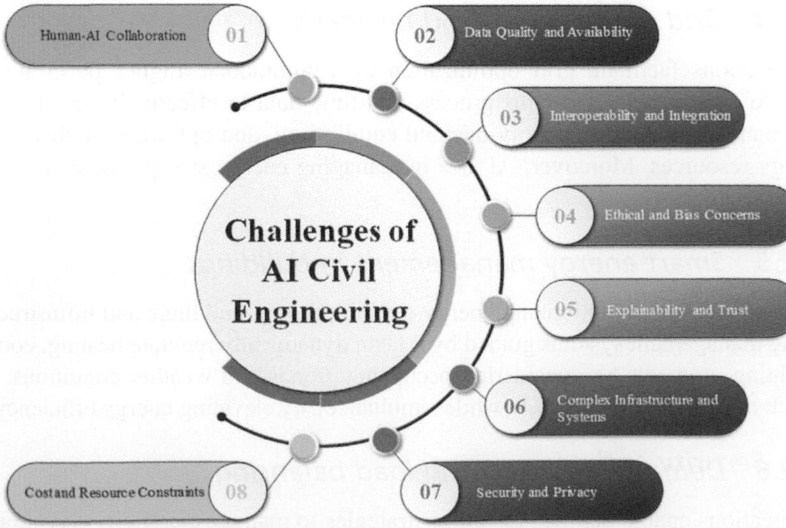

Figure 1.6 Challenges of AI in civil engineering.

1.4.1 Data quality and availability

AI algorithms are heavily dependent on substantial quantities of high-quality data for both their training and decision-making processes. In civil engineering, obtaining comprehensive and accurate datasets can be challenging due to the diverse and dynamic nature of projects. Data collection, integration, and validation require significant efforts, and the lack of standardized data formats can hinder seamless collaboration.

1.4.2 Interoperability and integration

The integration of AI with existing civil engineering software and systems poses a significant challenge. Many AI solutions may not be designed to work harmoniously with legacy tools, leading to interoperability issues and potential data inconsistencies. Civil engineers must develop strategies to bridge this gap to ensure smooth AI adoption.

1.4.3 Ethical and bias concerns

AI algorithms have the potential to inadvertently reinforce biases that exist within the data used for their training, which can result in the generation of discriminatory outcomes during decision-making processes. In civil engineering, biased AI models could affect infrastructure planning, community development, and resource allocation. Addressing ethical considerations and ensuring fairness in AI models is paramount.

1.4.4 Explainability and trust

AI models frequently function as "black boxes," creating difficulties in comprehending the logic behind their decision-making processes. In civil engineering, where safety and reliability are crucial, the lack of model explainability can lead to skepticism and reluctance to adopt AI-driven solutions. Enhancing the transparency and interpretability of AI algorithms is crucial to building trust among stakeholders. Explainable AI is one solution to nudge the transparency and trust of AI and is blooming among researchers these days [173].

1.4.5 Complex infrastructure and systems

Civil engineering projects involve intricate systems with numerous variables and uncertainties. AI models that work well in controlled environments may struggle to handle the intricacies of real-world construction sites. Tailoring AI algorithms to account for the specific challenges in civil engineering is vital for accurate and practical outcomes.

1.4.6 Security and privacy

The use of AI in civil engineering often involves sharing sensitive data, including geospatial information and project details. Ensuring data security and protecting the privacy of individuals and organizations involved in construction projects is a critical concern that requires robust cybersecurity measures.

1.4.7 Human-AI collaboration

Successfully integrating AI into civil engineering requires a shift in mindset and work culture. Civil engineers must embrace the idea of collaborating with AI systems rather than perceiving them as replacements. Striking an accurate equilibrium between human expertise and AI assistance is essential for optimal project outcomes.

1.4.8 Cost and resource constraints

Implementing AI-driven solutions may involve significant upfront costs, especially for smaller firms or developing regions. Adequate infrastructure, hardware, and expertise to leverage AI may not always be readily available, limiting its widespread adoption.

Despite the challenges, the potential of AI in civil engineering remains immense. Addressing these obstacles will require collaboration among civil engineers, data scientists, policymakers, and industry stakeholders. By collectively working toward overcoming these challenges, the civil engineering community can harness the true power of AI, leading to more sustainable, efficient, and innovative construction practices for the betterment of society.

1.5 AI in civil engineering: A way forward

The future of civil engineering is set to be profoundly transformed by the integration of AI. As technology continues to advance, AI-driven applications are poised to revolutionize how infrastructure is designed, constructed, and managed. With the potential to address pressing challenges and unlock unprecedented opportunities, a glimpse of futuristic AI applications in civil engineering is now visible in real-time.

AI-driven robotics will spearhead a new era of autonomous construction. Self-driving construction vehicles and drones equipped with AI algorithms will work collaboratively to perform tasks with unparalleled precision and efficiency. These robots will navigate construction sites, lay bricks, weld, and perform intricate tasks, reducing reliance on human labor and enhancing safety. AI will play a pivotal role in designing smarter, more resilient infrastructure. By analyzing vast datasets and incorporating factors like climate change, urban growth, and sustainability goals, AI models will generate optimal designs that maximize performance and resource efficiency. From skyscrapers to bridges, AI-optimized structures will define the cities of the future.

AI-driven predictive maintenance will revolutionize how civil engineers manage and maintain infrastructure. Smart sensors and AI algorithms will monitor the health of bridges, roads, and buildings in real-time, detecting potential issues before they escalate. This proactive approach will prolong asset lifespans, minimize downtime, and save costs. AI-powered smart cities will become a reality, revolutionizing urban planning and management. AI algorithms will analyze data from IoT sensors, traffic cameras, and weather forecasts to optimize traffic flow, energy consumption, waste management, and public services. Smart cities will be more sustainable, resilient, and citizen-centric. Fig. 1.7 represents the future of AI in civil engineering.

AI applications will help civil engineers tackle climate change challenges. From flood prediction to coastal erosion management, AI-driven models will provide insights into the impacts of climate change on infrastructure. This data-driven approach will enable engineers to develop adaptive and resilient solutions. AI-enabled augmented reality will transform how civil engineers visualize and interact with designs. Engineers can use AI-generated virtual prototypes to walk through infrastructure projects, making real-time adjustments and detecting potential clashes before construction begins. This immersive experience will enhance collaboration and reduce errors. AI-powered safety systems will improve job site safety. Wearable devices and AI algorithms will continuously monitor workers' movements and environmental conditions, issuing alerts for potential hazards. These AI-driven safety measures will prevent accidents and protect the workforce.

AI is quite useful for additional classifications and determining the percentage of soil moisture content. ML may be used in the field of structural engineering to locate and quantify damage using sensory or visual data. AI applications can be used to enhance output by cutting down on idle time and determining the ideal moisture content and maximum dry density for concrete. A significant area is using image recognition to correctly monitor a site while taking security and dangerous working conditions into account. AI in identifying the gaps and the supplies needed to complete the job as soon as possible is another opportunity. AI for predicting trip times and improving signs in transportation engineering. BIM is enabled with AI for effective infrastructure planning, design, and management. ANN is used to forecast concrete mix design features as well. Monitoring building site activities and forecasting cost changes based on raw material market pricing can be effectively tackled by using AI. Investigating foundation

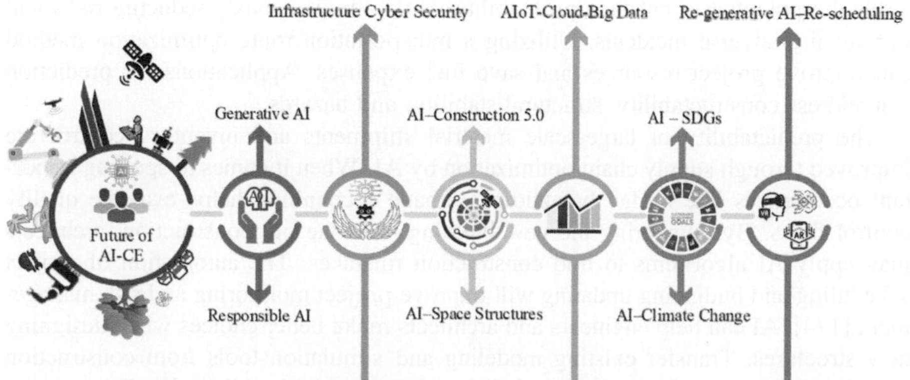

Figure 1.7 Future of AI in civil engineering.

settlement and slope stability is one more important area for AI in civil engineering. To monitor the building's structural health in real-time and provide notifications when and where repairs are needed using several AI techniques. Aiding in tidal forecasting by AI to help in maritime building is an advanced sector. Using AI in data analysis to reduce mistakes in the project is essential. AI can contribute as part of project management, create site layouts, and forecast hazards. In foundation engineering, a remedy for damage caused by prestressed concrete pile drives must be devised. To address complex difficulties at various phases of the project, AI is the best tool. To make judgments on the subject of design, AI can play a vital role. AI can be used in the fields of construction waste management and smart material handling. AI for expert cost monitoring and optimization in the work system is also an important area.

AI has several drawbacks in civil engineering as well. Technological advancements are accompanied by a rise in expenses. AI adoption in the construction industry necessitates periodic software upgrades. Similarly, another facet of technological invasion is the lack of human employment prospects. Robotic functions under AI are more suited to replace human labor. Construction employment is declining significantly, and the remaining workforce will be impacted further. While AI is effective in mitigating potential hazards on construction sites, its inherent limitation lies in its confined ability to perform tasks within its programmed scope. In contrast, skilled manual laborers can apply creative problem-solving approaches to tasks. The implementation of intricate algorithms tailored to the construction sector necessitates trained personnel and sufficient time. Undoubtedly, technological progress has facilitated various aspects of life. The construction industry, traditionally less influenced by software interventions, is experiencing a sudden transformation. Anticipating further developments in AI and ML concepts is plausible in the years ahead!

AI adoption has been progressively expanding across all phases of the construction process, encompassing planning, design, and construction execution. The integration of AI within construction operations facilitates comprehensive project oversight, offering insights into risk mitigation, schedule adherence, structural soundness, and other critical considerations. Harnessing the capabilities of AI in the construction sector holds the potential to enhance profitability while simultaneously reducing risks and minimizing adverse incidents. Utilizing a transportation route optimization method can improve project resources and save fuel expenses. Applications for prediction can address constructability, structural stability, and hazards.

The predictability of large-scale material shipments and inventory control are improved through supply chain optimization by AI. When it comes to spotting important occurrences like bridge breakdowns, image recognition helps evaluate quality control flaws. By comparing the new building with the old construction, architects may apply AI algorithms to find construction mistakes. The automation of project scheduling and budgeting updating will improve project monitoring and risk management [174]. AI can help engineers and architects make better choices when designing new structures. Transfer existing modeling and simulation tools from construction planning to the actual practical work being done on the building site. To design a BIM system that includes everyone required to deploy the solution, develop an integrated and intelligent prototype solution.

Many repetitive and time-consuming operations, like transporting materials, setting rebar, constructing brick walls, and even erecting 3D-printed structures, are handled by programmed mechanical arms. With the aid of machines that are powered by digital data created in information models during the design process, architects can now do a number of jobs with higher accuracy. AI digitization lowers operating costs, eliminates human error, and increases productivity. Emerging technologies are currently altering the practice of architecture as well as its conventional design methodologies. Using innovative techniques like 3D printing, robots, cloud computing, and AI is part of this.

The future of civil engineering is bound to be shaped by the transformative potential of AI applications. From autonomous construction and smart infrastructure design to predictive maintenance and climate change adaptation, AI will revolutionize how civil engineers approach challenges and create sustainable, efficient, and resilient infrastructure for the world of tomorrow. Embracing these futuristic AI applications will usher in a new era of innovation, propelling the civil engineering industry into uncharted territories of progress and possibility.

1.6 Conclusion

The current era necessitates developing a positive attitude toward investigating AI applications in civil engineering so that even the most mundane tasks may be completed quickly and precisely. Responsible AI research is also gaining traction nowadays to nudge automation with safety. AI-powered sensors and data analytics enable real-time monitoring of infrastructure, predicting potential failures, and optimizing maintenance. AI and IoT integration streamlines construction workflows, enhances productivity, and improves safety. AI helps design energy-efficient buildings and eco-friendly construction practices, contributing to sustainable development. AI optimizes traffic flow, reduces congestion, and aids in designing smarter, more accessible cities. AI and IoT provide real-time data for disaster prediction, early warnings, and effective disaster response planning. Drones with AI-driven analytics autonomously inspect infrastructure and conduct maintenance tasks in hazardous areas. AI-enabled systems track the lifecycle of infrastructure, facilitating predictive maintenance and cost optimization. AI-driven video analytics enhance security at construction sites and critical infrastructure.

AI techniques like ML and DL are majorly used in civil engineering. In ML, algorithms learn from data to make predictions, optimize designs, and perform PR tasks. In DL, neural networks process complex data and enable advanced tasks like image recognition and natural language processing. Focusing on AIoT, AI combines with IoT devices to gather real-time data, enabling data-driven decision-making. High-quality data is essential for training accurate AI models, but data availability can be limited. Integrating AI with existing systems may be challenging due to compatibility issues. Ensuring AI compliance with regulations and ethical considerations is crucial, especially in safety-critical applications. There may be a shortage of skilled professionals proficient in both AI and civil engineering.

AI can optimize construction processes, reducing costs and project timelines. AI enables proactive maintenance, reducing downtime and extending infrastructure lifespan. AI can promote environmentally friendly designs and construction practices, supporting the SDGs. AI-driven analytics enhance safety measures, reducing accidents and improving worker well-being. The future of AI in civil engineering is very bright. AI-driven autonomous machinery will revolutionize construction processes and improve safety. AI-powered urban planning will lead to the development of smarter, more sustainable cities. AI will play a crucial role in creating resilient infrastructure that can withstand various environmental challenges. Engineers will work closely with AI systems to enhance decision-making and innovation in civil engineering projects.

In the near future, the AI in civil engineering will have "responsible AI," as "generative AI" is developing presently at a very good pace. As AI continues to advance, it will profoundly impact civil engineering and allied domains, shaping a more sustainable, efficient, and resilient built environment for the future. Overcoming challenges and leveraging opportunities will be key to unlocking the full potential of AI in these fields. The ongoing digital transformation of the building industry is marked by the integration of AI. To unlock the sector's full potential, a comprehensive approach that incorporates technologies such as AI and ML throughout the entire engineering and construction lifecycle is imperative. This includes their application in design, preconstruction, construction, operations, and asset management phases. Companies that wish to maintain their competitive advantage should improve their technology as soon as possible with AI integration.

Eventually, there is no exaggeration in stating that "AI holds a future full of opportunities". However, in civil engineering along with all other domains the critical concern of the hour is… Can AI replace humans? Not sure! However, it is clear right now that someone who uses AI will undoubtedly replace someone who is not familiar with it. Wise engineers, architects, and professionals should work with AI for the welfare of society. The planet is still revolving at the same pace, but you now have new or updated information about AI and its significance in civil engineering, which helps in sprouting the world with technological advancements, and you can sustain yourself in the field rock-strong. Eventually, stakeholders need to remember that *it is better to learn & implement AI in Civil Engineering imperfectly than to oppose it with perfection.*

Notations

4D CAD	4D computer aided design
AL	Active learning
AAAI	American Association of Artificial Intelligence
AEC	Architecture, Engineering, and Construction
AI	Artificial intelligence
ANN	Artificial neural networks
BIM	Building information modeling

Cobots	Collaborative robots
CAD	Computer-aided design
CNN	Convolutional neural networks
CAP	Credit assignment path
DL	Deep learning
GPT	Generative pretrained transformer
GeoAI	Geospatial artificial intelligence
GIS	Geospatial information systems
IEQ	Indoor environmental quality
IVR	Interactive voice response
IoT	Internet of Things
ML	Machine learning
MV	Machine vision
MEP	Mechanical, electrical, and plumbing
NLP	Natural language processing
NDT	Nondestructive testing
PDE	Partial differential equations
PR	Pattern recognition
PHS	Public Health and Safety
QC	Quantum Computing
QDM	Quick decision making
SHM	Structure health monitoring
SCE	Sustainable Civil Engineering
SDGs	Sustainable Development Goals
UAV	Unmanned aerial vehicles
VR	Virtual reality
WSC	World Sustainability Council
XAI	Explainable artificial intelligence

References

[1] Y. Xu, X. Liu, X. Cao, C. Huang, E. Liu, S. Qian, J. Zhang, Artificial intelligence: a powerful paradigm for scientific research, The Innovation 2 (4) (2021).

[2] A. Kumar, N.R. Kapoor, H.C. Arora, A. Kumar, G. Saini, D.M. Nguyen, R. Shah, Smart cities: a step toward sustainable development, in: Smart Cities: Concepts, Practices, and Applications, CRC Press, Boca Raton and London, 2022, pp. 1−43.

[3] N.R. Kapoor, A. Kumar, A. Kumar, A. Kumar, H.C. Arora, Prediction of Indoor Air Quality Using Artificial Intelligence. Machine Intelligence, Big Data Analytics, and IoT in Image Processing: Practical Applications, 2023, pp. 447−469.

[4] N.R. Kapoor, A. Kumar, T. Alam, A. Kumar, K.S. Kulkarni, P. Blecich, A review on indoor environment quality of Indian school classrooms, Sustainability 13 (21) (2021) 11855.

[5] S.M. Zayed, G.M. Attiya, A. El-Sayed, E.E.D. Hemdan, A review study on digital twins with artificial intelligence and internet of things: concepts, opportunities, challenges, tools and future scope, Multimedia Tools and Applications (2023) 1−27.

[6] N. Díaz-Rodríguez, J. Del Ser, M. Coeckelbergh, M.L. de Prado, E. Herrera-Viedma, F. Herrera, Connecting the dots in trustworthy Artificial Intelligence: from AI principles,

ethics, and key requirements to responsible AI systems and regulation, Information Fusion (2023) 101896, https://doi.org/10.1016/j.inffus.2023.101896.
[7] S.M. Harle, Advancements and challenges in the application of artificial intelligence in civil engineering: a comprehensive review, Asian Journal of Civil Engineering (2023) 1−18, https://doi.org/10.1007/s42107-023-00760-9.
[8] I.G. Smith, J. Oliphant, Use of a knowledge-based system for civil engineering site investigations, in: Second International Conference on the Application of Artificial Intelligence to Civil and Structural Engineering, 1991, pp. 105−112.
[9] A. Kaplan, M. Haenlein, Siri, Siri, in my hand: who's the fairest in the land? On the interpretations, illustrations, and implications of artificial intelligence, Business Horizons 62 (1) (2019) 15−25, https://doi.org/10.1016/j.bushor.2018.08.004.
[10] K. El Bouchefry, R.S. de Souza, Learning in big data: introduction to machine learning, in: Knowledge Discovery in Big Data from Astronomy and Earth Observation, Elsevier, 2020, pp. 225−249, https://doi.org/10.1016/B978-0-12-819154-5.00023-0.
[11] L. Floridi, AI and its new winter: from myths to realities, Philosophy and Technology 33 (2020) 1−3, https://doi.org/10.1007/s13347-020-00396-6.
[12] F.H. Hsu, IBM's deep blue chess grandmaster chips, IEEE micro 19 (2) (1999) 70−81, https://doi.org/10.1109/40.755469.
[13] J.Y. Sung, L. Guo, R.E. Grinter, H.I. Christensen, "My Roomba is Rambo": intimate home appliances, in: UbiComp 2007: Ubiquitous Computing: 9th International Conference, UbiComp 2007, Innsbruck, Austria, September 16-19, 2007, Springer Berlin Heidelberg, 2007, pp. 145−162, https://doi.org/10.1007/978-3-540-74853-3_9. Proceedings 9.
[14] J.K. Sørensen, J. Hutchinson, Algorithms and public service media, in: Public Service Media in the Networked Society: RIPE@ 2017, Nordicom, 2018, pp. 91−106. https://vbn.aau.dk/en/publications/algorithms-and-public-service-media.
[15] D.A. Ferrucci, Introduction to "this is watson", IBM Journal of Research and Development 56 (3.4) (2012) 1, https://doi.org/10.1147/JRD.2012.2184356.
[16] M.I.P. Nasution, S.D. Andriana, P.D. Syafitri, E. Rahayu, M.R. Lubis, Mobile device interfaces illiterate, in: 2015 International Conference on Technology, Informatics, Management, Engineering & Environment (TIME-E), IEEE, September, 2015, pp. 117−120, https://doi.org/10.1109/TIME-E.2015.7389758.
[17] S. Pal, N. Das, A. Sarkar, N. Begam, L. Bhowmik, S.B. Pal, Desktop voice assistant using artificial intelligence, American Journal of Electronics and Communication 3 (4) (2023) 12−17, https://doi.org/10.15864/ajec.3403.
[18] N. Slonim, Y. Bilu, C. Alzate, R. Bar-Haim, B. Bogin, F. Bonin, R. Aharonov, An autonomous debating system, Nature 591 (7850) (2021) 379−384, https://doi.org/10.1038/s41586-021-03215-w.
[19] M.Y. Chaudhary, Augmented reality, artificial intelligence, and the re-enchantment of the world: with Mohammad Yaqub Chaudhary, "augmented reality, artificial intelligence, and the re-enchantment of the world"; and William Young, "Reverend robot: automation and Clergy, Zygon 54 (2) (2019) 454−478, https://doi.org/10.1111/zygo.12521.
[20] Q.V. Pham, D.C. Nguyen, T. Huynh-The, W.J. Hwang, P.N. Pathirana, Artificial intelligence (AI) and big data for coronavirus (COVID-19) pandemic: a survey on the state-of-the-arts, IEEE Access 8 (2020) 130820−130839, https://doi.org/10.1109/ACCESS.2020.3009328.
[21] Sharkov, G. Harnessing the Potential of AI against Covid-19 through the Lens of Cybersecurity: Challenges, Tools, and Techniques.

[22] N.R. Kapoor, A. Kumar, A. Kumar, A. Kumar, K. Kumar, Transmission probability of SARS-CoV-2 in office environment using artificial neural network, IEEE Access 10 (2022) 121204−121229, https://doi.org/10.1109/ACCESS.2022.3222795.
[23] N.R. Kapoor, A. Kumar, A. Kumar, D.A. Zebari, K. Kumar, M.A. Mohammed, M.A. Albahar, Event-specific transmission forecasting of SARS-CoV-2 in a mixed-mode ventilated office room using an ANN, International Journal of Environmental Research and Public Health 19 (24) (2022) 16862, https://doi.org/10.3390/ijerph192416862.
[24] A. Haleem, M. Javaid, R.P. Singh, An era of ChatGPT as a significant futuristic support tool: a study on features, abilities, and challenges, Bench Council Transactions on Benchmarks, Standards and Evaluations 2 (4) (2022) 100089, https://doi.org/10.1016/j.tbench.2023.100089.
[25] N.R. Kapoor, A. Kumar, A. Kumar, A. Kumar, H.C. Arora, H. Jahangir, Quantum computing for indoor environmental quality: a leapfrogging technology, in: Handbook of Research on Quantum Computing for Smart Environments, IGI Global, 2023, pp. 191−216, https://doi.org/10.4018/978-1-6684-6697-1.ch011.
[26] J.M. Spector, S. Ma, Inquiry and critical thinking skills for the next generation: from artificial intelligence back to human intelligence, Smart Learning Environments 6 (1) (2019) 1−11, https://doi.org/10.1186/s40561-019-0088-z.
[27] K.L. Siau, Y. Yang, Impact of Artificial Intelligence, Robotics, and Machine Learning on Sales and Marketing, 2017.
[28] D. Zhang, M. Xie, Artificial intelligence's digestion and reconstruction for humanistic feelings, in: 2018 International Seminar on Education Research and Social Science (ISERSS 2018), Atlantis Press, Paris, 2018.
[29] P. Thaichon, S. Quach (Eds.), Artificial Intelligence for Marketing Management, Taylor & Francis, 2022.
[30] L. Cominelli, D. Mazzei, D.E. De Rossi, SEAI: social emotional artificial intelligence based on Damasio's theory of mind, Frontiers in Robotics and AI 5 (2018) 6, https://doi.org/10.3389/frobt.2018.00006.
[31] P.M. Nadkarni, L. Ohno-Machado, W.W. Chapman, Natural language processing: an introduction, Journal of the American Medical Informatics Association 18 (5) (2011) 544−551, https://doi.org/10.1136/amiajnl-2011-000464.
[32] J. de Almeida Barbosa Franco, A.M. Domingues, N. de Almeida Africano, R.M. Deus, R.A.G. Battistelle, Sustainability in the civil construction sector supported by industry 4.0 technologies: challenges and opportunities, Infrastructures 7 (3) (2022) 43, https://doi.org/10.3390/infrastructures7030043.
[33] J.M.D. Delgado, L. Oyedele, A. Ajayi, L. Akanbi, O. Akinade, M. Bilal, H. Owolabi, Robotics and automated systems in construction: understanding industry-specific challenges for adoption, Journal of Building Engineering 26 (2019) 100868, https://doi.org/10.1016/j.jobe.2019.100868.
[34] R. Jain, R. Kasturi, B.G. Schunck, Machine Vision, vol. 5, McGraw-hill, New York, 1995, pp. 309−364.
[35] C. Steger, M. Ulrich, C. Wiedemann, Machine Vision Algorithms and Applications, John Wiley & Sons, 2018.
[36] W.E. Snyder, H. Qi, Machine Vision, vol. 1, Cambridge University Press, 2004.
[37] A. Kumar, N. Mor, Prediction of accuracy of high-strength concrete using data mining technique: a review, in: Proceedings of International Conference on IoT Inclusive Life (ICIIL 2019), NITTTR Chandigarh, Springer Singapore, India, 2020, pp. 259−267, https://doi.org/10.1007/978-981-15-3020-3_24.

[38] A. Kumar, N. Mor, An approach-driven: use of artificial intelligence and its applications in civil engineering, Artificial Intelligence and IoT: Smart Convergence for Eco-friendly Topography (2021) 201–221, https://doi.org/10.1007/978-981-33-6400-4_10.
[39] Z. Li, J. Yoon, R. Zhang, F. Rajabipour, W.V. Srubar III, I. Dabo, A. Radlińska, Machine learning in concrete science: applications, challenges, and best practices, Npj Computational Materials 8 (1) (2022) 127, https://doi.org/10.1038/s41524-022-00810-x.
[40] S.R. Vadyala, S.N. Betgeri, J.C. Matthews, E. Matthews, A review of physics-based machine learning in civil engineering, Results in Engineering 13 (2022) 100316, https://doi.org/10.1016/j.rineng.2021.100316.
[41] A.B. Nassif, I. Shahin, I. Attili, M. Azzeh, K. Shaalan, Speech recognition using deep neural networks: a systematic review, IEEE Access 7 (2019) 19143–19165, https://doi.org/10.1109/ACCESS.2019.2896880.
[42] T. Hastie, R. Tibshirani, J. Friedman, T. Hastie, R. Tibshirani, J. Friedman, Overview of supervised learning, The Elements of Statistical Learning: Data Mining, Inference, and Prediction (2009) 9–41.
[43] P. Cunningham, M. Cord, S.J. Delany, Supervised learning, in: Machine Learning Techniques for Multimedia: Case Studies on Organization and Retrieval, Springer Berlin Heidelberg, Berlin, Heidelberg, 2008, pp. 21–49, https://doi.org/10.1007/978-3-540-75171-7_2.
[44] F. Schwenker, E. Trentin, Pattern classification and clustering: a review of partially supervised learning approaches, Pattern Recognition Letters 37 (2014) 4–14, https://doi.org/10.1016/j.patrec.2013.10.017.
[45] H.B. Barlow, Unsupervised learning, Neural Computation 1 (3) (1989) 295–311, https://doi.org/10.1162/neco.1989.1.3.295.
[46] Z. Ghahramani, Unsupervised learning, in: Summer School on Machine Learning, Springer Berlin Heidelberg, Berlin, Heidelberg, 2003, pp. 72–112, https://doi.org/10.1007/978-3-540-28650-9_5.
[47] G. James, D. Witten, T. Hastie, R. Tibshirani, J. Taylor, Unsupervised learning, in: An Introduction to Statistical Learning: With Applications in Python, Springer International Publishing, Cham, 2023, pp. 503–556, https://doi.org/10.1007/978-3-031-38747-0_12.
[48] X.J. Zhu, Semi-supervised Learning Literature Survey, 2005. http://digital.library.wisc.edu/1793/60444.
[49] M.F.A. Hady, F. Schwenker, Semi-supervised learning, Handbook on Neural Information Processing (2013) 215–239, https://doi.org/10.1007/978-3-642-36657-4_7.
[50] J.E. Van Engelen, H.H. Hoos, A survey on semi-supervised learning, Machine Learning 109 (2) (2020) 373–440, https://doi.org/10.1007/s10994-019-05855-6.
[51] Z.H. Zhou, Semi-supervised learning, Machine Learning (2021) 315–341, https://doi.org/10.1007/978-981-15-1967-3_13.
[52] L.P. Kaelbling, M.L. Littman, A.W. Moore, Reinforcement learning: a survey, Journal of Artificial Intelligence Research 4 (1996) 237–285, https://doi.org/10.1613/jair.301.
[53] R.S. Sutton, A.G. Barto, Reinforcement Learning: An Introduction, MIT press, 2018.
[54] M.A. Wiering, M. Van Otterlo, Reinforcement learning, Adaptation, Learning, and Optimization 12 (3) (2012) 729.
[55] B. Shneiderman, Design lessons from AI's two grand goals: human emulation and useful applications, IEEE Transactions on Technology and Society 1 (2) (2020) 73–82, https://doi.org/10.1109/TTS.2020.2992669.
[56] AI with MATLAB. https://www.mathworks.com/discovery/artificial-intelligence.html.

[57] B.B. Benuwa, Y.Z. Zhan, B. Ghansah, D.K. Wornyo, F. Banaseka Kataka, A review of deep machine learning, International Journal of Engineering Research in Africa 24 (2016) 124–136.
[58] N.R. Kapoor, J.P. Tegar, Human comfort indicators pertaining to indoor environmental quality parameters of residential buildings in Bhopal, International Research Journal of Engineering and Technology 5 (2018) 2395.
[59] B. Manzoor, I. Othman, S. Durdyev, S. Ismail, M.H. Wahab, Influence of artificial intelligence in civil engineering toward sustainable development—a systematic literature review, Applied System Innovation 4 (3) (2021) 52, https://doi.org/10.3390/asi4030052.
[60] H. Salehi, R. Burgueño, Emerging artificial intelligence methods in structural engineering, Engineering Structures 171 (2018) 170–189, https://doi.org/10.1016/j.engstruct.2018.05.084.
[61] M. Ahmadi, A.G. Lonbar, A. Sharifi, A.T. Beris, M. Nouri, A.S. Javidi, Application of Segment Anything Model for Civil Infrastructure Defect Assessment, 2023, https://doi.org/10.1109/ISEN.2023.3240092 arXiv preprint arXiv:2304.12600.
[62] K. Smarsly, K. Lehner, D. Hartmann, Structural health monitoring based on artificial intelligence techniques, in: Computing in Civil Engineering, 2007, pp. 111–118, https://doi.org/10.1061/40937(261)14.
[63] L. Souza, P. Savoikar, Need for artificial intelligence in geotechnical earthquake engineering, in: Earthquake Geotechnics: Select Proceedings of 7th ICRAGEE 2021, 2022, pp. 425–434, https://doi.org/10.1007/978-981-16-5669-9_35.
[64] Y. Xie, M. Ebad Sichani, J.E. Padgett, R. DesRoches, The promise of implementing machine learning in earthquake engineering: a state-of-the-art review, Earthquake Spectra 36 (4) (2020) 1769–1801, https://doi.org/10.1177/8755293020919419.
[65] F. Xu, H. Uszkoreit, Y. Du, W. Fan, D. Zhao, J. Zhu, Explainable AI: a brief survey on history, research areas, approaches and challenges, in: Natural Language Processing and Chinese Computing: 8th CCF International Conference, NLPCC 2019, Dunhuang, China, October 9–14, 2019, Proceedings, Part II 8, Springer International Publishing, 2019, pp. 563–574, https://doi.org/10.1007/978-3-030-32236-6_51.
[66] S. Koutsourelakis, J.H. Prévost, G. Deodatis, Risk assessment of an interacting structure–soil system due to liquefaction, Earthquake Engineering and Structural Dynamics 31 (4) (2002) 851–879.
[67] A. Asatiani, P. Malo, P.R. Nagbøl, E. Penttinen, T. Rinta-Kahila, A. Salovaara, Challenges of explaining the behavior of black-box AI systems, MIS Quarterly Executive 19 (4) (2020) 259–278.
[68] D. Pratt, M. Sansalone, Impact-echo signal interpretation using artificial intelligence, Materials Journal 89 (2) (1992) 178–187. https://www.concrete.org/publications/internationalconcreteabstractsportal/m/details/id/2265.
[69] A. Kumar, H.C. Arora, N.R. Kapoor, M.A. Mohammed, K. Kumar, A. Majumdar, O. Thinnukool, Compressive strength prediction of lightweight concrete: machine learning models, Sustainability 14 (4) (2022) 2404, https://doi.org/10.3390/su14042404.
[70] D.M. Dimiduk, E.A. Holm, S.R. Niezgoda, Perspectives on the impact of machine learning, deep learning, and artificial intelligence on materials, processes, and structures engineering, Integrating Materials and Manufacturing Innovation 7 (2018) 157–172, https://doi.org/10.1007/s40192-018-0117-8.
[71] K. Guo, Z. Yang, C.H. Yu, M.J. Buehler, Artificial intelligence and machine learning in design of mechanical materials, Materials Horizons 8 (4) (2021) 1153–1172, https://doi.org/10.1039/D0MH01451F.

[72] S.O. Abioye, L.O. Oyedele, L. Akanbi, A. Ajayi, J.M.D. Delgado, M. Bilal, A. Ahmed, Artificial intelligence in the construction industry: a review of present status, opportunities and future challenges, Journal of Building Engineering 44 (2021) 103299, https://doi.org/10.1016/j.jobe.2021.103299.

[73] M. Pan, Y. Yang, Z. Zheng, W. Pan, Artificial intelligence and robotics for prefabricated and modular construction: a systematic literature review, Journal of Construction Engineering and Management 148 (9) (2022) 03122004, https://doi.org/10.1061/(ASCE)CO.1943-7862.0002324.

[74] N.R. Kapoor, A. Kumar, H.C. Arora, A. Kumar, Structural health monitoring of existing building structures for creating green smart cities using deep learning, in: Recurrent Neural Networks, CRC Press, 2022, pp. 203–232. https://www.taylorfrancis.com/chapters/edit/10.1201/9781003307822-15/structural-health-monitoring-existing-building-structures-creating-green-smart-cities-using-deep-learning-nishant-raj-kapoor-aman-kumar-harish-chandra-arora-ashok-kumar.

[75] A. Kumar, J.S. Rattan, N.R. Kapoor, A. Kumar, R. Kumar, Structural health monitoring of existing reinforced cement concrete buildings and bridge using nondestructive evaluation with repair methodology, Advanced Technology of Building Construction and Structural Analysis 87 (2021), https://doi.org/10.5772/intechopen.101473.

[76] S. Sharma, H.C. Arora, A. Kumar, D.P.N. Kontoni, N.R. Kapoor, K. Kumar, A. Singh, Computational intelligence-based structural health monitoring of corroded and eccentrically loaded reinforced concrete columns, Shock and Vibration (2023), https://doi.org/10.1155/2023/9715120.

[77] Y. Pan, L. Zhang, Roles of artificial intelligence in construction engineering and management: a critical review and future trends, Automation in Construction 122 (2021) 103517, https://doi.org/10.1016/j.autcon.2020.103517.

[78] C. Candrian, A. Scherer, Rise of the machines: delegating decisions to autonomous AI, Computers in Human Behavior 134 (2022) 107308, https://doi.org/10.1016/j.chb.2022.107308.

[79] Q.P. Ha, L. Yen, C. Balaguer, Robotic autonomous systems for earthmoving in military applications, Automation in Construction 107 (2019) 102934, https://doi.org/10.1016/j.autcon.2019.102934.

[80] A. Aljawder, W. Al-Karaghouli, The adoption of technology management principles and artificial intelligence for a sustainable lean construction industry in the case of Bahrain, Journal of Decision Systems (2022) 1–30, https://doi.org/10.1080/12460125.2022.2075529.

[81] J. Dumrak, S.A. Zarghami, The Role of Artificial Intelligence in Lean Construction Management. Engineering, Construction and Architectural Management, 2023, https://doi.org/10.1108/ECAM-02-2022-0153.

[82] A. Lekan, A. Clinton, E. Stella, E. Moses, O. Biodun, Construction 4.0 application: industry 4.0, internet of things and lean construction tools' application in quality management system of residential building projects, Buildings 12 (10) (2022) 1557, https://doi.org/10.3390/buildings12101557.

[83] M. Regona, T. Yigitcanlar, B. Xia, R.Y.M. Li, Opportunities and adoption challenges of AI in the construction industry: a PRISMA review, Journal of Open Innovation: Technology, Market, and Complexity 8 (1) (2022) 45, https://doi.org/10.3390/joitmc8010045.

[84] J. Ko, J. Ajibefun, W. Yan, Experiments on Generative AI-Powered Parametric Modeling and BIM for Architectural Design, 2023, https://doi.org/10.48550/arXiv.2308.00227 arXiv preprint arXiv:2308.00227.

[85] E. Touloupaki, T. Theodosiou, Energy performance optimization as a generative design tool for nearly zero energy buildings, Procedia Engineering 180 (2017) 1178–1185, https://doi.org/10.1016/j.proeng.2017.04.278.
[86] A. Zarzycki, Parametric BIM as a Generative Design Tool, ACSA, Boston, 2012, in: https://www.acsa-arch.org/proceedings/Annual%20Meeting%20Proceedings/ACSA.AM.100/ACSA.AM.100.90.pdf.
[87] C. Soddu, New naturality: a generative approach to art and design, Leonardo 35 (3) (2002) 291–294. https://muse.jhu.edu/article/19876/pdf.
[88] G. Fobiri, I. Musonda, F. Muleya, Reality capture in Construction Project Management: a review of opportunities and challenges, Buildings 12 (9) (2022) 1381, https://doi.org/10.3390/buildings12091381.
[89] S. Sarkar, J. Maiti, Machine learning in occupational accident analysis: a review using science mapping approach with citation network analysis, Safety Science 131 (2020) 104900, https://doi.org/10.1016/j.ssci.2020.104900.
[90] T. Birkstedt, M. Minkkinen, A. Tandon, M. Mäntymäki, AI governance: themes, knowledge gaps and future agendas, Internet Research 33 (7) (2023) 133–167, https://doi.org/10.1108/INTR-01-2022-0042.
[91] M.R. Davahli, The Last State of Artificial Intelligence in Project Management, 2020, https://doi.org/10.48550/arXiv.2012.12262 arXiv preprint arXiv:2012.12262.
[92] M. Martínez-Rojas, N. Marín, M.A. Vila, The role of information technologies to address data handling in construction project management, Journal of Computing in Civil Engineering 30 (4) (2016) 04015064, https://doi.org/10.1061/(ASCE)CP.1943-5487.0000538.
[93] L. Zhang, Y. Pan, X. Wu, M.J. Skibniewski, Artificial Intelligence in Construction Engineering and Management, Springer, Singapore, 2021, pp. 95–124, https://doi.org/10.1007/978-981-16-2842-9.
[94] M. Attalla, T. Hegazy, Predicting cost deviation in reconstruction projects: artificial neural networks versus regression, Journal of Construction Engineering and Management 129 (4) (2003) 405–411, https://doi.org/10.1061/(ASCE)0733-9364(2003)129:4(405).
[95] S. Shoar, N. Chileshe, J.D. Edwards, Machine learning-aided engineering services' cost overruns prediction in high-rise residential building projects: application of random forest regression, Journal of Building Engineering 50 (2022) 104102, https://doi.org/10.1016/j.jobe.2022.104102.
[96] E. Yildirim, Text-to-image generation ai in architecture, Art and Architecture: Theory, Practice and Experience 97 (2022).
[97] S. Cousins, AI takes on city design [artificial intelligence-architecture], Engineering and Technology 16 (11) (2021) 34–37, https://doi.org/10.1049/et.2021.1104.
[98] C.M. Eastman, Architectural CAD: a ten year assessment of the state of the art, Computer-Aided Design 21 (5) (1989) 289–292, https://doi.org/10.1016/0010-4485(89)90034-1.
[99] B.R. Hunde, A.D. Woldeyohannes, Future prospects of computer-aided design (CAD)—A review from the perspective of artificial intelligence (AI), extended reality, and 3D printing, Results in Engineering 14 (2022) 100478, https://doi.org/10.1016/j.rineng.2022.100478.
[100] R. Sacks, T. Bloch, M. Katz, R. Yosef, Automating design review with artificial intelligence and BIM: state of the art and research framework, in: ASCE International Conference on Computing in Civil Engineering 2019, American Society of Civil Engineers, Reston, VA, June, 2019, pp. 353–360. https://ascelibrary.org/doi/abs/10.1061/9780784482421.045.

[101] S. Alizadehsalehi, A. Hadavi, J.C. Huang, From BIM to extended reality in AEC industry, Automation in Construction 116 (2020) 103254, https://doi.org/10.1016/j.autcon.2020.103254.
[102] Spacemaker, AI Software for Urban Development, is Acquired by Autodesk for $240M. https://techcrunch.com/2020/11/17/spacemaker-ai-software-for-urban-development-is-acquired-by-autodesk-for-240m/.
[103] Autodesk Forma: Cloud-Based Software for Early-Stage Planning and Design https://www.autodesk.in/products/forma/overview?term=1-YEAR&tab=subscription.
[104] World Economic Forum, The Future of Jobs Report 2020, 2020 (Retrieved from Geneva).
[105] AI is Putting our Jobs as Architects Unquestionably at Risk. https://www.dezeen.com/2023/02/13/ai-architecture-jobs-risk-neil-leach-opinion/.
[106] What is the Future Role of Architects in the Age of AI and Data? https://www.archdaily.com/995781/what-is-the-future-role-of-architects-in-the-age-of-ai-and-data.
[107] J. Khakurel, B. Penzenstadler, J. Porras, A. Knutas, W. Zhang, The rise of artificial intelligence under the lens of sustainability, Technologies 6 (4) (2018) 100, https://doi.org/10.3390/technologies6040100.
[108] Japan's Daiwa House Industry Is Using Generative Design to Retool Urban Housing. https://redshift.autodesk.com/articles/daiwa-house-industry.
[109] Unleash the Power of AI that is revolutionizing architectural design. https://www.architectandinteriorsindia.com/insights/unleash-the-power-of-ai-that-is-revolutionizing-architectural-design.
[110] C. Lee, S. Shin, R.R. Issa, Rationalization of free-form architecture using generative and parametric designs, Buildings 13 (5) (2023) 1250, https://doi.org/10.3390/buildings13051250.
[111] A. Baghbani, T. Choudhury, S. Costa, J. Reiner, Application of artificial intelligence in geotechnical engineering: a state-of-the-art review, Earth-Science Reviews 228 (2022) 103991, https://doi.org/10.1016/j.earscirev.2022.103991.
[112] W. Zhang, H. Li, Y. Li, H. Liu, Y. Chen, X. Ding, Application of deep learning algorithms in geotechnical engineering: a short critical review, Artificial Intelligence Review (2021) 1−41, https://doi.org/10.1007/s10462-021-09967-1.
[113] R. Abduljabbar, H. Dia, S. Liyanage, S.A. Bagloee, Applications of artificial intelligence in transport: an overview, Sustainability 11 (1) (2019) 189, https://doi.org/10.3390/su11010189.
[114] Y. Liu, L. Liu, W.P. Chen, Intelligent traffic light control using distributed multi-agent Q learning, in: 2017 IEEE 20th International Conference on Intelligent Transportation Systems (ITSC), IEEE, October, 2017, pp. 1−8, https://doi.org/10.1109/ITSC.2017.8317730.
[115] I.H. Sarker, Ai-based modeling: techniques, applications and research issues towards automation, intelligent and smart systems, SN Computer Science 3 (2) (2022) 158. https://link.springer.com/article/10.1007/s42979-022-01043-x.
[116] J. Bharadiya, Artificial intelligence in transportation systems A critical review, American Journal of Computing and Engineering 6 (1) (2023) 34−45, https://doi.org/10.47672/ajce.1487.
[117] M. Mnyakin, Applications of ai, IoT, and cloud computing in smart transportation: a review, Artificial Intelligence in Society 3 (1) (2023) 9−27. https://researchberg.com/index.php/ai/article/view/108.
[118] the impact of artificial intelligence on the future of workforces in the European union and the united states of America. https://www.whitehouse.gov/wp-content/uploads/2022/12/TTC-EC-CEA-AI-Report-12052022-1.pdf.

[119] W. Zhou, J. Whyte, R. Sacks, Construction safety and digital design: a review, Automation in Construction 22 (2012) 102−111, https://doi.org/10.1016/j.autcon.2011.07.005.
[120] S.K. Baduge, S. Thilakarathna, J.S. Perera, M. Arashpour, P. Sharafi, B. Teodosio, P. Mendis, Artificial intelligence and smart vision for building and construction 4.0: machine and deep learning methods and applications, Automation in Construction 141 (2022) 104440, https://doi.org/10.1016/j.autcon.2022.104440.
[121] Y. Pan, L. Zhang, Integrating BIM and AI for smart construction management: current status and future directions, Archives of Computational Methods in Engineering 30 (2) (2023) 1081−1110, https://doi.org/10.1007/s11831-022-09830-8.
[122] Z. Babović, B. Bajat, V. Đokić, F. Đorđević, D. Drašković, N. Filipović, S. Zak, Research in computing-intensive simulations for nature-oriented civil-engineering and related scientific fields, using machine learning and big data: an overview of open problems, Journal of Big Data 10 (1) (2023) 1−21, https://doi.org/10.1186/s40537-023-00731-6.
[123] M. Živković, M. Žujović, J. Milošević, 3D-printed Architectural Structures Created Using Artificial Intelligences: A Review of Techniques and Applications, 2023, https://doi.org/10.20944/preprints202307.1826.v1.
[124] I. Rojek, J. Dorożyński, D. Mikołajewski, P. Kotlarz, Overview of 3D printed exoskeleton materials and opportunities for their AI-based optimization, Applied Sciences 13 (14) (2023) 8384, https://doi.org/10.3390/app13148384.
[125] A. Salzano, M. Intignano, C. Mottola, S.A. Biancardo, M. Nicolella, G. Dell'Acqua, Systematic literature review of open infrastructure BIM, Buildings 13 (7) (2023) 1593, https://doi.org/10.3390/buildings13071593.
[126] F. Zhang, A.P. Chan, A. Darko, Z. Chen, D. Li, Integrated applications of building information modeling and artificial intelligence techniques in the AEC/FM industry, Automation in Construction 139 (2022) 104289, https://doi.org/10.1016/j.autcon.2022.104289.
[127] J.P. Zhang, Z.Z. Hu, BIM-and 4D-based integrated solution of analysis and management for conflicts and structural safety problems during construction: 1. Principles and methodologies, Automation in Construction 20 (2) (2011) 155−166, https://doi.org/10.1016/j.autcon.2010.09.013.
[128] D.M. Botín-Sanabria, A.S. Mihaita, R.E. Peimbert-García, M.A. Ramírez-Moreno, R.A. Ramírez-Mendoza, J.D.J. Lozoya-Santos, Digital twin technology challenges and applications: a comprehensive review, Remote Sensing 14 (6) (2022) 1335, https://doi.org/10.3390/rs14061335.
[129] A. Keshmiry, S. Hassani, M. Mousavi, U. Dackermann, Effects of environmental and operational conditions on structural health monitoring and non-destructive testing: a systematic review, Buildings 13 (4) (2023) 918, https://doi.org/10.3390/buildings13040918.
[130] A. Almusaed, I. Yitmen, A. Almssad, Reviewing and Integrating AEC Practices into Industry 6.0: Strategies for Smart and Sustainable Future-Built Environments, 2023, https://doi.org/10.20944/preprints202308.0860.v1.
[131] Y. Gao, H. Li, G. Xiong, H. Song, AIoT-informed digital twin communication for bridge maintenance, Automation in Construction 150 (2023) 104835, https://doi.org/10.1016/j.autcon.2023.104835.
[132] W. Bronner, H. Gebauer, C. Lamprecht, F. Wortmann, Sustainable AIoT: how artificial intelligence and the internet of things affect profit, people, and planet, Connected Business: Create Value in a Networked Economy (2021) 137−154, https://doi.org/10.1007/978-3-030-76897-3_8.

[133] A. Aliahmadi, H. Nozari, J. Ghahremani-Nahr, AIoT-based sustainable smart supply chain framework, International Journal of Innovation in Management, Economics and Social Sciences 2 (2) (2022) 28−38, https://doi.org/10.52547/ijimes.2.2.28.
[134] M. De Filippo, S. Asadiabadi, J.S. Kuang, D.K. Mishra, H. Sun, AI-Powered Inspections of Facades in Reinforced Concrete Buildings, 2023. https://www.hkie.org.hk/hkietransactions/upload/2023-02-03/THIE-2020-0023.pdf.
[135] C. Haas, M. Skibniewski, E. Budny, Robotics in civil engineering, Computer-Aided Civil and Infrastructure Engineering 10 (5) (1995) 371−381, https://doi.org/10.1111/j.1467-8667.1995.tb00298.x.
[136] H.W. Choi, H.J. Kim, S.K. Kim, W.S. Na, An overview of drone applications in the construction industry, Drones 7 (8) (2023) 515, https://doi.org/10.3390/drones7080515.
[137] K. Lnc Prakash, S.K. Ravva, M.V. Rathnamma, G. Suryanarayana, AI applications of drones, Drone Technology: Future Trends and Practical Applications (2023) 153−182, https://doi.org/10.1002/9781394168002.ch7.
[138] W. Li, C.Y. Hsu, GeoAI for large-scale image analysis and machine vision: recent progress of artificial intelligence in geography, ISPRS International Journal of Geo-Information 11 (7) (2022) 385, https://doi.org/10.3390/ijgi11070385.
[139] H. Wang, L. Zhang, H. Luo, J. He, R.W.M. Cheung, AI-powered landslide susceptibility assessment in Hong Kong, Engineering Geology 288 (2021) 106103, https://doi.org/10.1016/j.enggeo.2021.106103.
[140] C.V. Ekeanyanwu, I.F. Obisakin, P. Aduwenye, N. Dede-Bamfo, Merging GIS and machine learning techniques: a paper review, Journal of Geoscience and Environment Protection 10 (9) (2022) 61−83, https://doi.org/10.4236/gep.2022.109004.
[141] M. Chen, C. Claramunt, A. Çöltekin, X. Liu, P. Peng, A.C. Robinson, G. Lü, Artificial intelligence and visual analytics in geographical space and cyberspace: research opportunities and challenges, Earth-Science Reviews (2023) 104438, https://doi.org/10.1016/j.earscirev.2023.104438.
[142] S. Dastgheibifard, M. Asnafi, A review on potential applications of unmanned aerial vehicle for construction industry, Sustainable Structure and Materials 1 (2) (2018) 44−53, https://doi.org/10.26392/SSM.2018.01.02.044.
[143] P. Pradhananga, M. ElZomor, G. Santi Kasabdji, Identifying the challenges to adopting robotics in the US construction industry, Journal of Construction Engineering and Management 147 (5) (2021) 05021003, https://doi.org/10.1061/(ASCE)CO.1943-7862.0002007.
[144] N. Emaminejad, R. Akhavian, Trustworthy AI and robotics: implications for the AEC industry, Automation in Construction 139 (2022) 104298, https://doi.org/10.1016/j.autcon.2022.104298.
[145] F. Escrig, Expandable space structures, International Journal of Space Structures 1 (2) (1985) 79−91, https://doi.org/10.1177/026635118500100203.
[146] M. Chen, R. Goyal, M. Majji, R.E. Skelton, Review of space habitat designs for long term space explorations, Progress in Aerospace Sciences 122 (2021) 100692, https://doi.org/10.1016/j.paerosci.2020.100692.
[147] R. Doyle, T. Kubota, M. Picard, B. Sommer, H. Ueno, G. Visentin, R. Volpe, Recent research and development activities on space robotics and AI, Advanced Robotics 35 (21−22) (2021) 1244−1264, https://doi.org/10.1080/01691864.2021.1978861.
[148] Y. Xu, W. Pei, W. Hu, A current overview of the biological effects of combined space environmental factors in mammals, Frontiers in Cell and Developmental Biology 10 (2022) 861006, https://doi.org/10.3389/fcell.2022.861006.

[149] N. Leach, 3D printing in space, Architectural Design 84 (6) (2014) 108−113, https://doi.org/10.1002/ad.1840.
[150] G. Cesaretti, E. Dini, X. De Kestelier, V. Colla, L. Pambaguian, Building components for an outpost on the Lunar soil by means of a novel 3D printing technology, Acta Astronautica 93 (2014) 430−450, https://doi.org/10.1016/j.actaastro.2013.07.034.
[151] A. Obili, S. Ramachandran, L. Pattabiraman, A comprehensive study of space debris & its removal using nanobots, in: AIP Conference Proceedings, AIP Publishing, June 2023, https://doi.org/10.1063/5.0144950.
[152] G. Viavattene, E. Devereux, D. Snelling, N. Payne, S. Wokes, M. Ceriotti, Design of multiple space debris removal missions using machine learning, Acta Astronautica 193 (2022) 277−286, https://doi.org/10.1016/j.actaastro.2021.12.051.
[153] E. Lagona, S. Hilton, A. Afful, A. Gardi, R. Sabatini, Autonomous trajectory optimisation for intelligent satellite systems and space traffic management, Acta Astronautica 194 (2022) 185−201, https://doi.org/10.1016/j.actaastro.2022.01.027.
[154] D.J. Dijk, D.F. Neri, J.K. Wyatt, J.M. Ronda, E. Riel, A. Ritz-De Cecco, C.A. Czeisler, Sleep, performance, circadian rhythms, and light-dark cycles during two space shuttle flights, American Journal of Physiology - Regulatory, Integrative and Comparative Physiology (2001), https://doi.org/10.1152/ajpregu.2001.281.5.R1647.
[155] F.E. Garrett-Bakelman, M. Darshi, S.J. Green, R.C. Gur, L. Lin, B.R. Macias, F.W. Turek, The NASA Twins Study: a multidimensional analysis of a year-long human spaceflight, Science 364 (6436) (2019) eaau8650, https://doi.org/10.1126/science.aau8650.
[156] A. Russo, G. Lax, Using artificial intelligence for space challenges: a survey, Applied Sciences 12 (10) (2022) 5106, https://doi.org/10.3390/app12105106.
[157] M.J. Mashala, T. Dube, B.T. Mudereri, K.K. Ayisi, M.R. Ramudzuli, A systematic review on advancements in remote sensing for assessing and monitoring land use and land cover changes impacts on surface water resources in semi-arid tropical environments, Remote Sensing 15 (16) (2023) 3926, https://doi.org/10.3390/rs15163926.
[158] Y.A. Nanehkaran, B. Chen, A. Cemiloglu, J. Chen, S. Anwar, M. Azarafza, R. Derakhshani, Riverside landslide susceptibility overview: leveraging artificial neural networks and machine learning in accordance with the United Nations (UN) sustainable development goals, Water 15 (15) (2023) 2707, https://doi.org/10.3390/w15152707.
[159] S.M. Islam, Drones on the Rise: Exploring the Current and Future Potential of UAVs, 2023 arXiv preprint arXiv:2304.13702. https://doi.org/10.48550/arXiv.2304.13702.
[160] V. Kumar, H.M. Azamathulla, K.V. Sharma, D.J. Mehta, K.T. Maharaj, The state of the art in deep learning applications, challenges, and future prospects: a comprehensive review of flood forecasting and management, Sustainability 15 (13) (2023) 10543, https://doi.org/10.3390/su151310543.
[161] C.E. Richards, A. Tzachor, S. Avin, R. Fenner, Rewards, risks and responsible deployment of artificial intelligence in water systems, Nature Water (2023) 1−11, https://doi.org/10.1038/s44221-023-00069-6.
[162] D. Vallejo-Gómez, M. Osorio, C.A. Hincapié, Smart irrigation systems in agriculture: a systematic review, Agronomy 13 (2) (2023) 342, https://doi.org/10.3390/agronomy13020342.
[163] Y. Gamal, A. Soltan, L.A. Said, A.H. Madian, A.G. Radwan, Smart Irrigation Systems: Overview, IEEE Access, 2023, https://doi.org/10.1109/ACCESS.2023.3251655.
[164] R. Anjum, F. Parvin, S.A. Ali, Machine Learning Applications in Sustainable Water Resource Management: A Systematic Review. Emerging Technologies for Water Supply, Conservation and Management, 2023, pp. 29−47, https://doi.org/10.1007/978-3-031-35279-9_2.

[165] E.K. Nti, S.J. Cobbina, E.A. Attafuah, L.D. Senanu, G. Amenyeku, M.A. Gyan, A.R. Safo, Water Pollution Control and Revitalization Using Advanced Technologies: Uncovering Artificial Intelligence Options towards Environmental Health Protection, Sustainability and Water Security, Heliyon, 2023, https://doi.org/10.1016/j.heliyon.2023.e18170.

[166] C. Irrgang, N. Boers, M. Sonnewald, E.A. Barnes, C. Kadow, J. Staneva, J. Saynisch-Wagner, Towards neural Earth system modelling by integrating artificial intelligence in Earth system science, Nature Machine Intelligence 3 (8) (2021) 667–674, https://doi.org/10.1038/s42256-021-00374-3.

[167] Y. Xiang, Y. Chen, J. Xu, Z. Chen, Research on sustainability evaluation of green building engineering based on artificial intelligence and energy consumption, Energy Reports 8 (2022) 11378–11391, https://doi.org/10.1016/j.egyr.2022.08.266.

[168] M.S. Pathan, E. Richardson, E. Galvan, P. Mooney, The role of artificial intelligence within circular economy activities—a view from Ireland, Sustainability 15 (12) (2023) 9451, https://doi.org/10.3390/su15129451.

[169] E.M. Ditria, C.A. Buelow, M. Gonzalez-Rivero, R.M. Connolly, Artificial intelligence and automated monitoring for assisting conservation of marine ecosystems: a perspective, Frontiers in Marine Science 9 (2022) 918104, https://doi.org/10.3389/fmars.2022.918104.

[170] D. Perera, O. Seidou, J. Agnihotri, H. Mehmood, M. Rasmy, Challenges and Technical Advances in Flood Early Warning Systems (FEWSs). Flood Impact Mitigation and Resilience Enhancement, 2020. https://www.intechopen.com/chapters/72571.

[171] H. Jain, R. Dhupper, A. Shrivastava, D. Kumar, M. Kumari, AI-enabled strategies for climate change adaptation: protecting communities, infrastructure, and businesses from the impacts of climate change, Computational Urban Science 3 (1) (2023) 25, https://doi.org/10.1007/s43762-023-00100-2.

[172] A. Kumar, K. Kumar, N.R. Kapoor, Optimization of renewable energy sources using emerging computational techniques, in: Sustainable Developments by Artificial Intelligence and Machine Learning for Renewable Energies, Elsevier, 2022, pp. 187–236, https://doi.org/10.1016/B978-0-323-91228-0.00012-4.

[173] A.B. Arrieta, N. Díaz-Rodríguez, J. Del Ser, A. Bennetot, S. Tabik, A. Barbado, F. Herrera, Explainable Artificial Intelligence (XAI): concepts, taxonomies, opportunities and challenges toward responsible AI, Information Fusion 58 (2020) 82–115, https://doi.org/10.1016/j.inffus.2019.12.012.

[174] Y. Himeur, M. Elnour, F. Fadli, N. Meskin, I. Petri, Y. Rezgui, A. Amira, AI-big data analytics for building automation and management systems: a survey, actual challenges and future perspectives, Artificial Intelligence Review 56 (6) (2023) 4929–5021, https://doi.org/10.1007/s10462-022-10286-2.

Application of artificial intelligence in sustainable construction: A secret eye toward the latest civil engineering techniques

2

Navdeep Mor[1], Pawan Kumar[2], Madhu[3], Gopal Lal Jat[4] and Ankush Kumar[5]
[1]Department of Civil Engineering, Guru Jambheshwar University of Science and Technology, Hisar, Haryana, India; [2]Department of Computer Science and Engineering, Guru Jambheshwar University of Science and Technology, Hisar, Haryana, India; [3]School of Chemistry, University of Hyderabad, Hyderabad, Telangana, India; [4]Department of Electrical Engineering, PEC (Deemed to be University), Chandigarh, India; [5]Department of Public Health Engineering, (GOH), Hisar, Haryana, India

2.1 Introduction

In order to improve the living standards of humans, there is a subsequent need for the construction of various types of infrastructure. Construction in any field, like roads, bridges, buildings, dams, flyovers, underpasses, shipping, and all other related items required huge data and strategies to get a fruitful output. As the data in the industries related to construction is massive and complex in terms of volume and variables, critical thinking is required to make better decisions. With such a difficult task to perform and the limitations of the human mind, advanced engineering is required to fulfill the desired objectives. What may be called a better technique or accurate technique is being used by the world of work, which can perform the tasks in time with the best accuracy.

For the requirement of accuracy up to that level, engineering alone is not sufficient to handle the said important task, as manual analysis involves time-consuming, costly, and manpower-oriented tasks with the possibility of errors. Rapid construction needs additional manpower to assist. Today's era is technology-oriented, and a accordingly number of features are available that can assist the human brain to reduce the stress related to multiple tasks involved in construction, save time, minimize loss, and ensure the overall completion of sustainable construction. To solve the issues related to the complex problems of construction industries, an advanced intelligent technique called artificial intelligence (AI) is being used by multiple stakeholders to improve the growth of society and the nation. In order to better understand the use of AI in construction, it is important to understand the concept of AI and its type along with the major areas in which AI is being used frequently.

Artificial Intelligence Applications for Sustainable Construction. https://doi.org/10.1016/B978-0-443-13191-2.00012-2
Copyright © 2024 Elsevier Ltd. All rights are reserved, including those for text and data mining, AI training, and similar technologies.

AI can be defined as the science of making intelligent and smart machines. Nowadays, AI has taken the ability and efficiency of the human brain to the next level. It has reduced human effort. Initially, it is very important to understand the concept, type, and technologies that are related to AI. And to begin with, let us start with the definitions available in the world of work, which are as follows:

1. "AI can be defined as the ability of an advanced digital computer or similar robot controlled by a computer in order to perform any particular task effectively and efficiently."
2. "AI can be defined as the machine completing the tasks that were previously assumed to be done only with intelligent brains."
3. "AI is a set of incorporated instructions to process tasks that involve learning/reasoning and accordingly self-correcting themselves to have better efficient results using machines, especially computers."
4. "AI is simply the capability of machines to perform similar but efficient intelligent human behavior."

All the above-mentioned definitions are correct, but ultimately, the results can be compared when processed using AI and human brains.

In the previous 2 decades, an increase in the applications of AI can be seen very easily in the field of civil engineering. AI basically includes machine learning (ML) and deep learning (DL). Here, ML can be found as a system of small, deep architectures, where the deep simply refers to the increase in the number of layers in the input data itself. The depth of any deep model is governed by the number of actual layers between input and output data. Deep techniques presently being used contain thousands of layers to process the data automatically; on the other hand, ML contains only one or two layers. The relationship between AI, ML, and DL is presented in Fig. 2.1.

Figure 2.1 Relationship between artificial intelligence, machine learning, and deep learning.

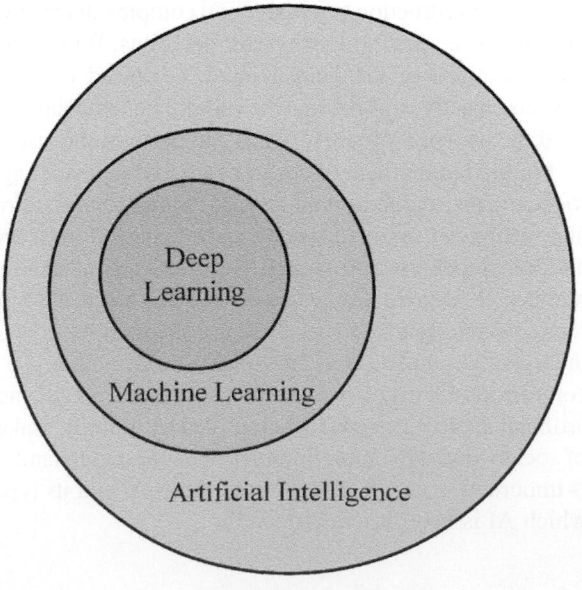

2.1.1 Working mechanism of artificial intelligence

ML simplifies the task or operations; in other words, it simply reduces human effort with better efficiency. This can be better understood by taking an appropriate example. If the task in construction is to find out the presence of any defect, one of the ML algorithms might work in this direction through a tree of questions like "how old the building or infrastructure is." Then, "Okay," a similar exercise will be executed subsequently by the number of questions in one branch. In a similar way, a number of iterations will be done. If we select yes, then another branch will be selected and another set of questions will be executed. The process will continue until we get a fruitful optimized result with the best efficiency. It's like a game of multiple questions kids might have played, with a difference in the case of ML: that such questions are automatically generated to get equal to the desired result.

In the case of already-inbuilt or already-incorporated commands, the AI will reflect some selected commands and accordingly ensure the task is executed. Some of the tasks that can be fixed are related to sales, marketing, the legal department, research and development, assurance of fixed quality, packing of items, entry into prohibited areas, a particular section that needs follow-up action, any time-bound specific activity, any warning or attention required, and the requirement of assets.

2.1.2 Types of AI

There are primarily two important categories in which AI can be classified, which are Type 1 and Type 2. Both categories are as follows:

Type 1: This particular category includes weak AI and strong AI. Both are discussed one by one.

Weak or narrow AI: This particular type focuses on a specific narrow task. This includes those types of machines that are not so intelligent as to complete their own work but are designed in such a way that they seem to be smart. It can be well understood by taking an example of a poker game in which a machine is able to beat human moves only that are already incorporated into the machine. In similar applications or games, all possible scenarios are required to be entered manually. All weakly incorporated AI ultimately contribute to the building of resilient AI.

Strong AI: In this category, only those machines are considered that can think and perform at their own level, like human beings. There are no specific examples in this category, but a lot of continuous efforts are being made by multiple stakeholders to build such advanced and brilliant machines.

Type 2 (based on functionalities): In this category, there are four possible categories, and one, by one all are discussed accordingly.

1. Reactive machines: The most basic form of AI comes under this category. The major component in this category is that it does not have the capacity to use previous memory, and accordingly, past memory cannot be used to perform future actions. The most accurate example in this category is the IBM chess program that beat the world's most brilliant player Garry Kasparov in the 1990s.

Table 2.1 Different ways of achieving artificial intelligence.

Artificial intelligence (AI)	Machine learning (ML)	Deep learning Supervised learning Unsupervized learning
	Natural language processing (NLP)	Content extraction Classification Machine translation Question answering Text generation
	Expert systems	—
	Vision	Image recognition Machine vision
	Speech	Speech to text Text to speech
	Planning	—
	Robotics	—

2. Limited memory: Some of the advanced AI-incorporated programs have the ability to use past experiences to make some future decisions. It can be well settled by taking the example of self-driving cars, which have the ability to make their own decisions like changing lanes, where to stop and move, and identifying any hazards or obstacles. These observations are not stored permanently but are utilized to make future decisions. Another example in this category is Apple's Chatbot Siri, which utilizes instant or recent commands and performs tasks accordingly.
3. Theory of mind: This particular type of AI is capable enough to understand the emotions, thoughts, and expectations of people and is able to interact socially. A lot of enhancement is being done in this category, but the final précize AI is not complete till now.
4. Self-awareness: This particular category is assumed to be almost equivalent to a human being due to its superintelligence and self-awareness, but multiple stakeholders are doing their best to build and design it so that it can act as a milestone in the field of AI.

AI can be achieved in many ways, but some of them are presented in Table 2.1.

The above-mentioned techniques can be used to improve the productivity of construction work. Overall, the type of technique to be used depends on the nature of the job and the type of work to be done.

2.1.2.1 Computer programming with and without incorporation of AI

All possible techniques of AI have been mentioned in Table 2.1. It is a well-settled fact that any smart activity related to technology requires the latest computer-based programming to execute the results. There is a massive history or background of AI that we need to understand, but the important key points related to AI programming are mentioned below:

1. AI is a modern concept and is a way that makes an arrangement of equivalence of manner in which an intelligent human thinks and a computer, software, or computer-controlled robot think intelligently, and as a result, it has the ability to reduce human efforts.

2. AI is programmed in such a way that it can study how the human brain thinks, learns, decides, and finally works in case of solving the problem faced by it, and based on that, it can develop intelligent mechanisms to perform the same.
3. Although it is similar to human mind efforts, there are three major points where programming with and without AI differs, and the same are presented in Table 2.2

In addition to this, AI is being used in multiple areas and sectors, as presented in Fig. 2.2 and Table 2.3.

Table 2.2 Programming with and without AI.

Sr. No.	Programming with AI	Programming without AI
1	Able to answer **generic** questions	Able to answer only **specific** questions
2	Able to absorb new modifications	If modification is done in the programming, its existing structure may completely change.
3	Modification is quick and easy and it will not result in any change of structure.	Modification is slow and difficult and as a result, it may adversely affect the program.

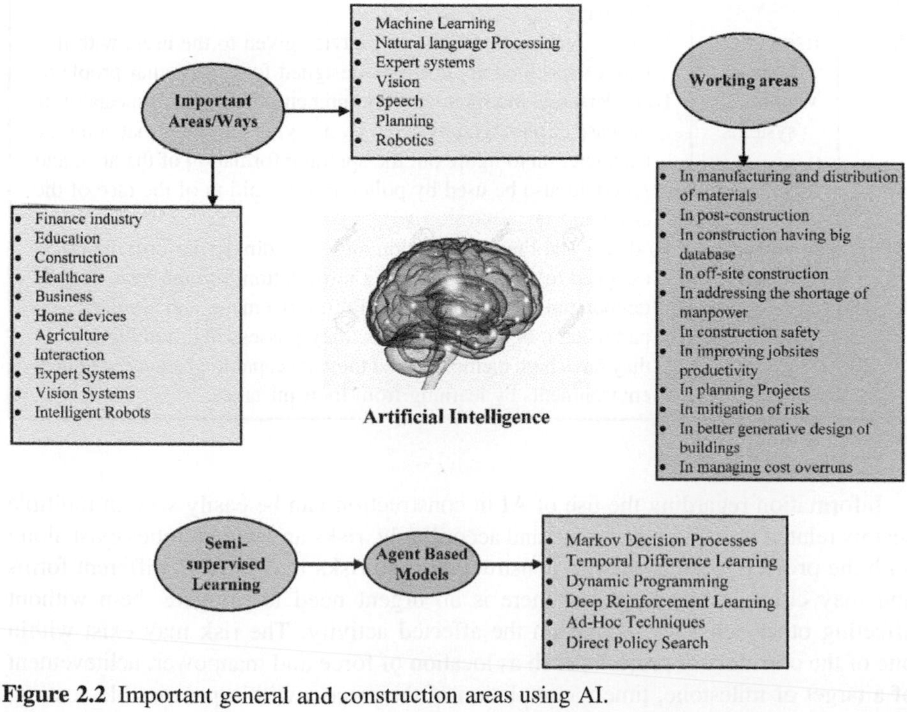

Figure 2.2 Important general and construction areas using AI.

Table 2.3 Sectors along with their working areas where AI can be used.

Sr. No.	Sector	Corridor up to where AI can be extended
1	Finance industry	To provide advice related to finances based on personally collected data
2	Education	To improve the learning process based on an automated grading system generated through accessing the performance of the students.
3	Healthcare	To assist customers with the problems related to the appointment schedule and processing of bills with the help of computer programming.
		To assist the doctors for better diagnosis and accordingly better answers to a specific problem.
4	Business	To improve the satisfaction level of customers by automating repetitive tasks based on the data collected for a better understanding of the needs of customers.
5	Home devices	To improve security and surveillance through smart home devices.
		To better navigate and travel.
		To enhance the music and media streaming especially in the case of video games like chess, poker, etc.
6	Agriculture	To improve productivity by consistently observing the growth and related aspects of crops.
7	Interaction	To interact with the computer effectively, which understands natural language.
8	Expert systems	To assist with explanations and advice given to the users with the help of specified applications designed for a particular problem.
9	Vision systems	To understand, interpret, and comprehend the inputs, especially in the case of photographs taken by a spying airplanes that could be used later on to figure out the spatial information of the area, and AI could also be used by police in recognition of the face of the criminal.
10	Intelligent robots	To detect real-time information and accordingly perform the tasks assigned to them by gathering information through heat, sound, temperature, noise, pressure, light, movement, and similar parameters. In addition to this, they possess high intelligence as they have high memories, and they are capable of adapting to new environments by learning from their mistakes.

Information regarding the use of AI in construction can be easily seen in multiple sectors related to civil engineering, and accordingly, risks and uncertainties exist along with the projects associated with construction. The risks may exist in different forms and may come at any stage, so there is an urgent need to mitigate them without affecting other activities relying on the affected activity. The risk may exist within one of the corridors of projects, such as location of force and manpower, achievement of a target or milestone, timely completion of the project within permissible costing, and overall management in the field of construction. ML, a branch of AI, is one of the most popular techniques being used in the domain of civil engineering. Some of the applications where AI can be used in the field of civil engineering are as follows:

1. AI can be used to classify the soil and accordingly estimate the moisture content of the soil without disturbing its strata.
2. The application of ML could be used to detect the damage and cracks in an existing building with the help of images and sensors to identify the exact location of defects along with the extent of damage, which could be used to take preventive measures accordingly within the domain of structural health monitoring.
3. To predict maximum dry density and optimum moisture content in the concrete.
4. To identify the gaps related to the requirement of raw materials to ensure timely completion of the tasks.
5. In the field of transportation engineering, AI could be used to predict time delays and road accidents; perform traffic simulations and correlate vehicle, human, and environmental-related activities in order to get the desired results.
6. In the field of green building, AI could be used to simulate energy consumption, effective building planning, and managing the building using building information modeling (BIM).
7. In the field of water resources, AI could be used to balance the water cycle.
8. In the field of hydraulics, AI could be used to detect the various activities going through the force of liquid pressure and various aspects of fluid.
9. In the case of the settlement of a foundation, AI could be used for soil stability to prevent failure.
10. In the case of marine-related work, AI could be used to forecast the tidal to aid according to the marine environment.
11. Artificial neural networks (ANNs) could be used to predict the properties of the design of concrete mixes.
12. To monitor the activity on the site to avoid a massive change in the cost.
13. To warn the building regarding repairs.
14. To reduce the errors in the project by opting for the option of automatic analysis of data or feedback.
15. AI could be used to solve the complicated problems that may arise at any stage of the project.
16. AI could be used to make decisions in the field of design.
17. AI could also be used in various aspects of environmentally friendly activities, especially waste management.

All above said domains/fields are available where AI could be used. But it is not possible to explore all those working areas. In the present investigation, the major focus has been on the use of AI in the field related to construction only.

2.1.3 Overall advanced construction using AI

The overall macro model showing the detailed procedure of how data analysis is being done is presented in Fig. 2.3. In this figure, it has been shown how input is being taken by the sensors, followed by data analysis to the prediction.

In addition to these, strengths, weaknesses, opportunities, and threats analyses can also be performed using AI, and accordingly, any required action can be taken. The detailed indication of tasks that can be ensured using AI is presented in Fig. 2.4.

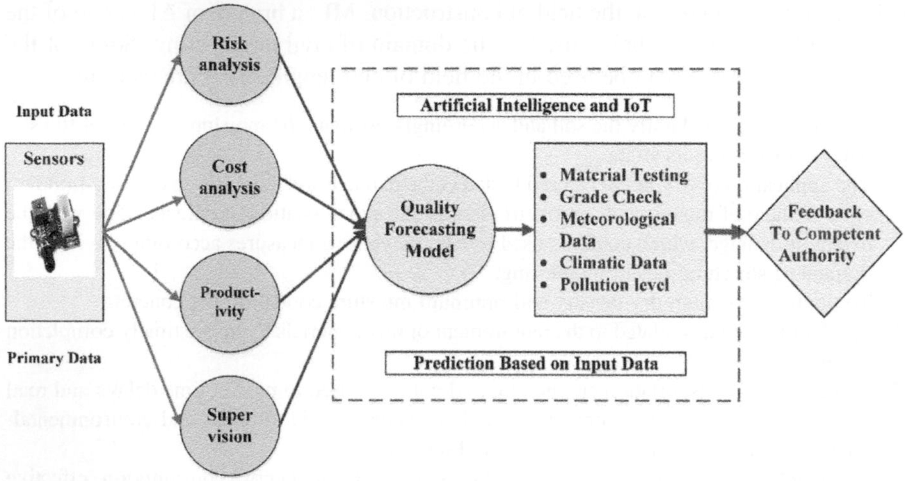

Figure 2.3 Macro model showing procedure of data analysis using AI in construction.

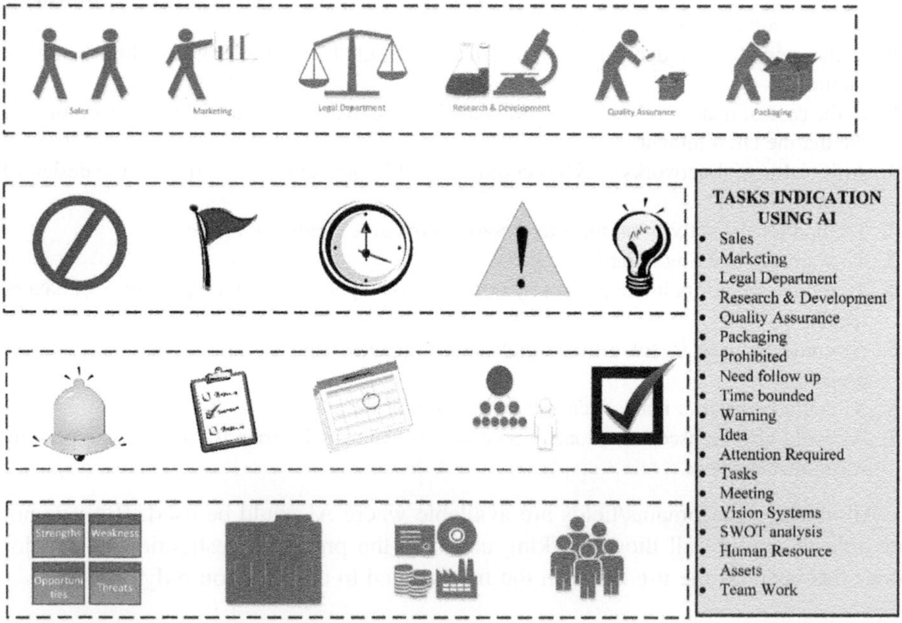

Figure 2.4 Important general indication of tasks using AI.

Construction has become one of the most resource-consuming tasks, and accordingly, it has become one of the largest polluters in the world. Moving forward in this direction, it is a matter of concern to deal with the construction-related work sustainably. To deal with this sustainable construction matter, multiple techniques and technologies are available in the world of work to minimize pollution and definitely

Table 2.4 Construction using Traditional and AI technique.

Sr. No.	Factor	Construction using traditional technique	Construction using AI technique
1	Cost	Overall comparatively low cost	Comparatively high cost in design stage
2	Design/Details of each task/activity	Comparatively more time consuming	Comparatively less time consuming
3	Risk analysis	Comparatively moderate risk analysis	Comparatively detailed risk analysis
4	Cost analysis	Comparatively moderate cost analysis	Comparatively detailed cost analysis
5	Supervision	Moderate supervision is required	High supervision is required
6	Team type	Moderate skilled team is required	Highly skilled team is required
7	Building design	Precise using traditional method	Detailed and precise using advanced 3D modeling
8	Productivity	Moderate to high range	High to improved range

utilize the resources efficiently. This section will explore this concept with a focus on the use of AI techniques in the field of construction, and to start with, let us understand the concept of sustainability.

The very first question that may arise is how AI can help stakeholders in smart, sustainable construction, and a detailed answer to this question has been mentioned above. The next major question is to discuss whether there will be any significant change if the concept of AI is used. In other words, it has become very important to differentiate between traditional ways of construction and construction using AI. The major difference between construction using traditional methods and construction using smart, advanced techniques is presented in Table 2.4.

2.1.4 Three pillars of sustainability

When we talk about sustainability, there are numerous available aspects in the market. In a nutshell, sustainable thought must comprise three pillars to fulfill the objectivity of sustainable construction. The three pillars of sustainable construction consist of:

1. Economic
2. Social
3. Environmental

The concept of all the above-mentioned pillars is essential to consider when there is a need for sustainable construction. Then, every country needs to find the purpose of sustainable development for itself. Generally, it has been found that countries focus more on construction than sustainable construction. With the growing demand for construction with limited resources, the concept of sustainable construction using some of the latest advanced techniques like AI and ML has become more popular.

The industries related to construction are resource-exhaustive industries. They have a significant and countable impact on the environment and, accordingly, a possibly greater impact on the environment in comparison to other available industries like agriculture, clothing, transport, manufacturing, etc. As per the latest available reports, construction-related work is contributing significantly to approximately 38% of total global energy consumption. It is also a well-settled fact that only the building-related sector accounts for about 55% of ultimate energy use. Considering the consumption and use of energy in this sector, the need for sustainable construction is being felt day by day by multiple stakeholders. In a nutshell, sustainable construction aims to provide better outcomes in terms of environmental-friendly construction using the energy, water, and materials in any infrastructure efficiently and ultimately by reducing the waste out of it and accordingly the environmental pollution. It also involves the improvement of social outcomes by helping people lead productive, better, healthy, and dignified lives. AI can be used in the following ways:

1. Stronger, better use of materials
2. More efficient use of energy
3. Smart planning with construction 3D printing
4. Construction Recycling
5. Controlling renewable energy generation
6. AI and sustainable building maintenance

2.1.5 Construction with certifications

After critically reviewing the research articles from various reputed sources, it has been found that different stakeholders/organizations are more concerned with reducing the impact on the environment due to excessive use of natural energy and resources in construction-related works, which ultimately causes pollution. The production of construction waste is a major issue that needs to be resolved on an urgent basis. This problem has led to the need for construction after considering environmental-related issues. This is being assured by various certifications like Leadership in Energy and Environmental Design (LEED), Building Research Establishment Environmental Assessment Method, ISO, and others (Green Rating for Integrated Habitat Assessment, Eco-housing, Comprehensive Assessment System for Built Environment Efficiency, building standard, Global Sustainability Assessment System (GSAS), passive house, SBTool, etc.). Out of these, LEED is the most popular one. These certifications have detailed guidelines to manage the projects and related construction works, which include work related to the construction of commercial as well as residential sites. These latest advanced techniques help in managing the resources in such a way that losses in construction could be reduced, thus improving efficiency.

2.1.5.1 AI in architecture, engineering and construction (AEC)

Whenever construction is to be made, architecture along with engineering comes along that needs to be studied in depth. Some of the available AI techniques applicable to Architecture, Engineering and Construction (AEC) and possible identified issues/problems where the concept of AI is approachable are presented in Table 2.5.

2.1.6 Methodology of working process of AI smart construction

After mentioning the need, requirement, type, and significant difference between construction using traditional and AI methods,there is a need to understand the working methodology of construction using AI. Now, to clarify this situation, the detailed methodology behind this style is presented in Fig. 2.5.

It helps us understand the background of the process and how AI is helpful in boosting the smart construction process. The mentioned techniques can help the companies to compete and stay one step ahead in comparison to other ones. Let us discuss each step of the detailed process for better understanding. The entire methodology is divided into five major components, which are as follows:

Table 2.5 AI techniques applicable in AEC.

Sr. No.	Some AI techniques applicable to AEC	Some AEC problems/issues/domains to which AI is applicable
1	Genetic algorithm	Optimization
2	Building information modeling	Project management
3	Genetic programming	Simulation
4	Neural network	Decision making
5	Case-based reasoning	Uncertainty
6	Expert system	Reliability analysis
7	Fuzzy logic/fuzzy set	Construction management
8	Deep learning	Structural health monitoring
9	Object oriented programming	Vibration control
10	Machine learning	Risk management
11	Harmony search	Planning
12	Firefly algorithm	Damage detection/assessment
13	Particle swarm optimization	Rehabilitation
14	Differential evolution	Forecasting
15	Evolutionary algorithm	Resource allocation
16	Support vector machine	Safety management/engineering
17	Adaptive neuro-fuzzy inference system	Large steel structures
18	Knowledge-based system	Inspection
19	Data mining	Life cycle assessment
20	Convolutional neural network	Scheduling

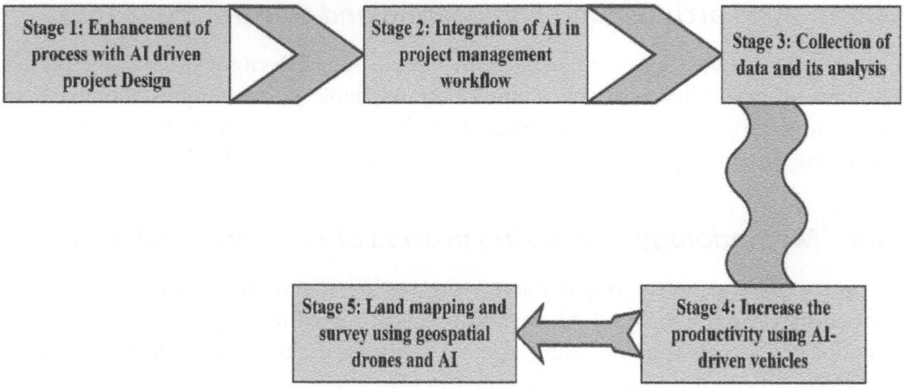

Figure 2.5 Methodology of working process of AI in smart construction.

Stage 1: Enhancement of processes with an AI-driven project design: The present process related to design and construction has become outdated and slow in handling, and as a result, the efficiency of work is quite slow. Thus, the incorporation of smart technology like AI, data analysis, and neural networks in the analysis of data related to buildings, materials, and nearby environments may help in the optimization of projects with better fruitful output.

Stage 2: Integration of AI in project management workflow: The tasks related to building or any infrastructure are comparatively tedious, more time-consuming, and more prone to errors in the case of manual work. Most of the time of managers or team leaders goes into assigning the work to the subleaders or workers and managing the records of the employees. The practice can be done accurately using the AI technique. Various tasks can be done with minimal or zero errors. An additional process of delegating tasks on the basis of data collected from different departments and employees can be ensured by AI without any time loss. This not only manages the streamline of the process related to workflow but also encourages the workers to work in their field of expertise.

Stage 3: Collection of data and its analysis: After the finalization of various tasks, the next step is to collect data from various units. Data nowadays is being collected through robots with cameras and AI-enabled construction equipment. They monitor the working process in detail and collect the data. The gathered data is then further processed for analysis based on which various decisions are made. Data can easily be collected in different formats as per requirements. All the collected data can be incorporated into a deep neural network to classify the progress of the project into various desired aspects. This analysis can help the stakeholder predict the problem at its initial stage and accordingly mitigate the upcoming subsequent possible major issues.

Stage 4: Increase productivity using AI-driven vehicles: We are aware of the fact that manual work results in comparatively less efficient output. In addition to this, the work progresses at a slower rate in comparison to the work progress at its initial stage due to the presence of uncalculated tasks and additional workload. Self-operated construction machines can be employed as they have advantages in terms of efficiency and quickness, and obviously they will not be tired and can do repetitive tasks. Some of the works where self-operated machines can be deployed are welding, the laying of bricks, and concrete work. Fully automated or semiautomated bulldozers can be easily found in companies for execution and prework. The only input they want is a specified task or activity to be performed, and after feeding the data, they start the task and stop when the work is finished. This can also be used to mitigate the human risk that may occur while performing such tasks.

Stage 5: Land mapping and survey using geospatial drones and AI: Land surveying and mapping are very tedious and time-consuming processes. The major problem here is the analysis of deep-detailed site aerial photographs, which may increase the possibility of errors in them. Using smart AI-enabled drones, the process of extracting information by the geographical information system gets simplified, and as a result, the possibility of errors gets reduced. AI also helps in tracking the progress of projects and problems associated with them in a better way, and accordingly, tentative solutions can be introduced.

2.1.6.1 Detailed working areas of artificial intelligence in construction

Some of the most frequently observed areas where AI can be used in the field of construction are mentioned below:

1. *AI could be used to prevent cost overruns*: It is a fact that most of the mega projects usually go over budget, despite the fact that the best project teams were employed. As big projects usually face unpredictable circumstances due to the involvement of multiple sectors and unavoidable circumstances, in such cases, the most precise prediction is required. AI may help the stakeholders predict cost overruns based on some of the most frequently concerned factors, like the size of the project, the type of contract, and the competence level of project managers. The most suitable technique for such cases is ANNs, which use historical construction data like the start and end of any completed activity as a base to predict and envision an actual and realistic timeline for the remaining activities of the projects. AI helps the staff to access the real-life training material remotely, which ultimately helps them to quickly enhance their knowledge and skills. Ultimately, the delivery of the project is expedited by minimizing the time taken to onboard the new resources.
2. *AI in better generative design of buildings*: Usually, the most common and frequently used BIM is the 3D model, which gives insights into design, planning, construction, and its management to engineers, architects, and professionals working in the field of construction. For a full plan and design of construction work, planning related to architecture, mechanical and electrical engineering, and other important plans related to plumbing (MEP) ought to be considered in 3D models. These models then help in selecting teams for these subsequent tasks. The major challenge in this phase is to ensure that different models do not hinder the activities of other teams. Usually, in order to perform these highly skilled-oriented tasks, different stakeholder groups use ML techniques in the form of AI-oriented power design in order to identify and mitigate clashes between different models generated for different subteams. Different iterations are performed based on the requirements of the user, and the task is performed until an ideal 3D model is obtained with optimized constraints.

3. *AI in mitigation of risk*: It is a well-settled fact that with construction, a lot of risk comes in different forms like quality, safety of manpower, time, and ultimately the cost of the project. It simply depends on the size and nature of the project. More risk arises in the case of larger projects due to the involvement of multiple subcontractors, whose work depends on the work done by other departments/contractors. In such cases, it becomes difficult to monitor them manually, but with the help of advanced AI techniques, each activity can be monitored deeply, and accordingly, a need-based priority can be assigned to avoid or minimize the risk on the job site. Ultimately resources can be utilized effectively in a limited time. AI automatically assigns priorities, is able to predict any risk, and could be handled timely by the managers or person concerned.
4. *AI in planning projects*: Nowadays, instead of manual work, the entire monitoring can be done easily by robots or advanced technologies using the concept of AI to solve the major issues related to delay and budget on construction projects. Robots automatically capture the 3D scans of the site and record the data, and that data acts as input and can be analyzed by a deep neural network that is capable enough to classify them into the desired categories. These categories can be managed separately by different competent authorities, and the prone area can be handled easily before the team faces the major issues. The upcoming era will use the concept of "reinforced learning," which is based on advanced algorithms. The said reinforced learning uses the trial-and-error method, and the method continues till it finds the best-optimized path for a properly planned project. It reduces the social and economic risk of the organization.
5. *AI in improving jobsites productivity*: Nowadays, many companies provide self-driven machinery capable enough to perform repetitive tasks with higher efficiency than humans. Some of the examples in this category are the pouring of concrete, welding, laying bricks, and work related to demolition. These instruments are autonomous and sufficient enough to perform the desired task effectively using the concept of AI. The time saved using such instruments could be utilized by the workers to do some extra activities, and thus the overall time required for the completion of the project can be reduced. Monitoring using cameras and facial recognition are some of the real-time examples that are being used by multiple stakeholders to improve productivity.
6. *AI in construction safety*: It is a generalized fact that usually, workers in the field of construction are more prone to risk in comparison to workers in other fields. AI is capable of assisting the security of the workers by taking on-site photographs at regular intervals and incorporating them into algorithms to produce efficient output that acts as an alarm and accordingly warns the managers about any possibility of mishap.
7. *AI in addressing the shortage of manpower*: The problems of low production in industry related to shortages of manpower at a particular site are another domain where AI can be used. The objectivity of such techniques increases as they are capable enough to analyze real-time data using data science, and as a result, the distribution of manpower at a deficient place can be ensured easily. The ML technique can also be used to distribute machinery and labor across the job site. For this, a robot can be used whose duty is to evaluate the progress of work assisted by manpower and machinery, and accordingly, need-based requirements can be suggested to the managers to ensure the completion of the project on time.
8. *AI in off-site construction*: Nowadays, companies are relying on need-based construction called off-site construction, which is assisted by robots having working mechanisms based on AI techniques. These robots are self-driven and autonomous and are designed to piece together different components of construction work, and the final piece-together work is ensured by human workers. A good example of this can be observed during the construction of infrastructure like walls. This ultimately improves quality by reducing human effort and saving time.

9. *AI in construction with a big database*: It is not even possible to manually design an infrastructure with a massive amount of data. Critical design and analysis of data can be ensured precisely using AI techniques. These advanced techniques have intelligent algorithms capable enough to give endless analysis in a minimum amount of time. In cases where prediction needs to be done through massive photographs captured through drones, mobile devices, and sensors and easily interpreted by AI, fruitful desired information can be extracted accordingly.
10. *AI in post-construction*: AI techniques are quite helpful even after construction work is done for monitoring the performance of a construction site. It can be done by using drone-based or sensor-based photographs, irrespective of the type of construction, i.e., whether it is a building, a road, or any similar type of construction. In a nutshell, it can be said that AI could be used to detect developing problems at a specific construction site. This can be used by multiple stakeholders working in this field, ultimately ensuring optimal safety and security.
11. *AI in manufacturing and distribution of materials*: AI assists people working in the construction field in predicting and forecasting the price of raw materials (gravel, sand, iron, etc.) and other similar elements. For such particular tasks, AI will study the historical data along with the evolution of other parameters that can have an effect on the price of the raw material. At a later stage, the next task of AI is to predict the price and the best suitable time to buy it. AI is capable enough to determine the requirements for a product, and accordingly, it will place an order for the desired product. Whenever the invoice is received by the system, AI initially verifies whether the invoice is correct or not before making the final decision to sell or decline it.

The present era demands the use of AI to maintain sustainable goals and minimize loss, cost, and human effort to increase productivity. Some of the most frequently used advanced techniques are robotics, the Internet of Things (IoT), and AI. As a reference, a few important objectives achieved by different authors using AI are presented in Table 2.6.

Table 2.6 AI used by some authors along with their objectives and conclusions.

Sr. No.	Name and year	Objective	Conclusion
1	Parveen [1]	AI in construction industry along with legal issues	AI could be used significantly to increase productivity.
2	Salehi and Burgueno [2]	To monitor and identify damage using AI	Neural network (NN) showed accurate results.
3	Patil [3]	To check applications of AI in construction	AI technique showed accurate results.
4	Schia et al. [4]	To check the impact of AI on human behavior in industry	A deep impact was found for betterment of industry.
5	Abioye [5]	To study the applications of AI in construction	AI can be used to mitigate multiple problems.
6	Harihanandh et al. [6]	To examine the use of AI to resolve the issues related to construction	Use of AI can improve the productivity of the project.
7	Rampini et al. [7]	To check the feasibility of AI in the construction industry	AI showed more effective results and could be used in the construction industry.

People working in this field can use cameras and robots to trace the overall progress of the project. Such techniques are also used for planning the routing of plumbing and electrical systems in upcoming buildings. AI especially focuses on tracking on-site interaction among workers, machinery, and related objects and accordingly warns managers and officials to take appropriate decisions for the detected potential issues related to productivity, construction, and safety. It can be said that AI is capable of predicting massive job losses. Based on the needs, AI can assign priority to the segment where more investment is required. Its impact can be seen at a later stage in the upcoming years and on a large scale in the long term.

2.2 Conclusions

The need for sustainable construction to reduce the damage to the environment that is occurring due to industries is nowadays a matter of concern for multiple stakeholders involved in the construction sector. Thus, sustainability practices have been adopted in recent years. The entire analysis was carried out in three phases, i.e., qualitative, meta-analysis, and quantitative analysis. From the critical study, it was deduced that project and material management were the major areas that had received the most attention. In addition to this, three pillars of construction considerations were also studied, and it was found that construction in this era is more concerned about the environmental pillar, and out of all certifications, LEED was the most popular one. The present study has focused on construction using sustainability, which could ultimately be helpful for the researchers doing their work in this field. It also covered the different methodologies related to environmental factors. This study also revealed some of the opportunities for future research in this emerging field, such as studies focusing on the identification of stakeholders for managing their needs after using the concept of sustainable construction and the development of concerned models keeping in view social issues.

References

[1] R. Parveen, Artificial intelligence in construction industry: legal issues and regulatory challenges, International Journal of Civil Engineering & Technology 9 (13) (2018) 957–962.

[2] H. Salehi, R. Burgueno, Emerging artificial intelligence methods in structural engineering, Engineering Structures 171 (2018) 170–189.

[3] A.G. Patil, Applications of artificial intelligence in construction management, International Journal of Religious Education 32 (03) (2019) 32–1541.

[4] M.H. Schia, The Introduction of AI in the Construction Industry and its Impact on Human Behavior, 2019.

[5] S.O. Abioye, et al., Artificial intelligence in the construction industry: a review of present status, opportunities and future challenges, Journal of Building Engineering 44 (2021) 103299.

[6] M. Harihanandh, P. Karthik, K. Murali, Application of artificial intelligence in construction project management, AIP Conference Proceedings 2385 (1) (2022) 100003.
[7] L. Rampini, A. Khodabakhshian, F. Re Cecconi, Artificial intelligence feasibility in construction industry, Computing in Construction 3 (2022) 1–8.

Further reading

[1] A. Kumar, N. Mor, An approach-driven: use of artificial intelligence and its applications in civil engineering, in: Artificial Intelligence and IoT: Smart Convergence for Eco-Friendly Topography, 2021, pp. 201–221.
[2] M. Nandal, N. Mor, H. Sood, An overview of use of artificial neural network in sustainable transport system, Computational Methods and Data Engineering: Proceedings of ICMDE 2020 1 (2021) 83–91.
[3] N. Mor, H. Sood, T. Goyal, Application of machine learning technique for prediction of road accidents in Haryana-A novel approach, Journal of Intelligent and Fuzzy Systems 38 (5) (2020) 6627–6636.
[4] A. Kumar, N. Mor, Prediction of accuracy of high-strength concrete using data mining technique: a review, in: Proceedings of International Conference on IoT Inclusive Life (ICIIL 2019), NITTTR Chandigarh, Springer Singapore, India, 2020, pp. 259–267.
[5] N. Mor, H. Sood, T. Goyal, A statistical model to prioritize selected Northern-Indian States/UT of India based on accident data, Journal of Discrcte Mathematical Sciences and Cryptography 23 (1) (2020) 305–312.
[6] H.C. Arora, S. Kumar, D.P.N. Kontoni, A. Kumar, M. Sharma, N.R. Kapoor, K. Kumar, Axial capacity of FRP-reinforced concrete columns: computational intelligence-based prognosis for sustainable structures, Buildings 12 (12) (2022) 2137.
[7] R.S. Suri, V. Dubey, N.R. Kapoor, A. Kumar, M. Bhushan, Optimizing the compressive strength of concrete with altered compositions using hybrid PSO-ANN, in: Information Systems and Management Science: Conference Proceedings of 4th International Conference on Information Systems and Management Science (ISMS) 2021, Springer International Publishing, Cham, November 2022, pp. 163–173.

Machine learning applications in the development of sustainable building materials to reduce carbon emission

3

Sikandar Ali Khokhar [1,2], *Mati Ullah Shah* [1], *Fazal Rehman* [1], *Hussnain Bilal Cheema* [1] and *Muhammad Usman* [1]
[1]School of Civil and Environmental Engineering (SCEE), National University of Sciences and Technology (NUST), Islamabad, Pakistan; [2]Bendcrete Construction Services (Pvt) Ltd., National Science and Technology Park (NSTP), Islamabad, Pakistan

3.1 Introduction

The construction sector is one of the main consumers of natural resources. Among construction materials, concrete is the most commonly used material due to the easy availability of its ingredients, low cost compared to other alternative construction materials, and flexibility to adopt any desired shape. Concrete is comprised of aggregates, cement, and water [1]. The role of aggregates in concrete is as an inert filler material, while cement and water provide binding properties due to the hydration process. Around 50 billion tons of aggregates are extracted annually for concrete production [2]. While ordinary portland cement (OPC) consumption is 3.5 billion tons per annum worldwide in the construction sector [3]. Around 46% of raw materials extracted globally are consumed in the manufacturing of concrete. The environmental impact of concrete production is significant in terms of CO_2 emissions. Concrete contributes to the production of around 7%−8% of the world's CO_2 emissions [4]. Cement is the main CO_2-contributing ingredient in concrete. In OPC production, 850 kg of CO_2 is emitted per ton of cement produced [5]. The current annual cement production is 3.5 billion tons, which is expected to reach 4.4 billion tons annually in 2050 to meet increasing urbanization demands.

The world population is anticipated to reach about 9.8 billion by 2050, out of which 7 billion will be living in urban settings. In the current urban infrastructure, concrete is the most favorable construction material and will continue to be a top choice as a primary building material in the coming years. Keeping in view the negative environmental impacts of concrete production, there is an immense need for the world to think about the production of environmentally friendly, sustainable concrete. It can be achieved by using composites that have a minimal environmental impact, are more durable and sustainable, and contain a low content of cement compared to conventional concrete. It is likely impossible to discourage concrete construction in the

Artificial Intelligence Applications for Sustainable Construction. https://doi.org/10.1016/B978-0-443-13191-2.00002-X
Copyright © 2024 Elsevier Ltd. All rights are reserved, including those for text and data mining, AI training, and similar technologies.

future built environment. However, it can be made sustainable by using recycled materials, industrial waste, etc. A significant amount of work has been done by researchers to make concrete composites more resilient, sustainable, and eco-friendly by using industry wastes and by-products, construction, and demolition wastes, and adopting the most optimal mix proportion according to requirements. Therefore, the future urban built environment can be more resilient, sustainable, and eco-friendly if conventional concrete is partially or fully replaced with sustainable composites.

Sustainable construction materials should be cheap, environment-friendly, durable, and have qualities like cement in the long run. Conventional concrete composites can be converted into sustainable green composites by adopting four main approaches [6] (Fig. 3.1): (i) Replacing or introducing new ingredients in the traditional mix; (ii) using durable and high-performance materials; (iii) using optimal mix design; and (iv) prolonging the lifecycle of materials by timely rectifying the damages by formulating an optimize maintenance strategy. Conventional composites can be converted into sustainable composites by using one or more strategies combined. The coming paragraphs will highlight comprehensively how sustainable composites can be achieved by adopting these strategies.

In the first approach, sustainable composites can be achieved by replacing or adding new materials in the traditional mix, which can be further classified into three categories. First, the replacement of cement with supplementary cementitious materials (SCMs) such as fly ash, blast furnace slag, rice husk ash, silica fume, bentonite, waste brick powder, clay, ceramic product waste, etc. Cement is the main contributor to CO_2 emissions in concrete construction. Therefore, the partial or full replacement of cement

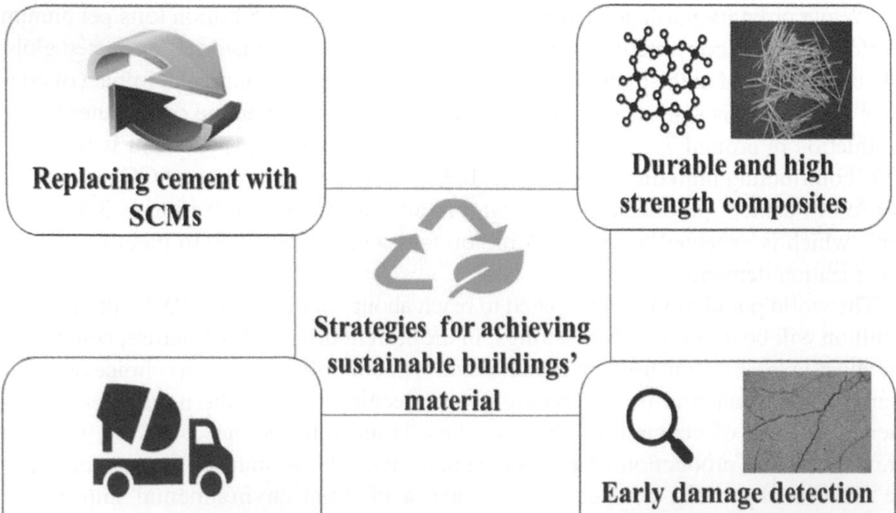

Figure 3.1 Four main approaches for sustainable building materials.

to produce concrete can lead to a substantial reduction in CO_2 emissions. The SCMs have the potential to replace cement and have been proven over time by researchers as a partial replacement for cement. The SCM acts as a pozzolanic material to give cementitious properties. Replacing cement with some percentage of pozzolanic materials decreases the cost of concrete. In addition to that, durability is enhanced. Second, to make the traditional mix sustainable, design low-water-to-cement ratio concrete and replace the coarser particles of cement with limestone powder. Here is the role of limestone powder as an inert filler: Studies have proved that coarser particles of cement can be replaced by limestone powder up to 5% without causing any detrimental effect on the performance of the concrete [7]. Thus, using inert filler smartly results in a sustainable, cheap, eco-friendly, and durable composite. Third, replacing the natural aggregate with crumb rubber and recycled aggregate (RA) results in a sustainable concrete composite that conserves natural resources. Worldwide construction and demolition waste (CDW) is around three billion tons per year [8]. Deposing the CDW in landfills causes major problems for solid waste management departments across the globe. However, recycling aggregates from CDW and using them as construction materials not only conserves natural resources but also reduces the cost of dumping them in landfills.

The second approach for achieving sustainable building materials is to use high-strength durable concrete. Adopting this approach will reduce the number of repairs and maintenance cycles of the structure. Apart from that, the expected life of the structure will increase. Hence, this approach significantly reduces CO_2 associated with repairs and maintenance activities and makes the concrete structure more sustainable. High-strength and durable concrete can be achieved by incorporating nanoparticles, fibers, superplasticizers, etc. into the traditional mix design. The incorporation of nano-silica greatly reduces the porosity of concrete and makes it less vulnerable to damages caused due to freeze and thaw action, while at the same time increasing the density, hence improving the mechanical properties. The fiber incorporation in the mix improves the response of concrete against tensile cracks and imparts ductile behavior. The different types of fibers being used in concrete are glass fibers, polypropylene fibers, carbon fibers, and steel fibers. Whereas superplasticizers improve the workability and flowability of concrete without adding excessive water, adding superplasticizers to concrete improves its fresh, rheological, and hardened state properties.

In the third approach, sustainable composites are attained by designing an optimal mixture recipe containing all the ingredients in an optimal ratio and capable of the required desired properties. In the traditional concrete mixture, while designing, the parameters that are taken into account are the water-cement ratio, coarse aggregate size, fineness modulus of sand, slump, strength, and durability. The main drawback of conventional mix design is the lack of consideration of CO_2 emission aspects. However, in designing sustainable composite building materials, the associated CO_2 emissions from the process cannot be ignored. For designing sustainable concrete, each ingredient should be added in an optimal content for reduce CO_2 emissions while still achieving the desired properties as per the requirements. But for designing an optimal mixture recipe, knowledge regarding the embodied CO_2 associated with the material is a must. Table 3.1 presents the embodied CO_2 associated with different ingredients of

Table 3.1 Embodied carbon associated with concrete ingredients.

Material	Embodied carbon (Kg CO_2/Kg)	References
Ordinary Portland cement (OPC)	0.83	[9]
Coarse aggregate	0.0062	[10]
Fine aggregate	0.0025	[11]
Water	0.0003	[12]
Fly ash	0.010	[13]
Blast furnace slag	0.019	[13]
Rice husk ash	0.173	[14]
Clay/ceramic wastes	0.004−0.01	[15]
Metakaolin	0.33	[16]
Silica fume	0.014	[11]
Recycle aggregate	0.004	[15]
Superplasticizer	0.72	[9]
Limestone powder	0.0034	[15]

concrete. Maximum CO_2 is released due to cement in concrete production (0.83 kg of CO_2 per kg of cement). However, by making an optimal mixture recipe and replacing cement with SCMs, CO_2 emissions can be significantly reduced.

By implementing the three approaches discussed above in the construction sector, sustainable building materials can be attained successfully. However, there is still some space left to improve the sustainability of the building materials. The building materials can be more sustainable if their performance is continuously monitored. Because continuous careful monitoring of concrete structures enables the stakeholders to identify the damages timely if occurred any and adopt the necessary procedures to resolve them. Therefore, the fourth approach, which is the timely detection of damages in the concrete structures and rectifying them by adopting optimal maintenance techniques, is very important for sustainable construction. Because the lifetime maintenance and repair costs of concrete structures are generally greater than the original cost of construction. The repeated repairs and maintenance cycles have not only financial impacts but also bad environmental impacts. However, the timely rectification of the damages by adopting optimal maintenance techniques increases the expected service life of the structure at a lower cost, reduces the probability of uncertain failures in the structure, and reduces CO_2 emissions by reducing extensive repair and maintenance work. In sustainable development, this strategy has a great role and cannot be ignored. For example, in the current world scenario, climate change has influenced every walk of life. The extreme weather events due to climate change have increased the deterioration process in the existing concrete infrastructure, and their expected service lives have been reduced. If the deterioration in a structure is not resolved timely, it will increase the repair and maintenance costs. Globally, around 22 billion dollars are annually consumed for the repair and maintenance of roads, bridges, and rail infrastructure. It is expected that in the coming years, this cost will be further increased. However, adopting this strategy for making materials sustainable and durable can significantly cause financial benefits by reducing repair costs.

For the successful development of sustainable composite building materials for reducing carbon emissions using the above four strategies, there is a need for advanced models for true estimation of the material properties. Because the classical methods to estimate the material properties are not reliable and viable options due to the complexities involved in the problems. Apart from that, classical solutions are time-consuming, non-economical, inefficient, and have varying empirical relationships for estimations, and those relations hardly predict correct results due to the non-linearities of the data. However, with the latest progress in the fields of machine learning (ML) and artificial intelligence (AI), their applications extended beyond the computer sciences and software engineering domains. Recently, the applications of ML have increased drastically in civil engineering fields. Researchers are engaged in the development of predictive models for material properties using ML [17–23]. Many studies have shown that these ML predictive models are more reliable in predicting material properties than classical empirical relationship predictions [24–26]. Because in the classical method, it is assumed that data follow the stochastic model, ML-based models do not adopt the stochastic approach. While ML-based predictive models are capable of identifying complex non-linear relationships among different variables and can predict the properties of materials accurately. In short, ML-based models are efficient, fast, and smart. Fig. 3.2 presents a general comparison between classical and ML-based models for data predictions. Therefore, in the conclusion to the above discussions, it is clear that using ML-based models is a reliable and viable option to develop sustainable composite building materials with low carbon emissions.

The significance of this work lies in its focus on the application of ML techniques for the design and development of sustainable composites with the objective of reducing CO_2 emissions. This chapter provides a practical framework to develop environmentally friendly materials. The remaining chapter is organized as follows: Section 02 contains the basic steps involved in the development of the ML models. Section 3 focuses on the preceding studies that employed ML models to develop predictive mimickers for the fresh properties, mechanical properties, and durability aspects of concrete. Section 04 highlights the ML models for formulating optimal mixture recipes to attain sustainable composite materials. Section 05 presents the application of ML for the detection of early damages in structures to prolong service life. Section 06 introduces the concept of multiobjective optimizations for sustainable development. In section 07, insightful and concise conclusions are drawn by highlighting the importance of the development of sustainable composite materials using ML.

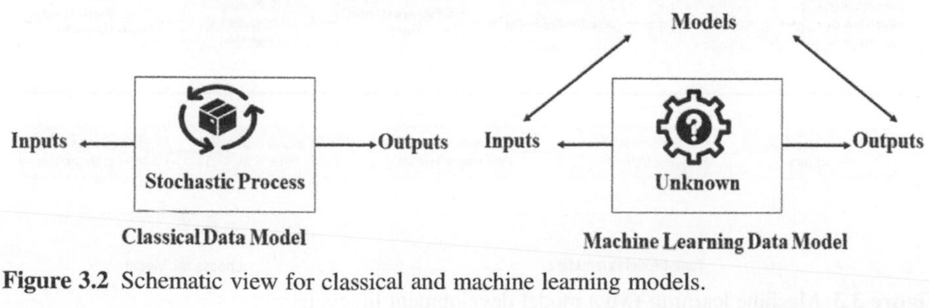

Figure 3.2 Schematic view for classical and machine learning models.

3.2 Steps involved in developing an ML model

The development of the ML model involves a series of steps that must be followed in the correct sequence to build an efficient mimicker [27]. The ML Model Lifecycle refers to the process of identifying source data, developing models, deploying them, and maintaining them. At a high level, the full set of operations may be divided into four major categories: Initial data analysis, developing the ML model, tuning model parameters, and finalizing the model [28]. The lifecycle of the ML model is graphically presented in Fig. 3.3.

As shown in Fig. 3.3, the development of a predictive ML model is an iterative process to finalize the optimum algorithm with the highest accuracy. To achieve a high-accuracy predictive model, the following steps are involved:

3.2.1 Initial data analysis

3.2.1.1 Data collection

The initial step for developing a predictive model is data collection and organization. This is a time-consuming task, as the data must be comprehensive and representative of the system being modeled [29]. Once the data is collected, it is fed into the ML algorithm. The algorithm then analyzes the data and produces a predictive model. Finally, the predictive model is validated to ensure that it accurately represents the system. Data collection is therefore the first and most important step in developing a predictive model using ML.

3.2.1.2 Data processing and cleaning

Data preprocessing is a crucial step in predictive modeling, and it generally includes four main operations on data: cleaning, transformation, reduction, and integration [30,31]. ML algorithms cannot be trained on data that is not properly processed because that would introduce biases to the models. Fig. 3.4 elaborates on the main attributes of data pre-processing.

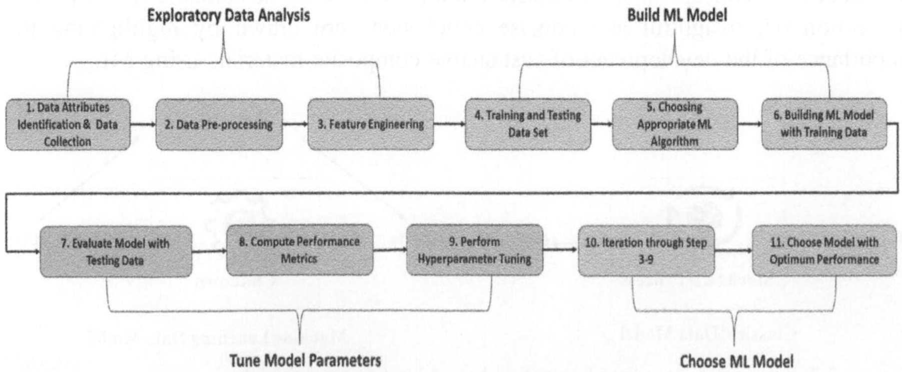

Figure 3.3 Machine learning (ML) model development lifecycle.

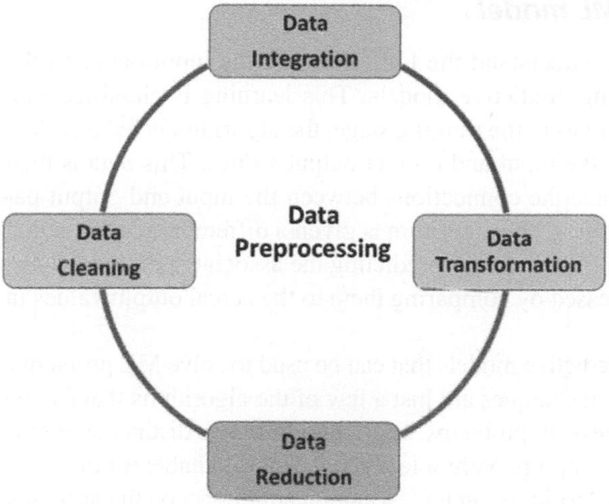

Figure 3.4 The main attributes of data pre-processing for ML.

The process of identifying and fixing errors and anomalies in a dataset is known as data cleaning. Data transformation is the process of changing data into a format that ML algorithms can understand. Data reduction is the procedure to shrink the dataset without sacrificing its quality. The process of merging numerous datasets into one is known as data integration.

3.2.1.3 Feature engineering

Training predictive models or ML algorithms requires a lot of data. However, much of the data collected is noise, and some of the parameters in the dataset may not contribute significantly to the model's performance [26]. Having a huge amount of data also slows down the training process and makes the model slower. This irrelevant input may potentially cause the model to learn erroneously. To avoid these problems, it is important to be aware of the sources of noise in our data and to remove them before training the model. Additionally, we should only use relevant columns of data in our training set to prevent the machine from learning from irrelevant information. By following these simple steps, the surety is achieved that the ML models are trained on high-quality data and perform accurately.

In predictive modeling, the process known as feature selection involves the careful selection of a subset of input variables (features) to use in an ML algorithm, as shown in Fig. 3.5 [32]. The goal is to select only the most relevant features while excluding features that are not predictive or that add noise to the data.

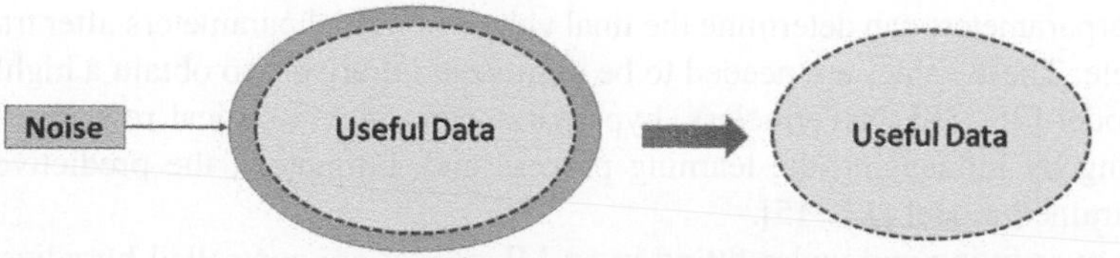

Figure 3.5 Feature selection to remove noise from the dataset.

3.2.2 Development of ML model

ML algorithms are employed to understand the links between the input data and the intended output while developing predictive models. This learning is classified into two stages: training and testing. During the training stage, the algorithm is fed a collection of data that comprises both the input and correct output values. This data is then used by the algorithm to recognize the connections between the input and output parameters [28]. During the testing step, the algorithm is given a different set of data that solely contains input values and is tasked with predicting the associated output values. The predictions' accuracy is assessed by comparing them to the actual output values in the test data [28].

There are a wide variety of predictive models that can be used to solve ML problems. ANN, CART, and deep learning techniques are just a few of the algorithms that can be applied to classification and regression problems. Each algorithm has distinct strengths and weaknesses, so it is crucial to appropriately select the one most suitable for the specified task at hand. The choice of algorithm can have a significant impact on the accuracy of the predictive model. In general, more sophisticated algorithms like deep learning tend to outperform simpler ones like regression methods, but they are also more computationally expensive and require more data to train. As a result, choosing a suitable ML algorithm is a critical step in the predictive modeling process [33,34]. Following are a few of the factors that must be considered for the selection of the ML model.

- Memory requirements
- Interpretability
- Training and prediction time
- Data format
- Linearity of data
- The number of data points and features

Recently, ML models have been applied for the prediction of various mechanical, durability, and rheological properties of sustainable construction materials. The selection of the ML model shows a significant impact on the performance of the predictive model. For example, for complex datasets involving a variety of parameters, like fiber-reinforced concrete, ultra-high-performance concrete (UHPC) gradient boosting algorithms provide better accuracy. For relatively simple datasets, simple regression models and regression trees provide satisfactory results.

3.2.3 Hyperparameter tuning

Hyperparameters are parameters that define and control the learning process itself, and as such, they are said to be external to the predictive model being learned. The values of hyperparameters can determine the final values of model parameters after training is complete. These values are needed to be optimized iteratively to obtain a highly accurate model [35–38]. Nevertheless, hyperparameters play a critical role in predictive modeling by influencing the learning process and, ultimately, the predictive power of the trained model [13–15].

The over-fitting and under-fitting in an ML model are controlled by adjusting the hyperparameters. It is important to note that optimal hyperparameters typically vary across different datasets.

Some of the main hyperparameters of ML models are as under.

- Batch size
- Size of pooling
- The ratio of the split between train and test
- Choice of activation function
- Optimization algorithm
- The number of hidden layers
- The choice of loss or cost function of the model
- Activation units in each layer
- The number of iterations
- The rate of dropout
- Kernel or filter size in convolutional layers
- Learning rate in optimization algorithms

3.2.4 Validation of ML model

In the field of ML, "model validation" means the process of evaluating a trained model using a testing data set [39–41]. This evaluation takes place after the model has been trained. The training data set is derived from the same source as the testing data set, but the two sets of data are kept separate for testing purposes. The primary objective of utilizing the testing data set is to assess the extent to which a trained model can generalize its results. The validation process is explained with the help of a flowchart in Fig. 3.6. Flowchart illustrating the sequential steps for constructing, training, validating, and selecting an optimal model for a given task. In model construction, the most common practice is that the data is divided into three sets: The training set,

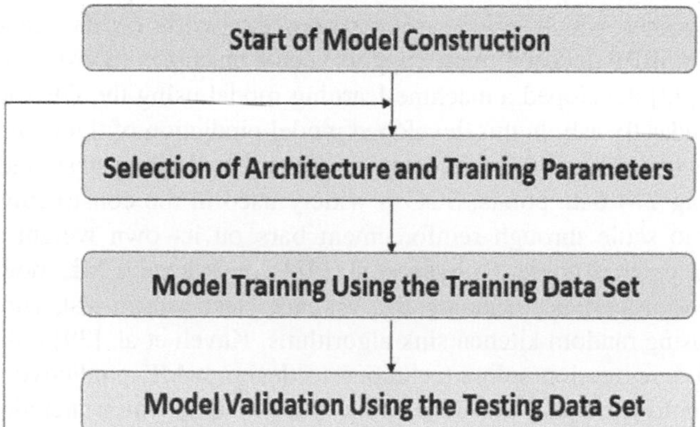

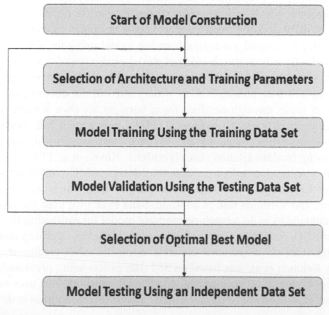

Figure 3.6 The flowchart for validation of the ML model.

the validation set, and the testing set. The training set commonly consists of 70% of the total data set, while the remaining 30% of data is divided equally into a validation/testing set (15%) and an independent set (15%). The ML model that gives the best results and accurate predictions on training and testing sets is selected. Then the selected ML model's performance is verified through an independent set of data.

Finally, the ML model's performance for training, testing, and validation is assessed and compared in the form of statistical metrics. These parameters provide us with information about the performance of the generated ML model. These statistical metrics include mean square error (MSE), mean absolute error, root MSE, coefficient of determination (R2), and Pearson correlation coefficient (R).

3.3 ML-based prediction mimickers for sustainable materials

3.3.1 Rheological and fresh properties

The fresh properties are important parameters of the concrete, which are indirectly linked to the other mechanical properties as well. Rheological properties should have satisfactory flowability; otherwise, it would lead to honeycombing, cold jointing, segregation, and bleeding. To avoid issues while pumping, especially in high-rise buildings, transporting, compacting, and placing, concrete should have an acceptable rheology. In addition to that, rheological parameters become more important while working with special types of concrete, e.g., self-compacting concrete (SCC), 3D printing concrete, and UHPC.

In order to avoid experimentally tedious work and a time-consuming approach to finding the rheological properties, many researchers developed ML models using different approaches, which include artificial neural networks (ANN), gene expression programming (GEP), decision trees, support vector machines (SVM), etc.

Balf et al. [42] developed a machine-learning model using the DEA technique for SCC that includes fly ash. In this developed model prediction of three important rheological parameters of the SCC, including the funnel test, slump test, and L-box test, was done using 114 data points. SCC is widely used in the construction sector due to its ability to settle through reinforcement bars on its own weight without any compaction or external force. Sathyan et al. [43] constructed a ML model to predict the rheological properties, including the V funnel test, slump test, and J ring test, for the SCC using random kitchen sink algorithms. Kaveh et al. [39] utilized a multivariate adaptive regression spline technique to design a ML predictive model using 114 datapoints to anticipate the fresh properties of SCC, which included the slump test, L-box test, and V-funnel test. In addition, Saha et al. [40] used the support vector regression technique to build a machine-learning model for estimating the rheological parameters of SCC. Rehman et al. [41] developed a machine-learning model to forecast the slump value of geopolymer concrete using the ANN technique. The model developed by Rehman et al. was based on 384 data points using previously published literature. Geopolymer is an economical, eco-friendly concrete that uses no cement at all. Chandwani et al. used the ANN and genetic algorithms techniques to develop a ML model for the prediction of the slump value for ready-mix concrete. The developed model was based on 560 data samples and used seven input parameters to predict the slump value of the ready-mix concrete. Amin et al. [44] developed a machine-learning model for super-plasticized concrete to forecast the yield stress and plastic viscosity using an ANN and random forecast approach. These models were developed using 156 data samples and six input parameters to predict these two outputs. Super-plasticized concrete is a type of concrete that uses high-range water reducers to increase the flowability of the concrete. It is mainly used in heavy reinforcement sections and high-rise buildings to pump. Table 3.2 presents a summary of some other studies in which ML is used for predicting the fresh properties of cementitious materials.

[illegible].2 Mechanical properties

Mechanical properties are the most important parameters that are considered while designing the material. Mechanical properties include mainly compressive strength, tensile strength, flexural strength, elastic modulus, etc. The mechanical properties of the materials give an overview of the engineering performance of the material. Finding an optimum mix design is a process that is extensive, iterative, and unsustainable; therefore, researchers came up with the idea to implement AI in the construction industry. Several researchers made their contributions in integrating AI to find desired mix designs for different types of concrete, which are discussed below.

Gandomi et al. [49] formulated an ML model using GEP for predicting the shear strength of reinforced concrete based on 1942 data points. Chou [50] developed a ML model based on 1030 data points using the SVM approach to forecast the compressive strength of high-performance concrete (HPC).

Due to the brittle behavior of concrete, it has been an issue to resist dynamic loading during seismic activities; therefore, fibers were introduced in the mix in order to increase the ductility of the composite. Concrete with the inclusion of fiber is known as fiber-reinforced concrete. Fiber-reinforced concrete is gaining great attention for its enhanced performance, which is due to its bridging effect to control the width of the cracks. However, the mix design of the fiber-reinforced concrete became a bit complex due to the incorporation of the fibers and their properties, which include length, diameter, tensile strength, and elastic modulus of the fiber. Therefore, to find an optimum and sustainable mix design without going through extensive lab tests for the fiber-reinforced concrete, ML models have been developed by different researchers. Khokhar et al. [51] developed a ML-based predictive mimicker using five different approaches to predict the fracture behavior of fiber-reinforced concrete. In this predictive mimicker, five model techniques were used, which include the SVM, the ANN, the gaussian process of regression, the extreme gradient boosting tree (XGBoost), and the regression tree. A comparison of all the models has been performed in this study, and the XGBoost showed the best results with an accuracy of 98%. Similarly, Behnood et al. [52] used the SVM technique to develop an ML model for the prediction of the tensile strength of steel fiber-reinforced concrete. This model was developed based on 980 data points using four input parameters.

Furthermore, in order to promote sustainability, researchers started incorporating tire rubber or tire chips, which has been a critical pollution issue due to its slow

Table 3.2 Predictive models for the fresh properties.

Technique	Type of concrete	Input	Output	References
Support vector regression	SCC	W/b ratio, binder content, superplasticizer, fly ash, coarse aggregate, fine aggregate,	V-funnel, slump value, and L-box test results	[45]
ANN, RF	Super plasticized concrete	Cement, fine aggregate, water, superplasticizer, coarse gravel, medium coarse gravel	Yield stress, plastic viscosity	[44]
ANN	Geopolymer	FA type F, non-calcined clay, GGBFS, silica fume, rice husk ash (RHA), sodium silicate, sodium hydroxide quantity, sodium hydroxide molarity, superplasticizer, water quantity, sodium silicate (SiO_2/Na_2O) composition, fine aggregate, coarse aggregate	Slump value	[46]
Multivariate adaptive regression splines	SCC	Fly ash, binder content, water-binder ratio, superplasticizer, fine and coarse aggregates	L-box test, V funnel test, slump flow	[47]
Random forest	Alkali-activated concrete	Fine aggregate, fly ash, coarse aggregate, NaOH, Na_2SiO_3	Slump value	[48]
Data envelopment analysis	SCC	Superplasticizer, fine aggregates, coarse aggregates, fly ash replacement percentage, water–binder ratio, and the total binder content.	Funnel test, slump test, L box test	[42]
Extreme gradient boost	The super-plasticized cement paste	Cement, superplasticizer, water, test temperature	Yield stress, plastic viscosity	[43]

biodegradable properties. To decrease the amount of waste as well as its property to make concrete lightweight. To encourage the use of this rubber waste researchers provided easy access to find the desired mix design for the rubberized concrete. Bachir et al. [53] used the ANN approach to develop a forecasting model to predict compressive strength. Yoon [54] developed an ML model using the ANN to predict the compressive strength and elastic modulus of lightweight aggregate concrete. Due to the inclusion of different lightweight aggregate types, the mix design of the lightweight aggregate concrete becomes a bit complex. The main issues with the design, placement, and production of lightweight concrete are the high water-absorbing capacity, low strength of the lightweight aggregate, and reduced aggregate weight. Researchers developed ML models to predict the properties of lightweight aggregate concrete in order to avoid the extensive, iterative, and unsustainable process [54−56]. Ullah et al. [57] also developed a machine-learning model using the GEP technique to predict the compressive strength and density of sustainable lightweight foamed concrete. Zhang [58] also developed a machine-learning model to forecast the uniaxial compressive strength of lightweight SCC using a random forest approach. Jagadesh et al. [59] developed ML models based on different algorithms to predict the compressive strength of SCC made of RA. Rehman et al. [46] developed a ML model applying the ANN to predict the mechanical properties, including compressive strength, flexural strength, and elastic modulus of the geopolymer concrete. Awoyera et al. [60] have worked on the geopolymer SCC and developed an AI model using GEP and ANN to predict the strength properties, including compressive strength, split tensile strength, and flexural strength. To make the construction industry further sustainable, investigators started using recycled concrete as a coarse aggregate. Therefore, different ML models were developed in order to provide easy access for users to find an optimum mix design for RA concrete. Xu [61] developed a backpropagation neural network approach to predict the tensile strength of concrete containing RA. Furthermore, Golafshani [62] also built a support vector regression model of M to predict the elastic modulus of RA concrete. Samer et al. developed ML models for concrete containing RA and secondary raw materials [63]. Ammar et al. compared four different ML predictive models for the tensile and flexural strengths of 3D-printed concrete [19]. Table 3.3 presents a summary of some other studies in which ML is used for predicting the mechanical properties of cementitious materials.

3.3.3 Durability

The durability of the concrete structures mostly depends on the concrete mix proportion and the exposure conditions. According to the American Concrete Institute, the durability of concrete is defined as "the ability of the concrete to resist chemical attack, weathering action, abrasion, or any other process of deterioration and will maintain its original form, quality, and serviceability when exposed to its external environment" [75]. Durable concrete can be attained through proper design, proportioning, placement, finishing, testing, assessment, and curing. In sustainable construction, the durability design and estimation of the service life of the material hold great importance. The mechanisms that can affect the durability of the concrete are carbonation, chloride

Table 3.3 Detailed table of the ML predictive models for the mechanical properties.

Technique	Type of concrete	Input parameters	Outputs	References
Neural network	Structural lightweight concrete	Water, silica fume, cement, LW coarse aggregate, fine aggregate	Compressive strength (Comp-strg)	[64]
BPNN	Concrete using construction and demolition waste	Cement, mortar, aggregates, the maximum aggregate size of the fine and coarse aggregates, fineness modulus of fine and coarse aggregates, water absorption, w/c ratio, admixture, the ratio of recycled materials, age of testing	Comp-strg	[65]
BPNN, SVM,	High-performance concrete	Cement, fly ash, slag, coarse and fine aggregates, water, superplasticizer, testing age	Comp-strg	[49]
GEP	Reinforced concrete	Effective depth, beam width, compressive strength, shear span to depth ratio, longitudinal reinforcement ratio	Shear strength	[49]
M5	Concrete containing coarse recycled concrete aggregate	Coarse recycled concrete aggregate replacement ratio, bulk density of recycled concrete aggregate, aggregate-to-cement ratio, water absorption of coarse recycled concrete aggregate, water-to-cement ratio	Comp-strg	[66]
BPNN	Rubberized concrete	W/C ratio, fine aggregates, coarse aggregates, superplasticizer, crumb rubber, tire chips	Comp-strg	[53]
BPNN	Self-compacting concrete	Cement, fine aggregate, coarse aggregate, fly ash, limestone powder, slag, rice husk ash, silica fume, water, superplasticizers, viscosity modifying admixtures	Comp-strg	[67]
BPNN	Recycled aggregate concrete	Cement, cement type, specimen size, total aggregate to cement ratio, mass substitution rate of natural aggregate by the recycled aggregate, water to cement ratio, fine aggregate percentage, characteristic of coarse aggregate, constituents of recycled coarse aggregate, type and preparation methods of coarse aggregate	Elastic modulus	[68]
SVM	Steel fiber reinforced concrete	Compressive strength, water-to-binder ratio, fiber reinforcing index, age of the specimen	Tensile strength	[52]
ANN	Geopolymer	Fly ash, slag, water, superplasticizer. Sand, sodium hydroxide sodium silicate	Comp-strg	[69]

ANN	Fly ash-based geopolymer	Fly ash, fine aggregates, coarse aggregate, water, sodium hydroxide sodium silicate, molarity of sodium hydroxide, amount of Na_2O and SiO_2, curing age	Comp-strg	[56]
SVM, ANN	Conventional concrete	Cement, GGBFS, fly ash, water, superplasticizer, coarse aggregate, fine aggregate	Comp-strg at 28 days	[70]
TGAN	Ultra-high-performance concrete	Cement, silica fume, slag, fly ash, quartz powder, limestone powder, nano, silica, water	Comp-strg	[71]
ANN	Ultra-high-performance concrete, self-compacting concrete	Cement, W/B, W/C, W/P, FA/P, CA/P, HRWR/P, VMA/P (%), fly ash/P (%), MS/B	Comp-strg at 28 days, workability	[72]
ANN, SVR, CART, XGBoost	High-performance fiber-reinforced cementitious composites	C/B, fly ash/B, Slag/B, RHA/B, Limestone/B, Metakaolin/B, silica fume/B, Sand/B, Water/B, Superplastisizer/B, fiber volume, fiber length, fiber diameter, fiber elastic modulus	Comp-strg, tensile strength, tensile strain capacity	[26]
ANN	Ceramic waste-based concrete	Cement, waste, fine aggregate, coarse aggregate, water	Comp-strg	[73]
ANN	Fly- ash and bottom ash-based geopolymer concrete	Fly ash, bottom ash, coarse aggregate, fine aggregates, water, sodium hydroxide sodium silicate, molarity of sodium hydroxide, amount of Na_2O and SiO_2, L/P, curing age.	Comp-strg	[74]
ANN, CART, GPR, SVM, XGBoost	Fiber reinforced concrete	C/B, fly ash/B, Slag/B, RHA/B, Limestone/B, Metakaolin/B, silica fume/B, Sand/B, Water/B, Superplasticizer/B, fiber volume, fiber length, fiber diameter, fiber elastic modulus	Comp-strg, tensile strength, strain-hardening, tensile strain capacity	[46]
GEP	Lightweight foamed concrete	Cement, sand, w/c, foam	Density, Comp-strg,	[57]
ANN	Geopolymer concrete	Type F fly ash, GGBFS, non-calcined clay, silica fume, rice husk ash (RHA), sodium silicate, sodium hydroxide quantity, sodium hydroxide molarity, sodium silicate composition (SiO_2/Na_2O), water, superplasticizer, fine aggregate, coarse aggregate	Flexural strength, elastic modulus, slump value.	[46]

ion penetration, alkali-aggregate reactions, sulfate reactions, corrosion, etc. These mechanisms are very complex and affect concrete physically as well as chemically. Using simple empirical models for the prediction of the durability of the material is not accurate because no direct relation exists due to the involvement of complex processes in the deterioration mechanism of concrete. However, some recent studies showed that ML-based predictive models for estimation of the durability aspects of concrete, such as carbonation depth, chloride ion penetration depth, etc., are accurate and efficient. Therefore, some of the recent studies conducted for developing ML-based models for concrete durability are comprehensively discussed below.

Buenfeld et al. [76] developed an ML model using the ANN to predict the carbonation depth of concrete structures. The development of the predictive model is based on 39 parameters, which include cement composition, SCMs, concrete mix proportions, accelerated testing conditions, curing conditions, and exposed environmental regimes and conditions. Taffese et al. [77] built an optimized ML predictive model using algorithms including regression tree, bagged, and reduced bagged ensemble regression tree to forecast the carbonation depth of the concrete. 15 input parameters were considered for this study to predict the carbonation depth of the concrete. Furthermore, a comparison of the models was also performed, and it was found that the ensemble model made promising results with maximum accuracy. Considering chlorination as an important property of concrete, especially in marine structures, researchers also developed machine-learning models using different approaches to predict chlorine diffusion or chlorine permeability into concrete structures. Regarding this, Peng et al. [26] constructed a machine-learning model using particle swarm optimization (PSO) backpropagation and backpropagation neural networks for the prediction of chlorine diffusion in concrete structures. A total of eight input attributes were considered for the development of the predictive model. In the end, the comparison of the models was also completed, and the PSO backpropagation showed minimal error in predicting the chlorine diffusion. Gilan et al. [78] used hybrid support vector regression PSO techniques to develop machine-learning predictive models for the rapid chloride permeability test. This prototype was based on 100 data samples with four different recipe mix designs and seven different inputs of concrete, including cement, water, coarse aggregate, fine aggregate, metakaolin, days, and surface resistivity. A comparative study was also done between the hybrid support vector regression—PSO and adaptive neural-fuzzy inference systems, where hybrid support vector regression—PSO outperformed the adaptive neural-fuzzy inference system. In addition to that, Ghafoori et al. [72] created an ML method for predicting the rapid chloride permeability of self-consolidating concrete using four alternative neural network algorithms. A total of 72 samples were created in this investigation using 12 distinct mix designs and a variety of input parameters. In addition, Hodhod et al. [26] have created an ANN-based model to assess chloride diffusivity in HPC. The proposed model was based on 300 previously published data samples, considering four input parameters. Tao et al. developed predictive mimickers for the precise calculation of bond strength between corroded steel rebars and concrete using the hybrid ML approach [79]. Liu et al. developed ML models to accurately predict the creep in concrete containing secondary cementitious material [80]. Table 3.4 presents a summary of some other studies in which ML is used for predicting the durability properties of cementitious materials.

Table 3.4 Further machine learning models for durability properties.

Technique	Type of concrete	Input parameters	Output	References
Support vector machine,	Conventional concrete	Water-cement ratio, concrete strength, cement type, aggregate, admixture	Carbonation depth	[81]
ANN	Conventional concrete contains ground pozzolans	Water-binder ratio, percentage replacement, pozzolans types, aggregate-cement ratio, testing ages, compressive strength	Chloride permeability	[82]
ANN	Slag concrete	Binder content, slag percentage, water-binder ratio, slag acid index, slag fineness, carbon dioxide concentration, relative humidity, curing time, and exposure time	Carbonation depth	[83]
ANN	High-performance concrete	Cement content, water-binder ratio, fly ash or slag content, and curing age	Chloride diffusion coefficient	[84]
Gradient boosting regression tree	Recycled aggregate concrete	Metakaolin, slag, fly ash, silica fume, cement, gravel, water, water-binder ratio, water absorption of gravel, water absorption of recycled aggregate, recycled aggregate content, sand content, weighted density, superplasticizer, compressive strength, carbon dioxide content, exposure time	Carbonation depth	[85]
ANN	Conventional concrete	W/c ratio, superplasticizer to cement ratio, aggregate-cement ratio, cement type, curing conditions, and testing age	Chloride permeability	[86]

3.4 ML-based mimickers for optimized mix design

The reverse prediction of more than one parameter of the construction materials simultaneously is complex but critical in real-world engineering applications. It is crucial to concurrently design an appropriate mix to satisfy all of the properties required. Developing an ideal mix design that addresses our ecological concerns while also providing enhanced performance is a task that requires further research. Effective reverse prediction based on mechanical, rheological, economical, and sustainability aspects is much more difficult as compared to forward prediction. This approach of reverse prediction can help preserve the workforce and other resources. However, the majority of research to date has focused on the forward prediction of engineering properties using ML models to forecast the characteristic performance of construction materials from their constituents; however, predicting the constituents for the anticipated properties remains a challenge.

Compressive strength has been considered the most significant criterion for designing the concrete mix. Other variables, such as sustainability, have received little attention in the concrete sector and research. Hence, Naseri et al. [24] created a predictive model for mix design that predicts the optimum mix based on multiple different types of goal functions and provides the most sustainable, cost-effective, environmentally friendly, and least material-intensive mixes.

Several studies have been conducted on the mix design of conventional concrete. Gou et al. [87] developed a variational SVM using PSO to develop an accurate model in a multiinput, multioutput issue that estimates the ingredients from the strength and slump flow of the composite. In addition, Fan et al. [88] utilized the SVM technique for the backward prediction of the mix ingredients of conventional concrete. Furthermore, Ziolkowski and Niedostatkiewicz [81] created an ANN-based prediction model to estimate the compressive strength of ordinary concrete and translated ANN into an empirical mathematical equation. Jafari and Mahini [82] utilized GEP to construct equations for forecasting the compressive strength of a certain lightweight concrete combination.

Traditional single-objective approaches are incapable of producing optimal combinations for complicated materials such as fiber-reinforced concrete (FRC). In contrast to ordinary concrete, there are several output properties of FRC that influence its performance. Also, FRC requires relatively higher input parameters (including a vast variety of ingredients and fiber properties), as they affect the composite's behavior. Therefore, a comprehensive model was developed by Guo et al. [26] to predict the mechanical properties (compressive strength, tensile strength,and tensile strain) of the high-performance fiber-reinforced concrete. Further, Khokhar et al. [51] designed the model to predict the post-peak behavior (strain hardening or strain softening) of FRC. This helped to develop the complete fracture behavior of FRC. In this way, these models can be used to obtain an optimized and sustainable mix design for FRC iteratively.

Self-compacting mortars are commonly employed in the concrete industry to improve the mechanical and durability aspects of the composite. They can help to lessen the pollution caused by the landfilling of industrial byproducts. Most of them with low carbon emissions may decrease CO_2 emissions from the cement industry by manufacturing much more ecologically friendly materials; also, these components display higher performance in terms of strength and durability. Golafshani and Behnood [85] employed a hybrid model based on the Bailey-Borwein-Plouffe algorithm to construct a simple formula for forecasting the compressive strength and optimal mix design of silica fume concrete.

HPC and UHPC can help build more resilient and sustainable infrastructure. HPC and UHPC, on the other hand, are more environmentally friendly and have greater engineering performances. Tavares et al. [89] created an ML model that takes into account the sustainability of UHPC in terms of carbon footprints.

Geopolymer is also an environment-friendly and sustainable construction material that uses no cement at all. Geopolymers utilize industrial waste products such as fly ash and GGBS (ground granulated blast furnace slag) as binding materials. Rehman et al. [90] developed an ANN-based predictive model for the rheological and mechanical properties of geopolymer concrete. Similarly, many other studies were carried out for the prediction of different parameters of geopolymer concrete. These developed models can help ease the mixed design of the composite, which could assist in the commercial implementation of these green materials.

Yet, most research to date has focused on forward prediction and has utilized ML models to forecast the properties of materials from their constituents; however, predicting the ingredients for the required performance remains a challenge.

3.5 System health monitoring for damage detection

System health monitoring (SHM) refers to regular assessment activities for monitoring the conditions and characteristics of civil infrastructure [91,92]. The purpose of the SHM system is to monitor the health conditions of the structure compared to normal conditions. The workings of the SHM system for civil infrastructure can be classified into three categories [91]: (i) Identify the damages or changes in the structure; (ii) timely formulate a solid maintenance activity plan; and (iii) take appropriate action to bring the structure back to normal conditions. In civil engineering, the prime structure health index of the SHM system is crack detection. Because in civil structures the cracks weaken the structures over time and can manifest serious damages. Therefore, for the safety and maintenance of civil infrastructure, such as roads, rails, bridges, dams, culverts, buildings, tunnels, etc., cracks are mainly considered for determining the structural health state. The cracks in the structure are the initial signs of alarm that the structure is degrading. If the cracks are mitigated at the early stage of progression, it will stop excessive structural modifications that cracks can cause in the form of physical changes and loss of strength of material if not prevented early.

The cracks in civil structures can be identified and monitored through manual or automated procedures [92]. In the manual procedure, skilled labor is required for a periodical check of the structure to identify and monitor the cracks. This process is time-consuming and non-reliable, with huge chances of error, and decisions vary from person to person. Another main drawback of the procedure is that there is no visual recording of the cracks over the period, and it is very hard to pass judgment on the sensitivity of the cracks. On the other hand, in the automated procedures, images of the structure are taken periodically through automated installed cameras on the structure. Then taken images are regularly processed to identify and monitor the cracks. The images can be processed through traditional image processing techniques or using ML tools. The automated process for crack detection and monitoring is more reliable, robust, time-efficient, and cost-effective.

Crack detection through traditional image processing involves the following steps: (i) Image pre-processing; (ii) segmentation; (iii) feature extraction; and (iv) crack recognition. In image processing, multiple techniques are developed over time for crack detection. Some of the major techniques are edge information, morphology operations, pattern matching, digital image correlation, and statistical methods. The crack detection through the image process is quite impressive compared to the manual crack detection method. However, there are some drawbacks to this process. For example, many environmental conditions, such as dust, shadows, non-even light, and multiple background scenes, can trigger difficulty in detecting the crack accurately. Another main problem is that while designing the algorithms to detect cracks many factors need to be considered, such as crack and background characteristics, the camera's resolution and position, crack width, length, angle of inclination, etc. Compared to traditional image processing, ML is a powerful technique for crack detection and monitoring.

In ML methods, the primary objective is feature extraction and highlighting the areas in the image that contain cracks. Moussa and Hussain proposed a novel technique using ML for automatic crack detection in flexible pavement [93]. The technique consists of four main steps: (i) Segmentation; (ii) feature extraction; (iii) classification; and (iv) quantification of parameters. The graph-cut segmentation method was used to segment the image precisely into cracks and background regions. Using this method, the best balance of boundaries and region attributes can be achieved [93,94]. A SVM can be used to classify the crack and uncrack images. SVM can also classify cracks based on orientation and is capable of calculating the geometric values of cracks. Applying an ML-based crack segmentation procedure computes features based on color and texture, but first, remove the background elements [95]. Using these features, SVM can easily classify crack images. Many ML-based methods have been developed over time for crack detection in civil infrastructure. Some of the methods include deep belief networks, Markov-based methods, image binarization, recurrent neural networks, random forests, AdaBoost, etc. These methods are very efficient and capable of easily detecting cracks in the structures. The ML base models for image processing to detect cracks perform better than traditional image processing. Therefore, introducing these techniques in the SHM system for crack detection in structures will be a positive step in the direction of sustainable development.

3.6 Multiobjective optimal design

In the majority of real-world issues, more than one function must be optimized simultaneously. Although single-objective optimization problems have been studied, construction material design challenges often involve many objective functions. For instance, in the design of concrete mixtures, the strength-workability-cost dilemma must be addressed. This section examines the use of ML technologies for multiple objective optimization issues.

To progress toward more sustainable construction, fly ash, GGBFS, metakaolin, etc. may be used as a partial replacement for cement in concrete mixes, providing various engineering and economic benefits while also lowering energy consumption and carbon footprints. A multiobjective-based ML model can help to develop sustainable mixes with a lower amount of carbon footprint [96]. Kandiri et al. [93] employed the multiobjective to construct ANN models with the least amount of error and complexity, arriving at a Pareto front with 19 ANN models.

Similarly, geopolymer concrete is also an environment-friendly composite that does not require cement for its hydration. To ease its material processing, Rehman et al. [90] developed a multiobjective ANN-based model for predicting the mechanical and rheological properties of eco-friendly geopolymer concrete.

UHPFRC is used to strengthen and seismically retrofit concrete structures. For optimal mixture design of UHPFRC, Abella'n-Garc'a and Guzma'n-Guzma'n [97] established two prediction models for energy absorption capacity and maximum post-cracking strain. Similarly, Khokhar et al. [51] developed a multiobjective prediction model for the complete fracture behavior of FRC. To provide the model the capacity to discriminate the composite based on its post-peak behavior, it was trained with data for both strain-hardening and strain-softening FRC. This case's multiobjectives comprised compressive strength, tensile strength, tensile strain, and post-peak response.

This multiobjective technique can also be used for road maintenance. Matin et al. [98] used a hybrid of GEP and particle swarm optimization (GAPSO) for efficient road maintenance planning, which may reduce the life cycle cost of a road network while also improving pavement conditions. Cao et al. [99] also developed a multiobjective optimization model to maintain a low-noise pavement network system, which is a developing interest in the pavement management system. The model takes into account the average proximity level decrease, maintenance expenses, and greenhouse gas emissions.

3.7 Conclusion

This chapter presents a comprehensive overview of ML applications for the development of sustainable composite building materials to reduce greenhouse gas emissions. For making sustainable and low-carbon, eco-friendly building materials, four different approaches are discussed in detail. Adopting those approaches will make the construction more sustainable. However, the main problem is the need for an efficient advanced model that smartly estimates the properties of sustainable materials. The classical model cannot be used for making sustainable building materials due to the non-linearities and complexities involved in the data. While ML techniques are powerful tools to deal with data with non-linearities and complexities. ML-based models are efficient and perform well compared to traditional models. Many researchers studied ML-based mimickers and concluded that optimal mix proportions and their estimated properties based on these mimickers are very close to experimental results. Apart from that, ML applications have been studied by researchers for damage detection in structures and multiobjective optimization of mix design. Those studies' results are very promising. Therefore, using ML for obtaining optimal mix design and for damage detection of structures will make the building materials sustainable and eco-friendly.

ML is a very promising field, and it will revolutionize the construction sector shortly. However, there are still some problems that need to be addressed before ML completely takes over the traditional methods in the construction sector. Most of the studies these days are carried out by standalone research groups with a relatively narrow scope in terms of sustainable ingredients. The reproducibility of such models is still limited in terms of easier application in the construction industry. Moreover, most of the model studies lately primarily focus on concrete properties like slump, flow, strength, and some durability properties. Concrete being a heterogeneous material exhibits a standard variation in these properties, even if the samples are taken from the same batch of mixing. Since the construction industry mostly relies on empirical relations in terms of these properties from different codes, this trend is quite understandable. However, the reproducibility of such models across different regions with wide variation in the properties of concrete constituents. There is a need for wider collaboration and the creation of a global database or model where these deep learning algorithms can enrich the learning experience, and such databases can then be incorporated into different regional building codes for wider application.

Moreover, the application of ML in concrete applications has broadly used different ML tools as a black box, and little research is done to optimize the learning algorithm to suit the needs of the construction industry. Some interdisciplinary collaboration among researchers from a broader background will be critical in the development of more customized ML tools to better suit the needs of civil engineering practitioners. Such research is also needed to expand the horizons of ML to cover more useful parameters such as stress-strain behaviors, CO_2 emissions, etc., to automate the complete design process for sustainable concrete and structures.

References

[1] M.I. Khan, M.U. Shah, M. Usman, Experimental investigation of concrete properties using locally available coarse aggregates in Punjab, Pakistan, NUST Journal of Engineering Sciences 15 (1) (2022) 26–29, https://doi.org/10.24949/njes.v15i1.655.

[2] L. Gallagher, P. Peduzzi, Sand and Sustainability: Finding New Solutions for Environmental Governance of Global Sand Resources, 2019.

[3] F. Pelisser, A. Barcelos, D. Santos, M. Peterson, A.M. Bernardin, Lightweight concrete production with low Portland cement consumption, Journal of Cleaner Production 23 (1) (March 2012) 68–74, https://doi.org/10.1016/j.jclepro.2011.10.010.

[4] I.H. Shah, S.A. Miller, D. Jiang, R.J. Myers, Cement substitution with secondary materials can reduce annual global CO_2 emissions by up to 1.3 gigatons, Nature Communications 13 (1) (2022) 1–11, https://doi.org/10.1038/s41467-022-33289-7.

[5] M.U. Shah, M. Usman, M.U. Hanif, I. Naseem, S. Farooq, Utilization of solid waste from brick industry and hydrated lime in self-compacting cement pastes, Materials 14 (5) (2021) 1–23, https://doi.org/10.3390/ma14051109.

[6] H. Adel, M.I. Ghazaan, A.H. Korayem, Machine Learning Applications for Developing Sustainable Construction Materials, 2022, https://doi.org/10.1016/B978-0-323-90508-4.00002-2.

[7] D.P. Bentz, Replacement of 'coarse' cement particles by inert fillers in low w/c ratio concretes: II. Experimental validation, Cement and Concrete Research 35 (1) (2005) 185−188, https://doi.org/10.1016/j.cemconres.2004.09.003.

[8] A. Akhtar, A.K. Sarmah, Construction and demolition waste generation and properties of recycled aggregate concrete: a global perspective, Journal of Cleaner Production 186 (2018) 262−281, https://doi.org/10.1016/j.jclepro.2018.03.085.

[9] B. Chiaia, A.P. Fantilli, A. Guerini, G. Volpatti, D. Zampini, Eco-mechanical index for structural concrete, Construction and Building Materials 67 (2014) 386−392, https://doi.org/10.1016/j.conbuildmat.2013.12.090.

[10] M. Nisbet, M.G. Van Geem, Environmental life cycle inventory of Portland cement and concrete, World Cement 28 (4) (1997).

[11] G.P. Hammond, C.I. Jones, Embodied energy and carbon in construction materials, Proceedings of the Institution of Civil Engineers—Energy 161 (2) (2008) 87−98, https://doi.org/10.1680/ener.2008.161.2.87.

[12] M. Reiner, Technology, Environment, Resource and Policy Assessment of Sustainable Concrete in Urban Infrastructure, University of Colorado at Denver, 2007.

[13] P. Purnell, The carbon footprint of reinforced concrete, Advances in Cement Research 25 (6) (2013) 362−368, https://doi.org/10.1680/adcr.13.00013.

[14] L. Hu, Z. He, S. Zhang, Sustainable use of rice husk ash in cement-based materials: environmental evaluation and performance improvement, Journal of Cleaner Production 264 (2020) 121744, https://doi.org/10.1016/j.jclepro.2020.121744.

[15] B. Sizirici, Y. Fseha, C.-S. Cho, I. Yildiz, Y.-J. Byon, A review of carbon footprint reduction in construction industry, from design to operation, Materials (Basel) 14 (20) (October 2021), https://doi.org/10.3390/ma14206094.

[16] A. Adesina, Recent advances in the concrete industry to reduce its carbon dioxide emissions, Environmental Challenges 1 (November) (2020) 100004, https://doi.org/10.1016/j.envc.2020.100004.

[17] S. Rastbod, F. Rahimi, Y. Dehghan, S. Kamranfar, O. Benjeddou, M.L. Nehdi, An optimized machine learning approach for forecasting thermal energy demand of buildings, Sustainability 15 (1) (December 2022) 231, https://doi.org/10.3390/SU15010231.

[18] P. Hu, H. Aghajanirefah, A. Anvari, M.L. Nehdi, Combining artificial neural network and seeker optimization algorithm for predicting compression capacity of concrete-filled steel tube columns, Buildings 13 (2) (February 2023) 391, https://doi.org/10.3390/BUILDINGS13020391.

[19] A. Ali, et al., Machine learning-based predictive model for tensile and flexural strength of 3D-printed concrete, Materials 16 (11) (June 2023) 4149, https://doi.org/10.3390/ma16114149.

[20] F. Nejati, W.O. Zoy, N. Tahoori, P.A. Xalikovich, M.A. Sharifian, M.L. Nehdi, Machine learning method based on symbiotic organism search algorithm for thermal load prediction in buildings, Buildings 13 (3) (March 2023) 727, https://doi.org/10.3390/BUILDINGS13030727.

[21] C. Cakiroglu, K. Islam, G. Bekdasçbekdasç, M.L. Nehdi, Data-driven ensemble learning approach for optimal design of cantilever soldier pile retaining walls, Structures 51 (2023) 1268−1280, https://doi.org/10.1016/j.istruc.2023.03.109.

[22] J. Zeng, M. Gül, Q. Mei, A Computer Vision-Based Method to Identify the International Roughness Index of Highway Pavements, 2022, https://doi.org/10.1016/j.iintel.2022.100004.

[23] F. Hussain, S. Ali Khan, R.A. Khushnood, A. Hamza, F. Rehman, Machine learning-based predictive modeling of sustainable lightweight aggregate concrete, Sustainability 15 (1) (January 2023), https://doi.org/10.3390/su15010641.

[24] H. Naseri, H. Jahanbakhsh, P. Hosseini, F. Moghadas Nejad, Designing sustainable concrete mixture by developing a new machine learning technique, Journal of Cleaner Production 258 (2020) 120578, https://doi.org/10.1016/j.jclepro.2020.120578.

[25] X. Ke, Y. Duan, A Bayesian machine learning approach for inverse prediction of high-performance concrete ingredients with targeted performance, Construction and Building Materials 270 (2021) 121424, https://doi.org/10.1016/j.conbuildmat.2020.121424.

[26] P. Guo, W. Meng, M. Xu, V.C. Li, Y. Bao, Predicting mechanical properties of high-performance fiber-reinforced cementitious composites by integrating micromechanics and machine learning, Materials 14 (12) (2021), https://doi.org/10.3390/ma14123143.

[27] H. Wang, C. Ma, L. Zhou, A brief review of machine learning and its application, in: Proceedings—2009 International Conference on Information Engineering and Computer Science, ICIECS 2009, 2009, https://doi.org/10.1109/ICIECS.2009.5362936.

[28] B. Qian, et al., Orchestrating the development lifecycle of machine learning-based IoT applications: a taxonomy and survey, ACM Computing Surveys 53 (4) (2020), https://doi.org/10.1145/3398020.

[29] P. Amaral, J. Dinis, P. Pinto, L. Bernardo, J. Tavares, H.S. Mamede, Machine learning in software defined networks: data collection and traffic classification, in: Proceedings—International Conference on Network Protocols, ICNP, vol. 2016-December, No. NetworkML, 2016, pp. 91−95, https://doi.org/10.1109/ICNP.2016.7785327.

[30] C. Fan, M. Chen, X. Wang, J. Wang, B. Huang, A review on data preprocessing techniques toward efficient and reliable knowledge discovery from building operational data, Frontiers in Energy Research 9 (March) (2021) 1−17, https://doi.org/10.3389/fenrg.2021.652801.

[31] S.B. Kotsiantis, D. Kanellopoulos, Data preprocessing for supervised leaning, International Journal 1 (2) (2006) 1−7, https://doi.org/10.1080/02331931003692557.

[32] R.C. Chen, C. Dewi, S.W. Huang, R.E. Caraka, Selecting critical features for data classification based on machine learning methods, Journal of Big Data 7 (1) (2020), https://doi.org/10.1186/s40537-020-00327-4.

[33] S. Liu, X. Wang, M. Liu, J. Zhu, Towards better analysis of machine learning models: a visual analytics perspective, Visual Informatics 1 (1) (2017) 48−56, https://doi.org/10.1016/j.visinf.2017.01.006.

[34] W. Jin, Research on machine learning and its algorithms and development, Journal of Physics: Conference Series 1544 (1) (2020), https://doi.org/10.1088/1742-6596/1544/1/012003.

[35] J. Wu, X.Y. Chen, H. Zhang, L.D. Xiong, H. Lei, S.H. Deng, Hyperparameter optimization for machine learning models based on Bayesian optimization, Journal of Electronic Science and Technology 17 (1) (2019) 26−40, https://doi.org/10.11989/JEST.1674-862X.80904120.

[36] J. Wong, T. Manderson, M. Abrahamowicz, D.L. Buckeridge, R. Tamblyn, Can hyperparameter tuning improve the performance of a super learner?: a case study, Epidemiology 30 (4) (2019) 521−531, https://doi.org/10.1097/EDE.0000000000001027.

[37] P. Probst, A.L. Boulesteix, B. Bischl, Tunability: importance of hyperparameters of machine learning algorithms, Journal of Machine Learning Research 20 (2019) 1−32.

[38] E. Elgeldawi, A. Sayed, A.R. Galal, A.M. Zaki, Hyperparameter tuning for machine learning algorithms used for Arabic sentiment analysis, Informatics 8 (4) (2021) 1−21, https://doi.org/10.3390/informatics8040079.

[39] K. Jankowsky, U. Schroeders, Validation and generalizability of machine learning prediction models on attrition in longitudinal studies, International Journal of Behavioral Development 46 (2) (2022) 169−176, https://doi.org/10.1177/01650254221075034.

[40] F. Maleki, N. Muthukrishnan, K. Ovens, C. Reinhold, R. Forghani, Machine learning algorithm validation: from essentials to advanced applications and implications for degulatory certification and deployment, Neuroimaging Clinics of North America 30 (4) (2020) 433−445, https://doi.org/10.1016/j.nic.2020.08.004.

[41] A. Vabalas, E. Gowen, E. Poliakoff, A.J. Casson, Machine learning algorithm validation with a limited sample size, PLoS One 14 (11) (2019) 1−20, https://doi.org/10.1371/journal.pone.0224365.

[42] F.R. Balf, H.M. Kordkheili, A.M. Kordkheili, A new method for predicting the ingredients of self-compacting concrete (SCC) including fly ash (FA) using data envelopment analysis (DEA), Arabian Journal for Science and Engineering 46 (5) (May 2021) 4439−4460, https://doi.org/10.1007/s13369-020-04927-3.

[43] D. Sathyan, D. Govind, C.B. Rajesh, K. Gopikrishnan, G. Aswath Kannan, J. Mahadevan, Modelling the shear flow behaviour of cement paste using machine learning-XGBoost, in: Journal of Physics: Conference Series, IOP Publishing Ltd, September 2020, https://doi.org/10.1088/1742-6596/1451/1/012026.

[44] M.N. Amin, A. Ahmad, K. Khan, W. Ahmad, S. Ehsan, A.A. Alabdullah, Predicting the rheological properties of super-plasticized concrete using modeling techniques, Materials 15 (15) (August 2022), https://doi.org/10.3390/ma15155208.

[45] P. Saha, P. Debnath, P. Thomas, Prediction of fresh and hardened properties of self-compacting concrete using support vector regression approach, Neural Computing and Applications 32 (12) (June 2020) 7995−8010, https://doi.org/10.1007/s00521-019-04267-w.

[46] F. Rehman, S.A. Khokhar, R.A. Khushnood, ANN based predictive mimicker for mechanical and rheological properties of eco-friendly geopolymer concrete, Case Studies in Construction Materials 17 (December 2022), https://doi.org/10.1016/j.cscm.2022.e01536.

[47] A. Kaveh, T. Bakhshpoori, S.M. Hamze-Ziabari, M5' and mars based prediction models for properties of selfcompacting concrete containing fly ash, Periodica Polytechnica: Civil Engineering 62 (2) (March 2018) 281−294, https://doi.org/10.3311/PPci.10799.

[48] E. Gomaa, T. Han, M. ElGawady, J. Huang, A. Kumar, Machine learning to predict properties of fresh and hardened alkali-activated concrete, Cement and Concrete Composites 115 (January 2021), https://doi.org/10.1016/j.cemconcomp.2020.103863.

[49] A.H. Gandomi, A.H. Alavi, S. Kazemi, M. Gandomi, Formulation of shear strength of slender RC beams using gene expression programming, part I: without shear reinforcement, Automation in Construction 42 (2014) 112−121, https://doi.org/10.1016/J.AUTCON.2014.02.007.

[50] N. Chousidis, E. Rakanta, I. Ioannou, G. Batis, Mechanical properties and durability performance of reinforced concrete containing fly ash, Construction and Building Materials 101 (December 2015) 810−817, https://doi.org/10.1016/j.conbuildmat.2015.10.127.

[51] S.A. Khokhar, T. Ahmed, R.A. Khushnood, S.M. Ali, Shahnawaz, A predictive mimicker of fracture behavior in fiber reinforced concrete using machine learning, Materials 14 (24) (December 2021), https://doi.org/10.3390/ma14247669.

[52] A. Behnood, K.P. Verian, M. Modiri Gharehveran, Evaluation of the splitting tensile strength in plain and steel fiber-reinforced concrete based on the compressive strength,

Construction and Building Materials 98 (November 2015) 519–529, https://doi.org/10.1016/J.CONBUILDMAT.2015.08.124.

[53] View of Using Artificial Neural Networks Approach to Estimate Compressive Strength for Rubberized Concrete." https://pp.bme.hu/ci/article/view/11928/8035 (Accessed October 21, 2022).

[54] J.Y. Yoon, H. Kim, Y.J. Lee, S.H. Sim, Prediction model for mechanical properties of lightweight aggregate concrete using artificial neural network, Materials 12 (17) (August 2019) 2678, https://doi.org/10.3390/MA12172678, 2019, Vol. 12, Page 2678.

[55] T.M. Pham, et al., Dynamic compressive properties of lightweight rubberized geopolymer concrete, Construction and Building Materials 265 (December 2020), https://doi.org/10.1016/j.conbuildmat.2020.120753.

[56] A. Ahmad, et al., Prediction of geopolymer concrete compressive strength using novel machine learning algorithms, Polymers (Basel) 13 (19) (October 2021), https://doi.org/10.3390/polym13193389.

[57] H.S. Ullah, R.A. Khushnood, J. Ahmad, F. Farooq, Predictive modelling of sustainable lightweight foamed concrete using machine learning novel approach, Journal of Building Engineering 56 (September 2022), https://doi.org/10.1016/j.jobe.2022.104746.

[58] Y. Li, et al., Measurement and Statistics of Single Pellet Mechanical Strength of Differently Shaped Catalysts, 2000 [Online]. Available: www.elsevier.comrlocaterpowtec.

[59] P. Jagadesh, J. de Prado-Gil, N. Silva-Monteiro, R. Martínez-García, Assessing the compressive strength of self-compacting concrete with recycled aggregates from mix ratio using machine learning approach, Journal of Materials Research and Technology 24 (2023) 1483–1498, https://doi.org/10.1016/j.jmrt.2023.03.037.

[60] P.O. Awoyera, M.S. Kirgiz, A. Viloria, D. Ovallos-Gazabon, Estimating strength properties of geopolymer self-compacting concrete using machine learning techniques, Journal of Materials Research and Technology 9 (4) (July 2020) 9016–9028, https://doi.org/10.1016/J.JMRT.2020.06.008.

[61] J. Xu, X. Zhao, Y. Yu, T. Xie, G. Yang, J. Xue, Parametric sensitivity analysis and modelling of mechanical properties of normal- and high-strength recycled aggregate concrete using grey theory, multiple nonlinear regression and artificial neural networks, Construction and Building Materials 211 (June 2019) 479–491, https://doi.org/10.1016/J.CONBUILDMAT.2019.03.234.

[62] E.M. Golafshani, A. Behnood, Application of soft computing methods for predicting the elastic modulus of recycled aggregate concrete, Journal of Cleaner Production 176 (March 2018) 1163–1176, https://doi.org/10.1016/J.JCLEPRO.2017.11.186.

[63] S. Al Martini, R. Sabouni, A. Khartabil, T.G. Wakjira, M. Shahria Alam, Development and strength prediction of sustainable concrete having binary and ternary cementitious blends and incorporating recycled aggregates from demolished UAE buildings: experimental and machine learning-based studies, Construction and Building Materials 380 (2023) 131278, https://doi.org/10.1016/j.conbuildmat.2023.131278.

[64] M.M. Alshihri, A.M. Azmy, M.S. El-Bisy, Neural networks for predicting compressive strength of structural light weight concrete, Construction and Building Materials 23 (6) (June 2009) 2214–2219, https://doi.org/10.1016/j.conbuildmat.2008.12.003.

[65] A.T.A. Dantas, M. Batista Leite, K. de Jesus Nagahama, Prediction of compressive strength of concrete containing construction and demolition waste using artificial neural networks, Construction and Building Materials (38) (January 2013) 717–722, https://doi.org/10.1016/J.CONBUILDMAT.2012.09.026, vol. Complete.

[66] N. Deshpande, S. Londhe, S. Kulkarni, Modeling compressive strength of recycled aggregate concrete by artificial neural network, model tree and non-linear regression,

International Journal of Sustainable Built Environment 3 (2) (December 2014) 187–198, https://doi.org/10.1016/J.IJSBE.2014.12.002.
[67] P.G. Asteris, K.G. Kolovos, Self-compacting concrete strength prediction using surrogate models, Neural Computing and Applications 31 (January 2019) 409–424, https://doi.org/10.1007/S00521-017-3007-7.
[68] Z.H. Duan, S.C. Kou, C.S. Poon, Using artificial neural networks for predicting the elastic modulus of recycled aggregate concrete, Construction and Building Materials 44 (2013) 524–532, https://doi.org/10.1016/J.CONBUILDMAT.2013.02.064.
[69] S.K. John, A. Cascardi, Y. Nadir, M.A. Aiello, K. Girija, A new artificial neural network model for the prediction of the effect of molar ratios on compressive strength of fly ash-slag geopolymer mortar, Advances in Civil Engineering 2021 (2021), https://doi.org/10.1155/2021/6662347.
[70] K. O. Akande, T. O. Owolabi, S. Twaha, and S. O. Olatunji, "Performance Comparison of SVM and ANN in Predicting Compressive Strength of Concrete," Ver. I. [Online]. Available: www.iosrjournals.orgwww.iosrjournals.org88|.
[71] A. Marani, A. Jamali, M.L. Nehdi, Predicting ultra-high performance concrete compressive strength using tabular generative adversarial networks, Materials 13 (21) (November 2020) 1–24, https://doi.org/10.3390/ma13214757.
[72] B.K.R. Prasad, H. Eskandari, B.V.V. Reddy, Prediction of compressive strength of SCC and HPC with high volume fly ash using ANN, Construction and Building Materials 23 (1) (January 2009) 117–128, https://doi.org/10.1016/j.conbuildmat.2008.01.014.
[73] H. Song, A. Ahmad, K.A. Ostrowski, M. Dudek, Analyzing the compressive strength of ceramic waste-based concrete using experiment and artificial neural network (Ann) approach, Materials 14 (16) (August 2021), https://doi.org/10.3390/ma14164518.
[74] S. Aneja, A. Sharma, R. Gupta, D.Y. Yoo, Bayesian regularized artificial neural network model to predict strength characteristics of fly-ash and bottom-ash based geopolymer concrete, Materials 14 (7) (April 2021), https://doi.org/10.3390/ma14071729.
[75] P.K. Mehta, Concrete. Structure, Properties and Materials, 1986.
[76] N. Buenfeld, N.M. Hassanein, A.J. Jones, An Artificial Neural Network for Predicting Carbonation Depth in Concrete Structures, undefined, 1998.
[77] W.Z. Taffese, E. Sistonen, Machine learning for durability and service-life assessment of reinforced concrete structures: recent advances and future directions, Elsevier B.V. Automation in Construction 77 (May 01, 2017) 1–14, https://doi.org/10.1016/j.autcon.2017.01.016.
[78] S. Safarzadegan Gilan, H. Bahrami Jovein, A.A. Ramezanianpour, Hybrid support vector regression—particle swarm optimization for prediction of compressive strength and RCPT of concretes containing metakaolin, Construction and Building Materials 34 (September 2012) 321–329, https://doi.org/10.1016/j.conbuildmat.2012.02.038.
[79] T. Huang, et al., Modelling the interface bond strength of corroded reinforced concrete using hybrid machine learning algorithms, Journal of Building Engineering 74 (2023) 106862, https://doi.org/10.1016/j.jobe.2023.106862.
[80] Y. Liu, Y. Li, J. Mu, H. Li, J. Shen, Modeling and analysis of creep in concrete containing supplementary cementitious materials based on machine learning, Construction and Building Materials 392 (2023) 131911, https://doi.org/10.1016/j.conbuildmat.2023.131911.
[81] Z. Li, H. He, S. Zhao, Research on support vector machine's prediction of concrete carbonization, in: 2008 International Seminar on Business and Information Management, ISBIM 2008, IEEE Computer Society, 2008, pp. 319–322, https://doi.org/10.1109/ISBIM.2008.206.

[82] S. Inthata, W. Kowtanapanich, R. Cheerarot, Prediction of chloride permeability of concretes containing ground pozzolans by artificial neural networks, Materials and Structures/Materiaux et Constructions 46 (10) (October 2013) 1707–1721, https://doi.org/10.1617/s11527-012-0009-x.

[83] Y. Kellouche, B. Boukhatem, M. Ghrici, R. Rebouh, A. Zidol, Neural network model for predicting the carbonation depth of slag concrete, Asian Journal of Civil Engineering 22 (7) (November 2021) 1401–1414, https://doi.org/10.1007/s42107-021-00390-z.

[84] O.A. Hodhod, H.I. Ahmed, Developing an artificial neural network model to evaluate chloride diffusivity in high performance concrete, HBRC Journal 9 (1) (April 2013) 15–21, https://doi.org/10.1016/j.hbrcj.2013.04.001.

[85] I. Nunez, M.L. Nehdi, Machine learning prediction of carbonation depth in recycled aggregate concrete incorporating SCMs, Construction and Building Materials 287 (June 2021), https://doi.org/10.1016/j.conbuildmat.2021.123027.

[86] E. Güneyisi, M. Gesoğlu, T. Özturan, E. Özbay, Estimation of chloride permeability of concretes by empirical modeling: considering effects of cement type, curing condition and age, Construction and Building Materials 23 (1) (January 2009) 469–481, https://doi.org/10.1016/j.conbuildmat.2007.10.022.

[87] J. Gou, Z.W. Fan, C. Wang, W.P. Guo, X.M. Lai, M.Z. Chen, A minimum-of-maximum relative error support vector machine for simultaneous reverse prediction of concrete components, Computers and Structures 172 (2016) 59–70, https://doi.org/10.1016/j.compstruc.2016.05.003.

[88] Z. Fan, R. Chiong, Z. Hu, Y. Lin, A fuzzy weighted relative error support vector machine for reverse prediction of concrete components, Computers and Structures 230 (2020) 106171, https://doi.org/10.1016/j.compstruc.2019.106171.

[89] C. Tavares, X. Wang, S. Saha, Z. Grasley, Machine learning-based mix design tools to minimize carbon footprint and cost of UHPC. Part 1: efficient data collection and modeling, Cleaner Materials 4 (2022) 100082, https://doi.org/10.1016/j.clema.2022.100082.

[90] F. Rehman, S.A. Khokhar, R.A. Khushnood, ANN based predictive mimicker for mechanical and rheological properties of eco-friendly geopolymer concrete, Case Studies in Construction Materials 17 (October) (2022) e01536, https://doi.org/10.1016/j.cscm.2022.e01536.

[91] L. Long, M. Döhler, S. Thöns, Determination of structural and damage detection system influencing parameters on the value of information, Structural Health Monitoring 21 (1) (2022) 19–36, https://doi.org/10.1177/1475921719900918.

[92] R. Ali, J.H. Chuah, M.S.A. Talip, N. Mokhtar, M.A. Shoaib, Structural crack detection using deep convolutional neural networks, Automation in Construction 133 (September 2021) (2022) 103989, https://doi.org/10.1016/j.autcon.2021.103989.

[93] G. Moussa, K. Hussain, A new technique for automatic detection and parameters estimation of pavement crack. IMETI 2011—4th International Multi-Conference on Engineering and Technological Innovation, Proceedings 2, 2011, pp. 11–16, https://doi.org/10.13140/2.1.3191.2001.

[94] Y.Y. Boykov, M.-P. Jolly, Interactive graph cuts for optimal boundary and region segmentation of objects in N-D images, in: Proceedings Eighth IEEE International Conference on Computer Vision. ICCV 2001, vol. 1, July 2001, pp. 105–112, https://doi.org/10.1109/ICCV.2001.937505.

[95] S. Varadharajan, S. Jose, K. Sharma, L. Wander, C. Mertz, Vision for road inspection, in: IEEE Winter Conference on Applications of Computer Vision, March 2014, pp. 115–122, https://doi.org/10.1109/WACV.2014.6836111.

[96] T. Kim, S. Tae, S. Roh, Assessment of the CO_2 emission and cost reduction performance of a low-carbon-emission concrete mix design using an optimal mix design system, Renewable and Sustainable Energy Reviews 25 (2013) 729–741, https://doi.org/10.1016/j.rser.2013.05.013.

[97] J. Abellán-García, J.S. Guzmán-Guzmán, Random forest-based optimization of UHPFRC under ductility requirements for seismic retrofitting applications, Construction and Building Materials 285 (2021), https://doi.org/10.1016/j.conbuildmat.2021.122869.

[98] A. Gerami Matin, R. Vatani Nezafat, A. Golroo, A comparative study on using meta-heuristic algorithms for road maintenance planning: insights from field study in a developing country, Journal of Traffic and Transportation Engineering (English Edition) 4 (5) (2017) 477–486, https://doi.org/10.1016/j.jtte.2017.06.004.

[99] R. Cao, Z. Leng, J. Yu, S.C. Hsu, Multi-objective optimization for maintaining low-noise pavement network system in Hong Kong, Transportation Research Part D: Transport and Environment 88 (November) (2020) 102573, https://doi.org/10.1016/j.trd.2020.102573.

Application of machine learning models for the compressive strength prediction of concrete with glass waste powder

4

Miljan Kovačević [1], Ivanka Netinger Grubeša [2], Marijana Hadzima-Nyarko [3] and Emmanuel Karlo Nyarko [4]

[1]University of Pristina, Faculty of Technical Sciences, Mitrovica, Serbia; [2]University North, Varazdin, Croatia; [3]Faculty of Civil Engineering and Architecture Osijek, Josip Juraj Strossmayer University of Osijek, Osijek, Croatia; [4]Faculty of Electrical Engineering, Computer Science and Information Technology, Josip Juraj Strossmayer University of Osijek, Osijek, Croatia

4.1 Introduction

The annual production of waste glass (WG) in the world is estimated at almost 200 million tons [1]. It is estimated that only about 21% of that amount is recycled [2]. In the USA and the UK, less than half of the generated WG is recycled, and the highest degree of recycled WG is in the EU, which is 73% [2]. Significant differences in the quality of the collected WG represent a significant problem for its reuse. The emission of carbon dioxide during the production of portland cement, as well as the significant extraction of aggregates from nature, has a negative impact on the environment, and the use of WG can reduce this impact. WG can be used in concrete either as a partial replacement for commonly used aggregate [3–5] or as a partial replacement of cement [6–8] as a commonly used binder in concrete. However, one should be aware of the possibility of an alkali-silica reaction (ASR), which could occur when aggregate with a high content of silicate (such as WG) and cement with a high content of alkaline components are combined in a concrete mixture. According to Refs. [7,8], the possibility of ASR is minimized if finely ground WG is used in concrete. For this reason, the use of WG as a cement replacement in the production of concrete is being considered here. The mechanical characteristics of concrete with WG significantly depend on the chemical composition of individual components of WG as well as the size of the particles [7]. At the same time, the content of SiO_2 contained in WG plays a significant role in the characteristics of the concrete for which it is used, especially in compressive strength (CS). When the presence of chemical components (SiO_2, CaO, aluminum oxide (Al_2O_3), MgO, NaO_2, and Fe_2O_3) is present in a percentage higher than 70%, it is considered that such WG can represent a pozzolanic material according to ASTM [9]. The pozzolanic activity of such glass depends significantly on the size of the WG

Artificial Intelligence Applications for Sustainable Construction. https://doi.org/10.1016/B978-0-443-13191-2.00004-3
Copyright © 2024 Elsevier Ltd. All rights are reserved, including those for text and data mining, AI training, and similar technologies.

particles. It is considered that in order to ensure high pozzolanicity, the particle size should be in the range of 38−75 μm [10,11]. The performance of the prepared concrete can be significantly affected by the curing temperature. Elevated temperatures (e.g., 50°C) can speed up the cement hydration process and affect the strength of concrete treated in this way [12]. According to some studies, elevated temperatures can increase the CS of such concrete by up to 22% after 91 days [12].

In terms of modeling the CS of concrete with WG, it is a challenging task due to a correlation between the input components of such concrete, so the application of multidimensional linear regression gives poor results. In addition, using classic regression procedures, it is necessary to first select a model and determine the parameters of the model from experimental data, which introduces bias into the modeling process itself. Therefore, applying machine learning machine learning methods is appropriate in scenarios of such problems with multiple correlations of input variables.

There are numerous studies of concrete strength modeling with different cement supplements, except with WG [13−17]. When modeling the properties of concrete containing WG, there are a limited number of studies that have been carried out. Mirzahoseini et al. worked on a model for CS prediction using genetic programming and defined explicit models whose accuracy was expressed through absolute criteria RMSE = 5.33 MPa, MAE = 4.35 MPa, and whose correlation coefficient was 0.94 on a separate test data set [18]. The use of neural network models for the prediction of CS and tensile strength of concrete with WG was discussed by Ray et al. [19]. Ghorbani et al. [20] used artificial neural network models to model the dynamic characteristics of concrete with WG. Seghier et al. considered the application of Metaheuristic-based machine learning modeling of the CS of concrete containing WG [21]. The application of support vector regression (SVR), least-square support vector regression (LSSVR), adaptive neuro-fuzzy inference system (ANFIS), and multilayer perceptron neural network (MLP) in the development of prediction models were analyzed. The hybrid LSSVR-MPA model is proposed as the optimal model that is recommended. Golafhani and Kashani worked on the application of multi-objective automatic regression models of different complexity in predicting the strength of concrete with WG, and the accuracy of the obtained models expressed through the correlation coefficient was in the range from 0.8569 to 0.9154 on the test data set [22]. The use of artificial neural networks as an optimal model for predicting the CS of concrete with WG is recommended by Ahmad et al [23]. The following variables were considered: curing time, water/binder ratio, cement content, sand content, WG content, and coarse aggregate. The use of the SVM method with the application of different kernel functions in CS concrete with glass powder-based admixture was analyzed by Harish and Janardhan [24]. As an optimal kernel function, the research recommends the application of the erbf function. Ghosh and Ransinchung worked on researching the application of the linear model, regression trees, RF method, and SVM in the prediction of the CS of concrete that has additions of WG [25]. They recommend the RF model as the optimal model for the application. Sun et al. [26] considered the application of the RF model for the prediction of flexural strength (FS) and ASR expansion of WG concrete. As a recommendation for the optimal model, the RF model was given, which showed good agreement with the experimental data with values of 0.9545 for FS and 0.9416 for ASR expressed through the correlation coefficient.

This chapter investigated the application of different methods and algorithms of machine learning, namely regression trees, ensembles based on regression trees, support vector machines (SVM) models, and models of Gaussian process regression (GPR), was investigated. To the knowledge of the authors of this chapter, the application of GPR models has not been used for modeling the CS of concrete with WG.

4.2 Methods

4.2.1 Machine learning methods based on decision trees

Decision trees represent a class of supervised algorithms that are used to create classification trees and regression trees for solving classification problems and regression problems, respectively [27–29]. Both types of trees divide the prediction space into nonoverlapping regions. The idea is that instead of one global linear model that is valid for the entire problem area, the domain is divided into a larger number of subdomains, and then the corresponding model is defined in them.

Trees are built from the roots to the leaves (Fig. 4.1), i.e., by recursive binary division. In the beginning, all instances belong to one region. Then the space is divided using the so-called greedy approach so that in each iteration of the algorithm, the variable and the value of the splitting point are determined so as to have the smallest deviation from the output (target) variable that is being modeled. This approach is referred to as greedy because it considers the optimality of the selection of the variable and the value of the splitting point only in the current iteration and not globally. The processing of the variable is the value at which the splitting of the tree is performed using the following expression:

$$\min_{j,\, s}\left[\min_{c_1}\sum_{x_i\in R_1(j,\, s)}(y_i-c_1)^2+\min_{c_2}\sum_{x_i\in R_2(j,\, s)}(y_i-c_2)^2\right] \tag{4.1}$$

where $\widehat{c}_1 = average(y_i|x_i \in R_1(j,s))$ and $\widehat{c}_2 = average(y_i|x_i \in R_2(j,s))$.

a) b)

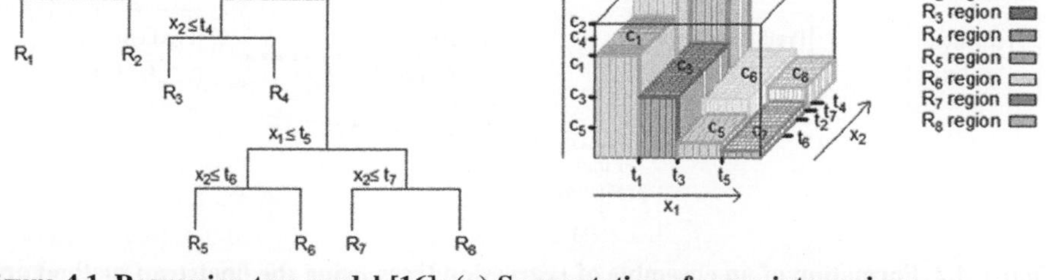

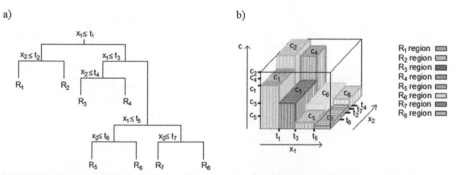

Figure 4.1 Regression tree model [16]: (a) Segmentation of space into regions, (b) averaging output values in regions.

In this way, the segmentation of the input space into M regions $R_1, R_2, \ldots, R_M$. is performed (Fig. 4.1a).

When a prediction is required for the test sample, the model gives a prediction that is equal to the mean value of all values of the output variable of the region to which that test sample belongs (Fig. 4.1b).

The division process continues until the defined division stop criterion is reached.

In cases of poor generalization when applying individual regression trees, ensemble learning methods based on trees can be applied. Among the ensemble learning methods, the TB algorithm [29], the RF [30], and the boosted trees (BT) [17] algorithm can be singled out.

The idea is to train a large number of regression trees, and the value of the unknown sample is determined by averaging the results of individual trees within the ensemble in the case of regression (Fig. 4.2). At the beginning, by applying the bootstrap method of resampling from the training set, B new sets are formed. The bootstrap resampling method involves creating new training sets by randomly drawing samples with replacements from the original set. Each regression tree is then formed based on one of the B newly obtained training sets.

When applying the RF algorithm to obtain uncorrelated trees, only a subset of randomly selected variables is used for splitting the tree, which in certain cases can improve generalization. For example, it is recommended that when analyzing regression problems, a subset of variables that is approximately a third of the total number of variables in the problem be used [30]. During each new iteration, a new random set of variables of the adopted size is selected, and the variable and the value of the splitting

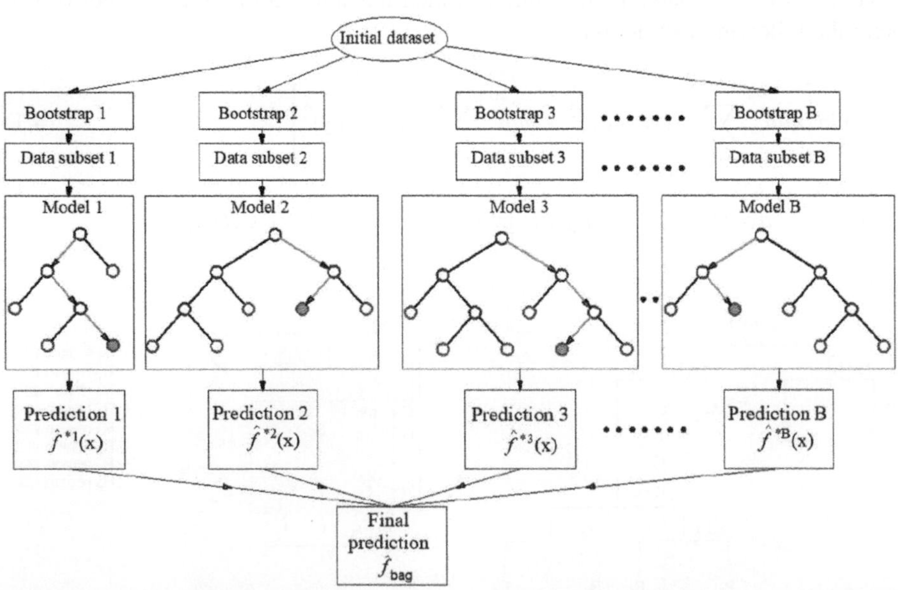

Figure 4.2 Formation of an ensemble of regression trees using the bootstrap method and prediction calculation [17].

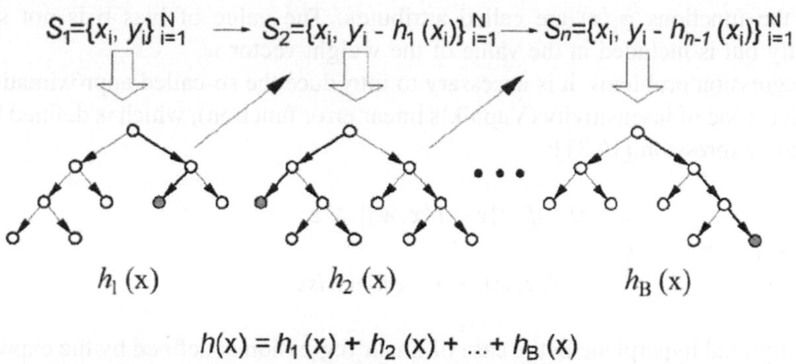

Figure 4.3 Regression tree ensemble creation using gradient boosting [17].

point are determined. The TB algorithm, like the RF algorithm, uses the Bootstrap method when creating individual trees of the ensemble. However, it uses the entire set of variables for potential splitting when creating trees.

With the gradient boosting method (Fig. 4.3), the BT model is created by training the first regression model on the entire training data set, and then a new set of data is created, which is obtained as the difference between the original target values and the predictions of the previously generated tree for the given values of the input variables. Next, the model trained on the thus-obtained data set is added to the previously generated tree. Then that procedure is iteratively repeated. The procedure is terminated by satisfying one of the defined stopping criteria. Unlike RF and TB models, BT models are trained sequentially, which can be a problem with a large amount of data for model training. Usually, when creating a BT model, the differences between the target value and the prediction are not taken in full but are multiplied with a reduction coefficient of less than 1 (learning rate) in order not to overtrain the model-while the model is gradually created.

4.2.2 Support vector machines for regression

The basic idea of the SVM method is the transformation of the input space and its application in the newly transformed space.

Let it be assumed that a given set of training data is composed of l parallel input and output values, on the basis of which one tries to determine the input-output dependence, i.e., mapping function f. The training data set $D = \{[x(i)], y(i) \in \mathbb{R}^n x\mathbb{R}, i = 1, \ldots, l\}$ consist of $(x_1, y_1), (x_2, y_2), \ldots, (x_l, y_l)$ where the inputs x (inputs) are n-dimensional vectors (inputs) $x \in \mathbb{R}^n$, and the responses of the system $y \in \mathbb{R}$ are continuous values. The SVM method considers an approximation function (Eq. 4.2) that has the following form [31]:

$$f(\mathbf{x}, \mathbf{w}) = \sum_{i=1}^{N} w_i \varphi_i(\mathbf{x}) \tag{4.2}$$

where the functions $\varphi_i(x)$ are called attributes. The value of bias b is not shown explicitly but is included in the value of the weight vector w.

In regression problems, it is necessary to introduce the so-called approximation error with ε-zone of insensitivity (Vapnik's linear error function), which is defined by the following expression [16,31]:

$$|y - f(\boldsymbol{x}, \boldsymbol{w})|_{\varepsilon} = \begin{cases} t0 \quad if \quad |y - f(\boldsymbol{x}, \boldsymbol{w})| \leq \varepsilon \\ |y - f(\boldsymbol{x}, \boldsymbol{w})| - \varepsilon \quad otherwise. \end{cases} \tag{4.3}$$

The optimal hyperplane in the case of linear regression is defined by the expression [16]:

$$z = f(\mathbf{x}, \mathbf{w}) = \mathbf{w}^{\mathbf{T}}\mathbf{x} + b = \sum_{i=1}^{l} (\alpha_i{}^* - \alpha_i)\mathrm{x_i}^{\mathrm{T}}\mathrm{x} + \mathrm{b}. \tag{4.4}$$

where α and α^* are Lagrangian multipliers obtained by applying Lagrangian optimization of the sum of empirical risk R^{ε}_{emp} and $\|w\|^2$ values simultaneously [31].

In the case of nonlinear regression (Fig. 4.4), expressions for linear regression can be used by replacing the scalar product $\mathrm{x_i^T x_j}$ with the kernel function $\mathrm{K(x_i, x_j)} = \Phi(\mathrm{x_i})^{\mathrm{T}}\Phi(\mathrm{x_j})$, i.e., similar to linear regression, the following is obtained for nonlinear regression:

$$w = \sum_{i=1}^{l} (\alpha_i{}^* - \alpha_i)\Phi(x_i) \tag{4.5}$$

$$f(x) = \sum_{i=1}^{l} (\alpha_i{}^* - \alpha_i)\mathrm{K(x_i, x)} + b. \tag{4.6}$$

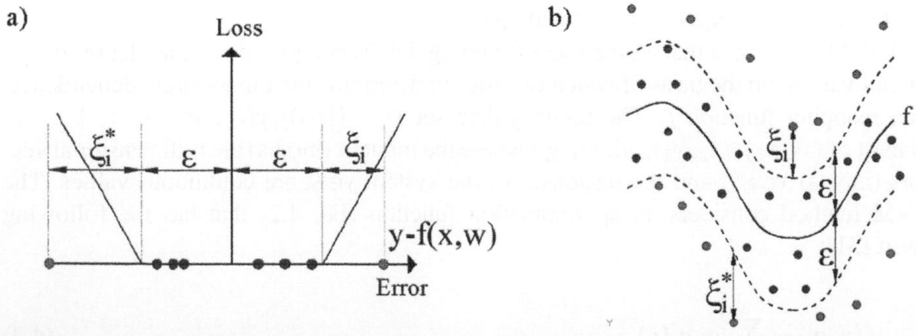

Figure 4.4 (a) Linear loss function with an ε-insensitivity zone, (b) SVM for nonlinear regression [16].

In this chapter, the application of linear, RBF, and sigmoid kernel functions is analyzed and defined by the following Eqs. (4.7—4.9), respectively [31]:

$$k\langle x_i, x\rangle = \langle x_i, x\rangle \quad (4.7)$$

$$k\langle x_i, x\rangle = exp\left(-\gamma\|x_i - x\|^2\right), \gamma > 0 \quad (4.8)$$

$$k\langle x_i, x\rangle = tanh(\gamma\langle x_i, x\rangle + r), \gamma > 0 \quad (4.9)$$

4.2.3 Gaussian process regression (GPR) for machine learning

A Gaussian process (GP) is a collection of random variables, any finite number of which have a joint Gaussian distribution [32]. GP can be applied to classification problems and regression problems. When applying GP for regression, it is assumed that the obtained value of the output variable that corresponds to an input vector, i.e., location, represents a random realization of a multidimensional Gaussian distribution. Since there are a large number of tested samples in the research, their output variables represent randomly realized values of the unknown function. All values of the output variables for the respective locations, as well as each of their subsets, follow a multidimensional Gaussian distribution.

In real situations, instead of observing the exact values of the functions, their values together with noise are assumed in the following form [32]:

$$y(X) = I_n f(X) + \epsilon_*, \quad (4.10)$$

where f is an unknown and needs to be estimated, y is the target variable, X is the input vector, and ε is the normally distributed additive noise, i.e., $\varepsilon \sim N(0, \sigma^2 I_n)$, and I_n is the identity matrix of size $n \times n$.

The following is assumed in the case when there are n total values of the output variable [32]:

$$y(y_1, \ldots, y_n)^T \sim N(\boldsymbol{\mu}, K). \quad (4.11)$$

where $\boldsymbol{\mu} = (\mu(x_1), \ldots, \mu(x_n))^T$ represents the mean value vector, and K is the $n\times n$ covariance matrix whose element (i,j) has the value $K_{ij} = k(x_i, x_j) + \sigma^2\delta_{ij}$. In this expression, δ_{ij} represents Kronecker's delta function [32].

Suppose that for a certain arbitrary point x^* it is necessary to determine the predicted value of the output variable, which is denoted by y^*. The value of the variable y^* will also, together with other values $y_1, \ldots, y_n$, follow a multidimensional Gaussian distribution, i.e.,

$$(y_1, \ldots, y_n, y^*)^T \sim N(\boldsymbol{\mu}^*, \Sigma), \quad (4.12)$$

where $\boldsymbol{\mu}^* = (\mu(x_1), \ldots, \mu(x_n), \mu(x^*))^T$, and the covariance matrix defined by:

$$\Sigma = \begin{bmatrix} K_{11} & K_{12} & \cdots & K_{1n} & K_{1*} \\ K_{21} & K_{22} & \cdots & K_{2n} & K_{2*} \\ \cdots & \cdots & \cdots & \cdots & \cdots \\ K_{n1} & K_{n2} & \cdots & K_{nn} & K_{n*} \\ K_{*1} & K_{*2} & \cdots & K_{*n} & K_{**} \end{bmatrix} = \begin{bmatrix} K & K^* \\ K^{*T} & K^{**} \end{bmatrix}, \tag{4.13}$$

where $K^* = (K(x^*, x_1), \ldots, K(x^*, x_n))^T$ and $K^{**} = K(x^*, x^*)$.

In order to find the distribution y^*, for the given $\mathbf{y} = (y_1, \ldots, y_n)^T$, it is necessary to find the conditional of the Eq. (4.4), which again represents the normal Gaussian distribution $N\left(\widehat{\mathbf{y}}^*, \widehat{\boldsymbol{\sigma}}^{*2}\right)$ where [32]:

$$\widehat{\mathbf{y}}^* = \mu(x^*) + K^{*T}K^{-1}(y - \mu), \tag{4.14}$$

$$\widehat{\boldsymbol{\sigma}}^{*2} = K^{**} + \sigma^2 - K^{*T}K^{-1}K^*. \tag{4.15}$$

One of the most important steps when modeling an unknown function using GPR is the adequate selection of covariance functions. There are covariance functions with one length scale parameter for all coordinate axes, i.e., input variables, and so-called ARD covariance functions with different length scale parameter values for each coordinate axis, i.e., input variables. The ARD squared exponential covariance function (Eq. 4.16) can be mentioned as one of the ARD covariance functions [32]:

$$k(x_p, x_q) = v^2\, exp\left[-\frac{1}{2}\sum_{i=1}^{n}\left(\frac{x_p^i - x_q^i}{r_i}\right)^2\right]. \tag{4.16}$$

The hyperparameters of the covariance functions $\{v, r_1, \ldots, r_n\}$ and the noise variance σ^2 can be determined by maximizing the following expression:

$$L(v, r_1, \ldots, r_n, \sigma^2) = -\frac{1}{2}\log\ \det K - \frac{1}{2}\,\mathrm{y}^T K^{-1}\mathrm{y} - \frac{n}{2}\log 2\pi. \tag{4.17}$$

4.3 Experimental dataset and model accuracy criteria

4.3.1 Experimental dataset

The experimental database consists of 70 samples on which the CS was experimentally determined, created by Mirzahosseini and Riding [18,33,34]. The CS was determined on cube-shaped test samples with dimensions of 50 mm, which contained glass waste powder that meets the requirements of the standard ASTM C109 (ASTM, 2012). Data on the percentage content of individual components representing the input variables of the model Al_2O_3, K_2O (potassium oxide), Cr_2O_3 (chromic oxide), and TiO_2 (titanium dioxide) were collected from all analyzed samples.

In addition, data on curing age (CA), curing temperature (T), and surface area (SA) of cementitious material for the analyzed samples were collected, which represents the next part of the input variables. The water-to-cementitious material ratio (w/cm) determined by ASTM C109 was 0.485 for all analyzed samples. The sand-to-cementitious material ratio specified by ASTM C109 for all analyzed samples was 2.75. Ordinary portland cement type I/II was used to make the samples. The gradation of glass waste powder particles and cement particles is shown in Ref. [13].

It can be seen from Fig. 4.5 that the problem of modeling CS is complex and that there is a mutual correlation of individual input variables as well as the correlation of input variables and output variables. The application of machine learning methods is suitable for modeling problems in this way.

Since all machine learning models predict within the ranges of the data (samples) within which they were trained, Table 4.1 provides data on the statistical properties of the samples. Detailed data of the experimental dataset can be found in the paper written by Mirzahosseini and Riding [18].

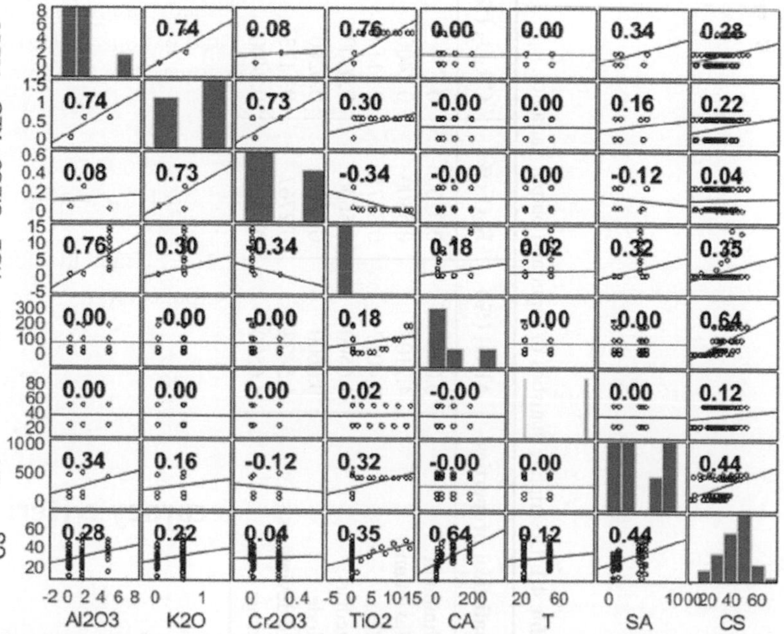

Figure 4.5 Correlation matrix of model variables.

Table 4.1 Descriptive statistics of input and output variables.

Statistical properties	Al_2O_3 (%)	K_2O (%)	CR_2O_3 (%)	TiO_2 (%)	CA (days)	T (°C)	SA (m^2/kg)	CS (MPa)
Minimum	0.0600	0.0200	0	0.0100	1	23	53	6.3100
Maximum	4.7100	0.5800	0.2400	14.2400	180	50	476	51.5800
Mean	1.4057	0.3371	0.1114	1.1571	61.4000	36.5000	237.4286	28.7429
Median	1.6500	0.5600	0.0200	0.1100	28	36.5000	126.0000	29.9350
Mode	0.0600	0.0200	0.0200	0.0100	1	23.0000	53.0000	39.2400
Std.deviations	1.5478	0.2767	0.1123	3.1582	67.8140	13.5975	175.6370	10.5086
Range	4.6500	0.5600	0.2400	14.2300	179	27	423	45.2700

4.3.2 Model accuracy criteria

After defining the input and output data of the model, it is necessary to divide the data into a data set for training the model and a data set for evaluating the model. It is necessary that the training data set contains representative data for the problem being modeled, and the evaluation data set must contain data with the same statistical characteristics as the training set.

In the research, the procedure of cross-validation of the model was applied. The procedure was chosen since it reduces the bias related to the random distribution of data for training and model training, which in principle should have the same statistical characteristics.

Fig. 4.6 illustrates the implementation procedure with 5-fold cross-validation on the considered model. To apply the cross-validation procedure shown in Fig. 4.6, it is necessary to randomly divide the data set into five disjoint subsets of the same size. One of the subsets is then removed, and the model is trained using the remaining four subsets. By applying the adopted criteria, the accuracy is then evaluated on the removed subset that was not used for training the model, that is, on the test data. The procedure is repeated 5 times on different data sets, where one subset is always removed. In this way, the entire data set is used to test the model. The accuracy of the model is determined by the mean value of the formed models according to the adopted criteria.

In this research, two absolute criteria, the root mean square error (RMSE) and mean absolute error (MAE), and two relative criteria, Pearson's linear correlation coefficient (R) and mean absolute percentage error (MAPE), were used to evaluate the accuracy of the model [31].

The mean square error (*MSE*) defined by Eq. (4.18) represents the criterion or measure that is most often used during model calibration, while the other criteria or measures are most often used in model evaluation.

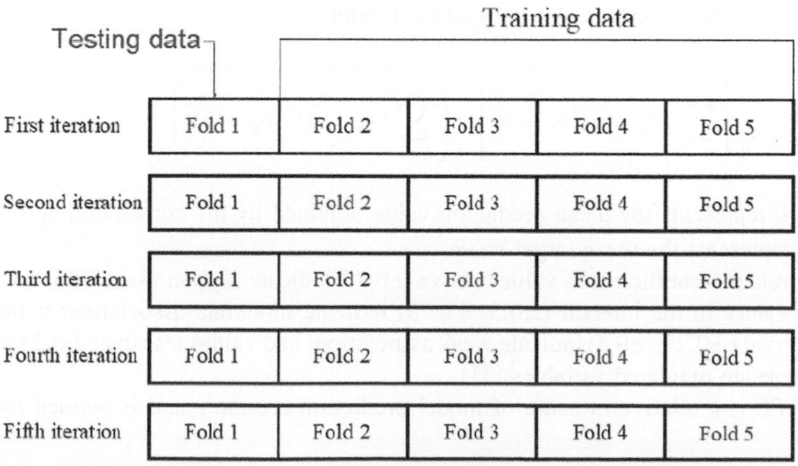

Figure 4.6 Procedure of applied cross validation (5-fold cross validation).

$$MSE = \frac{1}{N}\sum_{k=1}^{N}(d_k - o_k)^2. \tag{4.18}$$

where is

d_k – the actual value, i.e., the target value
o_k – output or prediction given by the model
N – number of training samples

The RMSE criterion is expressed in units of the output variable that is being modeled and is defined by Eq. (4.19),

$$RMSE = \sqrt{\frac{1}{N}\sum_{k=1}^{N}(d_k - o_k)^2}. \tag{4.19}$$

With the *RMSE* criterion, the difference between the target and forecast value is squared, then the average is found, and then the root of this value. Since the errors are squared before the average is found, the value of this measure is highly affected by large errors in the forecast, so it is a good indicator if you want to avoid large discrepancies between observed and forecasted values. The lower the value of this indicator, the better the model.

The *MAE* criterion is also expressed in units of the output variable and is defined by Eq. (4.20).

$$MAE = \frac{1}{N}\sum_{k=1}^{N}|d_k - o_k|. \tag{4.20}$$

Pearson's linear correlation coefficient *R* is defined by Eq. (4.21) and is a relative criterion for evaluating the accuracy of the model.

$$R = \sqrt{\left[\sum_{k=1}^{N}(d_k - \overline{d})(o_k - \overline{o})\right]^2 \cdot \left[\sum_{k=1}^{N}(d_k - \overline{d})^2(o_k - \overline{o})^2\right]^{-1}}. \tag{4.21}$$

where $\overline{o}$ represents the mean prediction value obtained by the corresponding model, and $\overline{d}$ represents the mean target value.

Correlation coefficient R values above ±0.75 indicate a good association of variables; values in the interval (±0.5, ±0.75) indicate moderate association; values in the interval (±0.25, ±0.5) indicate weak association; and values less than ±0.25 imply that these are unrelated variables [31].

MAPE is a relative measure of model prediction accuracy and is defined by Eq. (4.22),

$$MAPE = \frac{100}{N} \sum_{k=1}^{N} \left| \frac{d_k - o_k}{d_k} \right|. \tag{4.22}$$

The difference between the target and modeled values needs to be divided by the target value. The absolute value of this calculation is summed for all samples, then divided by the number of samples, and the resulting value is multiplied by 100. *MAPE* cannot be used when there are target values that are equal to zero [31].

4.4 Results and discussion

The chapter examines different models, from individual models of regression trees to ensembles composed of regression trees, SVM models, and different models based on Gaussian random processes.

With regression tree models, models of different structures and complexity were tested. The minimum number of data points per leaf was considered the main parameter of the regression trees. Matlab software was used to implement the model. The standard setting of the program generates trees that have a minimum number of data points per terminal leaf equal to 5.

The chapter examined different ensembles tree structures where the minimum number of data varied from the value 1 (when more complex and deeper trees are generated) to the value of the minimum number of data per leaf 10 (when shallow and less complex trees are generated). The results in terms of accuracy in relation to the defined criteria are given in Figs. 4.8 and 4.9. In the case of the RF and TB models, by creating an ensemble with 500 trees, the learning curves were saturated (Fig. 4.10).

In Table 4.2, it can be seen that according to the RMSE and R criteria, the optimal model has the minimum number of data points per terminal leaf 6, while according to the MAE and MAPE criteria, the optimal model has 5 data points per leaf. The model with a minimum leaf size equal to 5 is shown as a tree structure in Fig. 4.7. In further research, the possibility of creating more complex models in the form of ensembles of regression trees was examined.

For ensembles of regression trees, the RF algorithm, the TB algorithm, and models where the gradient boosting algorithm was applied (BT models) were considered.

In the RF and TB models, the accuracy is mainly influenced by three parameters, namely:

- The total number of trees,
- The minimum number of samples assigned to a terminal leaf (min leaf size),
- The number of randomly selected variables that will be used for selection when branching the tree (number of variables, RF only).

Models based on the TB algorithm differ from models based on the RF algorithm in that they take the entire set of input variables when generating the model, while models based on the RF algorithm take only a subset of the input variables of a certain size.

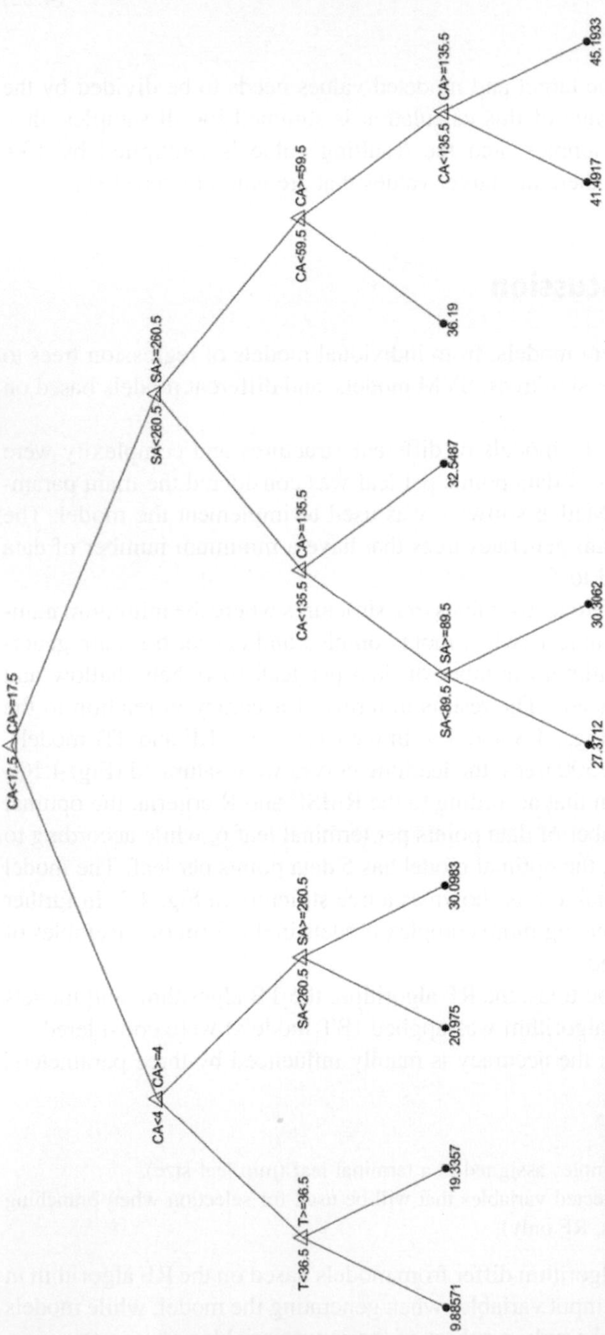

Figure 4.7 Optimal individual regression tree model.

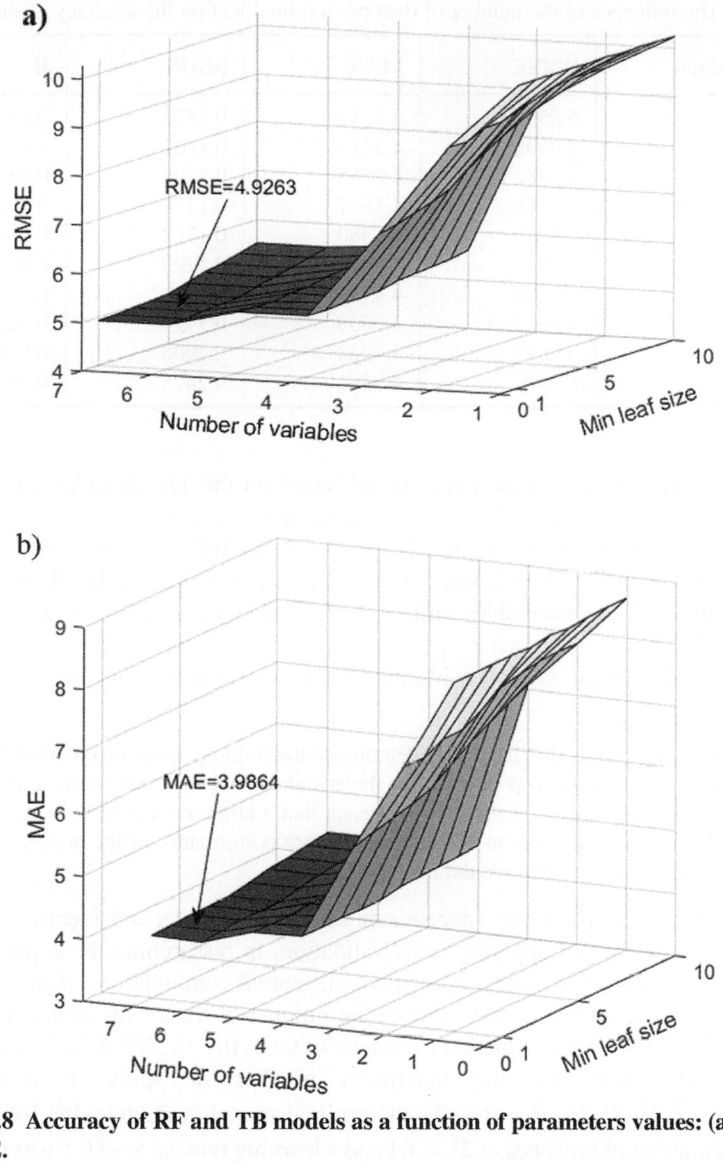

Figure 4.8 Accuracy of RF and TB models as a function of parameters values: (a) RMSE, (b) MAE.

Fig. 4.9, shows the dependence of the predictive accuracy of the model on its hyperparameters. It can be seen that the models with regard to the defined criteria have a higher accuracy when a larger number of variables are taken into account when creating the model.

Also, it can be seen that the model is more accurate when the minimum number of data points per sheet is smaller (Fig. 4.9). It was concluded that the optimal model

Table 4.2 The influence of the number of data per terminal leaf on the accuracy of the model.

Min leaf size	RMSE	MAE	MAPE	R
1	5.6598	4.3213	0.1802	0.8453
2	5.6598	4.3213	0.1802	0.8453
3	5.4387	4.2555	0.1790	0.8576
4	5.3375	4.1917	0.1774	0.8632
5	5.3789	4.1580	0.1762	0.8604
6	5.2971	4.1969	0.1790	0.8635
7	5.7109	4.4781	0.1957	0.8378
8	5.9312	4.6639	0.1995	0.8229
9	5.9643	4.7044	0.2008	0.8207
10	6.3479	4.9985	0.2425	0.7969

according to the RMSE criterion is a model based on the TB algorithm (number of trees = 500, number of variables = 7, minimum leaf size = 1).

According to other criteria, MAE, MAPE, and R, the RF model is optimal (number of trees = 500, number of variables = 6, minimum leaf size = 1). In all models, the learning curve of the ensemble with 500 regression trees satisfactorily saturates (Fig. 4.10).

With the BT method, the following model parameters were analyzed:

- Number of trees B,
- The learning rate is λ. This parameter determines the training speed of the model, and the right choice of this parameter depends on the specific problem under consideration. Small values of the parameter λ have the consequence that a large number of trees, i.e., a larger value of B is needed in order to obtain a satisfactory performance of the model.
- A maximum number of splits in the tree is d.

The value of these parameters depends on the problem under consideration, and the values are determined by applying cross-validation. In researching the application of the BT model using the grid search method, all mutual combinations of the three parameters were examined, i.e., the value of the number of trees with an upper limit of 100 trees, the values of the reduction parameter of 0.01, 0.1, 0.25, 0.5, 0.75, and 1.0, as well as the values for the maximum number of splits in a tree of $2^0 = 1, 2^1, 2^2, 2^3, 2^4, 2^5, 2^6 = 64$. As the optimal model, a model with 100 trees, a maximum number of branches of $2^6 = 64$ and a learning rate value of 0.5 was obtained (green line in Fig. 4.11).

Within the RF and TB methods, it is possible to rank the variables by building an ensemble model in the first step according to the data or samples available for training. The data obtained by the bootstrap method can be divided into the data used to build the model, so-called "in-bag data" and the data used to test the model "out-of-bag data". The data used for model calibration (in-bag data) are obtained by random sampling with replacement from the entire data set. Since it is sampling

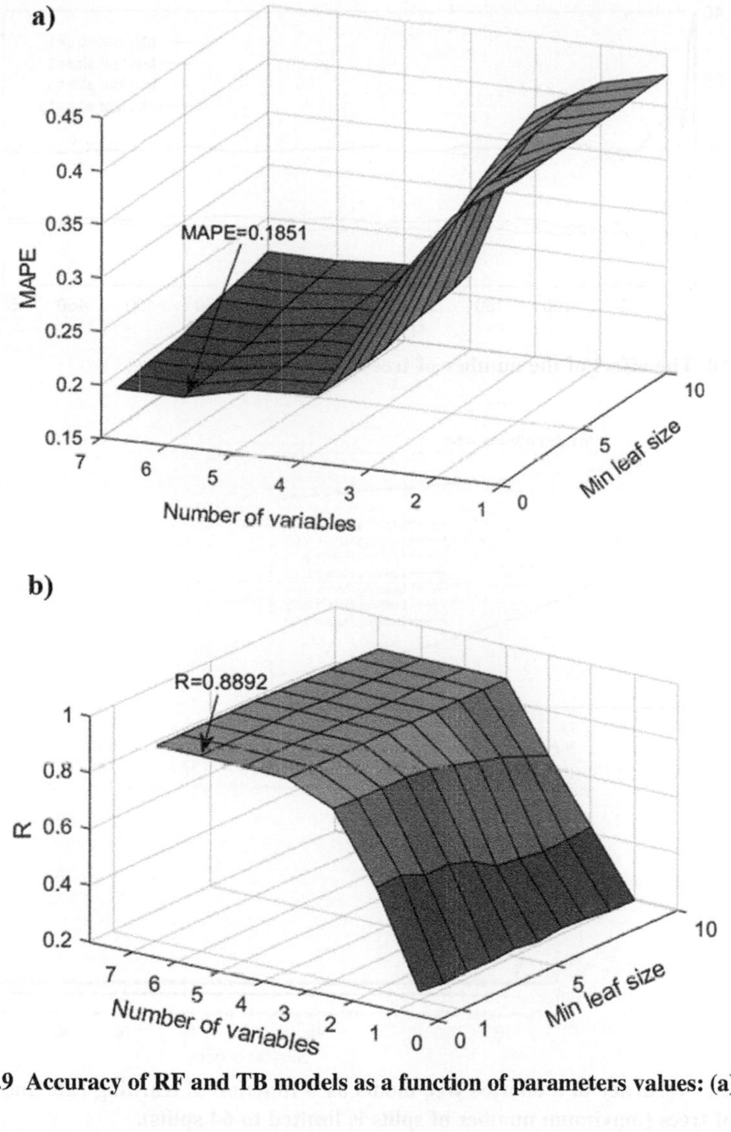

Figure 4.9 Accuracy of RF and TB models as a function of parameters values: (a) MAPE, (b) R.

with replacement, that data set will not contain identical data as the original data set because some data will be repeated. Data that are not selected can be used for model testing (out-of-bag data). The assessment of model accuracy can be obtained by using out-of-bag data by making predictions for the input values of that data, and since the target values are known for those data, it is possible to determine the error of the tree. By applying this procedure to all created trees, an ensemble error is obtained.

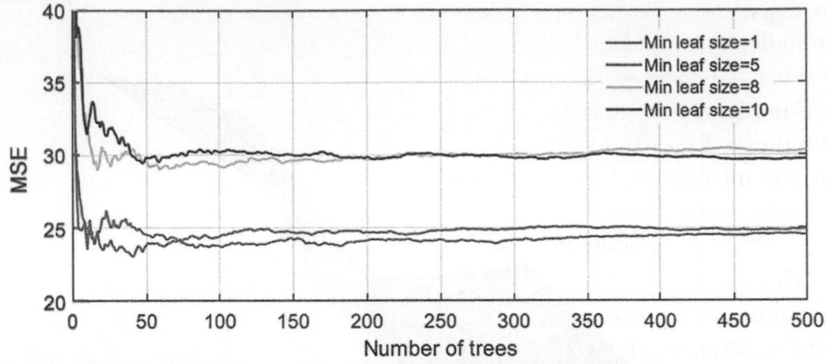

Figure 4.10 The effect of the number of trees on the learning curve.

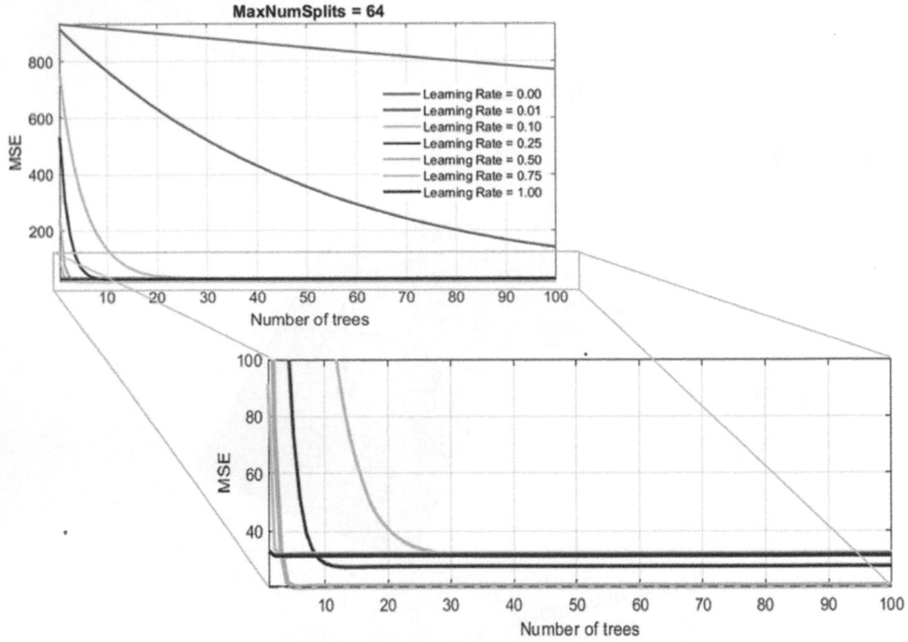

Figure 4.11 Accuracy of a boosted tree model as a function of learning rate and the number of trees (maximum number of splits is limited to 64 splits).

To determine the importance of the jth variable (within the RF and TB models), the values of the jth variable within the training data are permuted, and then the out-of-bag error is recalculated using the previously calibrated model and a prediction for such permuted data.

In order to determine the significance of the jth variable, it is necessary to determine the error or the difference between the prediction before and after the permutation of the jth variable for each individual calibrated tree, and then divide the obtained mean value by the standard deviation of the obtained differences. Higher values obtained in this way indicate the greater importance of that variable on the accuracy of the model [19,20] (Figs. 4.12 and 4.13).

When implementing the SVM method, three different kernel functions were analyzed: linear, RBF, and sigmoid. All values of input and output variables are first scaled to the interval [0, 1]. In all models, the values of the model parameters were analyzed using cross-validation, namely, in the case of the linear model, the penalty parameters C and ε, while in the case of the RBF and sigmoid kernel function, the parameters C, ε, and γ were determined.

The optimal values were determined using the grid search method during the initial search phase, and after identifying a wider area with lower MSE values, the search step was reduced and a detailed search was performed. The model with parameters corresponding to the smallest MSE value is adopted as optimal (Table 4.3).

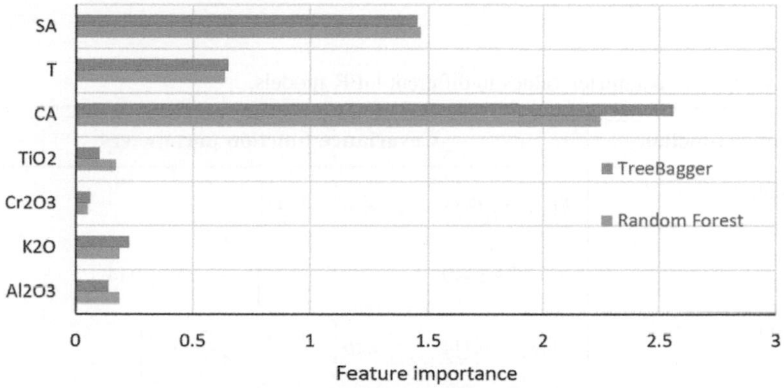

Figure 4.12 Importance of variables on the accuracy of the prediction model.

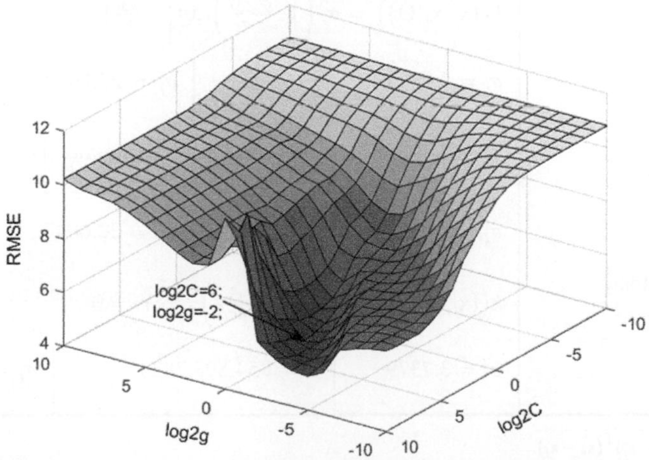

Figure 4.13 The search area determined during the rough parameter search procedure.

Table 4.3 Values of optimal kernel function parameters in the SVM method.

SVM linear	C = 0.8181	ε = 0.0833	/
SVM RBF	C = 48.6851	ε = 0.1304	γ = 0.2534
SVM sigmoid (default r = 0)	C = 17.9338	ε = 0.1243	γ = 0.0414

When applying GPR, a model was created where it was assumed that the function of the mean value of the GP was equal to zero. In order to satisfy such an assumption, the data was normalized to a mean value equal to zero and a standard deviation equal to one. Due to the significant influence of the type of covariance function, numerous function variants were analyzed: exponential, squared exponential, Matern 3/2, Matern 5/2, rational quadratic, as well as all ARD variants of the mentioned functions (Table 4.4 and Table 4.5). The parameters were determined by maximizing the Eq. (4.17).

Table 4.4 Optimal parameter values in different GPR models.

Covariance function	Covariance function parameters
Exponential	$k((\mathrm{x}_i,\mathrm{x}_j\|\Theta)) = \sigma_f^2\, exp\left[-\frac{1}{2}\frac{r}{\sigma_l^2}\right]$ $\sigma_l = 18.3349$ | $\sigma_f = 19.7944$
Squared exponential	$k((\mathrm{x}_i,\mathrm{x}_j\|\Theta)) = \sigma_f^2\, exp\left[-\frac{1}{2}\frac{(\mathrm{x}_i-\mathrm{x}_j)^{\mathrm{T}}(\mathrm{x}_i-\mathrm{x}_j)}{\sigma_l^2}\right]$ $\sigma_l = 3.7379$ | $\sigma_f = 15.5640$
Matern 3/2	$k((\mathrm{x}_i,\mathrm{x}_j\|\Theta)) = \sigma_f^2\left(1+\frac{\sqrt{3}r}{\sigma_l}\right)exp\left[-\frac{\sqrt{3}r}{\sigma_l}\right]$ $\sigma_l = 7.6493$ | $\sigma_f = 23.8407$
Matern 5/2	$k((\mathrm{x}_i,\mathrm{x}_j\|\Theta)) = \sigma_f^2\left(1+\frac{\sqrt{5}r}{\sigma_l}+\frac{5r^2}{3\sigma_l^2}\right)exp\left[-\frac{\sqrt{5}r}{\sigma_l}\right]$ $\sigma_l = 5.6377$ | $\sigma_f = 22.6047$
Rational quadratic	$k((\mathrm{x}_i,\mathrm{x}_j\|\Theta)) = \sigma_f^2\left(1+\frac{r^2}{2a\sigma_l^2}\right)^{-\alpha};\ r=0$ $\sigma_l = 3.7379$ | $a = 854{,}898.7823$ | $\sigma_f = 19.5640$

where $r = \sqrt{(\mathrm{x}_i-\mathrm{x}_j)^{\mathrm{T}}(\mathrm{x}_i-\mathrm{x}_j)}$.

Table 4.5 Optimal parameter values in different GPR ARD models.

Covariance function parameters						
σ_1	σ_2	σ_3	σ_4	σ_5	σ_6	σ_7
ARD exponential:						
$k((x_i, x_j\|\Theta)) = \sigma_f^2 \exp(-r);\ \sigma_F = 18.1057;\ r = \sqrt{\sum_{m=1}^{d} \frac{(x_{im} - x_{jm})^2}{\sigma_m^2}}$						
52.3136	197.3789	58.2961	7,946,201.7	1.0552	29.3206	21.4633
ARD squared exponential:						
$k((x_i, x_j\|\Theta)) = \sigma_f^2\, exp\left[-\frac{1}{2} \sum_{m=1}^{d} \frac{(x_{im} - x_{jm})^2}{\sigma_m^2}\right];\ \sigma_f = 10.2814$						
9.6532	4.1608	2587.9914	2.5851	0.0955	2.4321	2.2412
ARD matern 3/2:						
$k((x_i, x_j\|\Theta)) = \sigma_f^2 (1+\sqrt{3}\, r) exp[-\sqrt{3}\, r];\ \sigma_f = 11.6142$						
858.4592	6.4860	65,503.658	3.4459	0.1782	4.4078	3.0320
ARD matern 5/2:						
$k((x_i, x_j\|\Theta)) = \sigma_f^2 \left(1+\sqrt{5}\, r+\frac{5r^2}{3}\right) exp[-\sqrt{5}\, r];\ \sigma_f = 10.8887$						
7184.2135	5.1140	10,355.279	2.7106	0.1362	3.4309	2.5547
ARD rational quadratic:						
$k((x_i, x_j\|\Theta)) = \sigma_f^2 \left(1+\frac{1}{2\alpha} \sum_{m-1}^{d} \frac{(x_{im} - x_{jm})^2}{\sigma_m^2}\right)^{-\alpha};\ \alpha = 0.3623;\ \sigma_f = 12.1062$						
4323.5080	4.0886	421,224.11	2.2963	0.1197	3.0198	1.7910

where $r = \sqrt{\sum_{m=1}^{d} \frac{(x_{im} - x_{jm})^2}{\sigma_m^2}}$.

A comparative analysis of the accuracy of all analyzed models according to the defined accuracy criteria is given in Table 4.6. It can be seen that according to all the defined accuracy criteria, the optimal model is the one with the GP-ARD exponential covariance function. Other models based on the GP with the ARD covariance

Table 4.6 Comparative analysis of results of different machine learning models.

Model	RMSE	MAE	MAPE/100	R
Decision tree 1 (min leaf size = 5)	5.3789	4.1580	0.1762	0.8604
Decision tree 2 (min leaf size = 6)	5.2971	4.1969	0.1790	0.8635
TreeBagger	4.9263	4.0016	0.1863	0.8899
Random forest	4.9298	3.9864	0.1851	0.8892
Boosted trees	5.2172	4.1659	0.1895	0.8846
SVM lin. Kernel	6.8366	5.4855	0.2771	0.7578
SVM sigm. Kernel	6.8528	5.5012	0.2702	0.7573
GP exponential	6.6033	5.2378	0.2784	0.7774
GP Sq. exponential	6.0005	4.7230	0.2560	0.8185
GP Matern 3/2	6.2343	4.9933	0.2665	0.8024
GP Matern 5/2	6.1145	4.8653	0.2615	0.8108
GP rat. Quadratic	6.0005	4.7230	0.2560	0.8185
GP ARD exponential	**3.9833**	**3.0309**	**0.1146**	**0.9244**
GP ARD Sq. exponential	4.7122	3.6946	0.1427	0.8952
GP ARD Matern 3/2	4.1906	3.2243	0.1221	0.9166
GP ARD Matern 5/2	4.2894	3.3231	0.1250	0.9127
GP ARD rat. Quadratic	4.3584	3.3525	0.1266	0.9098

function, except for the model with the ARD squared exponential covariance function, achieve similar accuracies. However, all other models have worse accuracy according to all defined criteria (Table 4.6).

The optimal model with the ARD exponential covariance function can also be used to assess the importance of individual variables on the accuracy of CS prediction using length scale parameters (Fig. 4.14).

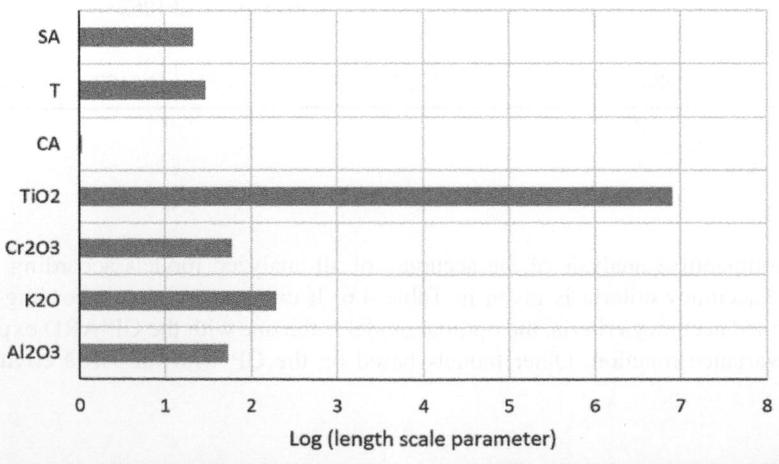

Figure 4.14 Log length scale parameters for optimal GPR ARD exponentials model.

When applying ARD covariance functions, it is characteristic that they have different length scale parameters for the input variables under consideration. Additionally, each parameter is correlated with the influence or importance of that variable on the output variable.

In the specific case, a smaller value of an individual parameter for a certain input variable indicates that the variable has a greater importance on the CS of concrete and vice versa (Fig. 4 14). As part of the research, the optimal model showed that the variable CA, which represents curing age, has the greatest importance. The variable TiO_2 has the least importance. The influence of other input variables is similar in terms of influence on the CS of concrete as an output variable.

Fig. 4.15 shows the values of the targeted and modeled compression strength values for their comparison. In the specific case, since it is a model that gives a model of the CS of concrete with glass additions at different times, it can be said that Fig. 4.15 shows a satisfactory match between the targeted and modeled values in absolute terms.

The predictive characteristics of the model can also be seen through the regression plot (Fig. 4.16), where the target values of the compression strength are represented on the horizontal axis and the modeled values obtained by the optimal model on the vertical axis. The correlation diagram of measured and modeled values shown in Fig. 4.16 indicates a satisfactory correlation between these values. In addition, the numerical value of the correlation coefficient R, which in this case is 0.9244, indicates that the obtained model can be considered a satisfactory prediction model.

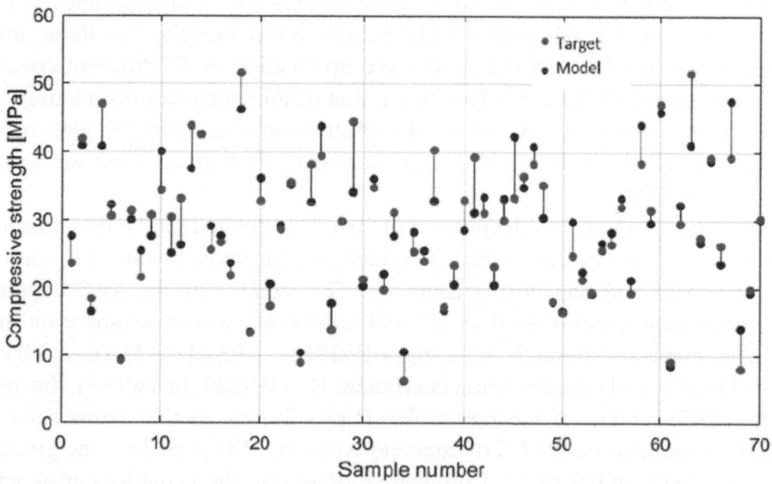

Figure 4.15 Predicted and target values for GPR ARD exponential model.

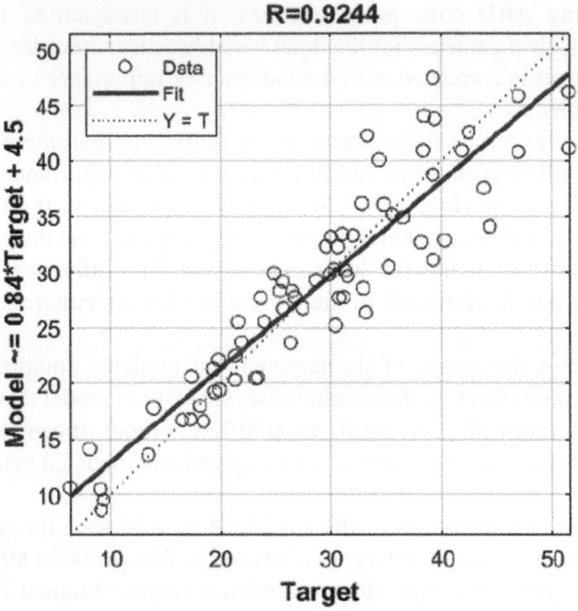

Figure 4.16 Regression plot for predicted and target values.

4.5 Conclusions

In this chapter, numerous machine learning algorithms were analyzed for predicting the CS of concrete with glass particles of the appropriate granulation. Models of individual regression trees were analyzed, as were models based on ensembles of regression trees using the TB, RF, and GB algorithms, SVM models with three different kernel functions, and GPR models where the application of 10 different covariance functions was analyzed. Research has shown that despite the recommendations stated in a large number of research papers for the application of ensemble models based on regression trees, it is possible to obtain greater accuracy with a significantly simpler model based on GPR.

To the author's knowledge, the method of GPR was applied for the first time in concrete with WG. In this way, a model was created that can serve as a basis for designing concrete with WG with satisfactory accuracy. The model with the ARD exponential covariance function was obtained as the optimal model, whose accuracy on the test data set was expressed through the criteria RMSE = 3.98 MPa, MAE = 3.03 MPa, MAPE = 11.46%, and the correlation coefficient R = 0.9244. In addition, the application of the ARD covariance function makes it possible to see the significance of the individual components on the CS of concrete with fine WG particles. The greatest influence of the variable CA ($\sigma_5 = 1.06$) can be observed; the variables curing temperature and the variable SA of cementitious materials ($\sigma_6 = 29.32$ and $\sigma_7 = 21.46$, respectively) have a lower and similar influence. The next group of influential variables is represented by the variables: percentage content of Al_2O_3 and percentage content of Cr_2O_3 ($\sigma_1 = 52.31$ and $\sigma_3 = 58.30$, respectively), while the variables that represent the percentage content of KiO_2 and TiO_2 ($\sigma_2 = 197.38$ and $\sigma_4 =$ 7,946,201.7), respectively, have the lowest influence.

References

[1] E.M. Golafshani, A. Kashani, Modeling the compressive strength of concrete containing waste glass using multi-objective automatic regression, Neural Computing and Applications (2022), https://doi.org/10.1007/s00521-022-07360-9.

[2] A. Siddika, A. Hajimohammadi, M.A.A. Mamun, R. Alyousef, W. Ferdous, Waste glass in cement and geopolymer concretes: a review on durability and challenges, Polymers 13 (13) (2021), https://doi.org/10.3390/polym13132071.

[3] E. Harrison, A. Berenjian, M. Seifan, Recycling of waste glass as aggregate in cement-based materials, Environmental Science and Ecotechnology 4 (2020) 100064, https://doi.org/10.1016/j.ese.2020.100064. ISSN 2666-4984.

[4] M. Malek, W. Lasica, M. Jackowski, M. Kadela, Effect of waste glass addition as a replacement for fine aggregate on properties of mortar, Materials 13 (14) (2020), https://doi.org/10.3390/ma13143189.

[5] S. Qaidi, H.M. Najm, S.M. Abed, Y.O. Özkılıç, H.A. Dughaishi, M. Alosta, M.M.S. Sabri, F. Alkhatib, A. Milad, Concrete containing waste glass as an environmentally friendly aggregate: a review on fresh and mechanical characteristics, Materials 15 (18) (2022), https://doi.org/10.3390/ma15186222.

[6] H. Du, K.H. Tan, Waste glass powder as cement replacement in concrete, Journal of Advanced Concrete Technology 12 (2014) 468−477.

[7] J. Ahmad, Z. Zhou, K.I. Usanova, N.I. Vatin, M.A. El-Shorbagy, A step towards concrete with partial substitution of waste glass (WG) in concrete: a review, Materials 15 (7) (2022), https://doi.org/10.3390/ma15072525.

[8] D. Grdić, N. Ristić, G. Topličić-Ćurčić, D. Đorđević, N. Krstić, Effects of addition of finely ground CRT glass on the properties of cement paste and mortar, Građevinar 72 (2020), https://doi.org/10.14256/JCE.25942018.

[9] ASTM D6868 Standard Specification for Biodegradable Plastics Used as Coatings on Paper and Other Compostable Substrates, ASTM International, West Conshohocken, PA, USA, 2017.

[10] R. Idir, M. Cyr, A. Tagnit-Hamou, Use of fine glass as ASR inhibitor in glass aggregate mortars, Construction and Building Materials 24 (2010) 1309−1312.

[11] Y. Shao, T. Lefort, S. Moras, D. Rodriguez, Studies on concrete containing ground waste glass, Cement and Concrete Research 30 (2000) 91−100.

[12] M. Mirzahosseini, K.A. Riding, Effect of curing temperature and glass type on the pozzolanic reactivity of glass powder, Cement and Concrete Research 58 (2014) 103−111.

[13] M. Hadzima-Nyarko, E.K. Nyarko, N. Ademović, I. Miličević, T.K. Šipoš, Modelling the influence of waste rubber on compressive strength of concrete by artificial neural networks, Materials 12 (4) (2019), https://doi.org/10.3390/ma12040561.

[14] M. Hadzima-Nyarko, E.K. Nyarko, H. Lu, S. Zhu, The Machine learning approaches for estimation of compressive strength of concrete, European Physical Journal Plus 135 (8) (2020), https://doi.org/10.1140/epjp/s13360-020-00703-2.

[15] M. Hadzima-Nyarko, S.H. Trinh, Prediction of compressive strength of concrete at high heating conditions by using artificial neural network-based Bayesian regularization, Journal of Science and Transport Technology 2 (1) (2022).

[16] M. Kovačević, S. Lozančić, E.K. Nyarko, M. Hadzima-Nyarko, Modeling of compressive strength of self-compacting rubberized concrete using machine learning, Materials 14 (15) (2021), https://doi.org/10.3390/ma14154346.

[17] M. Kovačević, S. Lozančić, E.K. Nyarko, M. Hadzima-Nyarko, Application of artificial intelligence methods for predicting the compressive strength of self-compacting concrete with class F fly ash, Materials 15 (12) (2022), https://doi.org/10.3390/ma15124191.

[18] M. Mirzahosseini, P. Jiao, K. Barri, K.A. Riding, A.H. Alavi, New machine learning prediction models for compressive strength of concrete modified with glass cullet, Engineering Computations 36 (3) (2019) 876−898, https://doi.org/10.1108/EC-08-2018-0348.

[19] S. Ray, M. Haque, T. Ahmed, T.T. Nahin, Comparison of artificial neural network (ANN) and response surface methodology (RSM) in predicting the compressive and splitting tensile strength of concrete prepared with glass waste and tin (Sn) can fiber, Journal of King Saud University - Engineering Sciences (2021), https://doi.org/10.1016/j.jksues.2021.03.006.

[20] B. Ghorbani, A. Arulrajah, G. Narsilio G, et al., Dynamic characterization of recycled glass-recycled concrete blends using experimental analysis and artificial neural network modeling, Soil Dynamics and Earthquake Engineering 142 (2020), https://doi.org/10.1016/j.soildyn.2020.106544.

[21] M.E.A. Ben Seghier, E.M. Golafshani, J.J.M. Arashpour, Metaheuristic-based machine learning modeling of the compressive strength of concrete containing waste glass, Structural Concrete (2023), https://doi.org/10.1002/suco.202200260.

[22] E.M. Golafshani, A. Kashani, Modeling the compressive strength of concrete containing waste glass using multi-objective automatic regression, Neural Computing and Applications 34 (2022) 17107−17127, https://doi.org/10.1007/s00521-022-07360−369.

[23] S.A. Ahmad, H.U. Ahmed, D.A. Ahmed, B.H.S. Hamah-ali, R.H. Faraj, S.K. Rafiq, Predicting concrete strength with waste glass using statistical evaluations, neural networks, and linear/nonlinear models, Asian Journal of Civil Engineering (2023), https://doi.org/10.1007/s42107-023-00692-4.

[24] Harish, P. Janardhan, Support vector machine in predicting epoxy glass powder mixed cement concrete, Materials Today: Proceedings 46 (Part 18) (2021) 9042−9046. https://www.sciencedirect.com/science/article/pii/S2214785321039857.

[25] G.D. Ayana Ghosh, R.N. Ransinchung, Application of machine learning algorithm to assess the efficacy of varying industrial wastes and curing methods on strength development of geopolymer concrete, Construction and Building Materials 341 (2022), https://doi.org/10.1016/j.conbuildmat.2022.127828.

[26] J. Sun, Y. Wang, X. Yao, Z. Ren, G. Zhang, C. Zhang, Machine-learning-aided prediction of flexural strength and ASR expansion for waste glass cementitious composite, Applied Sciences 11 (15) (2021), https://doi.org/10.3390/app11156686.

[27] T. Hastie, R. Tibsirani, J. Friedman, The Elements of Statistical Learning, Book, Springer, Berlin/Heidelberg, Germany, 2009.

[28] L. Breiman, H. Friedman, R. Olsen, C.J. Stone, Classification and Regression Trees", Book, Chapman and Hall/CRC, Wadsworth, OH, USA, 1984.

[29] L. Breiman, Bagging predictors, Machine Learning 24 (1996) 123−140.

[30] L. Breiman, Random forests, Machine Learning 45 (2001) 5−32.

[31] M. Kovačević, N. Ivanišević, P. Petronijević, V. Despotović, Construction cost estimation of reinforced and prestressed concrete bridges using machine learning, Građevinar 73 (1) (2021) 1−13, https://doi.org/10.14256/JCE.2738.2019.
[32] C.E. Rasmussen, C.K. Williams, Gaussian Processes for Machine Learning, Book, The MIT Press, Cambridge, MA, USA, 2006.
[33] M. Mirzahosseini, K.A. Riding, Influence of different particle sizes on reactivity of finely ground glass as supplementary cementitious materials (SCM), Cement and Concrete Composites 56 (2015) 95−105.
[34] M. Mirzahosseini, K.A. Riding, Effect of combined glass particles on hydration in cementitious systems, Journal of Materials Civil Engineering 27 (2015).

AI-based structural health monitoring systems

5

Ayoub Keshmiry[1], *Sahar Hassani*[2] *and Ulrike Dackermann*[3]
[1]Faculty of Civil Engineering, Shahrood University of Technology, Shahrood, Iran; [2]Centre for Infrastructure Engineering and Safety, School of Civil and Environmental Engineering, University of New South Wales, Sydney, NSW, Australia; [3]School of Civil and Environmental Engineering, University of New South Wales, Sydney, NSW, Australia

5.1 Introduction

Civil structures and infrastructure are essential and integral components of human society and are of high importance to the worldwide population and global economy. Due to the construction boom after World War II, a large proportion of critical infrastructure is approaching the end of its service life [1]. Since the cost and labor involved in replacing these structures far exceed the available resources, structural health monitoring (SHM) systems can help to extend their service lives, ensuring the structures' integrity and safety. Furthermore, SHM improves the reliability of current asset management strategies and reduces the possibility of financial and life losses due to structural failure [2,3]. SHM systems determine the current state of a structure by continuously measuring and analyzing structural responses and by providing periodic updates on the structure's capacity to continue serving [4]. In a typical SHM system, a variety of sensors are installed at different locations of the structure, enabling real-time monitoring of the global structure or of specific local components [5]. An example of sensor measurements is the periodic recording of a structure's dynamic responses that are subsequently used to extract damage-sensitive features for structural health assessment. Statistical analysis is one of many approaches used to analyze the sensor measurements [6]. Following a disaster, such as a severe seismic event, measurement anomalies can be detected quickly, providing an evaluation of the structure's integrity and reliability and mitigating further damage [7,8]. Three key elements are included in an SHM system.

1. **Sensor network:** A structure is permanently equipped with a diverse network of sensors, enabling continuous and automated monitoring of the structure. This aspect distinguishes SHM from traditional nondestructive testing [9].
2. **Data transmission and processing:** Large volumes of data are continuously recorded by the sensors, requiring data transmission networks and on-board facilities for real-time processing [10].
3. **Data interpretation:** Using advanced algorithms, acquired data is analyzed and damage indices are derived, providing structural health information including damage type, location, and extent. Environmental and operational factors are typically considered and incorporated into the algorithms [11].

Artificial Intelligence Applications for Sustainable Construction. https://doi.org/10.1016/B978-0-443-13191-2.00008-0
Copyright © 2024 Elsevier Ltd. All rights are reserved, including those for text and data mining, AI training, and similar technologies.

The main components of an SHM system, including associated hardware and disciplines, are presented in Fig. 5.1. As it can be seen in the figure, SHM consists of different elements and involves expertise from a broad area of disciplines. Generally, SHM strategies fall into two categories [12].

1. **Model-driven:** In this type, physics-based models and inverse techniques are used to infer a structure's health state from measurement changes and involve the updating of model parameters. This strategy is defined as an inverse model-driven approach.
2. **Data-driven**: This type operates within a statistical pattern recognition paradigm. Rather than relying on physical modeling, data-driven SHM methods infer structural health directly from the measurement data.

The many benefits of SHM systems sparked the interest of a wide range of stakeholders and motivated the rapid growth of SHM research. Consequently, SHM systems have been implemented in a variety of structures, including high-rise buildings [13], bridges [14], towers [15], tunnels [16], and dams [17]. Recent advancements and cost reductions in sensor technologies, as well as data transmission and processing systems, have resulted in the application of large sensor networks and an ever-growing amount of big data requiring sophisticated, intelligent, and powerful computational analysis tools. This has ultimately led to the deployment of AI in the field of SHM.

AI emerged in the 1950s as part of computer science and has contributed significantly to the development of AI subfields including data mining [18], robotics [19], and pattern recognition [20]. Fig. 5.2 depicts various fields of AI and the categorization of machine learning (ML) algorithms. Many recent SHM applications incorporate AI, such as artificial neural networks (ANNs) [21], fuzzy logic algorithms [22], and knowledge-based systems [23].

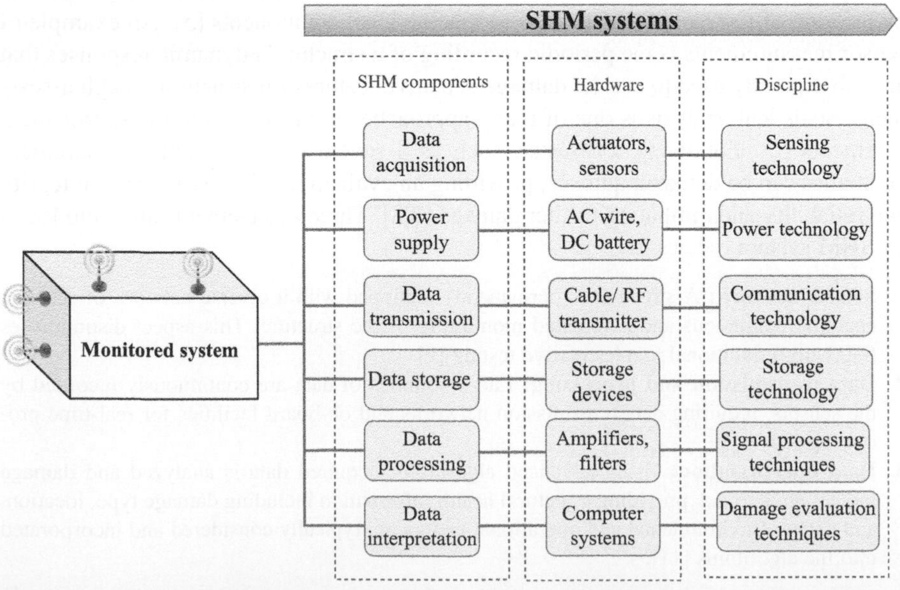

Figure 5.1 Main components and associated hardware and disciplines of an SHM system.

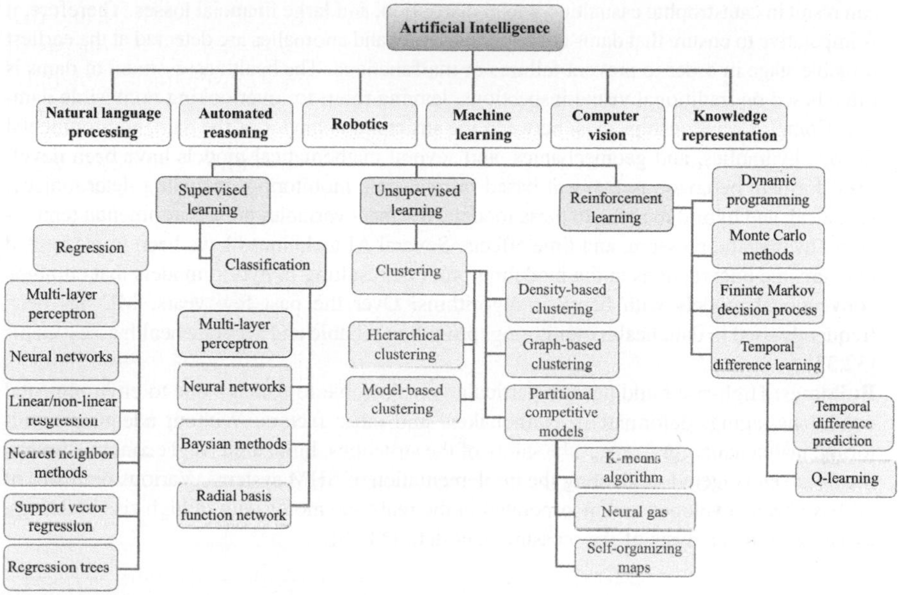

Figure 5.2 A taxonomy of AI systems.

5.2 Structural health monitoring applications

This section presents a brief description of typical SHM applications in civil and mechanical engineering structures and their incorporation of AI.

- **Bridges:** SHM systems have a long history of being installed in bridge structures for the continuous assessment of their health and factors causing structural degradation. These factors include aging materials, corrosion, mechanical overloading, and other degradation mechanisms. Different types of sensors are typically mounted on or even integrated into bridge components to gather quantitative data on the structural response as well as on environmental and operational conditions. The acquired data facilitates the real-time and automated feedback of the systems, enabling up-to-date health evaluations. The processing and interpretation of big data in real time presents a key challenge for bridge SHM, and AI is applied to address these demands [24,25].
- **Wind turbines:** To reduce the dependency on conventional energy sources, such as fossil fuels, much effort is invested in the development and expansion of sustainable energy sources. Wind turbines have become increasingly popular as the technology matures, and ever larger turbines are being developed to increase their efficiency and productivity to harvest more wind energy. To address wind turbine maintenance and repair issues, several approaches for the nondestructive testing and SHM of wind turbines have been developed and applied, such as thermal imaging [26], acoustic emission event detection [27], ultrasonic [28], eddy current [29], fiber optics, and modal-based approaches. Wind turbine monitoring accumulates large amounts of data, giving rise to the demand for ML for data analytics to determine and predict structural damage [30,31].
- **Dams:** Dams are crucial infrastructure that benefits the population by providing vital services such as drinking and irrigation water, as well as power generation. Undetected deterioration

can result in catastrophic casualties, social disruption, and large financial losses. Therefore, it is imperative to ensure that dams are operated safely and anomalies are detected at the earliest possible stage in order to prevent failures or malfunctions. The health assessment of dams is often based on traditional visual inspections, leaving room for overlooking nonvisible damage. Complex relationships exist between the structural behavior of the dam, environmental factors, hydraulics, and geomechanics, and several mathematical models have been developed for dam behavior assessment based on real-time monitoring, including deterministic, statistical, and hybrid models. In these models, the main variables are environmental temperature, hydrostatic pressure, and time effects. Several AI techniques have been implemented to overcome uncertainties in the modeling process, resulting in hybrid models that combine conventional models with heuristic algorithms. Over the past few years, ML has been frequently used in dam health monitoring, providing reliable and accurate health assessments [32,33].

- **Buildings:** High-rise buildings are particularly vulnerable to damage due to environmental conditions, aging, deformation, earthquakes, and other factors. Without adequate monitoring, maintenance, and repair, the safety of the structures, human life, and economic investments are endangered motivating the implementation of SHM systems. Various branches of AI have been developed and incorporated in the real-time monitoring of high-rise buildings, assisting in the analysis of the measurement data [34,35].

5.3 Artificial intelligence

As discussed earlier, artificial intelligence (AI) systems have been one of the focal points of the research community, and considerable attention has been paid in recent years to investigating their integration in SHM systems [36]. To assist in understanding these methods, this section presents an overview of the currently most popular AI techniques (i.e., ML and deep learning (DL)).

5.3.1 Machine learning (ML)

ML automates the process of building analytical models and integrates information or experience into machines or systems. In ML, a system's accuracy is enhanced by analyzing data structures and fitting them into models [37]. Machines can learn in a number of ways, including reinforcement learning, supervised learning (SL), and unsupervised learning. SL is utilized to provide a learning procedure with labeled data in order to address problems pertaining to regression and classification. As an application of SL in SHM, damage types and severity can be determined. Contrarily, unsupervised learning refers to learning from unlabeled data, that is, data sets with undefined outputs that can be grouped based on a certain pattern. It can be utilized as a means of detecting damage by clustering structural response data.

ML involves the loading of the input data, processing the data using a chosen algorithm, obtaining the output, and then determining whether to discontinue or restart the process based on the feedback received. A successful process concludes with a well-predicted and accurate result. In the context of AI, ML allows systems to automatically learn and enhance based on prior experience without explicit programming [38].

For clarification and a deeper understanding of the concept of ML, the following example is presented: Consider the objective of developing an AI-based SHM approach for determining whether a specific kind of damage is present in a structure. In order to train the system, large quantities of measurement samples must be available. The data of each test includes an input, which is a record of the field measurement, and an output, which states whether the structure has been diagnosed as damaged or not. In order to train the system, it is necessary for a structural engineer to provide a list of possible types of damage (damage categories). It is then possible to analyze a training dataset to extract data relevant to these damage types for each record sample. This process is known as feature extraction. The processed dataset contains information regarding the damage types identified in each recording as well as their associated outputs (i.e., whether a structure has been damaged or not). These outputs are then used to train an ML algorithm based on the processed dataset. As a result of the training process, the system will automatically be able to establish correlations between damage patterns and output (i.e., damaged or undamaged). After validation tests, any future structural state can be diagnosed using the trained system based on new structural records and trained damage types.

In the example provided above, a dataset consisting of human-labeled data must be provided to train the SL algorithms. Consequently, SL aims at finding the optimal mapping between inputs and outputs (or targets). Thus, it is necessary for SL to be supervised by a human, who assigns each data sample the correct target or label prior to training. Contrarily, unsupervised learning algorithms require only unlabeled input data. Unsupervised learning is operated to obtain practical information regarding its underlying structure by analyzing the data distribution. In general, the different tasks performed by ML techniques can be categorized as follows [39]:

- **Classification:** This task aims to identify the category in which the input falls. The presented example is a classification system where inputs are classified into "damaged" or "undamaged" categories.
- **Regression**: In this process, correlation models are established between input parameters and numerical outputs. Regression and classification differ with respect to the output format.
- **Prediction:** Prediction involves the prediction of the future value of a time series using a particular type of regression.
- **Clustering:** As the name implies, clustering is intended to divide the input dataset into clusters containing examples that are similar. As opposed to classification, regression, and prediction tasks, clustering is carried out using an unsupervised approach, such as self-organizing maps.

5.3.2 Deep learning (DL)

DL is a subset of ML within the AI context, referring to networks that are able to learn unsupervised from unstructured data. Feature extraction is necessary for various types of supervised and unsupervised ML algorithms to express the input data in terms of a predetermined number of representative features. It is essential to choose the right set of features that best reflect the properties of the input data. Once these features have been extracted, mapping them to desired outputs becomes relatively effortless using simple ML algorithms. This approach is practical in cases where it is possible to use feature extraction techniques to identify signal patterns, as demonstrated in the previous example (i.e., the list of damage types). It can be challenging for conventional feature extraction techniques to identify a representative set of features for training the AI system. Consider the example of a computer vision system used for the detection of vehicles in images for a better understanding of this concept. Several input-output samples are available for training the system. For each case, a traffic image of the street is the input, and the output is the location of the vehicles in the image. As a pretraining step before applying an ML algorithm, the input image must be represented by an array of features that represent significant clues regarding the location of vehicles in the input image. This approach may not be a practical solution considering that the appearance of vehicles in photographs is affected considerably by factors such as the viewing angle, camera position, occlusions from obstacles and other vehicles, light and shadow conditions, etc. While feature extraction technologies (or hand-crafted feature selection) may work in specific scenarios, they are unlikely to be practical in others. Accordingly, DL was established as an alternative to conventional feature extraction methods for complex ML applications. In this regard, DL, or representation learning, is a specific type of ML method that allows for the extraction of the optimal input representation directly from raw data, eliminating user intervention. To put it another way, DL algorithms are capable of learning not only how to correlate features with intended outputs but also how to extract features. It has been demonstrated that DL systems that have been appropriately trained can produce direct mappings from raw inputs to final outputs without requiring any prior extraction of features [40]. In this way, DL can provide an explanation of high-level and abstract features in terms of a hierarchy of simple and low-level learned features.

DL algorithms are able to handle complex tasks by breaking them into many simple tasks. According to some recent studies, it has been demonstrated that using learned features instead of hand-crafted features leads to considerably improved performance when it comes to tackling challenging tasks such as the classification of electrocardiogram beats [41], images [42], or object detection [43]. DL techniques can be divided into three main categories, namely, discriminative (supervised), generative (unsupervised), and hybrid learning algorithms. A taxonomy of DL techniques is presented in Fig. 5.3.

5.4 Hierarchy of ML algorithms

This section presents a brief general guideline of algorithm manipulation in ML for clarification and deeper understanding.

5.4.1 Input configuration

To select the most efficient ML algorithm, a better understanding of the data at the input stage is necessary. While some algorithms can be applied to small sample

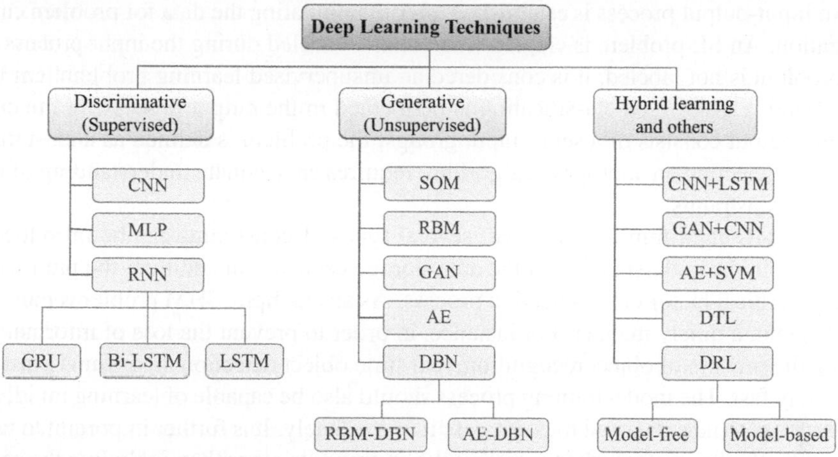

Figure 5.3 A taxonomy of DL techniques. *AE*, auto encoder; *CNN*, Convolutional neural network; *DBN*, deep belief network; *SOM*, self-organizing map; *DRL*, deep reinforcement learning; *DTL*, decision tree learning; *GAN*, generative adversarial network; *GRU*, gated recurrent unit; *LSTM*, long short-term memory; *MLP*, multilayer perceptrons; *RBM*, restricted Boltzmann machine; *RNN*, recurrent neural network.

sets, others require vast numbers of samples. Data manipulation can be summarized in the following steps:

1. **Know the data:**
   a. **Data statistics:** Averages and medians
   b. **Data visualization:** Density plots and histograms
2. **Clean the data:**
   a. **Dealing with missing data**
   b. **Outlier detection**
   c. **Data aggregation**
3. **Augment the data:**
   a. **Feature engineering**
   b. **Feature extraction**
   c. **Feature selection**

Before implementing an ML algorithm, it is essential to thoroughly understand and manipulate the data, employing analytical tools such as data visualization and statistics. Percentiles are used in data statistics to describe correlations and central tendencies based on the range, average, and median of the data. In addition, they provide an understanding of how data are related to one another. Data visualization typically consists of density plots and histograms to illustrate distributions of data, along with box plots to identify anomalies. As the next step, the data should be cleaned, involving the removal of outliers and missing values that can affect the accuracy of future predictions. As a final step, the data should be augmented or enriched in order to simplify interpretation, reduce dimensionality and redundancy, rescale some variables, and capture complex relationships.

An input-output process is established after manipulating the data for problem categorization. An SL problem is created when data is labeled during the input process. If the problem is not labeled, it is considered an unsupervised learning problem. On the other hand, a task-based classification is performed in the output process. In the case that the output consists of a set of input groups, the problem is defined as a clustering problem. Selecting an appropriate algorithm requires an adequate understanding of the problem constraints.

As an ML algorithm is developed, several types of constraints can be introduced, starting with the understanding of the data storage capacity. In addition, the prediction time is a critical factor in the selection process. As an example, SHM problems must be addressed in a timely manner. For instance, in order to prevent the loss of information during the process of object recognition, real-time object detection algorithms must be extremely fast. The model training process should also be capable of learning rapidly if it is exposed to new data and must process it immediately. It is further important to take into account other factors when selecting the appropriate algorithm, including the scale and accuracy of the model, computational overhead, preprocessing of the model, interactions, the number of features that are involved in learning, and the prediction of polynomial terms with greater complexity.

5.4.2 Output manipulation

Depending on the application, SHM outputs can vary for different problem tasks. These outputs include damage detection, settlement, damage classification, health index, predicting environmental factors (i.e., humidity, temperature, etc.), and object detection. A practical and accurate output must be provided at the end of the process. Otherwise, as a result of learning experience, the machine will be able to improve upon its performance in the future.

5.5 ML and DL applications in SHM

This section presents a description of several ML-based and DL-based approaches that have been applied in SHM applications. Over the past decades, various AI algorithms have been investigated for their use in SHM systems, including feedforward neural networks (FNNs), backpropagation neural networks (BPNNs), convolutional neural networks (CNNs), support vector machines (SVMs), and recurrent neural networks. Table 5.1 lists the most common AI algorithms used for SHM applications and a summary of their advantages and disadvantages.

5.5.1 CNNs

AI-based SHM systems can be trained to detect cracks in civil engineering structures quickly and accurately. As a result, engineers may be able to determine a structure's level of degradation and load-carrying capacity. In the assessment of the health and prediction of the remaining service life of structures, experts have often relied upon

Table 5.1 Advantages and disadvantages of AI-based approaches for SHM.

Approach	Applications in SHM	Advantages	Disadvantages	Type (supervised/ unsupervised)
CNNs	Image identification and classification damage detection damage localization	Large data capacity high learning capabilities highly reliable in image processing	Parameter tuning requirements and data demands result in overfitting and underfitting with high computational cost	Supervised
BPN	Degradation tracking damage detection damage identification	Rapid and straightforward analysis parameter tuning is not required easy implementation	Very sensitive to noisy data the matrix-based approach is utilized instead of a mini-batch, and the performance highly relies on the input data	Supervised
FNNs	Damage detection damage identification aging evaluation	Runs independently with a slight intermediary to ensure moderation simplified architecture highly suitable for nonlinear data	High computational and hardware requirements for handling large data sets require graphics processing units (GPUs) that require large amounts of data in order to function properly	Supervised
SVM	Damage classification degradation tracking damage quantification	Works efficiently with a clear margin of separation practical in high-dimensional spaces memory efficient	Indirect probability estimation weak performance for data with high noise pollution and time-consuming training for large data sets	Supervised

subjective assessments, which are further complicated by the difficulty of accessing critical structural areas. To address this issue, it is necessary to implement automated and intelligent crack detection methods independent of subjective operator assessment.

As a DL algorithm, CNNs are utilized in both descriptive and generative tasks, primarily image processing. It contains scripts for image and video recognition. A great deal of research is being conducted on image-based crack detection methods. As the technology improves, it will be able to address problems such as the identification of cracks with varying shapes and sizes, challenges with shading, concrete spalling, blemishes, and lighting conditions. Nevertheless, parameters must be carefully considered from input datasets to output.

In general, AI models are more successful if they are trained on a more comprehensive and extensive data set. To overcome the issue of lacking data, some techniques have been developed, such as data augmentation [44], whose main objective is the reduction of overfitting due to a limited and imbalanced set of training data.

The dropout technique [45] is another promising method for enhancing prediction accuracy, as it involves ignoring some units of the neural network randomly and temporarily in calculations. Several other considerations must be taken into account to achieve higher accuracy in processing image data, such as uncontrolled lighting conditions, image shooting distance, blurriness conditions, and shot angle.

Several CNN architectures and layouts have been proposed for the detection of cracks in asphalt pavements and concrete structures [46]. The investigated pretrained networks used different numbers of convolutional blocks, fully connected layers, and pooling layers. Additional features were added to the pretrained networks by transfer learning.

An alternative method of training a neural network to detect fatigue cracks in gusset plates has been developed by Dung et al. [47]. Using the output features of the VGG16 network architecture, the researchers were able to achieve the highest precision by fine-tuning the top layer of the VGG16 network architecture, previously trained on an image database called ImageNet. It was found that fine-tuning a fully connected layer, in combination with the convolutional layer on top of VGG16 and data augmentation, produced the best results for detecting cracks in structural elements.

Many proposals have been presented to identify the most robust crack detection algorithm by altering the architecture of CNNs, varying the number of convolutional blocks, increasing the pooling at the end of each convolutional block, increasing the activation layers, and normalizing the data to identify the most robust crack detection algorithm.

In addition to binary classifications (cracked or not), other research efforts explored rather complex structural relationships. There have been other innovative and practical solutions explored for monitoring schemes, such as the loosening of bolts [48], detection of efflorescence and spalling [49], the typology of cracks, and their length and width [50].

5.5.2 BPNNs

BPNN is an SL algorithm used to train multilayer perceptrons. Using the gradient descent technique, it primarily serves as a means of determining the minimal error function in the weight space. The learning problem is solved by the weight that minimizes the loss function.

According to Fan et al. [51], four different applications of ML were highlighted using BPNNs to detect and evaluate damage to structural elements of a building:

1. First, the location of damage was determined based on a changing ratio of modal strain energy on a reinforced concrete structure.
2. Second, the location and extent of damage were assessed using a simply supported beam and a finite element (FE) simulation based on the curvature of critical points in the beam.
3. Damage level identification was the third application conducted in a steel frame structure. Natural frequency ratios were used as inputs, with wind loads simulated as applied loads.
4. The Kewitte single-layer spherical reticulated shell was then analyzed by using a method for identifying the damage.

The use of ANNs may hold greater promise for inverse problems like the structural identification of large structures (for example, bridges), where in-situ measurement data are likely to be incomplete and often imprecise. As another application of BPNNs, in order to estimate the damage potential in truss joints, a bridge was subjected to an extensive campaign that evaluated parameters such as frequency, degrees of freedom, and mode shapes [52]. Routine monitoring is another area where the BPNNs have been applied. In Refs. [53,54] pile settlement was predicted based on pile displacement sequences and the tracking of normality as a function of their deviation. The methods above were effective in detecting damage in different types of structures with an adequate level of accuracy.

5.5.3 FNNs

FNNs and recurrent networks are two types of ANNs. Fig. 5.4 shows the classification of ANN. In FNNs, information flows forward from the input nodes to the output nodes in a single direction. Furthermore, the network does not contain any loops or cycles. In most cases, FNNs are used in SL circumstances where there is neither sequential nor time-dependent data to learn.

Some of the studies on the applications of FNNs in SHMs of various structures are presented below:

1. **Damage identification of steel frame structures** [55]:
   - **Input:** The first flexural modes were obtained using a FE model.
   - **Strategy:** Two main approaches were used in the study. In the first step, a healthy structure was calibrated, and then a damaged structure was identified after a seismic event.
   - **Output:** As a result, stiffness and the mass of the structure were calculated. The method's robustness was demonstrated by predicting damage in each story.
2. **Identification of damage locations and levels in a seven-story building** [56]:
   - **Input:** Modal properties of the structure
   - **Strategy:** First, a stochastic subspace system identification approach was used to determine a healthy structure's natural frequencies and mode shapes. A simplified structure model was derived based on natural frequencies and mode shapes, enabling the construction of various damage patterns based on varying stiffness terms. The modal properties of the structure and the damage patterns were used to train and build a neural network model.

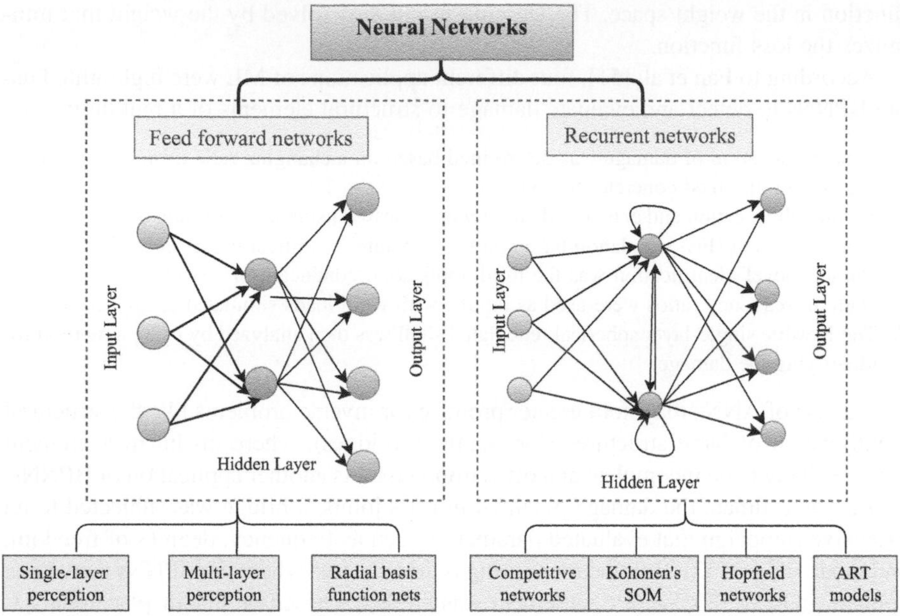

Figure 5.4 Neural network classification.

- **Output:** Damage patterns after critical events

Compared to traditional methods, this algorithm is simpler and more accurate, making it suitable for a wide range of applications, such as evaluating aging, stiffness, and long-term structural parameters in bridges [57]. Moreover, its efficiency has been demonstrated in determining radial dam displacements [58], leakage flow monitoring [59], and detecting dam pore pressure [60].

5.4 SVM

SVM is an SL algorithm employed in classification and regression problems; in addition, it is highly suitable for classification problems. In SVM algorithms, data are sorted into one of two categories, and then a map is generated based on the sorted data, maximizing the margins between the two categories. A kernel function allows the algorithm to perform both linear and nonlinear classifications. As data are mapped to a high-dimensional feature space, SVM can categorize data points regardless of whether the linear separation of data points is possible or not. First, it is necessary to identify a separator between the categories; then, the data needs to be transformed to represent the separator as a hyperplane. Some reported applications of SVM in SHM of various structures are described below.

1. **Vibration-based SHM** [61]:
   - **Input:** Vibration parameters
   - **Strategy:** The identification of damaged and undamaged cases was accomplished using a kernel-based SVM.
   - **Output:** Mapping of damage features

2. **Timber health monitoring** [62]:
   - **Input:** Vibration signals
   - **Strategy:** Analyzing the obtained waveform data using KNN and an SVM
   - **Output:** Damage detection, localization, and degradation tracking
3. **Characterization of gradually evolving structural deterioration** [63]:
   - **Input:** Force, acceleration, and time histories
   - **Strategy:** A Gaussian white noise excitation is assumed to be applied to the structure.
   - **Output:** Damage detection and structural health assessment

Some of the other applications of AI algorithms in SHM systems are presented in Table 5.2.

5.6 Future trends

As mentioned in this chapter, the problems previously addressed only using physical models are currently solved in different ways due to digital transformation. Implementing AI-based SHM and full utilization still demands considerable effort, despite current advancements. Integration is a crucial component of this process, and two key discussion areas have emerged.

- **Digital twins (DTs):** DTs refer to the full life cycle of an asset and the integration of all levels of information and simulation [75].
- **Physics-guided ML (PGML):** This part integrates data-driven learning with physics knowledge for consistent representation learning [76].

Observing recent results in AI in other domains and identifying the gaps in SHM, the following explanations are used to recommend future approaches for research in AI-based SHM.

5.6.1 Physics guided machine learning

ML models are not well suited to physical problems because of a lack of interpretability or theoretical base. This drawback leads to skepticism toward their sufficient practicality. Other drawbacks of such models include relying on big data and failing to generalize unseen scenarios, which may result in physical inconsistencies. However, physical-driven models struggle to account for historical conditions, uncertainties, and environmental constraints and rely on hypotheses and simplifications of boundary conditions. PGML is a rapidly growing domain that suggests merging data-driven and physics-driven models, benefiting from the optimal performance of both models.

5.6.2 Digital twins

DTs are integrated multiphysics and multiscale simulations of an actual system capable of effectively forecasting the system's health over time using the best physical models and data available. DTs can also create damage mitigation and performance improvement plans while accounting for system-associated uncertainties. DTs

Table 5.2 Applications of AI algorithms in SHM systems.

References	Problem	Model	Algorithm	Input
Maes et al. [64]	SHM based on natural frequencies	Railway bridge	Principle component analysis	Acceleration measurements of the deck and arches
Tibaduiza et al. [65]	Damage classification	Aluminum plate	Principle component analysis	Vibrational structural responses
Ibrahim et al. [66]	Identifying the status of buildings postevent	Building	K-nearest neighbours (KNN)	Accelerometer traces
Hasan and Kim [67]	Fault detection	Spherical tank	KNN	Time and frequency domain data
Worden and Cross [68]	Effects of environmental and operational conditions in SHM	Bridge	Treed Gaussian process	Modal parameters, humidity, and air and soil temperature.
Civera et al. [69]	Crack detection and localization	Beam	Treed Gaussian process	Mode shape curvatures
Barzegar et al. [70]	High-rate structural health monitoring	Hypersonic vehicles	RNNs	Experimental high-rate dynamic data
Eltouny and Liang [71]	Detecting and localizing damage in large-scale structures	10-story, 10- bay, numerical structure	RNNs	Structural responses
Dorafshan et al. [72]	Image-based crack detection	Concrete structure	CNNs	Images of damaged and undamaged concrete
Wang et al. [73]	Asphalt pavement cracking recognition	Pavement	CNNs	Manually processed images
Kim and Cho [74]	Automated detection technique for crack morphology	Concrete surface	CNNs	Images collected from the Internet

represent physical infrastructure in virtual form, enabling them to be updated in near real-time as new data are collected, integrate feedback, evaluate asset risks, and predict performance based on what-if scenarios. Originally introduced by Grieves and Vickers in 2003 [77], the concept only gained widespread popularity with Glaessgen and Stargel's famous article [78] in 2012. DTs are comprised of four main components.

1. Simulations for modeling the physics of the system.
2. Knowledge from specialists and prior experiences regarding the environment and the product variables.
3. Data of the physical twin.
4. Connectivity for linking the other elements and giving DT a means of evolving with updates in information.

In the context of SHM, DT is widely used in bridge SHM. A summary of a case study on applying DT to the SHM of a bridge is presented below.

Ye et al. [79] investigated four domains of research.

1. **Real-time data management using building information modeling (BIM):** Using SHM datasets, real-time sensor data and associated bridge behavior were visualized within a dynamic BIM environment. This operation can assist in the identification of anomalies in the data due to various factors, such as anomalous behavior and faulty sensors. Moreover, it can be utilized in visualizing strain/stress distribution and evolution.
2. **Physics-based approaches:** A 3D FE model was generated, and its predictions were verified by strain measurements using fiber optic sensors both spatially and temporally. This information can be used in establishing a performance baseline for achieving data-informed asset management and long-term condition monitoring.
3. **Data-driven approaches:** A refined moving average algorithm was first applied to sensor data from an 80-sensor fiber Bragg grating network for removing both short-term and long-term environmental trends and extracting individual train-passage events. Linear dynamic (statistical) modeling was subsequently used to detect anomalies in the data and for short-term forecasting. In addition, a streaming model was developed that was capable of real-time updating.
4. **Data-centric engineering (DCE) approaches:** Lastly, data-driven and physics-based approaches were integrated into a DCE framework. In a DCE approach, the physics-based model was integrated and balanced with monitoring data from various sources to model the bridge response. This approach was utilized to combine multiple data sources, minimize systematic errors, and quantify underlying uncertainties.

5.7 Conclusion

Modern SHM systems are continuously generating vast amounts of data. Challenges associated with the storage, processing, analysis, and interpretation of such amounts of data led to the implementation of AI techniques, especially ML algorithms, in SHM systems. Over the past decade, AI techniques have evolved rapidly, with unprecedented capabilities in data mining and analytics. These features are particularly attractive for SHM and have resulted in considerable attention and advancements in AI-based SHM technology. Despite its advantages, it has certain limitations as well.

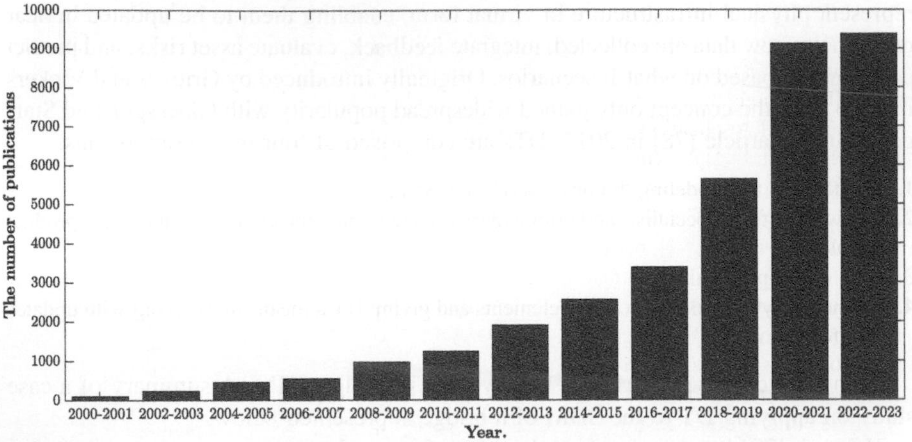

Figure 5.5 The number of articles in SHM based on AI
All data have been derived from Google Scholar in October 2022.

Among these limitations are the need for greater interpretability and transparency, bias and discrimination, over- and underfitting, adequate computational resources, causation, ethical considerations, and poor data quality. A thorough understanding of these limitations is necessary for the development and effective use of ML algorithms.

As it can be seen in Fig. 5.5, since 2000, there has been a significant increase in research on AI technologies for system monitoring. Nevertheless, there are a number of challenges remaining faced by AI-based SHM, including the selection and configuration of the most appropriate AI technique based on the problem, data generation for adequate data training, suitable software and hardware for efficient system operation, and the time-consuming and unaffordable operation of AI-based SHM systems.

References

[1] U. Dackermann, J. Li, R. Rijal, K. Crews, A dynamic-based method for the assessment of connection systems of timber composite structures, Construction and Building Materials 102 (2016) 999−1008.

[2] U. Dackermann, W.A. Smith, M.M. Alamdari, J. Li, R.B. Randall, Cepstrum-based damage identification in structures with progressive damage, Structural Health Monitoring 18 (1) (2019) 87−102.

[3] C.R. Farrar, K. Worden, An introduction to structural health monitoring, Philosophical Transactions of the Royal Society A: Mathematical, Physical & Engineering Sciences 365 (1851) (2007) 303−315.

[4] S. Hassani, M. Mousavi, A.H. Gandomi, A Hilbert transform sensitivity-based model-updating method for damage detection of structures with closely-spaced eigenvalues, Engineering Structures 268 (2022) 114761.

[5] U. Dackermann, Y. Yu, E. Niederleithinger, J. Li, H. Wiggenhauser, Condition assessment of foundation piles and utility poles based on guided wave propagation using a network of tactile transducers and support vector machines, Sensors 17 (12) (2017) 2938.

[6] H.-P. Chen, Structural Health Monitoring of Large Civil Engineering Structures, 2018.
[7] S. Hassani, M. Mousavi, A.H. Gandomi, Structural health monitoring in composite structures: a comprehensive review, Sensors 22 (1) (2021) 153.
[8] M.P. Limongelli, M. Çelebi, Seismic Structural Health Monitoring: From Theory to Successful Applications, Springer, 2019.
[9] S. Hassani, U. Dackermann, A systematic review of advanced sensor technologies for non-destructive testing and structural health monitoring, Sensors 23 (4) (2023) 2204.
[10] C. Cremona, J. Santos, Structural health monitoring as a big-data problem, Structural Engineering International 28 (3) (2018) 243−254.
[11] A. Keshmiry, S. Hassani, M. Mousavi, U. Dackermann, Effects of environmental and operational conditions on structural health monitoring and non-destructive testing: a systematic review, Buildings 13 (4) (2023) 918.
[12] Model-based vs. data-driven approaches for anomaly detection in structural health monitoring: A case study, in: A. Moallemi, A. Burrello, D. Brunelli, L. Benini (Eds.), 2021 IEEE International Instrumentation and Measurement Technology Conference (I2MTC), IEEE, 2021.
[13] J.P. Amezquita-Sanchez, H. Adeli, Synchrosqueezed wavelet transform-fractality model for locating, detecting, and quantifying damage in smart highrise building structures, Smart Materials and Structures 24 (6) (2015) 065034.
[14] S. Hassani, M. Mousavi, Z. Sharif-Khodaei, Smart Bridge Monitoring. The Rise of Smart Cities, Elsevier, 2022, pp. 343−372.
[15] T.-H. Yi, H.-N. Li, X.-D. Zhang, Sensor placement on Canton Tower for health monitoring using asynchronous-climb monkey algorithm, Smart Materials and Structures 21 (12) (2012) 125023.
[16] S. Bhalla, Y. Yang, J. Zhao, C. Soh, Structural health monitoring of underground facilities−technological issues and challenges, Tunnelling and Underground Space Technology 20 (5) (2005) 487−500.
[17] S. Oliveira, A. Alegre, Seismic and structural health monitoring of dams in Portugal, in: Seismic Structural Health Monitoring, Springer, 2019, pp. 87−113.
[18] X. Li, W. Yu, S. Villegas, Structural health monitoring of building structures with online data mining methods, IEEE Systems Journal 10 (3) (2015) 1291−1300.
[19] J. Zhou, H. Xiao, W. Jiang, W. Bai, G. Liu, Automatic subway tunnel displacement monitoring using robotic total station, Measurement 151 (2020) 107251.
[20] A. Entezami, H. Sarmadi, B. Behkamal, S. Mariani, Big data analytics and structural health monitoring: a statistical pattern recognition-based approach, Sensors 20 (8) (2020) 2328.
[21] S. Kiakojoori, K. Khorasani, Dynamic neural networks for gas turbine engine degradation prediction, health monitoring and prognosis, Neural Computing & Applications 27 (8) (2016) 2157−2192.
[22] H. Pragalath, S. Seshathiri, H. Rathod, B. Esakki, R. Gupta, Deterioration Assessment of Infrastructure Using Fuzzy Logic and Image Processing Algorithm, 2018.
[23] S. Wang, S.A. Zargar, F.-G. Yuan, Augmented reality for enhanced visual inspection through knowledge-based deep learning, Structural Health Monitoring 20 (1) (2021) 426−442.
[24] B.T. Svendsen, G.T. Frøseth, O. Øiseth, A. Rønnquist, A data-based structural health monitoring approach for damage detection in steel bridges using experimental data, Journal of Civil Structural Health Monitoring 12 (1) (2022) 101−115.
[25] A. Sofi, J.J. Regita, B. Rane, H.H. Lau, Structural health monitoring using wireless smart sensor network−an overview, Mechanical Systems and Signal Processing 163 (2022) 108113.

[26] Thermal imaging for monitoring rolling element bearings, in: R. Schulz, S. Verstockt, J. Vermeiren, M. Loccufier, K. Stockman, S. Van Hoecke (Eds.), 12th International Conference on Quantitative InfraRed Thermography (QIRT 2014), 2014. http://www.ndt.net.

[27] J. Tang, S. Soua, C. Mares, T.-H. Gan, An experimental study of acoustic emission methodology for in service condition monitoring of wind turbine blades, Renewable Energy 99 (2016) 170−179.

[28] S.P. Weaver, C.D. Hein, T.R. Simpson, J.W. Evans, I. Castro-Arellano, Ultrasonic acoustic deterrents significantly reduce bat fatalities at wind turbines, Global Ecology and Conservation 24 (2020) e01099.

[29] J. Lian, Y. Zhao, C. Lian, H. Wang, X. Dong, Q. Jiang, et al., Application of an eddy current-tuned mass damper to vibration mitigation of offshore wind turbines, Energies 11 (12) (2018) 3319.

[30] V.B. Sharma, K. Singh, R. Gupta, A. Joshi, R. Dubey, V. Gupta, et al., Review of structural health monitoring techniques in pipeline and wind turbine industries, Applied System Innovation 4 (3) (2021) 59.

[31] F.P. García Márquez, P.J. Bernalte Sánchez, I. Segovia Ramírez, Acoustic inspection system with unmanned aerial vehicles for wind turbines structure health monitoring, Structural Health Monitoring 21 (2) (2022) 485−500.

[32] A. Sivasuriyan, D.S. Vijayan, R. Munusami, P. Devarajan, Health assessment of dams under various environmental conditions using structural health monitoring techniques: a state-of-art review, Environmental Science and Pollution Research (2021) 1−12.

[33] Y. Xu, H. Huang, Y. Li, J. Zhou, X. Lu, Y. Wang, A three-stage online anomaly identification model for monitoring data in dams, Structural Health Monitoring 21 (3) (2022) 1183−1206.

[34] A. Sivasuriyan, D.S. Vijayan, W. Górski, Ł. Wodzyński, M.D. Vaverková, E. Koda, Practical implementation of structural health monitoring in multi-story buildings, Buildings 11 (6) (2021) 263.

[35] J. Wang, Y. Fu, X. Yang, An integrated system for building structural health monitoring and early warning based on an Internet of things approach, International Journal of Distributed Sensor Networks 13 (1) (2017), 1550147716689101.

[36] S. Hassani, M. Mousavi, A.H. Gandomi, A mode shape sensitivity-based method for damage detection of structures with closely-spaced eigenvalues, Measurement 190 (2022) 110644.

[37] C. Huyen, Designing Machine Learning Systems, O'Reilly Media, 2022.

[38] Z.H. Zhou, S. Liu, Machine Learning, Springer Nature Singapore, 2021.

[39] O. Avci, O. Abdeljaber, S. Kiranyaz, M. Hussein, M. Gabbouj, D.J. Inman, A review of vibration-based damage detection in civil structures: from traditional methods to Machine Learning and Deep Learning applications, Mechanical Systems and Signal Processing 147 (2021) 107077.

[40] Deep learning-based unsupervised methods for real-time condition monitoring of structures: a state-of-the-art survey, in: M. Mousavi, A.H.H. Gandomi (Eds.), Health Monitoring of Structural and Biological Systems XVI, SPIE, 2022.

[41] M.M. Al Rahhal, Y. Bazi, H. AlHichri, N. Alajlan, F. Melgani, R.R. Yager, Deep learning approach for active classification of electrocardiogram signals, Information Sciences 345 (2016) 340−354.

[42] Y. Zhang, X. Sun, K.J. Loh, W. Su, Z. Xue, X. Zhao, Autonomous bolt loosening detection using deep learning, Structural Health Monitoring 19 (1) (2020) 105−122.

[43] Z.-Q. Zhao, P. Zheng, S.-T. Xu, X. Wu, Object detection with deep learning: a review, IEEE Transactions on Neural Networks and Learning Systems 30 (11) (2019) 3212−3232.

[44] L. Perez, J. Wang, The effectiveness of data augmentation in image classification using deep learning, arXiv preprint arXiv:171204621 (2017).

[45] Y. Gal, Z. Ghahramani, A theoretically grounded application of dropout in recurrent neural networks, Advances in Neural Information Processing Systems 29 (2016).
[46] K. Gopalakrishnan, S.K. Khaitan, A. Choudhary, A. Agrawal, Deep convolutional neural networks with transfer learning for computer vision-based data-driven pavement distress detection, Construction and Building Materials 157 (2017) 322–330.
[47] C.V. Dung, H. Sekiya, S. Hirano, T. Okatani, C. Miki, A vision-based method for crack detection in gusset plate welded joints of steel bridges using deep convolutional neural networks, Automation in Construction 102 (2019) 217–229.
[48] Q. Han, Y. Pan, D. Yang, Y. Xu, CNN-based bolt loosening identification framework for prefabricated large-span spatial structures, Journal of Civil Structural Health Monitoring 12 (3) (2022) 517–536.
[49] Structural defects classification and detection using convolutional neural network (CNN): a review, in: P. Arafin, A. Billah (Eds.), Canadian Society of Civil Engineering Annual Conference, Springer, 2023.
[50] S. Sony, K. Dunphy, A. Sadhu, M. Capretz, A systematic review of convolutional neural network-based structural condition assessment techniques, Engineering Structures 226 (2021) 111347.
[51] Developing situation and research advances of structural damage detection using BP network, in: J. Fan, Y. Yuan, X. Cao (Eds.), 2016 4th International Conference on Machinery, Materials and Computing Technology, Atlantis Press, 2016.
[52] M. Mehrjoo, N. Khaji, H. Moharrami, A. Bahreininejad, Damage detection of truss bridge joints using Artificial Neural Networks, Expert Systems with Applications 35 (3) (2008) 1122–1131.
[53] Application of machine learning method in bridge health monitoring, in: J. Peng, S. Zhang, D. Peng, K. Liang (Eds.), 2017 Second International Conference on Reliability Systems Engineering (ICRSE), IEEE, 2017.
[54] A study on the application of GA-BP neural network in the bridge reliability assessment, in: J. Yang, J. Zhou, F. Wang (Eds.), 2008 International Conference on Computational Intelligence and Security, IEEE, 2008.
[55] M.P. González, J.L. Zapico, Seismic damage identification in buildings using neural networks and modal data, Computers & Structures 86 (3–5) (2008) 416–426.
[56] C.-M. Chang, T.-K. Lin, C.-W. Chang, Applications of neural network models for structural health monitoring based on derived modal properties, Measurement 129 (2018) 457–470.
[57] S. Soyoz, M.Q. Feng, Long-term monitoring and identification of bridge structural parameters, Computer-Aided Civil and Infrastructure Engineering 24 (2) (2009) 82–92.
[58] F. Salazar, M. Toledo, E. Oñate, R. Morán, An empirical comparison of machine learning techniques for dam behaviour modelling, Structural Safety 56 (2015) 9–17.
[59] The performance of the neural networks to model some response parameters of a buttress dam to environment actions, in: A. Popovici, C. Ilinca, T. Ayvaz (Eds.), Proceedings of the 9th ICOLD European Club Symposium, Venice, Italy, 2013.
[60] V. Ranković, A. Novaković, N. Grujović, D. Divac, N. Milivojević, Predicting piezometric water level in dams via artificial neural networks, Neural Computing & Applications 24 (5) (2014) 1115–1121.
[61] Vibration-based support vector machine for structural health monitoring, in: H. Pan, M. Azimi, G. Gui, F. Yan, Z. Lin (Eds.), International Conference on Experimental Vibration Analysis for Civil Engineering Structures, Springer, 2017.
[62] Timber Health Monitoring using piezoelectric sensor and machine learning, in: R. Oiwa, T. Ito, T. Kawahara (Eds.), 2017 IEEE International Conference on Computational

Intelligence and Virtual Environments for Measurement Systems and Applications (CIVEMSA), IEEE, 2017.

[63] M.M. Alamdari, N. Khoa, P. Runcie, J. Li, S. Mustapha, Characterization of gradually evolving structural deterioration in jack arch bridges using support vector machine, in: Maintenance, Monitoring, Safety, Risk and Resilience of Bridges and Bridge Networks, CRC Press, 2016, p. 555.

[64] K. Maes, L. Van Meerbeeck, E. Reynders, G. Lombaert, Validation of vibration-based structural health monitoring on retrofitted railway bridge KW51, Mechanical Systems and Signal Processing 165 (2022) 108380.

[65] D.A. Tibaduiza, L.E. Mujica, J. Rodellar, Damage classification in structural health monitoring using principal component analysis and self-organizing maps, Structural Control and Health Monitoring 20 (10) (2013) 1303−1316.

[66] A. Ibrahim, A. Eltawil, Y. Na, S. El-Tawil, A machine learning approach for structural health monitoring using noisy data sets, IEEE Transactions on Automation Science and Engineering 17 (2) (2019) 900−908.

[67] M.J. Hasan, J.-M. Kim, Fault detection of a spherical tank using a genetic algorithm-based hybrid feature pool and k-nearest neighbor algorithm, Energies 12 (6) (2019) 991.

[68] K. Worden, E. Cross, On switching response surface models, with applications to the structural health monitoring of bridges, Mechanical Systems and Signal Processing 98 (2018) 139−156.

[69] M. Civera, C. Surace, K. Worden, Detection of cracks in beams using treed Gaussian processes, Structural Health Monitoring & Damage Detection 7 (2017) 85−97. Springer.

[70] V. Barzegar, S. Laflamme, C. Hu, J. Dodson, Ensemble of recurrent neural networks with long short-term memory cells for high-rate structural health monitoring, Mechanical Systems and Signal Processing 164 (2022) 108201.

[71] K.A. Eltouny, X. Liang, Large—scale structural health monitoring using composite recurrent neural networks and grid environments, Computer—Aided Civil and Infrastructure Engineering (2022).

[72] S. Dorafshan, R.J. Thomas, M. Maguire, Comparison of deep convolutional neural networks and edge detectors for image-based crack detection in concrete, Construction and Building Materials 186 (2018) 1031−1045.

[73] K.C. Wang, A. Zhang, J.Q. Li, Y. Fei, C. Chen, B. Li, Deep learning for asphalt pavement cracking recognition using convolutional neural network, Airfield and Highway Pavements (2017) 166−177.

[74] B. Kim, S. Cho, Automated vision-based detection of cracks on concrete surfaces using a deep learning technique, Sensors 18 (10) (2018) 3452.

[75] M. Batty, Digital Twins, Sage Publications, London, England, 2018, pp. 817−820.

[76] Z. Zhang, C. Sun, Structural damage identification via physics-guided machine learning: a methodology integrating pattern recognition with finite element model updating, Structural Health Monitoring 20 (4) (2021) 1675−1688.

[77] M. Grieves, J. Vickers, Digital Twin: Mitigating Unpredictable, Undesirable Emergent Behavior in Complex Systems. Transdisciplinary Perspectives on Complex Systems, Springer, 2017, pp. 85−113.

[78] The digital twin paradigm for future NASA and US Air Force vehicles, in: E. Glaessgen, D. Stargel (Eds.), 53rd AIAA/ASME/ASCE/AHS/ASC Structures, Structural Dynamics and Materials Conference 20th AIAA/ASME/AHS Adaptive Structures Conference 14th AIAA, 2012.

[79] C. Ye, L. Butler, B. Calka, M. Iangurazov, Q. Lu, A. Gregory, et al., A Digital Twin of Bridges for Structural Health Monitoring, 2019.

Application of ensemble learning in rock mass rating for tunnel construction

6

Denise-Penelope N. Kontoni [1,2], Mahdi Shadabfar [3] and Jiayao Chen [4]
[1]Department of Civil Engineering, School of Engineering, University of the Peloponnese, Patras, Greece; [2]School of Science and Technology, Hellenic Open University, Patras, Greece; [3]Center for Infrastructure Sustainability and Resilience Research, Department of Civil Engineering, Sharif University of Technology, Tehran, Iran; [4]Key Laboratory for Urban Underground Engineering of the Education Ministry, Beijing Jiaotong University, Beijing, China

6.1 Introduction

Relating the most fundamental geological features of rock masses and their quality together, the engineering rock mass classification system is the pillar of various experimental design approaches commonly utilized in rock engineering [1−3]. In these systems, engineering observations, measurements, and experiences are used to determine how the said features correlate with certain qualities of rock masses, which can prove very useful for geologists, designers, and engineers [4,5]. However, since geological structures are difficult to predict and complex in nature, it could be challenging to build a suitable classification system of rock mass for tunneling in rock masses, especially considering the frequency with which such a system must be utilized in a rock tunneling process [6,7]. Such a system should have the capability of presenting a precise evaluation of the rock mass qualities in the tunnel face prior to the subsequent tunneling phase based on the data compiled from multiple sources.

There are a variety of different empirical and theoretical models for rock classification, which include rock mass rating (RMR) [8,9], rock mass index (RMi) [10], the rock mass quality Q-system (Q) [11,12], the geological strength index (GSI) [13], etc. Falling into multifactor quantitative classification schemes, empirical rock classification models are extensively used for design and planning purposes in engineering projects. RMR is one such classification system widely employed in rock tunnel engineering [4,14,15]. RMR is a hundred-mark system functioning based on six main factors, namely joint and bedding spacing (JS), joint condition (JC), groundwater condition (GW), discontinuity orientation relative to the opening axis, and uniaxial compressive strength (UCS), which are assigned a rating representing how they impact the stability of the tunnel [16]. RMR provides a convenient yet effective way for evaluating the rock mass quality wherein the tunnel is bored. Many engineers use RMR to determine what excavation methods and support systems would be most suitable for a

Artificial Intelligence Applications for Sustainable Construction. https://doi.org/10.1016/B978-0-443-13191-2.00007-9
Copyright © 2024 Elsevier Ltd. All rights are reserved, including those for text and data mining, AI training, and similar technologies.

given tunneling project. Nevertheless, such formulaic/framed classification models tend to face the following challenges:

(1) In the case of missing data, the remedial methods are limited, and the computational power is weakened significantly.
(2) Major differences exist between the parameters in different experimental models, and the assessment metrics used in a real project may not perfectly fit a specific system.
(3) The acquisition of mechanical parameters is costly and time-consuming, which poses a challenge for tunneling projects with high-frequency construction stages.

It is, therefore, crucial to find an implicit method that can not only summarize the critical parameters from the real tunnel project but also benchmark the empirical formulas with the predicted values [17]. Traditional field survey methods like compass measurement could be not only time-consuming but also unsafe and challenging to perform as needed, and they may also inject their errors and biases into the data [18–20]. Hence, the application of emergent noncontact approaches, such as light detection and ranging (LIDAR) and photogrammetry, is preferred to extract the required data from rock masses (water inflow, weak interlayers, rock structure, discontinuity, etc.) [6,7,21,22]. Using these technologies, engineers can perform the needed in-situ data collection with greater efficiency and convenience. Thus, noncontact approaches are more recommended than contact ones [23].

Besides empirical and theoretical methods, researchers have shown increasing interest in using machine learning (ML) in rock mass classification [24–27]. Flexible in nature, it has been confirmed that these methods have high effectiveness in solving various engineering problems, especially complex nonlinear ones. Most of these methods function as a "black box", finding a correlation between model predictions and actual measurements. However, despite the advancement of such ML methods, empirical models are still more popular because of how easily they can be implemented [28,29]. Moreover, ML models are susceptible to poor training due to unavoidable noise in the available data, which can negatively affect their prediction accuracy [30]. There are three types of challenges in working with single-hypothesis ML algorithms, namely representation challenge, computational challenge, and statistical challenge, which are associated with a high level of bias, high computational variance, and high variance, respectively [31]. It is possible to overcome such challenges using ensemble learning algorithms. This approach involves building a set of hypotheses and then putting them to a "vote" for each new prediction rather than searching for the best hypothesis for explicated data [32]. Ensemble models provide more accurate predictions but require more computation [33]. In any case, these methods are gradually becoming popular choices for benchmarking formulaic systems.

In ML, using hyperparameters, the model architecture and its learning process are shaped [34]. The learning performance of an ML model tends to depend strongly on how its hyperparameters are chosen and set. Thus, an automatic hyper-parameter optimization process can significantly facilitate the development of an ML model in the following ways: (1) reducing how much human effort needs to be put into deploying the model, which is essential for practical purposes because it might be necessary to use various hyper-parameter settings for different datasets [35]; (2) making the ML model perform better by giving it the best hyper-parameter setting (the one most suitable for the specified objectives) [36]; (3) improving the reproducibility of the results by giving a firm structure to how hyper-parameters are set (compared to manual setting, which depends on human choices), which also allows for a fair assessment of how different models perform in a specific application compared to each other [37]. Thus, a hyper-parameter optimization approach might critically play a role in the prediction performance of the ML-based model.

According to the above, this work aimed at providing a systematic approach for using ML algorithms in rock mass quality assessments. The subjects covered in this study include how the ML algorithm is selected, how optimum hyper-parameters are determined, and how model robustness is improved, plus a sensitivity analysis. Since it has been shown that integrated classification and regression trees (CARTs) perform ideally while maintaining reasonable interpretability in such applications, this study uses a tree-based ML model, that is, the gradient boosting regression tree (GBRT), for its tests. To make the process less dependent on brute-force search and rule of thumb, the ML model is tuned using a Bayesian hyper-parameter optimization approach [34] called tree-structured Parzen estimator (TPE) [38]. When using TPE, the prediction error of k-fold cross-validation (CV) sets is utilized as its fitness function. A sensitivity analysis is also performed to determine how the model performs when given unobserved data and which parameters are more essential for its outputs.

The organization of the study is as follows: the presented methodology (i.e., GBRT) is first explained in Section 6.2, and then it proceeds to describe the TPE hyper-parameter optimization method. Section 6.3 describes how the multisource database is compiled from the data obtained from noncontact and contact techniques. In Section 6.4, the method is compared with different algorithms with regard to prediction performance, and a comprehensive analysis of test results is provided. The method results for a case study of a real tunnel are discussed in Section 6.5. Finally, Section 6.6 presents the conclusion and discusses promising avenues for future work.

6.2 Methodology

To determine how the ML method performs in providing RMR predictions based on various factors, an integrated boosting approach is applied [33,39,40]. Classification and regression trees (CARTs) used as a root algorithm in the development of the ensemble model, that is, the gradient boosting regression tree (GBRT). In order to boost, the position of the CARTs is set in such a way that each CART reduces the error of the one that precedes it, or, in other words, learns from the previous trees to achieve a lower residual error.

GBRT enhances the traditional CART by boosting [41,42]. This approach revolves around merging several weaker models into a single stronger consensus model instead of developing a new optimized model (Fig. 6.1). Using GBRT, a prediction p is produced based on a model (F) trained by least-squares regression (t). Then, the following forward-stagewise strategy is used to improve the model with an additive estimator:

$$F_t(x) = F_{t-1}(x) + \gamma_t h_t(x) \tag{6.1}$$

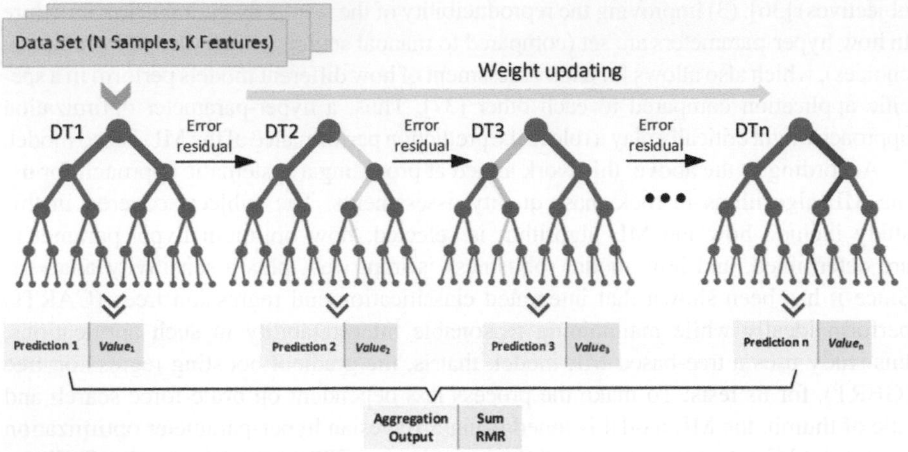

Figure 6.1 Flowchart of GBRT model.

where γ_t denotes the learning rate, $h_t(x)$ denotes the weak learners, and F_t is the GBRT model containing t CARTs. In each boosting iteration t, a new CART will be added to the GBRT model. Thus, the following relation is used to calculate h:

$$F_t(x) = F_{t-1}(x) + h(x) = p \tag{6.2}$$

$$h(x) = p - F_{t-1}(x) \tag{6.3}$$

Next, the predicted p is approximated using a weighted sum function *h(x)* as given by:

$$p = \sum_{i=1}^{T} \gamma_i h_i(x) + const \tag{6.4}$$

In this equation, T denotes the total tree number. Taking an empirical risk minimization approach, the GBRT model employs each CART variable that has the greatest reductive impact on loss function L (minimizing the final L). Thus, the function $F(x)$ that minimizes L is approximated as follows:

$$F_0(x) = \underset{\gamma}{\operatorname{argmin}} \sum_{i=1}^{n} L(p_i, \gamma) \tag{6.5}$$

Here, $F_0(x)$ denotes a constant function.

$$F_t(x) = F_{t-1}(x) + \underset{h \in H}{\operatorname{argmin}} \sum_{i=1}^{n} L(y_i, F_{t-1}(x_i) + (x_i)) \tag{6.6}$$

The minimization process is conducted by the steepest descent method as shown below:

$$F_t(x) = F_{t-1}(x) - \gamma_t \sum_{i=1}^{n} \nabla_{F_{t-1}} L(y_i, F_{t-1}(x_i)) \tag{6.7}$$

$$\gamma_m = \underset{\gamma}{\operatorname{argmin}} \sum_{i=1}^{n} L\left(p_i, F_{t-1}(x_i) - \gamma \frac{\partial L(p_i, F_{t-1}(x_i))}{\partial F_{t-1}(x_i)}\right) \tag{6.8}$$

It has been shown that the use of relatively small learning rates results in the model gaining higher generalizability compared to boosting without shrinkage [41]. However, this imposes extra computational cost because it requires more CARTs and makes it necessary to fine-tune other parameters that affect the structure and complexity of the model, like the maximum number of splits and the number of trees T.

6.2.1 Hyper-parameter tuning

In this chapter, a Bayesian optimization (BO) approach called TPE is used for automatic hyper-parameter optimization [38]. Before presenting mathematical formulations, it should be explained that authors assume it is more costly to evaluate the function $f : X \to R$ than to conduct a TPE approximation for function f. The place where the true f is predicted is considered to be the point x^* where the model M is maximized. The criterion used for performance comparisons is the expected improvement (EI), which is chosen since it is intuitive and effective for this purpose [38]. Given by Eq. (6.1), this parameter is the expectation that $f(x)$ will improve beyond a specific level, like y^* under model M:

$$EI(x) = \int_{-\infty}^{+\infty} \max((y^*, y), 0).P_M((y|x))dy \tag{6.9}$$

The TPE approach can be used to define model M easily and search for the local hyper-parameter configuration. For the approximation of model M, $P_M(x|y)$ and $P_M(x)$ are indirectly referred to as express TPE models $P_M(y|x)$, where $P_M(x|y)$ is defined as follows:

$$P_M(x|y) = \begin{cases} g(x) & \text{if } y \geq y^* \\ l(x) & \text{if } y < y^* \end{cases} \tag{6.10}$$

In this equation, $l(x)$ *and* $g(x)$ represent the density estimates obtained based on observations and show the probability of the presence of the hyper-parameter set x in the equivalent groups prespecified by y^*. To simplify the EI optimization in the TPE algorithm, $P_M(x|y)$ is parametrized as $P_M(y)P_M(x|y)$.

$$EI_{y^*}(x) = \int_{-\infty}^{y^*} (y^* - y).P_M(y|x)dy = \int_{-\infty}^{y^*} \frac{P_M(x|y).P_M(y)}{P_M(x)} dy \tag{6.11}$$

By defining $\gamma = p(y < y^*)$ and $p(x) = f_R P_M(x|y).P_M(y)dy = \gamma l(x) + (1 - \gamma)g(x)$.

Then, Eq. (6.13) is calculated as follows:

$$\int_{-\infty}^{y^*} (y^* - y).P_M(y|x)dy = l(x) \int_{-\infty}^{y^*} (y^* - y).P_M(y)dy = \gamma.y^* l(x) - l(x) \int_{-\infty}^{y^*} P_M(y)dy \tag{6.12}$$

The value of $EI_{y^*}(x)$ is finally computed by Eq. (6.14):

$$EI_{y^*}(x) = \frac{\gamma.y^*\, l(x) - l(x) \int_{-\infty}^{y^*} P_M(y)dy}{\gamma(x) + (1 - \gamma).g(x)} \propto \left(\gamma + \frac{g(x)}{l(x)}(1 - \gamma)\right)^{-1} \tag{6.13}$$

As shown in Eq. (6.14), with the tree structure consisting of *g* and *l*, it is easy to obtain a great number of candidates based on *l* and determine how they perform based on *g(x)/l(x)*. The algorithm in each iteration returns the candidate *x** with the greatest EI. Fig. 6.2 shows the process of optimization with TPE.

6.2.2 Evaluation measures

For a reliable evaluation of predictive models, it is important to conduct this evaluation with appropriate performance measures. For a comprehensive performance evaluation of the ML models, evaluations of this study are conducted with four widely used performance measures, i.e., root mean square error (RMSE), coefficient of determination (R), mean absolute percentage error (MAPE), and mean absolute error (MAE). The formulas of these four indicators are given below [43,44]:

$$RMSE = \sqrt{\frac{1}{n}\sum_{i=1}^{n}(m_i - p_i)^2} \tag{6.14}$$

$$R = \sqrt{1 - \frac{\sum (p_i - \widehat{p_i})^2}{\sum (p_i - \overline{p_i})^2}} \tag{6.15}$$

$$MAPE = \frac{1}{n}\sum_{i=1}^{n}\left|\frac{m_i - p_i}{m_i}\right| \times 100\% \tag{6.16}$$

$$MAE = \frac{1}{n}\sum_{i=1}^{n}|m_i - p_i| \tag{6.17}$$

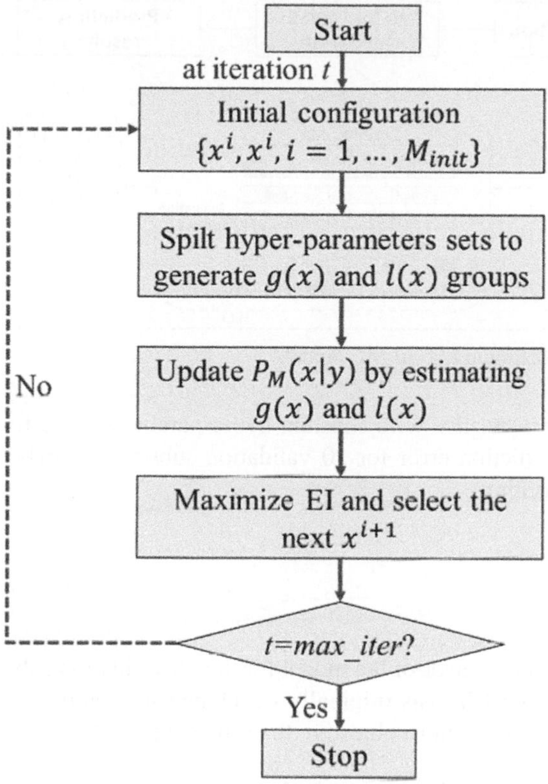

Figure 6.2 Flowchart of TPE optimization procedure.

where *m* denotes the measured RMR value, *p* denotes the predicted value of RMR, and *n* denotes the total number of datasets. Lower values of MAE, MAPE, and RMSE indicate better performance of a model. The values of these three indicators range from 0 to ∞. The fourth measure, R, describes the proportion of the variance in the results of the ML models, in the range from 0 to 1.

6.2.3 K-fold cross-validation

ML models are developed in three main stages: training, validation, and testing. Widely used in model validation [45], the K-fold CV algorithm involves random partitioning of the original dataset into *k* subsamples with equal size and using the k-1 subsets for training and the rest for validation, as shown in Fig. 6.3. This CV procedure will be repeated k times (hence, k folds), and each time one of the k subsamples will be taken for the validation process. Since previous studies have recommended setting k to 10 [46], this study also uses a 10-fold CV.

In each iteration of the 10-fold CV, the target ML models with fixed hyper-parameters are trained 10 times with nine randomly chosen subsets of the data, and the remaining subset is used for validation. For each training-followed-by-testing iteration, the CV error is calculated, and on such a basis, the related parameters are fine-

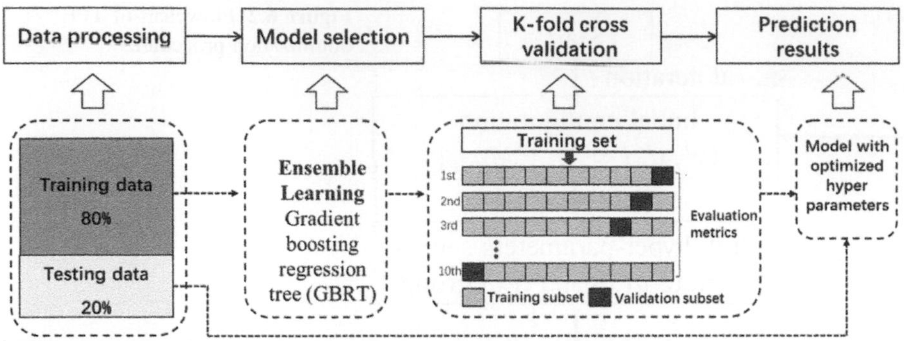

Figure 6.3 Flowchart of determining optimum hybrid ML model.

tuned gradually. The ML model performance with constant hyper-parameters is estimated in terms of their average prediction error for 10 validation subsets—namely, the fitness function in the TPE algorithm,

$$\text{Fitness} = \frac{1}{10}\sum_{i=1}^{10} MAE_i \tag{6.18}$$

In this formula, MAE_i is the prediction error of the model for the ith validation subset. It should be noted that the k-fold CV was originally developed for improving robustness and avoiding overfitting and is more objective than the simple CV.

6.3 Establishment of a multisource database

To demonstrate that the geostatistical approach is applicable in the prediction of RMR for rock tunnel engineering, we used the project of the Mengzi-Pingbian freeway in Yunnan, China, as a case study. This project is situated on the periphery of the South China plate to the north of the southern part of the Honghe fault and the east of the southern part of the Xiaojiang fault. Stratum lithology and landforms govern the hydrogeological conditions at the tunnels of this project, which are mostly of tectonic denudation type on terrain with tall mountains and great height differences. The project location is mostly engulfed in a Cambrian slate with low water permeability that serves as an aquiclude.

6.3.1 Database compilation

Developing a cost-effective and reliable approach for automatic RMR prediction during dynamic drilling and blasting processes is a troublesome yet important task. It is a complex process from multisource data acquisition and feature extraction to comprehensive analysis. The simplified flowchart of the process of building the multisource dataset is illustrated in Fig. 6.4. It is possible to divide this process into three stages:

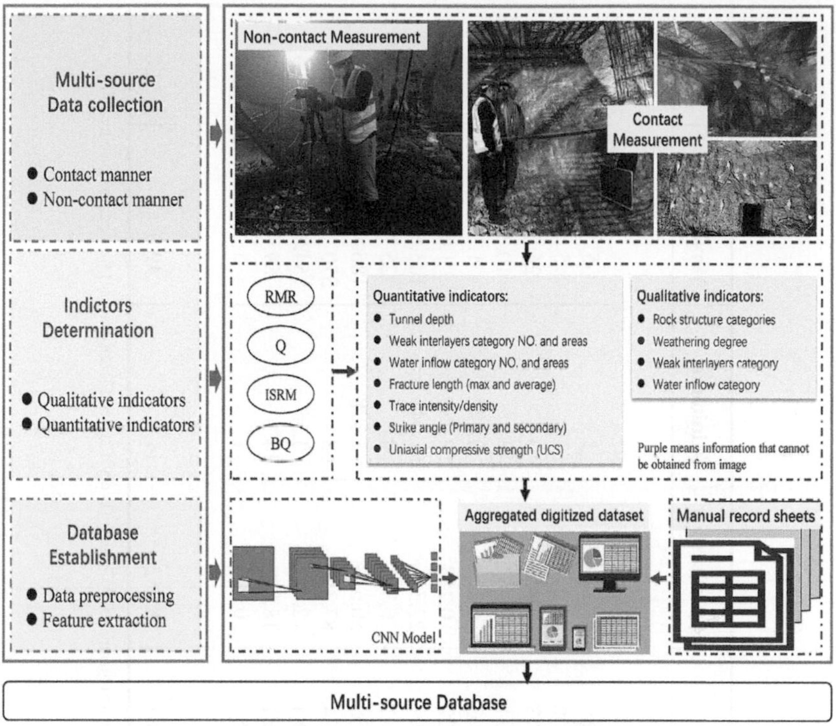

Figure 6.4 Compilation of multisource datasets in a rock tunnel.

multisource data gathering, indicator determination, and compilation of the database [47]. Here, a convolutional neural network (CNN)-based approach is used on the rock tunnel face images for a quantitative segmentation of the main features, including the water inflow (area and statistical number), weak interlayers (area and statistical number), fracture length (mean and max), density and intensity of discontinuity trace, and joint dip angle. A CNN is also used for automatic classification based on the categories of weak interlayers, water inflow, and rock surface structure. Other information, including UCS, weathering degree, and depth of tunnel for each tunnel face, was manually gained from the strata. Overall, multisource heterogeneous data was compiled from tunnel faces for RMR prediction. Because of space limitations, the database extraction process will not be described in detail.

6.3.2 Data description

To build a comprehensive database, the basic information of over 133 tunnel faces was included in the database for RMR prediction. Additional data is continuously collected from the site to increase the total scale. As explained, 13 input variables with the greatest impact on rock mass quality were chosen for RMR prediction. The min, max, and mean values of these inputs, as well as the model outputs, are presented in Table 6.1,

Table 6.1 Definition and distribution of variables.

Symbol	Variable	Unit	Mean	Standard	Minimum	Maximum	Median
Input							
RSC	Category of rock structure	–	2.489	1.229	1	5	2
GWC	Category of groundwater	–	2.406	1.2	1	5	2
GWA	Area of groundwater	m^2	8.493	7.826	0	26.593	7.824
WIA	Area of weak interlayer	m^2	3.521	2.752	0	10.33	3.051
TL_a	Length of trace (average)	m	0.503	0.23	0.022	1.122	0.522
TL_{max}	Length of trace (maximum)	m	0.922	0.258	0.283	1.637	0.928
TD	Trace density	–	40.309	23.346	1.046	103.54	36.605
TI	Trace intensity	–	7.801	6.383	0	25.101	6.275
PTA	Primary trace angle	°	97.705	50.636	3.378	186.849	113.309
H	Depth of tunnel	m	221.624	87.352	68	370	217
UCS	Uniaxial compressive strength	MPa	20.084	12.247	3.55	54.89	17.5
TS	Tunnel strike	°	61.481	18.827	14.1	89.9	63.6
WD	Weathering degree	–	3.15	0.996	1	5	3
Output							
RMR	Rock mass rating	–	32.083	8.952	11.0	49.0	32.0

Table explanations: RSC, type of structure: (1) block, (2) layered, (3) mosaic, (4) fragmentation, (5) granular. GWC, state: (1) dry, (2) wet, (3) dripping, (4) flowing, (5) gushing. WD, state: (1) unweathered, (2) slightly, (3) moderately, (4) highly weathered, (5) decomposed.

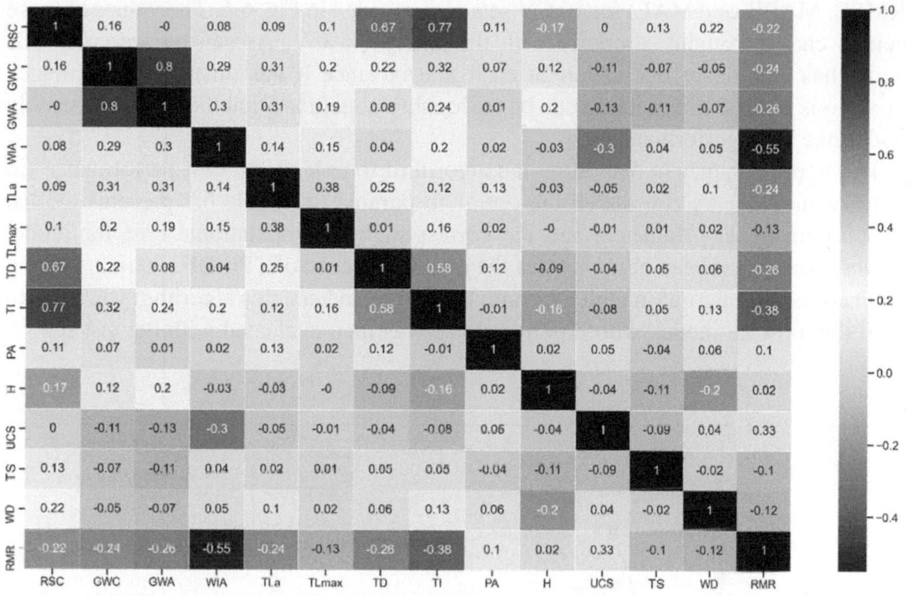

Figure 6.5 Correlation matrix of multisource variables for prediction of RMR.

along with some other information. Fig. 6.5 shows how input variables and outputs are correlated in a matrix analysis chart.

The resulting database was used to build the TPE-GBRT ML-based model. In this process, the database was randomly classified into two subsets, one for training (93/133) and the other for testing (40/133). First, using the training dataset, the optimal set of model parameters was chosen according to the average mean-square error delivered by them. After determining the best parameters, the trained ML model on this dataset and then used the test dataset to determine the ML model's performance with regard to the evaluated evaluation metrics on an independent part of the database.

6.4 Test results

To simulate real-world conditions, the available data were classified into two subsets: testing and training. To check the robustness of the models, the tests were repeated 10 times, and the test results (values of evaluation metrics) were averaged to weaken the impact of how the data were partitioned. Using the hyperopt package [48] and the sklearn package [49] on Python 3.7, implemented the ML models and the Bayesian hyper-parameter optimization (TPE) method, and they were executed on a workstation with a 3.70 GHz Intel i7-8700K CPU, 32 GB of RAM, and the Windows 10 operating system.

6.4.1 Hyper-parameter tuning

After using the training set for hyper-parameter tuning, testing was conducted by using CV on the chosen validation set. The values of the three performance measures

(RMSE, MAPE, and MAE) for 10 CV sets are provided in Fig. 6.6. The values of these metrics change roughly similarly with the change in CV. As can be seen, the ML model has significant fluctuation in each matrix since it has unstable performance for various RMRs. In general, it can be reasonably concluded that the hybrid ensemble model has ideal performance.

The hyper-parameters that allow the algorithm to exhibit the best performance (in terms of metrics) are considered the optimum parameters. Table 6.2 presents the list of optimum hyper-parameters and the corresponding computational cost for model convergence. It has been observed that the operation speed of TPE-RF is quick. Having obtained the ML algorithm with optimum hyper-parameters, this algorithm was used to build the ultimate model, whose performance was then evaluated with the test set.

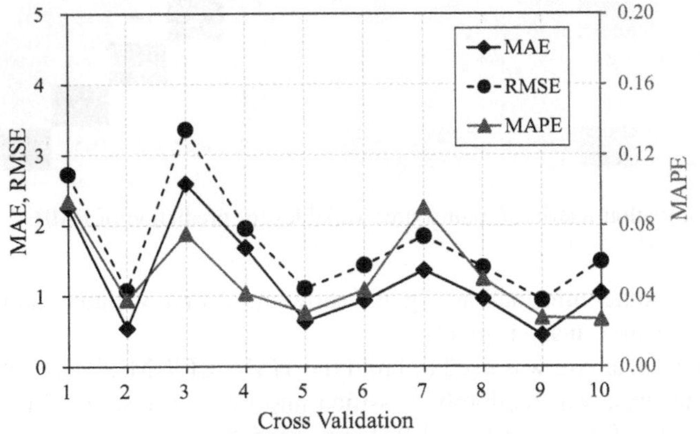

Figure 6.6 Evaluation metrics for TPE-GBRT algorithm.

Table 6.2 TPE-optimized hyper-parameter of the proposed ML algorithm.

Algorithm: GBRT					
Hyper-parameters to be optimized	**Space**	**Step**	**Optimal value**	**Fitness**	**Tuning time (s)**
Tree numbers of (tree_num)	[10, 1000]	10	370	0.256	1083
Maximum features (max_features)	[1, 15]	1	3		
Maximum depth (max_depth)	[5, 50]	1	19		
Minimum number of samples required to split an internal node (min_samples_split)	[2, 11]	1	5		
Maximum number of samples required to split an internal node (max_samples_split)	[1, 11]	1	2		
Rate of learning (learning_rate)	[0.001, 0.1]	–	0.09		

6.4.2 Performance analysis

The results obtained on the test and training sets with the tuned hyper-parameters are plotted in Fig. 6.7. Please note that all of these predictions were obtained with only a several-second-long execution of the trained ML model. Table 6.3 reports the values of evaluation metrics (RMSE, MAPE, MAE, and R) for testing and training sets, which provides a comprehensive performance evaluation of the models. For the training set, the predictions obtained with TPE-GBRT exhibit a high constituency with the measurements with R ~ 1. The good prediction performance of TPE-GBRT is also reflected in the low values of RMSE, MAPE, and MAE. Generally, the greater performance of the hybrid ensemble ML model is shown by the results. For the testing set, the ensemble algorithm has somewhat higher R and lower MAE, RMSE, and MAPE, but generally exhibits excellent performance, reflected in how almost all data points have fallen on a straight line with a slope of 1 (P = M line). Notably, the TPE tuning algorithm has enhanced the overall performance of the ML model in both the training and testing processes.

For further analysis, the RMR prediction performance of the hybrid TPE-GBRT model was plotted on the Taylor diagram [50] displayed in Fig. 6.8. This diagram visualizes the RMSE, R, and SD of the predictions in both the testing and training stages. Here, using SD and R, the comparability of predictions and measurements is quantified, and using RMSE, it is described how much predictions deviate from measurements. In Fig. 6.8, the points labeled "actual" are the measured RMR values. Ideally, R and RMSE should be very close to 1 and 0, and the SD of the predictions should be close to that of the measurements. As can be seen, TPE-GBRT exhibits great performance for both testing and training data.

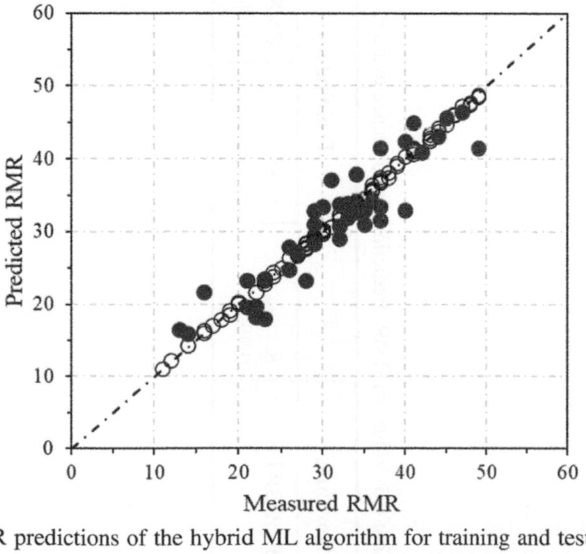

Figure 6.7 RMR predictions of the hybrid ML algorithm for training and testing sets.

Table 6.3 Values of RMR prediction performance evaluation metrics for training and test sets.

Algorithms	Training set				Test set			
	R	MAE	RMSE	MAPE	R	MAE	RMSE	MAPE
TPE-GBRT	0.9995	0.2557	0.3947	0.85%	0.9220	2.7520	3.3676	9.67%

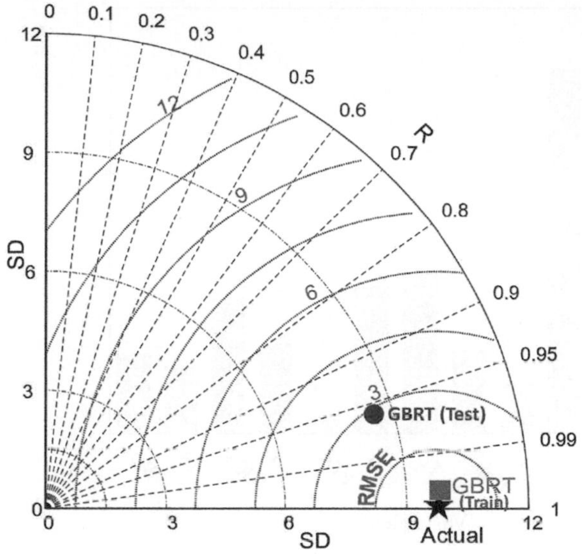

Figure 6.8 Taylor diagrams of the models for testing and training sets.

6.5 Discussion

6.5.1 Sensitivity analysis

Sensitivity analyses can be conducted by using the variable importance measure (VIM) as a measure of how much different inputs contribute to the output of a model [51]. Here, the importance of variables in TPE-GBRT is specified by investigating how much the outputs are influenced by the Gini index of the input variables' variations. Then, these Gini indices are normalized using the Sklearn package to reach the final results in relation to the importance of variables. It should be noted that the variables that trigger larger changes in VIM tend to impose a greater impact on the model's prediction capacity [52].

The VIM values obtained for different input variables are illustrated in Fig. 6.9. As this figure shows, the variables with the greatest level of sensitivity for RMR prediction are WIA, GWA, and RSC, with VIM values of 0.2053, 0.1157, and 0.1471. RMR predictions are also significantly influenced by GWC and USC, with VIM values of 0.0681 and 0.0759. While TD, TI, and RMR are well correlated, the same cannot be said for the other three discontinuity trace parameters, namely PTA, TL_{max}, and TL_a. In summary, TPE-GBRT can produce accurate RMR predictions, and the significance of variables in this model matches what is expected in practice. In future works, the authors will try to improve the TPE-GBRT algorithm for better RMR prediction and also compile a larger database from different rock tunneling projects to improve its validity, reliability, and applicability in different situations.

6.5.2 Comparison of hyper-parameter optimization approaches

Optimization techniques are broadly divided into two categories: approximation approaches and exact approaches. Exact optimization approaches involve using brute force

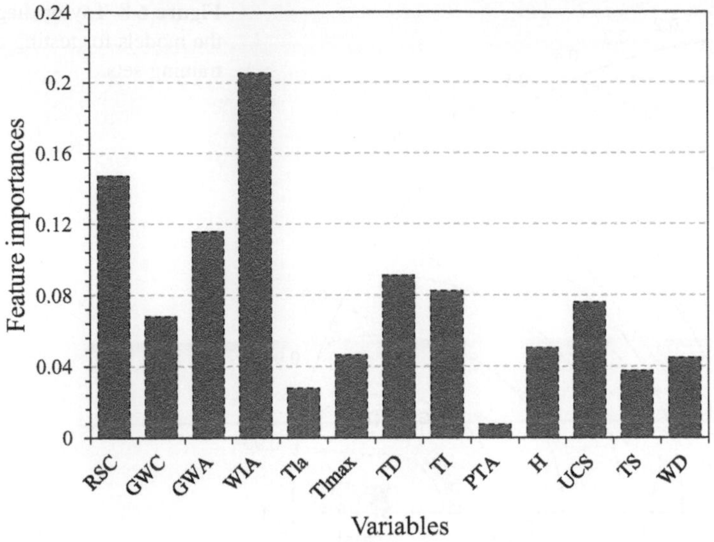

Figure 6.9 Variable importance of RMR by the use of TPE-GBRT.

methods, such as grid search (GS) and random search (RS) [38,53]. Approximation-type optimization approaches are themselves subdivided into two subgroups: model-based and metaheuristic-based methods [54]. The first subgroup includes a wide range of evolutionary algorithms, e.g., genetic algorithm (GA) and particle swarm optimization, which can be employed to optimize hyper-parameters [55,56]. In the second subgroup, the performance of hyper-parameters is approximated by sequentially built models based on historical measurements, and the best hyper-parameters are selected accordingly.

For a comprehensive analysis, the hyper-parameters of the GBRT model were also tuned with three commonly used optimization approaches, namely RS, GS, and GA. In this part of the study, the model performance is measured in terms of the convergent fitness of the optimization methods and their training duration. For the elimination of randomness, tests were repeated 10 times, followed by averaging the results, providing the findings presented in Fig. 6.10. As this figure shows, TPE has significantly better computation time than the other three optimization methods. GS has worse fitness performance than the other three methods, but there is not much difference between these three methods in this respect. Overall, model-based methods outperform benchmark methods in terms of computational costs. This is perhaps because of the ability of approximation-type optimization algorithms to quickly find near-optimal solutions for nondeterministic polynomial hard problems, whereas exact optimization algorithms tend to perform poorly on these problems since their execution time increases exponentially with the problem size [54].

6.6 Case study

To investigate the generalization ability of TPE-GBRT in a real tunnel, it was used in a case study of the Lengquan tunnel, which is a highly fractured rock tunnel in the same project (Fig. 6.11a). This case study was conducted by using the images taken from

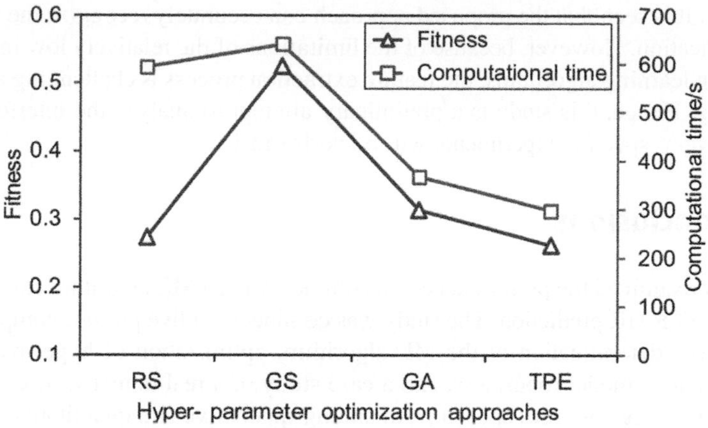

Figure 6.10 Computational time and fitness value for comparative optimization strategies.

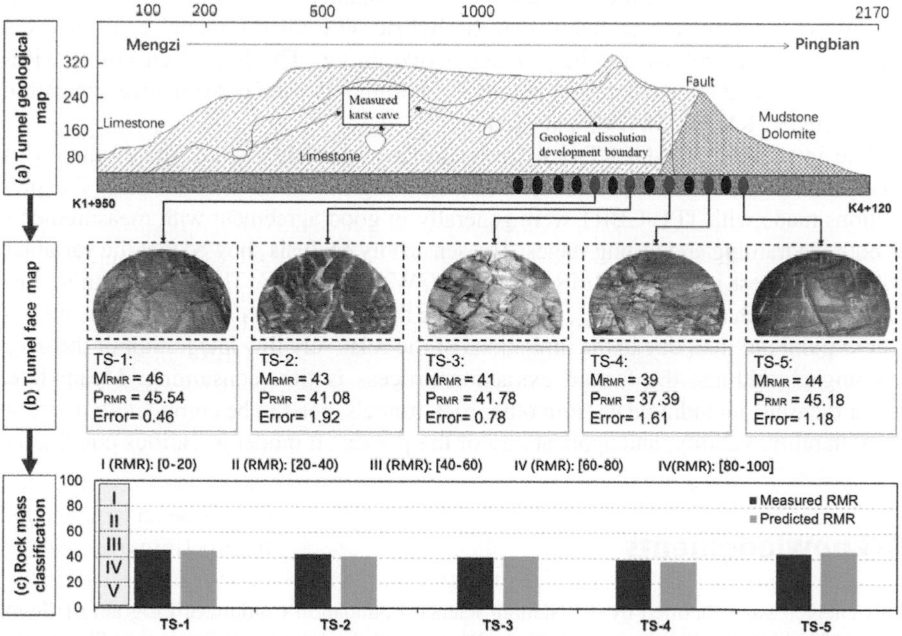

Figure 6.11 Variable importance for RMR prediction with TPE-GBRT represented in three distinct stages: (a) Tunnel geological map; (b) Tunnel face map; (c) Rock mass classification.

five tunnel face sections (TS-1, TS-2, TS-3, TS-4, and TS-5), which are illustrated in Fig. 6.11b, for RMR prediction. The aforementioned 13 quantitative and qualitative indicators were extracted by deep learning-based algorithms and in-field measurements. The predicted RMR values of the applied five tunnel sections are obtained by using the pretrained TPE-GBRT algorithm. The measured and predicted RMR values are compared in Fig. 6.11c, which shows great agreement between them. Meanwhile, based on the RMR classification criterion [9], the corresponding classification results of rock mass quality, which are converted from the RMR value, are shown in

Fig. 6.11c. It is seen that the proposed approach can accurately recognize the rock tunnel classification. However, because of the limitations of the relatively low integration of the deep learning algorithms, the feature extraction process is challenging and time-consuming. Hence, this study is a preliminary attempt to analyze the interfering factors, and more specific experiments will be performed.

6.7 Conclusions

This study examined the performance of a commonly used ML algorithm called GBRT in the area of RMR prediction. The study was conducted in five phases: compilation of the database, determination of the ML algorithm, optimization of hyper-parameters, improvement of model robustness, and a case study of a real tunnel section.

13 multisource variables chosen from among qualitative and quantitative rock features as input in the compilation of a database for RMR prediction. Using a BO method known as TPE, the hyper-parameters of the presented TPE-GBRT algorithm were optimized. For TPE, the mean prediction error obtained in 10-fold CV was used as the fitness function. It was found that the hybrid TPE-GBRT model can make the ML model more robust and help it avoid overfitting. The hybrid ensemble TPE-GBRT was found to be competent in terms of commonly used performance evaluation indicators (R, RMSE, MAPE, and MAE).

The intelligent TPE-GBRT-based model can accurately capture the evolution of measurements with a low prediction error and a high correlation coefficient. The predictions made with TPE-GBRT were generally in good agreement with measurements in both the training and testing stages. The sensitivity analysis showed that the variables with the greatest impact on RMR are WIA, GWA, and RSC. The discontinuity trace parameters, such as PTA, TL_{max}, and TL_a, have ignorable scores. Moreover, it is important to point out that, due to the limitations of the relatively low integration of the deep learning algorithms, the feature extraction process is time-consuming. Meanwhile, larger datasets, including data from other rock tunnels, need to be compiled to improve the reliability, validity, and applicability of the presented model in various conditions.

Acknowledgments

This chapter was supported by the Natural Science Foundation Committee Program of China (No. 51778474), the Key Innovation Team Program of the Innovation Promotion Plan of the Ministry of Science and Technology of China (No. 2016RA4059), as well as the Science and Technology Project of Yunnan Provincial Transportation Department (No. 25 of 2018).

References

[1] B. Singh, R.K. Goel, Rock Mass Classification: A Practical Approach in Civil Engineering, Elsevier, 1999.

[2] S. Tzamos, A. Sofianos, Extending the Q system's prediction of support in tunnels employing fuzzy logic and extra parameters, International Journal of Rock Mechanics and Mining Sciences 43 (2006) 938−949, https://doi.org/10.1016/j.ijrmms.2006.02.002.
[3] S. Tzamos, A. Sofianos, A correlation of four rock mass classification systems through their fabric indices, International Journal of Rock Mechanics and Mining Sciences 44 (2007) 477−495, https://doi.org/10.1016/j.ijrmms.2006.08.003.
[4] C.O. Aksoy, M. Geniş, G.U. Aldaş, V. Özacar, S.C. Özer, Ö. Yılmaz, A comparative study of the determination of rock mass deformation modulus by using different empirical approaches, Engineering Geology 131 (2012) 19−28, https://doi.org/10.1016/j.enggeo.2012.01.009.
[5] H. Rehman, A.M. Naji, J.-j. Kim, H. Yoo, Extension of tunneling quality index and rock mass rating systems for tunnel support design through back calculations in highly stressed jointed rock mass: an empirical approach based on tunneling data from Himalaya, Tunnelling and Underground Space Technology 85 (2019) 29−42, https://doi.org/10.1016/j.tust.2018.11.050.
[6] J. Chen, T. Yang, D. Zhang, H. Huang, Y. Tian, Deep learning based classification of rock structure of tunnel face, Geoscience Frontiers 12 (1) (2021) 395−404, https://doi.org/10.1016/j.gsf.2020.04.003.
[7] X. Li, J. Chen, H. Zhu, A new method for automated discontinuity trace mapping on rock mass 3D surface model, Computers and Geosciences 89 (2016) 118−131, https://doi.org/10.1016/j.cageo.2015.12.010.
[8] Z.T. Bieniawski, Engineering classification of jointed rock masses, Transactions of the South African Institution of Civil Engineers 15 (1973) 335−344.
[9] Z. Bieniawski, The rock mass rating (RMR) system (geomechanics classification) in engineering practice, STP48461S, in: L. Kirkaldie (Ed.), Rock Classification Systems for Engineering Purposes, ASTM International, 1988.
[10] A. Palmstrøm, Characterizing rock masses by the RMi for use in practical rock engineering: Part 1: the development of the Rock Mass index (RMi), Tunnelling and Underground Space Technology 11 (1996) 175−188, https://doi.org/10.1016/0886-7798(96)00015-6.
[11] N. Barton, Suggested methods for the quantitative description of discontinuities in rock masses, ISRM, International Journal of Rock Mechanics and Mining Sciences and Geomechanics Abstracts 15 (1978) 319−368, https://doi.org/10.1016/0148-9062(78)91472-9.
[12] N. Barton, R. Lien, J. Lunde, Engineering classification of rock masses for the design of tunnel support, Rock Mechanics 6 (1974) 189−236, https://doi.org/10.1007/BF01239496.
[13] E. Hoek, Strength of rock and rock masses, ISRM News Journal 2 (1994) 4−16.
[14] C. Aksoy, Review of rock mass rating classification: historical developments, applications, and restrictions, Journal of Mining Science 44 (2008) 51−63, https://doi.org/10.1007/s10913-008-0005-2.
[15] H. Rehman, W. Ali, A.M. Naji, J.-J. Kim, R.A. Abdullah, H.-k. Yoo, Review of rock-mass rating and tunneling quality index systems for tunnel design: development, refinement, application and limitation, Applied Sciences 8 (2018) 1250, https://doi.org/10.3390/app8081250.
[16] A. Palmström, Combining the RMR, Q, and RMi classification systems, Tunnelling and Underground Space Technology 24 (2009) 491−492, https://doi.org/10.1016/j.tust.2008.12.002.
[17] T.-T. Wang, T.-H. Huang, Anisotropic deformation of a circular tunnel excavated in a rock mass containing sets of ubiquitous joints: theory analysis and numerical modeling, Rock Mechanics and Rock Engineering 47 (2014) 643−657, https://doi.org/10.1007/s00603-013-0405-8.

[18] Y. Ge, H. Tang, D. Xia, L. Wang, B. Zhao, J.W. Teaway, H. Chen, T. Zhou, Automated measurements of discontinuity geometric properties from a 3D-point cloud based on a modified region growing algorithm, Engineering Geology 242 (2018) 44–54, https://doi.org/10.1016/j.enggeo.2018.05.007.
[19] M. Vöge, M.J. Lato, M.S. Diederichs, Automated rockmass discontinuity mapping from 3-dimensional surface data, Engineering Geology 164 (2013) 155–162, https://doi.org/10.1016/j.enggeo.2013.07.008.
[20] K. Zhang, W. Wu, H. Zhu, L. Zhang, X. Li, H. Zhang, A modified method of discontinuity trace mapping using three-dimensional point clouds of rock mass surfaces, Journal of Rock Mechanics and Geotechnical Engineering 12 (3) (2020) 571–586, https://doi.org/10.1016/j.jrmge.2019.10.006.
[21] J. Chen, D. Zhang, H. Huang, M. Shadabfar, M. Zhou, T. Yang, Image-based segmentation and quantification of weak interlayers in rock tunnel face via deep learning, Automation in Construction 120 (2020b) 103371, https://doi.org/10.1016/j.autcon.2020.103371.
[22] G. Gigli, N. Casagli, Semi-automatic extraction of rock mass structural data from high resolution LIDAR point clouds, International Journal of Rock Mechanics and Mining Sciences 48 (2011) 187–198, https://doi.org/10.1016/j.ijrmms.2010.11.009.
[23] R. García-Luna, S. Senent, R. Jurado-Piña, R. Jimenez, Structure from Motion photogrammetry to characterize underground rock masses: experiences from two real tunnels, Tunnelling and Underground Space Technology 83 (2019) 262–273, https://doi.org/10.1016/j.tust.2018.09.026.
[24] M. Galende-Hernández, M. Menéndez, M. Fuente, G. Sainz-Palmero, Monitor-While-Drilling-based estimation of rock mass rating with computational intelligence: the case of tunnel excavation front, Automation in Construction 93 (2018) 325–338, https://doi.org/10.1016/j.autcon.2018.05.019.
[25] D.J. Lary, A.H. Alavi, A.H. Gandomi, A.L. Walker, Machine learning in geosciences and remote sensing, Geoscience Frontiers 7 (2016) 3–10, https://doi.org/10.1016/j.gsf.2015.07.003.
[26] H.N. Rad, Z. Jalali, H. Jalalifar, Prediction of rock mass rating system based on continuous functions using Chaos–ANFIS model, International Journal of Rock Mechanics and Mining Sciences 73 (2015) 1–9, https://doi.org/10.1016/j.ijrmms.2014.10.004.
[27] Z. Şen, B.H. Sadagah, Modified rock mass classification system by continuous rating, Engineering Geology 67 (2003) 269–280, https://doi.org/10.1016/S0013-7952(02)00185-0.
[28] S. Finlay, Multiple classifier architectures and their application to credit risk assessment, European Journal of Operational Research 210 (2011) 368–378, https://doi.org/10.1016/j.ejor.2010.09.029.
[29] S. Lessmann, B. Baesens, H.-V. Seow, L.C. Thomas, Benchmarking state-of-the-art classification algorithms for credit scoring: an update of research, European Journal of Operational Research 247 (2015) 124–136, https://doi.org/10.1016/j.ejor.2015.05.030.
[30] H.I. Erdal, Two-level and hybrid ensembles of decision trees for high performance concrete compressive strength prediction, Engineering Applications of Artificial Intelligence 26 (2013) 1689–1697, https://doi.org/10.1016/j.engappai.2013.03.014.
[31] T.G. Dietterich, Machine Learning for Sequential Data: A Review, Joint IAPR International Workshops on Statistical Techniques in Pattern Recognition (SPR) and Structural and Syntactic Pattern Recognition (SSPR), Springer, 2002, pp. 15–30, https://doi.org/10.1007/3-540-70659-3_2.
[32] I. Brown, C. Mues, An experimental comparison of classification algorithms for imbalanced credit scoring data sets, Expert Systems with Applications 39 (2012) 3446–3453, https://doi.org/10.1016/j.eswa.2011.09.033.

[33] J. Elith, J. Leathwick, T. Hastie, A working guide to boosted regression trees, Journal of Animal Ecology 77 (2008) 802−813, https://doi.org/10.1111/j.1365-2656.2008.01390.x.
[34] M. Feurer, J.T. Springenberg, F. Hutter, Initializing Bayesian hyperparameter optimization via meta-learning, in: Twenty-Ninth AAAI Conference on Artificial Intelligence, 2015, https://doi.org/10.1609/aaai.v29i1.9354.
[35] R. Kohavi, G.H. John, Automatic parameter selection by minimizing estimated error, in: Machine Learning Proceedings 1995, Proceedings of the Twelfth International Conference on Machine Learning, Tahoe City, California, July 9−12, 1995, Elsevier, 1995, pp. 304−312, https://doi.org/10.1016/B978-1-55860-377-6.50045-1.
[36] J. Snoek, H. Larochelle, R.P. Adams, Practical Bayesian optimization of machine learning algorithms, in: Advances in Neural Information Processing Systems 25: 26th Annual Conference on Neural Information Processing Systems 2012, NIPS 2012, Lake Tahoe, NV, United States, December 3−6 2012, 2012, pp. 2951−2959 (Advances in Neural Information Processing Systems; vol. 4).
[37] J. Bergstra, D. Yamins, D. Cox, Making a science of model search: hyperparameter optimization in hundreds of dimensions for vision architectures, in: International Conference on Machine Learning, PMLR, 2013a, pp. 115−123.
[38] J. Bergstra, Y. Bengio, Random search for hyper-parameter optimization, Journal of Machine Learning Research 13 (2012) 281−305.
[39] G. James, D. Witten, T. Hastie, R. Tibshirani, Tree-based methods, in: An Introduction to Statistical Learning, Springer, New York, NY, 2021, pp. 327−365, https://doi.org/10.1007/978-1-0716-1418-1_8. Springer Texts in Statistics.
[40] G. Wang, J. Ma, L. Huang, K. Xu, Two credit scoring models based on dual strategy ensemble trees, Knowledge-Based Systems 26 (2012) 61−68, https://doi.org/10.1016/j.knosys.2011.06.020.
[41] J.H. Friedman, Greedy function approximation: a gradient boosting machine, Annals of Statistics (2001) 1189−1232.
[42] J.H. Friedman, Stochastic gradient boosting, Computational Statistics and Data Analysis 38 (2002) 367−378, https://doi.org/10.1016/S0167-9473(01)00065-2.
[43] R. Boddy, G.L. Smith, Statistical Methods in Practice: For Scientists and Technologists, Wiley Online Library, 2009.
[44] R.J. Hyndman, A.B. Koehler, Another look at measures of forecast accuracy, International Journal of Forecasting 22 (2006) 679−688, https://doi.org/10.1016/j.ijforecast.2006.03.001.
[45] M. Stone, Cross-validatory choice and assessment of statistical predictions, Journal of the Royal Statistical Society: Series B 36 (1974) 111−133, https://doi.org/10.1111/j.2517-6161.1976.tb01573.x.
[46] R. Kohavi, A Study of Cross-Validation and Bootstrap for Accuracy Estimation and Model Selection, Ijcai, Montreal, Canada, 1995, pp. 1137−1143, in: https://www.ijcai.org/Proceedings/95-2/Papers/016.pdf.
[47] M. Zhou, J. Chen, H. Huang, D. Zhang, S. Zhao, M. Shadabfar, Multi-source data driven method for assessing the rock mass quality of a NATM tunnel face via hybrid ensemble learning models, International Journal of Rock Mechanics and Mining Sciences 147 (2021) 104914, https://doi.org/10.1016/j.ijrmms.2021.104914.
[48] J. Bergstra, D. Yamins, D.D. Cox, Hyperopt: a python library for optimizing the hyperparameters of machine learning algorithms, in: Proceedings of the 12th Python in Science Conference, 2013, pp. 13−19, https://doi.org/10.25080/Majora-8b375195-003.
[49] B. Komer, J. Bergstra, C. Eliasmith, Hyperopt-sklearn: automatic hyperparameter configuration for Scikit-learn, ICML workshop on AutoML, in: Proceedings of the 13th Python in Science Conference (SCIPY 2014), 2014, pp. 32−37, https://doi.org/10.25080/Majora-14bd3278-006.

[50] K.E. Taylor, Summarizing multiple aspects of model performance in a single diagram, Journal of Geophysical Research: Atmospheres 106 (2001) 7183–7192, https://doi.org/10.1029/2000JD900719.

[51] A. Hapfelmeier, T. Hothorn, K. Ulm, C. Strobl, A new variable importance measure for random forests with missing data, Statistics and Computing 24 (2014) 21–34, https://doi.org/10.1007/s11222-012-9349-1.

[52] L. Breiman, J. Friedman, C.J. Stone, R.A. Olshen, Classification and Regression Trees, CRC Press, Boca Raton, 1984.

[53] C.-L. Huang, M.-C. Chen, C.-J. Wang, Credit scoring with a data mining approach based on support vector machines, Expert Systems with Applications 33 (2007) 847–856, https://doi.org/10.1016/j.eswa.2006.07.007.

[54] Y. Xia, C. Liu, Y. Li, N. Liu, A boosted decision tree approach using Bayesian hyper-parameter optimization for credit scoring, Expert Systems with Applications 78 (2017) 225–241, https://doi.org/10.1016/j.eswa.2017.02.017.

[55] J. Kennedy, R. Eberhart, Particle swarm optimization, in: Proceedings of ICNN'95-International Conference on Neural Networks, IEEE, 1995, pp. 1942–1948, https://doi.org/10.1109/ICNN.1995.488968.

[56] D. Whitley, A genetic algorithm tutorial, Statistics and Computing 4 (1994) 65–85, https://doi.org/10.1007/BF00175354.

AI-based framework for Construction 4.0: A case study for structural health monitoring

7

Anas Alsharo, Samer Gowid, Mohammed Al Sageer, Amr Mohamed and Khalid Kamal Naji
College of Engineering, Qatar University, Doha, Qatar

7.1 Introduction

The construction sector significantly contributes to the economic, social, and sustainable growth of any country. Globally, the value of the construction industry in 2021 was estimated at 7.28 trillion USD and was predicted to have a compound average growth rate of 7.3% and to reach 14.41 trillion USD by 2030 [1]. The estimated values reflect the significance and size of the construction industry. Hence, it is vital to equip the sector with the latest state-of-the-art physical and digital technology trends to promote a positive shift in the efficient management of construction projects. However, the construction industry is known to lag behind in this respect, being slow to adapt to new technologies. This gradual shift toward digital transformation can be linked to the different roles and objectives of the various participants in a construction project, in addition to the unique nature of the construction industry [2]. The lack of an efficient adoption of digitization technology within this sector can explain the challenges associated with construction projects, which include time overruns, poor or inadequate collaboration, inefficient resource management, and adverse effects on cost [3].

The flaws of the architecture, engineering, and construction (AEC) industry have led researchers, industry professionals, and software developers to collaborate to create innovative solutions to bridge these gaps. This collaboration resulted in the birth of building information modeling (BIM) [4]. The concept of BIM was first presented in 1992 by van Nederveen and Tolman, who proposed an approach to building different models that represent different aspects of building elements to describe the building in a simple yet efficient way [5]. BIM is an innovative concept that is built on the creation of single or multiple digital (virtual) three-dimensional (3D) models of the structure. These virtual models contain precise construction information on the structure's geometry, material details, element interactions, costs, and code requirements. These models work as a backbone for the construction process throughout its various phases [6]. The various stakeholders involved in a construction project can access the same virtual environment used to build the digital 3D model. Edits can be applied to the digital model, and all the teams involved in the construction project can access and explore these edits and update the construction plan accordingly.

Artificial Intelligence Applications for Sustainable Construction. https://doi.org/10.1016/B978-0-443-13191-2.00013-4
Copyright © 2024 Elsevier Ltd. All rights are reserved, including those for text and data mining, AI training, and similar technologies.

The enhanced collaboration in BIM-enabled projects has led to a significant shift in the running of construction projects, with fewer physical and logistical clashes, more efficient cost estimation and planning, fewer time overruns, and adequate resource planning [6,7]. Despite the fact that BIM leverages digital technology to boost efficiency in managing construction projects throughout its full service life, it lacks the ability to provide real-time information on the as-built construction. In other words, these models are passive, providing detailed construction information on the preconstruction stage of the structure in a collaborative environment but lacking the real-time connection with the structure necessary to update the modeled construction information [8]. This drawback led to the development of integration schemes between physical structures equipped with sensors and digital models to provide data sources and feedback mediums that together support and update digital models to provide precise, up-to-date, and informative representations of actual structures. These are known as digital twin (DT) schemes [9].

A DT, as the name implies, aims to build a digital replica (twin) of a structure to allow different participants in the construction process to obtain detailed information and feedback from a system that is equipped with sensors. This converts the passive BIM model into an active one [10]. The DT concept, in terms of building an active digital replica of a structure, was first proposed by aerospace engineers and researchers [11] with the goal of building a link between a digital finite element model and a physical model of an aerospace structure. The driving force for adopting this approach in their study was the fact that digital models built at the design stage, which are computationally expensive, fail to give an accurate representation of the actual behavior and state of a structure under variable operational conditions. Accordingly, as the efficient operation and maintenance of any civil engineering structure cannot be satisfied by relying solely on BIM models, the exhaustive task of acquiring precise and up-to-date data from the structure should also be included to allow data-driven decisions to be taken with regard to maintenance or rehabilitation [12]. In the context of facility management (FM), DTs are considered to be creative schemes that overcome the drawbacks of reliance on off-line design information and construction documents or notes by providing an accessible and easy-to-explore twin of the structure [9]. Hence, moving toward active digitization would result in significant savings in time, construction and operation costs, and effort in the life cycle management of buildings and structures [12].

Researchers in the field of AEC have conducted intensive research on the implementation and use cases of a vast array of technology trends in the construction industry. The literature is rich with studies exploring the use cases, benefits, and challenges of BIM and DT in the construction industry [4–17], sensing technologies and state-of-the-art in structural health monitoring (SHM) [18–23], additive manufacturing and its influences in the construction industry [24–28], artificial intelligence (AI) implementation and innovation in the AEC sector [29–34], and employing blockchain in construction [13,14]. Other studies have explored the use of technology tools such as unmanned aerial vehicles (UAVs) and their role in boosting monitoring and data collection in construction projects [16,35–38], as well as the use of virtual reality (VR), augmented reality (AR), and extended reality (XR) in supporting the design, construction, and monitoring of structures [21,39–42]. However, the architecture, engineering, construction, and operation (AECO) sector works with products that are complex and contain large-scale variability in size, cost, required resources, manufacturing and operational procedures, and other related factors [12]. This calls for integrated, fully connected (Fc) schemes that leverage the use of up-to-date technology tools and advancements at all stages of the construction process to achieve high-quality deliverables and outcomes rather than segmented or inefficient utilization of technology. The need for such an approach has influenced the shift toward a broad scheme referred to as Construction 4.0 [43].

Construction 4.0 is a technology-driven holistic framework that aims to employ disruptive technologies at all phases of construction and in all sectors of the AECO industry to build an enhanced support system that ensures the achievement of the highest level of desired outcomes in terms of sustainability, quality, time, cost, and environmental impact [44]. Construction 4.0 was first proposed as a concept and plan by the Roland Berger company in 2016 [45]. In the magazine Think:Act, the authors mentioned that although 93% of construction industry participants believed in the high impact of technology implementation on the industry, less than 6% of construction companies employed the relevant digital tools in their work. It was thus of fundamental importance to move toward a plan that bridged this gap in the market. It is worth mentioning that Construction 4.0 did not evolve separately as a concept but was an extension of the famous and well-known Industry 4.0 framework [43]. As Construction 4.0 is an adoption of the theoretical base of Industry 4.0, a discussion of Industry 4.0's definition, characteristics, benefits, challenges, driving technologies, and use cases is useful to give a clear idea of Construction 4.0.

Industry 4.0 as a concept is related to the fourth stage of industrial evolution and development, also known as the Fourth Industrial Revolution [46]. This evolution involves the employment of emerging and advanced technologies in a way that ensures efficient connectivity between machines, human-machine collaboration, and machine-to-machine communication in order to build a robust connected network for data sharing, acquisition, and transfer [47]. Creating a technology-driven connected environment results in enhanced control and monitoring of industrial systems supported by real-time data acquisition, enhanced quality of shared data, and data-supported decision-making. Such efficient connectivity will positively influence the economy, human quality of life, and environmental sustainability [48]. To provide a clear idea of how the Fourth Industrial Revolution came about, it is useful to summarize the first three industrial revolutions, as follows [49].

- The First Industrial Revolution was characterized by mechanization—the replacement of human power and workforces with machines. In this industrial phase, a single operator was assigned to monitor machines and the manufacturing process and acquire data on them.
- The Second Industrial Revolution evolved with the development of electricity. In this stage, manufacturing processes underwent a shift characterized by continuous operation, scaled-up production, and an enhanced ability to monitor the production process.
- The Third Industrial Revolution was driven by the use of computers and information technology (IT), which allowed process automation and enhanced production process control. Moreover, programmable devices and controllers boosted the efficiency of monitoring and data acquisition techniques.

Notice that each evolution in the industrial process was driven by the need to overcome the challenges and problems exhibited by the previous phase. Accordingly, the fourth stage of industrial evolution is a natural development toward enhanced connectivity and monitoring, the efficient perception of the physical environment, and guided decision-making. The characteristics of the Industry 4.0 phase can be summarized as follows [44]: (1) Large-scale and optimized digitization; (2) automation of production and monitoring processes; (3) connectivity and collaboration; (4) enhanced logistical efficiency; (5) constructive exchange of data; (6) modernized industrial business models; (7) personalized production; and (8) an increase in digital services and products. In terms of connectivity and collaboration, the idea of Industry 4.0 is to build horizontal connections to link different business participants, different business models within an industry, and different business models locally, regionally, and overseas. Additionally, vertical connectivity is desired to build a smart and technology-driven integration between the different layers within the industrial process in order to support production time, enhance quality, and overcome production challenges [47].

Construction 4.0 represents an implementation paradigm of Industry 4.0 concepts and principles in the construction industry. The positive impact of the Industry 4.0 transformation has led researchers and professionals to adopt a similar scheme to tackle the persistent challenges of the construction industry, such as project delays, cost overruns, inconsistent quality, a lack of interconnected relationships between different industry participants, and a lack of robust control and monitoring schemes in construction projects [50]. Hence, the following presents a detailed discussion of Construction 4.0's definition, characteristics, benefits and challenges, and driving technologies. The immersive and AI-based technologies and their possible implementations in the construction industry will also be elaborated.

The remainder of this chapter is organized as follows: Section 7.2 highlights the definitions and characteristics of the Construction 4.0 paradigm, in addition to the benefits and challenges that motivate technologists and researchers to develop solutions in this area. Section 7.3 summarizes the key enabling technologies that shape this new generation of construction solutions. Section 7.4 summarizes the application use cases linked to the different phases of Construction 4.0, such as the design, construction, and operation phases. Section 7.5 discusses the key frameworks and tools used for the digital transformation of construction assets and provides a high-level comparison between them. This is followed by a summary of the key AI techniques and neural network architectures leveraged for data and model analysis. Finally, section 7.7 discusses in detail an important case study in which ImageNet was used to analyze the health of construction assets based on images of buildings captured by drone-mounted cameras.

7.2 Construction 4.0 definition and characteristics

Construction 4.0, as a relatively new scheme, can be defined in different ways according to the context and desired goal.

Nevertheless, these different definitions all agree that Construction 4.0 is about digitization and transformation. Roland Berger, in Ref. [45], defined Construction 4.0 as the realization, among construction industry participants, of the importance of digitization and the progress of its implementation. This was followed by the definition of the four key elements of Construction 4.0: (1) Data digitization; (2) self-organized and automated systems; (3) connectivity of different activities; and (4) accessibility to data and digital systems. Sawhney et al. [51] defined Construction 4.0 as the combined effect of trends and advancements in physical and digital technologies in transforming construction practice. This definition considers cyber-physical systems (CPS) as the key element in the digital transformation of the construction industry. CPS are environments in which virtual and physical systems communicate and cooperate by means of sensors, actuators, and other data exchange mechanisms [12,38]. This framework facilitates the transformation of physical systems (existing assets) into digital assets, the creation of digital assets for new physical assets, and the transformation of these digital assets into physical assets via technology drivers. In other words, the framework enables a digitization platform for existing and planned assets. Incorporating advances in construction production (e.g., prefabrication, precast construction, and additive manufacturing) into CPS is the key element of the Construction 4.0 framework. Moreover, the incorporation of digital technologies such as machine learning (ML), data science, cloud computing, blockchain, and other related technologies adds an additional layer to the transformation process. Forcael et al. [43] presented Construction 4.0 as a combination of construction industry digitization and construction industry industrialization.

7.2.1 Construction 4.0 benefits and challenges

The construction industry works with large-scale projects and products, and there are complex interrelationships between the different participants of the industry. However, the large size of the industry means that small changes and improvements in any process or procedure will result in significant improvements in terms of cost, quality, and impact on society [52]. Accordingly, any improvement of any scale will uplift the construction industry and improve different aspects of quality of life. With the framework proposed in Ref. [51], wide-ranging advantages and benefits are expected to be obtained in:

- **Innovation**: Construction 4.0 works as the main driving force for building innovation in the construction industry. Toh and Park [22] mentioned that, based on global reports, many changes are required to take place in the AEC industry. One of the main changes required is sensible innovation. It is believed that the lack of innovation in the construction industry stems from organizational and institutional challenges [14]. Construction 4.0 elements function as a catalyst in reinforcing innovation within construction organizations.
- **Sustainability**: The efficient utilization of digitization technology and the continuous monitoring of physical systems through CPS allow for improved tracking and utilization of resources and materials. Moreover, it allows for efficient energy use. The implementation of the Industry 4.0 paradigm (and subsequently Construction 4.0) will result in improved social, economic, and environmental sustainability [46].

- **Cost**: Improved connectivity with physical systems, enhanced monitoring of different elements of the construction process, and innovative digital twinning can facilitate the precise estimation of productivity, applied and stocked materials, and progress tracking. This will result in improved cost control and savings. Moreover, it will support providing stakeholders with sufficient high-quality data to help them make data-informed decisions. Moreover, utilizing CPS (e.g., UAVs and robots) can help with a wide range of monitoring tasks that require time and expenditure [35].
- **Time**: The time element can be improved in various ways through the adoption of Construction 4.0 principles. In terms of construction industrialization, advanced construction methods (e.g., 3D printing, off-site manufacturing, and precast technologies) can result in considerable time savings. Moreover, abundant real-time and precise data on construction sites gathered through a digital-physical connection can help with the early detection of issues that might produce time delays.
- **Safety**: Connectivity and real-time data acquisition are key elements of Construction 4.0. In addition to the cost and time savings enabled by the elements of Construction 4.0, they can also significantly improve site safety. The physical and digital technologies adopted by Construction 4.0 play an important role in providing detailed and informative data that can help safety teams monitor safety at both the personal and project levels.
- **Quality**: The adoption of new technologies in the AECO industry can result in improved quality in the design, planning, construction, and operation of structures. Quality can be ensured by controlling and managing the construction process at different stages through a unified technology-driven platform. Construction 4.0 tools provide Fc digital and physical systems that can support this goal. Moreover, the improvement of time, cost, productivity, and safety elements can result in subsequent improvements in quality.
- **Coordination and collaboration**: The extensive use of BIM, Common Data Environments (CDEs), and DT in construction projects improves collaboration and coordination between the different players in construction projects. Collaboration can support project management, enhance the accessibility of project-related data, improve transparency and trust among different participants, and positively influence the time and cost of the project [3].
- **Image and reputation of the construction industry**: The construction industry is known for being reluctant to adopt emergent techniques and technologies [23,53]. The industry is also known for its difficult working conditions and environment. Adopting technology in the industry will help to improve the industry's image and reputation [54]. Additionally, new technologies can be utilized to automate strenuous construction tasks and reduce workers' exposure to tough site environments [52]. This will lead to a more appealing and modernized industry.

Arabshahi et al. [23] performed an intensive literature review on the different types of sensing technology trends used in construction sites and their respective benefits. Their summary of the benefits of utilizing such technologies, based on feedback from construction stakeholders, included cost reductions, quality enhancement, improved safety, enhanced FM, improved and data-informed decision-making, technical enhancement, project management and delivery enhancement, and an improvement in workers' satisfaction in the workplace. Begić and Galić [52] mentioned the benefits of employing additive manufacturing as one of the catalysts in Construction 4.0. The use of such technology can result in a wide range of improvements in the construction industry, including new structural designs and concepts, improved construction precision, reduced material waste, increased safety levels for workers, the ability to mix and utilize different types of materials, and a reduced need for skilled construction labor.

The wide range of benefits that accompany the adoption of the Construction 4.0 framework and the digital transformation of the industry are countered by various challenges and difficulties with organizational, financial, technical, social, or legal roots, which could slow or prevent transformation [53,55,56]. The main challenges of employing Industry 4.0 principles in Construction 4.0 can be linked to the following issues:

- **Government support**: Governments are considered the largest clients in the construction market. However, the construction requirements of such a client are based, in general, on traditional tendering schemes. This makes the adoption of innovative and emergent technologies by construction companies difficult or slow [44]. Moreover, there is a lack of awareness among construction companies of the positive impact of Construction 4.0 on workforce and organizational development [56]. This issue affects the willingness of stakeholders in the construction industry (including governments) to support digital transformation.
- **Financing challenges**: The efficient utilization of the enabling technologies of Construction 4.0 requires large initial investments and substantial digitization costs [46]. Additionally, different stakeholders in the construction industry consider the Industry 4.0 paradigm to be costly to adopt [54]. This makes the key players in the construction industry reluctant to adopt and invest in emerging technologies and digital trends.
- **Technical challenges**: Construction 4.0 is about the intensive integration of different technological elements of the construction industry. This requires equipping the construction workforce with newly developed skills to lead the implementation of new technologies [44]. Moreover, the digitization of construction projects and construction information requires secure IT infrastructure to ensure the cybersecurity of digitized data [51]. A lack of cybersecurity can affect stakeholders' trust in digitization.
- **Legal and contractual challenges:** The relationship between the different parties involved in a construction project is managed by contractual and legal agreements. The terms and clauses of these agreements can assign uncertain and vague responsibilities to the different players in the project [56]. Hence, the presentation of new paradigms and technologies in the construction industry requires the definition of clear and specific regulations that organize the new paradigm and clearly define responsibilities. For example, BIM implementation faces regulatory issues related to digital data ownership and involves responsibility for any gaps and errors in the digital model [54].
- **Complex nature of the construction industry:** The construction industry is complex in terms of the processes and overlapping responsibilities of the different participants in a construction project [54]. Moreover, construction projects are longitudinally fragmented processes, meaning that different construction projects are not linked to each other. Construction teams switch from one project to another, and the required knowledge and skills are different for each project. This results in a lack of built-up knowledge within construction companies and teams and a subsequent slow progression toward technology implementation [57].

7.3 Construction 4.0 enabling technologies

Uplifting the construction industry and redeeming the image of construction require the efficient employment of emerging and state-of-the-art digital and physical technology tools. The implementation of advanced technology aims to support tasks related to the design, construction, management, and operation of construction projects. Supporting construction tasks can be achieved by using emerging technologies to fulfill tasks that are usually accomplished via conventional, time-intensive, dangerous, or inefficient techniques [17]. Different emerging technologies can be employed to support construction projects at different stages. The technology drivers for Construction 4.0 can be summarized as follows:

7.3.1 Building information modeling (BIM)

Using BIM, a 3D graphical replica of a structure can be constructed that contains information on its geometry, materials, and specifications, as well as cost-related data [13]. Manzoor et al. [58] conducted an intensive literature survey on the digital technologies used in the AEC industry and concluded that BIM is considered the leading digital technology in this industry. BIM helps facilitate the design, construction, procurement, management, and maintenance of buildings [14]. The virtual digital model of the structure can be accessed by the various different participants in a project, allowing them to collaborate and make decisions based on data derived from a common source. This helps reduce clashes and reinforce time, cost, quality, and collaboration improvements [6].

7.3.2 Robots and automation

Robots are programmed physical systems that can be actuated to perform different tasks and functions [51]. Manzoor et al. [58] concluded that robotics is considered the second-most important technology in construction, following BIM. Robots play a key role in system automation. The construction industry is labor-intensive and demands skilled laborers. Robots can revolutionize the industry and provide automation solutions that reduce dependency on labor and enhance productivity and quality [59]. Moreover, robots can be employed to implement tasks in dangerous and inaccessible areas [49].

7.3.3 Additive manufacturing (3D printing)

Additive manufacturing is the construction of 3D objects in a segmented process by laying ultrathin layers of a filament to produce a precise and scale object [50]. Additive manufacturing is considered to play an important role in the transformation of construction. It also forms a key component of planetary construction [26,28]. Additive manufacturing supports the elements of Construction 4.0 by improving time, cost, labor, sustainability, safety, and other elements in construction projects [25].

7.3.4 Artificial intelligence (AI)

AI is the science of building smart machines that can employ state and environmental data to learn, build knowledge, interact, and assist decision-making [33]. AI can assist in the design, construction, monitoring, and FM stages of construction projects [50]. The ability of AI to assist efficiently in data-driven decisionmaking has led different stakeholders to consider it for various applications in the manufacturing and construction industries. AI has the potential for wide application in the construction industry as a revolutionary solution for various different challenges facing the industry [29–31,33,34].

7.3.5 Unmanned aerial vehicles

UAVs are remote-controlled aerial vehicles that can be equipped with control units, inertial measurement units, different types of sensors, and Global Positioning System receivers. Additionally, various visual, thermal, and environmental data-capturing sensors can be attached to UAVs by users. LiDAR (light detection and ranging) equipment can also be attached to drones. UAVs, with their proven capabilities, provide an efficient replacement for the manned aerial vehicles commonly used to perform various different tasks [60]. One of the main advantages of UAVs is their wide field of view or coverage area [35]. This makes UAVs a competitive technology for a variety of surveying, monitoring, mapping, and other critical tasks. Furthermore, UAV technologies have exhibited significant and rapid growth, and their affordability has increased [36]. This allows their efficient use and offers the benefits of their wide-ranging capabilities without imposing overly high costs on construction companies.

Other driving technologies for Construction 4.0 include VR, AR, XR, cloud computing, blockchain, big data analytics and data science, cybersecurity, deep learning (DL) ML, CDEs, sensors, the Internet of Things (IoT), radio frequency identification, laser scanners, natural language processing, and data mining (see Refs. [50,51] for details). Fig. 7.1 describes the Construction 4.0 framework with its related enabling technologies.

7.4 Construction 4.0 use cases and applications

Technologies adopted with the implementation of Construction 4.0 can span different applications and use cases to provide long-term solutions for known and repeated faults in the construction industry. To help build a better understanding of the different use cases of Construction 4.0 technologies, the following list links them to the design, construction, management, operation, and maintenance phases of construction projects.

7.4.1 Design phase

The Construction 4.0 framework has great potential to tackle the challenges that AEC designers face at the design stage. BIM, for example, can help designers create a virtual model of a new asset and consolidate all related geometric, material, specification, and cost information for all structural and nonstructural components of the asset in one

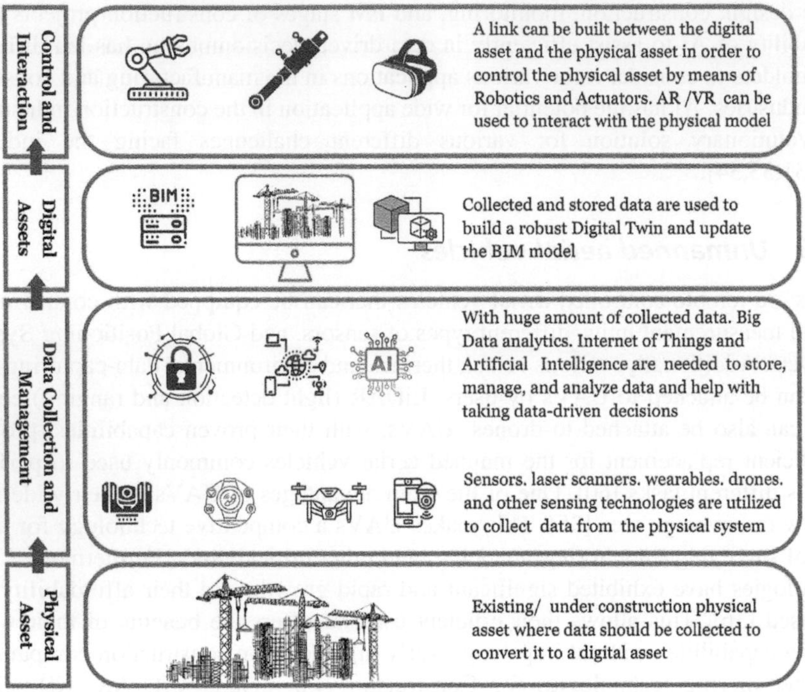

Figure 7.1 Construction 4.0 framework and enabling technologies.

digital model [15]. Moreover, BIM models can create both physical and analytical structural models, which can be transferred to structural analysis software programs to reduce the time and effort needed to prepare the input data for analysis [6]. UAVs, on the other hand, can help in surveying and mapping by creating 3D topographical models based on low-altitude visual data [60]. This can efficiently reduce the time and cost of site surveying required to develop topographical models and contour maps.

7.4.2 Construction phase

Various digital and physical technologies can assist engineers and decision-makers during the construction phase. UAVs can be used for construction site monitoring and progress assessment [36]. This significantly reduces the cost of conventional and human-based inspections and permits the assessment of locations that are inaccessible or difficult for humans to access. Additionally, using visual data and LiDAR, UAVs can be employed on construction sites to produce 3D models of construction materials using visual data and LiDAR in order to enhance estimates of the quantity of the materials [61]. Moreover, Nanayakkara et al. [62] reviewed the significance of employing smart contracts and blockchain with a focus on resolving payment issues. They concluded that these technologies can significantly reduce challenges related to payments and financial transactions in the construction industry. With the rapid development of data acquisition technologies and sensors and their implementation in construction technology, big data can serve as a major facilitator in sharing, storing, and analyzing the data on the life cycle of the structure [52]. AI also plays a key role in the construction phase. In Ref. [63], an AI-based approach to enhancing safety on construction sites is proposed.

7.4.3 Operation phase

New assets have the potential to be BIM-enabled and to be designed in a way that allows the implementation of new technologies within their life cycles. However, there is little or no availability of construction information for existing, old, and heritage structures. Precise structure-related information such as geometry, architectural layouts, materials, and installed building services is necessary for FM. UAVs and laser scanners have made significant contributions to the construction of 3D models of existing facilities in order to enhance FM. Soliman et al. [16] presented a case study where BIM, UAVs, and laser scanners were used to build a precise 3D model of an old structure to help facility managers make data-informed decisions related to the structure. Construction 4.0 technologies are also vital in SHM. Al-Sabbag et al. [42] proposed a structural health assessment technique using robots, AI, visual and depth sensors, mixed reality tools, and computer algorithms to enable human-machine collaboration that could enhance structural health assessments.

The implementation and utilization levels of Construction 4.0 technologies in the different construction phases are not uniform across the different technologies. El Jazzar et al. [64] summarized graphically the utilization level of each Construction 4.0 technology at each construction stage. The implementation level and the envisioned utilization of each technology are presented in Fig. 7.2. The figure demonstrates that some technologies are highly utilized at different stages, and some of the technologies exhibit low or nonexistent utilization.

7.5 Modeling frameworks and tools

In this section, we highlight and compare the usage of three well-known postprocessing drone photogrammetry platforms, namely DJI Terra, Pix4D, and Bentley ContexCapture. These platforms are used to transform assets into digital forms by surveying areas using various different sensors, as discussed in previous sections. Each platform has its own workflow and produces a unique outcome to help decision-makers in critical sectors such as public safety, construction, infrastructure, and agriculture.

DJI Terra [65] was developed by the DJI company, a well-known drone manufacturer. This platform is used to build 2D or 3D models of assets, which enable general inspections in agriculture and electricity, point cloud accuracy optimization, and highly detailed inspections in many fields. This platform's strength is that it is the most compatible platform for data collected by a DJI drone: its workflow uses the photos'

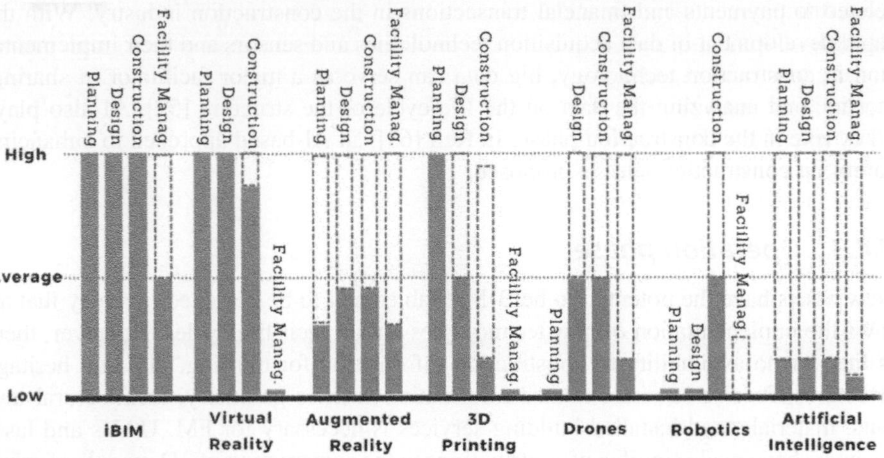

Figure 7.2 Current utilization level and envisioned roadmap for construction 4.0 technologies (green bars show the current utilization level of the technology and the dashed line shows the envisioned use of the technology).

geotagged data, and it automatically generates high-quality output digital assets. The platform also allows flight mission plans to be created and uploaded to execute professional-level missions for various purposes. This allows the user to plan, execute, and process data holistically using one platform, which guarantees the best quality outcomes. The weakness of DJI Terra is that the process of setting the digital transformation parameters can require an exhaustive process of trial-and-error and that there may be no deterministic set of parameters that guarantees the successful generation of a 2D or 3D model consistently and without error. Therefore, we have tried to incorporate the time needed to set these parameters into the comparison we conducted (see Fig. 7.3).

Pix4D Mapper [66], developed by the Pix4D company, has many great features that deliver high-quality models in numerous output formats, which makes projects more informative in multiple industries. Its workflow is somewhat more advanced compared to DJI Terra, and attention is required to make selections at each step; the quality of the

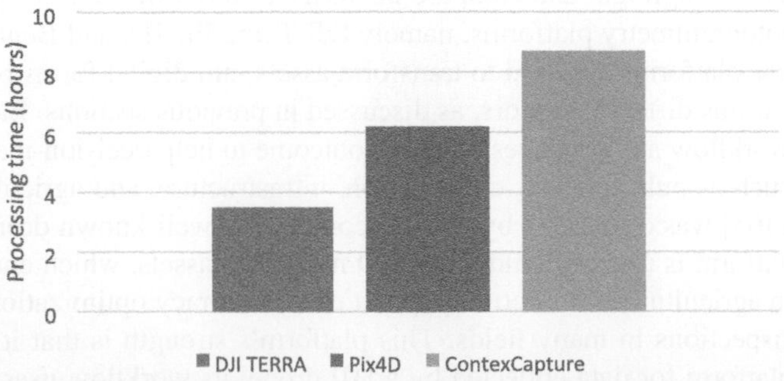

Figure 7.3 Comparing the duration of process workflows for modeling platforms.

reconstruction and the processing time required depend on the chosen parameters. The strength of this platform is that it creates a quality report that can be used to verify the success of processing and evaluate the processed data for better enhancement. It also provides the option to edit some viewing parameters after processing the model. The weakness of Pix4D Mapper is that it takes more time to process the models than DJI Terra and generates lower-quality output. The tool also seems to have some incompatibility issues with drone output data, which may affect the quality of the output or cause the loss of some parts of the models produced by it.

The ContextCapture [67] Master platform is built by Bentley Systems Company. This platform provides only 3D reconstruction and orthophoto or digital surface model (DSM) reconstruction types, enabling the user to create 3D models with very advanced properties that require a high level of attention to build the correct model. These properties include:

- Spatial framework: This defines the spatial reference system, region of interest, and tiling. It also includes resolution settings for an orthophoto/DSM.
- Geometry constraints: This allows the use of existing 3D data to control the reconstruction and avoid reconstruction errors.
- Reference model: This is the reconstruction sandbox; it stores the model in a native format, which is progressively completed throughout the process. This reference model is the model to which retouches and reconstruction constraints are applied and from which future productions are derived.
- Processing settings: This sets the geometric precision level (high, extra, etc.) and other reconstruction settings. It also offers a list of production formats.

The strength of the ContexCapture Master platform is that it creates an effective model if the right parameters are set; it also allows numerous properties to be selected if there is a need for more focus on specific details of the model.

The weakness of the ContexCapture platform is that the spatial framework and processing settings can be edited only before starting production, and if these are not set incorrectly, the model will not be correctly viewed and a new reconstruction process will need to be begun.

A comparison of the preparation and processing times of the three platforms indicated that DJI Terra was able to complete the workflow and process the model approximately 2–3 times faster than the other two platforms.

Furthermore, when comparing the quality of the data processed on the three platforms, DJI Terra was found to be able to provide the highest quality outcome for the same data, as shown in Fig. 7.4.

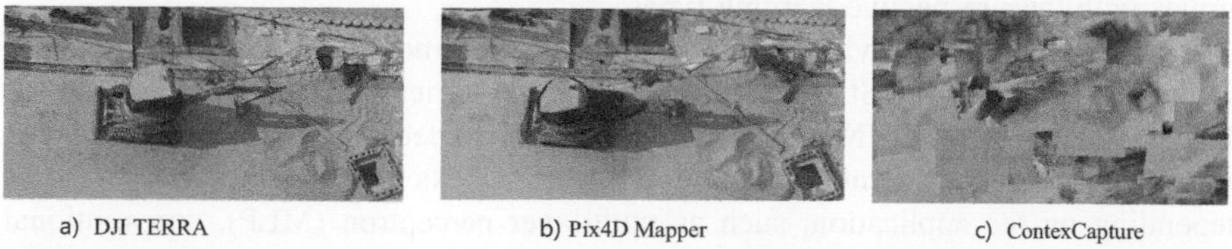

Figure 7.4 Comparing processing quality outcomes for the three platforms (a) DJI Terra, (b) Pix4D Mapper, and (c) ContexCapture.

From Fig. 7.4, it can be observed that the first higher-quality outcome was built by DJI Terra. This was composed faster than Pix4D Mapper, which produced a high-quality outcome after the general parameters were set but took more time to produce the output. The last lower-quality output was produced by ContexCapture Master after all parameters had been set correctly. ContexCapture Master took much more time to process the model than DJI Terra and Pix4D Mapper did, using the same computer. However, it is worth mentioning that the data were collected using a DJI drone, which gave DJI Terra an advantage in building an accurate 3D model as the drone was more compatible with this postprocessing platform than the other platforms.

7.6 Artificial intelligence in construction 4.0

The digital transformation of construction assets and the big data generated as a result of data acquisition through IoT devices necessitate the leverage of smart techniques for data analysis rather than reliance on manual/visual inspection of data by humans or primitive programs. AI and ML techniques provide robust architectures for pattern recognition, classification, and problem identification, which can help optimize construction operations and management [68]. A deeper and more comprehensive survey regarding the design of pervasive AI techniques for different IoT applications can be found in Ref. [69].

Generally, AI allows computers to imitate human intelligence in order to perceive and understand problems. The term was first coined in 1956, initially referring to rule-based systems and expert systems, and later branched out to ML systems. Hence, ML is a subfield of AI that focuses on learning from data to predict, classify, or recognize sophisticated unseen patterns. DL, on the other hand, is a subfield of ML that utilizes different architectures of artificial neural networks [70] to design ML models that learn patterns from datasets; this will be the main focus of the discussion in the following section. ML techniques can be largely categorized as supervised, unsupervised, or reinforcement learning techniques. Supervised ML techniques rely on data with well-known labels that identify the outcome of data samples and hence provide clear, solved examples to guide the ML algorithm to learn and identify unseen patterns. Unsupervised ML techniques rely on data with no specific outcome labels, meaning that the data can at best be clustered into groups with similar features to help understand and identify their taxonomy. Reinforcement learning techniques do not necessarily rely on datasets but instead utilize agents that form perceptions through direct interactions with the environment. Fig. 7.5 summarizes the different categories of AI techniques with their respective learning types.

In this section, we provide a brief overview of the most common DL techniques used in Construction 4.0. Stacked consecutive layers of neurons are commonly termed deep neural networks (DNNs), which provide different deep structures to analyze data patterns in different applications. The structures of DNNs may take on different forms depending on the application, such as multilayer perceptron (MLP), convolutional neural networks (CNN), and recurrent neural networks (RNN).

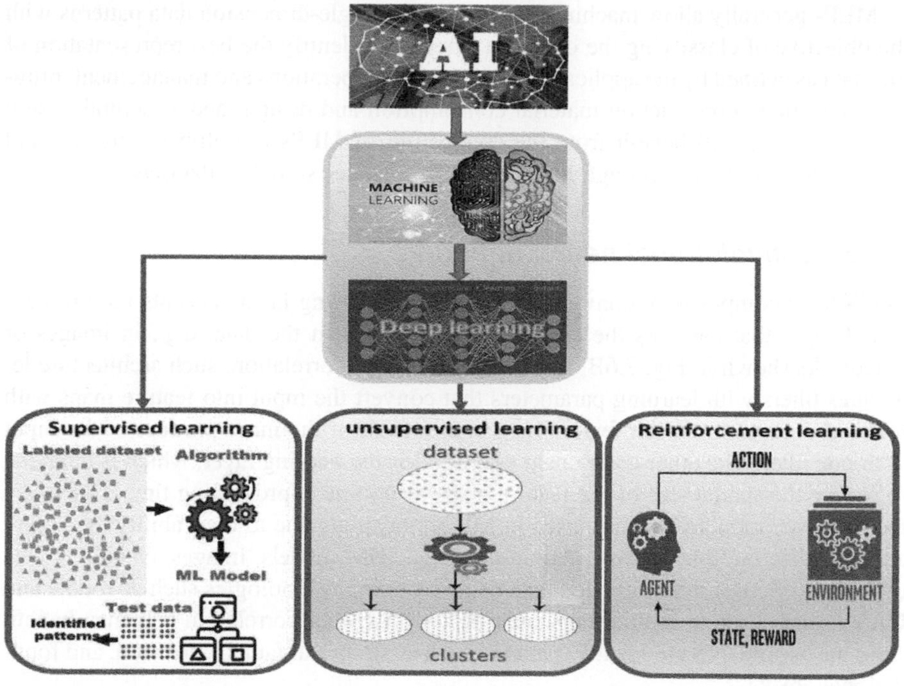

Figure 7.5 Categorization of AI, ML, and DL with their respective learning types.

7.6.1 Multilayer perceptron

A neural network with layers of single-dimension neurons, stacked in a feed-forward manner, is called a feed-forward network or a MLP, as shown in Fig. 7.6A. The layers are FC to each other, causing the output from each layer (x_i) to be linearly weighted ($y_i = w_i x_i$) and scaled using an activation function ($x| |i + 1 = (y_i)$) before being fed to the next layer [71]. MLPs are trained using an acquired labeled dataset to estimate the optimal weights ($\forall i$) that minimize the loss function ($L(w)$), which represents the difference between the estimated labels using the weights and the existing labels in the dataset.

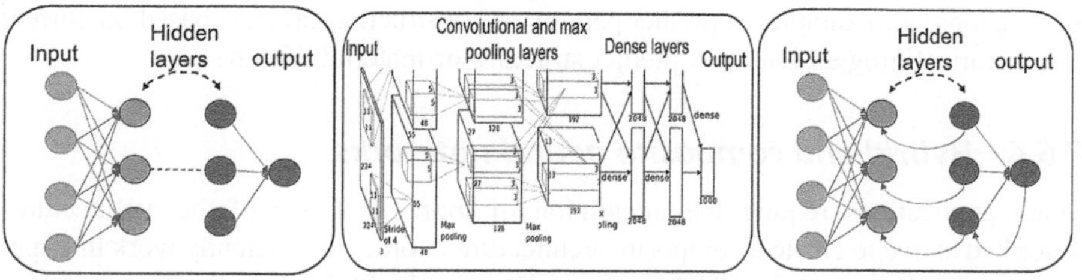

a) Multilayer Perceptron b) Convolutional Neural Networks c) Recurrent Neural Networks

Figure 7.6 DL neural network architectures: (a) multilayer perceptron; (b) convolutional neural networks; (c) recurrent neural networks.

MLPs generally allow machines to understand single-dimension data patterns with the objective of classifying the data into labels that identify the best representation of this data as defined by the application. Construction operations and management, progress and safety, construction material consumption and design, and structural design and analysis can all benefit from the application of MLPs to optimize, predict, and identify best strategies through historical best practices stored in datasets.

7.6.2 Convolutional neural networks

In CNNs, a composite of convolutional and max-pooling layers accepts multidimensional input that captures the spatial context hidden in the data, e.g., in images or videos. As shown in Fig. 7.6B, to capture the spatial correlation, such architecture leverages filters with learning parameters that convert the input into feature maps with reduced dimensions. Each feature map is the result of the inner product of the input with one filter. The other component of a CNN is the pooling layer, which is intended to reduce the spatial size of the feature maps to minimize processing time. CNNs are the most common architecture used in ML applications due to their ability to capture and visualize multidimensional input data, e.g., BIM models, images, or videos. The collection of construction-related data using enabling technologies such as robots and UAVs often leads to multidimensional data with spatial correlation that may benefit from the use of CNNs to identify patterns such as structural faults, corrosion, and foundational cracks.

7.6.3 Recurrent neural networks

RNNs are efficient at detecting and identifying temporal correlations among sequential input samples. To achieve this, the structure of RNNs can support backward connections from one layer to the previous layer to help influence the current estimated values based on the previous history of values through a recursive scheme, as shown in Fig. 7.6c. Several different variations of RNNs exist, the most common being the long short-term memory (LSTM) architecture. The LSTM architecture is based on short-term memory blocks that are used to create a long-term memory of the input sequence; hence, the project's future output values are based on historical sequences.

RNNs are ideal for predicting a projected future based on sequential, progressive processes and milestones. Such a paradigm can be significantly utilized in construction applications, for example, to predict periods of construction projects based on current and historical progression or to predict structure or material lifetimes.

7.6.4 Hybrid and composite neural networks

Many applications require the integration of more than one of the architectures described above to create a composite architecture in order to efficiently work in a specific context. For example, CNNs are often integrated with Fc dense layers to produce single-dimensional output from a multidimensional input dataset. MLPs with a bottleneck hidden layer logically divide the architecture into two parts, namely an encoder and a decoder, and are hence often called stacked autoencoder architectures. The encoder is used to reduce the dimensions of the input data by generating low-dimension feature vectors, while the decoder can reverse this operation to reconstruct the original features. This concept is often used for data compression and to remove data redundancy. Another revolutionary composite architecture is the generative adversarial network, which promotes the interconnection of two types of neural networks, namely the generator and discriminator networks. While the generator tries to generate innovative features that resemble the features in the dataset, the discriminator tries to provide feedback to the generator about how discriminant these new features are by reconstructing the original features from the newly generated features [71].

7.6.5 Transfer learning

Transfer learning is the process of training a pretrained network to learn new patterns in new data. This is useful because it takes advantage of the knowledge provided by a pretrained network to avoid the need to define a network architecture and train it from scratch. Leveraging this concept, we can utilize many well-established pretrained ML models for different applications, such as the GoogLeNet, ImageNet, ResNet-40, Visual Geometry Group 16 (VGG16), and VGG19 networks [69,71–73]. In the following case study, transfer learning was implemented by applying and adapting the knowledge gained by the pretrained network, namely ImageNet, to our image dataset.

7.7 AI-based concrete column base cover localization and degradation detection: A case study

This case study focuses on automated SHM to detect the degradation of a concrete column base cover and the application of the DL training models ResNet-50, GoogLeNet, and VGG19 with seven different network configurations. The results of this study are based on the findings of the research work introduced in Ref. [73]. Many FM organizations are currently investigating automated technologies for the efficient and timely monitoring of civil structures with reduced labor costs. The main factors causing concrete deterioration involve the corrosion of embedded metals, freeze-thaw deterioration, and chemical attacks due to groundwater, erosion, fire/heat, concrete spalling, and reinforcing bars (REBARs) [74,75]. All of the above can cause serious damage to the structure as a whole, which may also result in loss of property and public injury. In this case study, we studied a cost-effective approach to automated structural monitoring based on a combination of UAV imaging and a state-of-the-art DL framework using a CNN-based architecture, namely ResNet-50. This technique was developed for the detection and localization of the major types of column defects. Concrete spalling and rebar exposure are among the major concrete column defects, as shown in Fig. 7.7. These defects may lead to loss of property, public injury, and damage to the whole structure, for example, through damage to rebars. It is necessary to repair these defects promptly, as the repair costs and risks significantly increase with time.

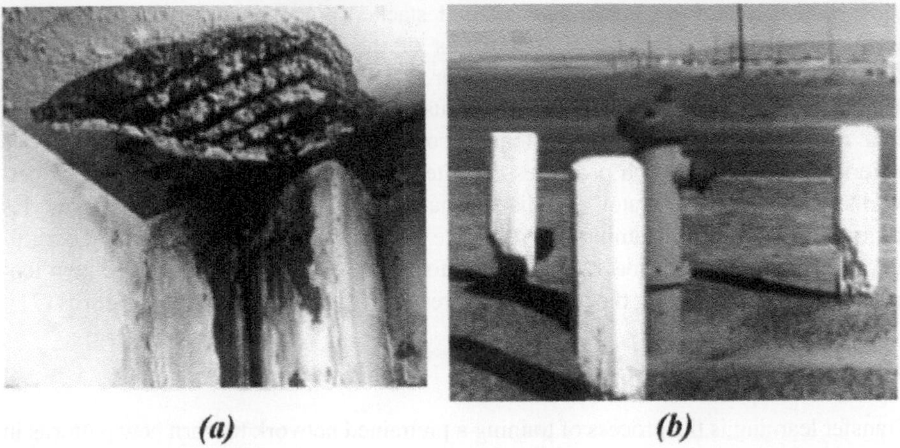

Figure 7.7 Causes of concrete spalling: (a) corrosion of reinforcing steel; (b) concrete posts suffering from a sulfate attack.

Zhao and Zhou [76] developed and tested a fully CNN network to detect various cracks, spalling, and holes in concrete covers. While the proposed network yielded good results, detectability against surrounding objects was not reported. Perez et al. [77] proposed a DL network to detect building mold, staining, and paint deterioration, wherein they transferred learning to the VGG16. Bhavani et al. [78] introduced a CNN network that achieved 81% accuracy when detecting building damage and predicting the robustness of the repaired mortar. The network is based on the widely known ResNet-50 pretrained network. The AlexNet and GoogLeNet models are among the best-performing pretrained models, with proven image classification effectiveness. These networks were used to detect cracks in highways and the locations of illegal buildings, respectively, in Refs. [79,80]. Spalling and rebar damage are the most common concrete defects. GoogLeNet, AlexNet, and ResNet-50 have been used to develop efficient concrete defect detection algorithms. However, there is a research gap concerning the proposal of a hybrid algorithm for the localization and binary health condition classification of concrete column base covers. Concrete columns are usually surrounded by aggregates and soil of the same color as the concrete, and sometimes with a texture that is close to that of concrete spalling. This makes the classification process more challenging. There is no specific algorithm that can guarantee precision, and thus this particular classification problem requires investigation to determine the best-performing technique and input option.

7.7.1 Methodology

High-resolution digital images of columns were taken as the input, and the outputs were the barcode, location, and health condition of each column base cover. This produced a list of defective column locations along with a digital image of each defective column base. The defect identification algorithm did not consider the severity of

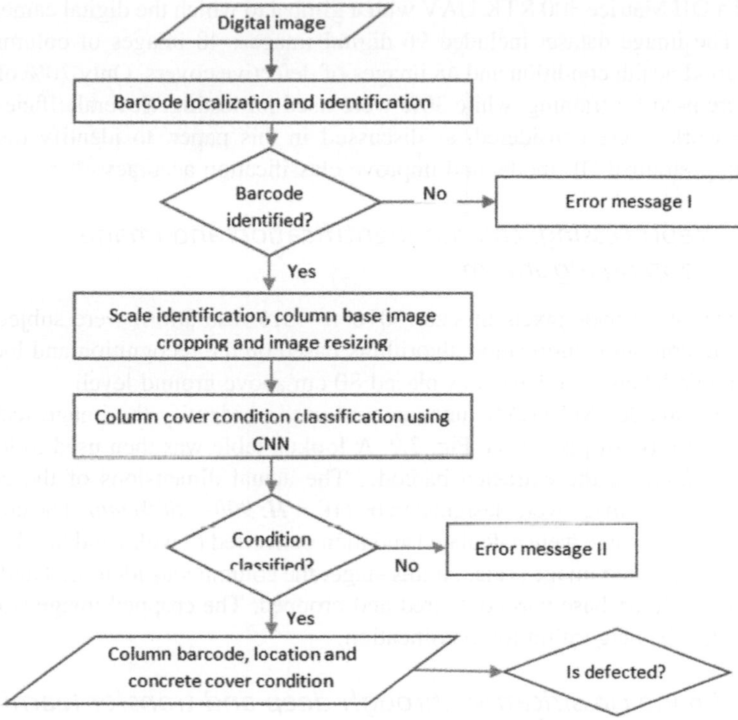

Figure 7.8 Automatic AI-based defect detection algorithm for concrete column base covers.

spalling, as immediate maintenance would be required in all cases, regardless of severity. However, the images of the defective column base covers would allow a maintenance team to make decisions on maintenance priorities and act accordingly.

An automatic AI-based defect detection algorithm for concrete column base covers is shown in Fig. 7.8. The input is in the form of a picture or video. It is then subjected to a barcode identification process to provide information on the column number, location, and image scale. The image scale is calculated by dividing the width/height of the barcode label in pixels by the actual width/height of the barcode label (140 mm). The algorithm resizes the image to pass through a pretrained CNN image classification algorithm for decision-making. If the CNN network fails to classify the image, an "Error message II" is sent to the maintenance team to provide them with information about the column's location as well as the classification of the fault. In cases of successful classification, the algorithm sends the column's barcode, location, and concrete cover health condition to the maintenance team.

7.7.1.1 Dataset

The data were collected using a digital wireless HD 24-megapixel camera. For indoor image acquisition, the camera must be either fixed or attached to a robotic platform; we opted for the fixed deployment for our experiment. For outdoor image acquisition, the team used a DJI Matrice 300 RTK UAV with a gimbal to which the digital camera was attached. The image dataset included 96 digital images: 48 images of column-base covers in good health condition and 48 images of defective covers. Only 70% of those images were used for training, while 30% were used for testing. Several efficient pre-trained networks were considered, as discussed in this paper, to identify the best-performing pretrained DL model and improve classification accuracy.

7.7.1.2 Preprocessing: column identification and image cropping algorithm

Input images of 24 megapixels in size (6016 × 4016 resolution) were subjected to cropping and column identification algorithms based on the recognition and localization of barcode labels. The label was placed 80 cm above ground level.

The "readbarcode" MATLAB function was used to identify the image scale and barcode in the form of pixels; see Fig. 7.9. A lookup table was then used to localize the column based on the extracted barcode. The actual dimensions of the column base in the cropped image were designed to be ***(W × H: 900 × 800) mm***. The cropping dimensions of the image were calculated and then converted to width and height pixels based on the identified image scale. At this stage, the column was identified and localized, and the column base was contoured and cropped. The cropped image was then passed to the CNN algorithm for classification.

7.7.1.3 Image classification through deep and transfer learning using ImageNet

CNNs are widely used in similar defect detection applications, particularly the GoogLeNet, ResNet-40, and VGG19 networks. As the data passes through CNNs, the

Figure 7.9 (a) Column barcode identification and localization process; (b) column base image cropping object detection.

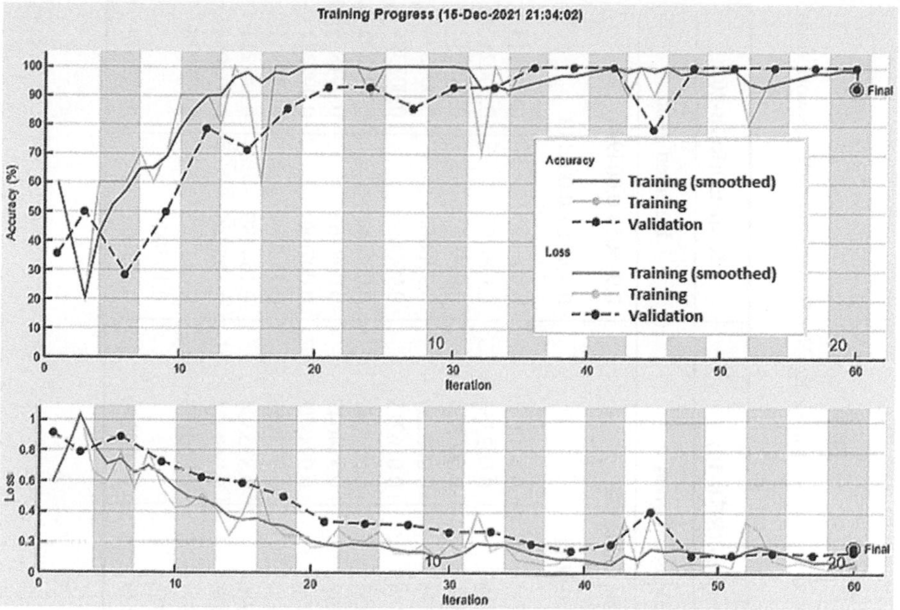

Figure 7.10 Training progress of the AI-based defect detection algorithm using Resnet50 network and gray images as inputs (classification accuracy of 92.5%, MATLAB Figure).

depth of the input image (n) increases through the convolution layers, and the width and height decrease through the pooling layers. This improves the processing speed as well as the precision of the results. We used an ImageNet pretrained model and employed transfer learning by adopting Fc dense layers at the output of the model to customize the outcome based on our application. The network had an image input size of 224-by224 pixels [3].

Various CNN network configurations were investigated to identify the network with the best binary classification performance (i.e., either good or defective concrete cover condition). A ResNet-50 pretrained model was used for binary classification and yielded a correct image classification accuracy of 92.5% (see Fig. 7.10 and Table 7.1). The certainty level was assessed using the maximum and root mean square error (RMSE) scores given to the wrong class (wrong condition); all classification scores are given in Tables 7.1 and Table 7.2. A 100% certainty score could be achieved only if a 0 was given to the wrong class. All network configurations using different input image types yielded a classification performance of 100%. However, the certainty of image classification was different. VGG19 demonstrated the highest certainty, with both RMSE and max error values less than 1%. RGB (red-green-blue) and grayscale images were used as inputs to the CNN network. The classification rate remained unchanged at 100%, and the conversion of RGB images to grayscale images improved the classification certainty, reducing the maximum error score from 12.77% to 0.87% and the RMSE from 4.81% to 0.29% (Figs. 7.10–7.13).

Table 7.1 Summary of the prediction scores of the proposed algorithm using different configurations for the ResNet-50 pretrained DL network (max. score value (probability) = 1, false classification in bold).

	Network	**Resnet50(Grayscale)**		**Resnet50(RGB-10)**		**Resnet50(RGB-20)**	
s. no	**Classification**	**Defected good**		**Defected good**		**Defected good**	
1	Defected	0.999777 0.000223		0.997401 0.002599		0.995054 0.004946	
2	Defected	0.996647 0.003353		0.9989 0.0011		0.998887 0.001113	
3	Defected	0.996065	0.003935	0.9989	0.0011	0.998887	0.001113
4	Defected	0.997214	0.002786	0.997646	0.002354	0.99979	0.00021
5	Defected	0.999807	0.000193	0.999736	0.000264	0.99979	0.00021
6	Defected	0.999033	0.000967	0.999736	0.000264	0.996606	0.003394
7	Defected	0.999033	0.000967	0.983785	0.016215	0.996606	0.003394
8	Good	0.157906	0.842094	0.445953	0.554047	0.008165	0.991835
9	Good	0.157906	0.842094	0.445953	0.554047	0.008165	0.991835
10	Good	0.100307	0.899693	0.005074	0.994926	0.007689	0.992311
11	Good	**0.654957**	**0.345043**	0.026837	0.973163	0.154789	0.845211
12	Good	0.143291	0.856709	0.124696	0.875304	0.154789	0.845211
13	Good	0.143292	0.856708	0.039926	0.960074	0.195046	0.804954
14	Good	0.447808	0.552192	0.039841	0.960159	0.011282	0.988718
	Max. error	0.654957		0.445953		0.195046	
	RMSE	0.323043	0.002282	0.244138	0.006299	0.111022	0.002674

Table 7.2 - Summary of the prediction scores of the proposed algorithm using GoogLeNet and VGG19 pretrained DL networks (max. score value (probability = 1).

	Network	GoogLeNet (RGB)		GoogLeNet (Grayscale)		VGG19 (RGB)		VGG19 (Grayscale)	
s. no	**Classification**	**Defected**	**Good**	**Defected**	**Good**	**Defected**	**Good**	**Defected**	**Good**
1	Defected	0.969243	0.030757	0.997918	0.002082	0.999836	0.00014	0.99174	0.007883
2	Defected	0.969243	0.030757	0.99951	0.00049	0.869828	0.127292	1	2.53E-09
3	Defected	0.929899	0.070101	0.999455	0.000545	0.999943	5.63E-05	0.999998	5.38E-07
4	Defected	0.980153	0.019847	0.93933	0.06067	0.99996	3.89E-05	1	8.81E-08
5	Defected	0.991365	0.008635	0.93933	0.06067	1	1.07E-08	0.999996	1.42E-06
6	Defected	0.77988	0.220121	0.996861	0.003139	0.996789	0.000976	0.999989	6.3E-06
7	Defected	0.929097	0.070904	0.951068	0.048932	0.998301	0.00046	0.999964	7.71E-06
8	Good	0.004074	0.995926	0.040438	0.959562	7.31E-08	0.99999	8.26E-08	0.999999
9	Good	0.004074	0.995926	0.00663	0.99337	2.65E-07	0.999962	3.39E-08	0.999998
10	Good	0.001514	0.998486	0.041824	0.958176	7.68E-07	0.999998	0.008741	0.989347
11	Good	0.0124	0.9876	0.041824	0.958176	2.34E-07	0.999991	5.11E-07	0.999994
12	Good	0.075749	0.924252	0.002568	0.997432	9.58E-07	0.999889	3.71E-10	0.999999
13	Good	0.271405	0.728595	0.077213	0.922787	1.03E-08	0.999986	1.02E-08	0.99999
14	Good	0.271404	0.728596	0.077213	0.922787	5.06E-07	0.999999	4.57E-07	0.999997
	Max. error	0.220121		0.077213		0.127292		0.008741	
	RMSE	0.147961	0.093163	0.049437	0.037361	5.2E-07	0.048114	0.003304	0.002979

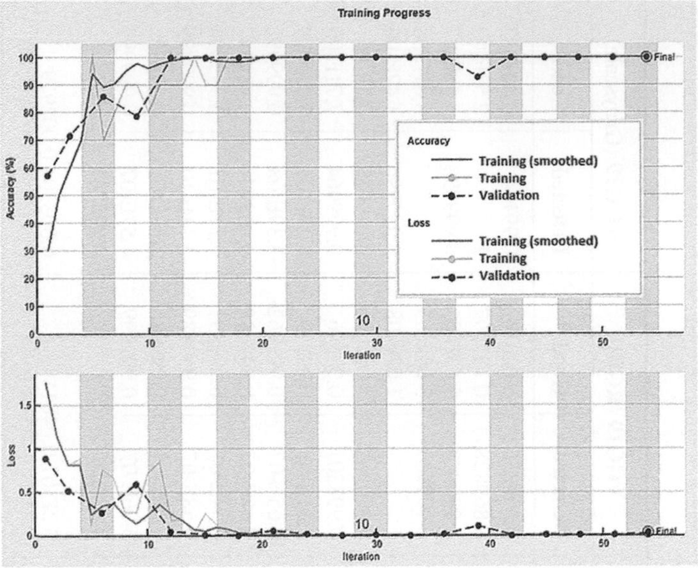

Figure 7.11 Training progress of the AI-based defect detection algorithm using VGG19 network and grayscale images as inputs (classification accuracy of 100%, MATLAB figure).

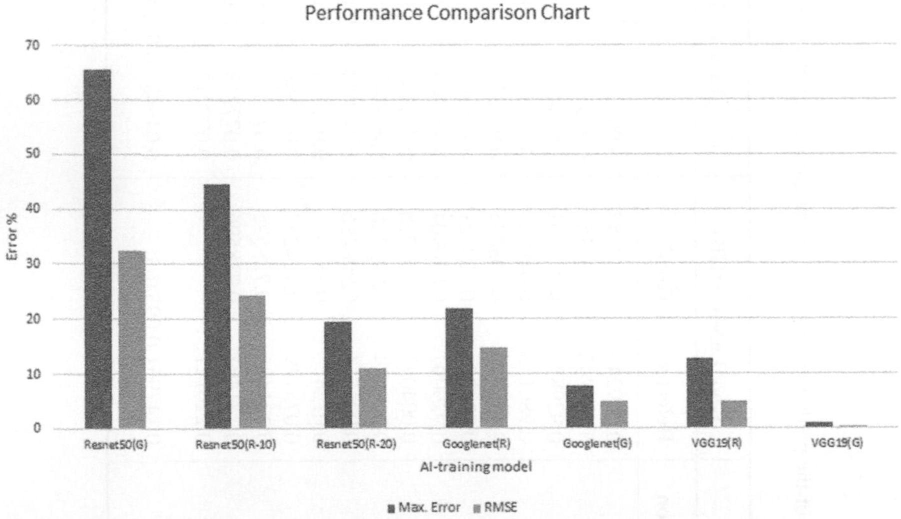

Figure 7.12 Performance comparison chart for the studied DL-pretrained algorithms.

7.7.2 Results and discussion

The different networks yielded different classification performances and score certainty percentages. The studied neural networks were evaluated based on the classification score produced by the softmax layer, as this layer returned a score that

Figure 7.13 An example of image classification of four column covers taken from a site in Qatar.

represents the probability for each classification class (out of 1), with a total summed score value of 1 for all classes. All network configurations using different input image types yielded a performance of 100%, except the ResNet-50 network when trained using grayscale images; this only yielded a classification performance of 92.5% (see line 11 in Table 7.1). The VGG19 pretrained network yielded a correct image classification score of 100% when trained using grayscale images. The RMSE and maximum image classification/prediction scores were used for the classification performance evaluation. VGG19 produced an RMSE value and a classification score of 0.33% and 0.87%, respectively. This resulted in an average certainty of correct classifications as low as 99.63%.

7.8 Conclusion

In this chapter, we discussed the application of AI to facilitate interconnections between CPS and BIM as part of the Construction 4.0 paradigm. We shed light on the benefits and challenges of such systems and how AI can help to reduce construction costs and time through data and/or model analysis. We also summarized three of the common frameworks used for the digital transformation of construction assets and provided a quick comparison of the typical processing times needed for each tool, starting with setting the processing parameters for model construction. We then detailed a case study in which AI was applied to the detection and localization of concrete column base covers for proactive structure health monitoring using CNN DL architecture. With the maturity of the presented tools and frameworks, the digital transformation of construction assets is clearly becoming more efficient and feasible. Tools such as DJI Terra have been proven efficient with minimum processing time for model construction, although more tool stability is needed for consistent results given a predefined set of parameters. The results of the AI-based case study also highlight the successful detection and identification of column base covers with accuracy reaching up to 99.63%.

Acknowledgments

This work was made possible by NPRP grant # NPRP12C-0828-190023 from the Qatar National Research Fund (a member of Qatar Foundation). The findings achieved herein are solely the responsibility of the authors.

References

[1] Ltd, R. and M. (n.d.). Construction market by construction type, by Sector - Global Opportunity Analysis and Industry Forecast 2022-2030. Research and Markets - Market Research Reports - Welcome. Retrieved September 1, 2022, from https://www.researchandmarkets.com/reports/5519701/constructionmarket-by-construction-type-by.

[2] J. Koeleman, M.J. Ribeirinho, D. Rockhill, S. Erik, Gernot Strube, Decoding digital transformation in construction, Capital Projects and Infrastructure Practice (2019).

[3] A. Nikas, A. Poulymenakou, P. Kriaris, Investigating antecedents and drivers affecting the adoption of collaboration technologies in the construction industry, Automation in Construction 16 (5) (2007) 632−641.

[4] M.K. Karen, N. Douglas, Building Information Modeling BIM in Current and Future Practice. Hoboken, New Jerse, 2014, pp. 59−62.

[5] G.A. van Nederveen, F.P. Tolman, Modelling multiple views on buildings, Automation in Construction 1 (3) (1992) 215−224.

[6] C.M. Eastman, C. Eastman, T. Paul, R. Sacks, K. Liston, BIM Handbook: A Guide to Building Information Modeling for Owners, Managers, Designers, Engineers and Contractors, John Wiley & Sons, 2011.

[7] J. Lou, W. Lu, X. Fan, A review of BIM data exchange method in BIM collaboration, in: International Symposium on Advancement of Construction Management and Real Estate, Springer, Singapore, 2020, pp. 1329−1338.

[8] S. Tang, D.R. Shelden, C.M. Eastman, P. Pishdad-Bozorgi, X. Gao, A review of building information modeling (BIM) and the internet of things (IoT) devices integration: present status and future trends, Automation in Construction 101 (2019) 127−139.

[9] C.C. Menassa, From BIM to digital twins: a systematic review of the evolution of intelligent building representations in the AEC-FM industry, Journal of Information Technology in Construction 26 (5) (2021) 58−83.

[10] S.H. Khajavi, N.H. Motlagh, A. Jaribion, L.C. Werner, H. Jan, Digital twin: vision, benefits, boundaries, and creation for buildings, IEEE Access 7 (2019) 147406−147419.

[11] E.J. Tuegel, A.R. Ingraffea, T.G. Eason, S. Michael Spottswood, Reengineering aircraft structural life prediction using a digital twin, International Journal of Aerospace Engineering 2011 (2011).

[12] J.M.D. Delgado, L. Oyedele, Digital twins for the built environment: learning from conceptual and process models in manufacturing, Advanced Engineering Informatics 49 (2021) 101332.

[13] C. Coupry, S. Noblecourt, P. Richard, D. Baudry, D. Bigaud, BIM-based digital twin and XR devices to improve maintenance procedures in smart buildings: a literature review, Applied Sciences 11 (15) (2021) 6810.

[14] R. Davies, C. Harty, Implementing 'Site BIM': a case study of ICT innovation on a large hospital project, Automation in Construction 30 (2013) 15−24.

[15] S.T. Matarneh, M. Danso-Amoako, S. Al-Bizri, M. Gaterell, M. Rana, Building information modeling for facilities management: a literature review and future research directions, Journal of Building Engineering 24 (2019) 100755.
[16] K. Soliman, K. Naji, M. Gunduz, O. Tokdemir, F. Faqih, T. Zayed, BIM-based facility management models for existing buildings, Journal of Engineering Research 10 (1A) (2022) 21−37.
[17] D.-G.J. Opoku, S. Perera, R. Osei-Kyei, M. Rashidi, Digital twin application in the construction industry: a literature review, Journal of Building Engineering 40 (2021) 102726.
[18] J. Paek, C. Krishna, R. Govindan, J. Caffrey, S. Masri, A wireless sensor network for structural health monitoring: performance and experience, in: The Second IEEE Workshop on Embedded Networked Sensors, 2005. EmNetS-II, IEEE, 2005, pp. 1−9.
[19] H. Sohn, C.R. Farrar, N.F. Hunter, W. Keith, Structural health monitoring using statistical pattern recognition techniques, Journal of Dynamic Systems, Measurements and Control 123 (4) (2001) 706−711.
[20] J.M.W. Brownjohn, Structural health monitoring of civil infrastructure, Philosophical Transactions of the Royal Society A: Mathematical, Physical & Engineering Sciences 365 (1851) (2007) 589−622.
[21] Z.A. Al-Sabbag, C.M. Yeum, S. Narasimhan, Interactive defect quantification through extended reality, Advanced Engineering Informatics 51 (2022) 101473.
[22] G. Toh, J. Park, Review of vibration-based structural health monitoring using deep learning, Applied Sciences 10 (5) (2020) 1680.
[23] M. Arabshahi, Di Wang, J. Sun, P. Rahnamayiezekavat, W. Tang, Y. Wang, X. Wang, Review on sensing technology adoption in the construction industry, Sensors 21 (24) (2021) 8307.
[24] Y.W.D. Tay, B. Panda, S.C. Paul, N.A. Noor Mohamed, M.J. Tan, K.F. Leong, 3D printing trends in building and construction industry: a review, Virtual and Physical Prototyping 12 (3) (2017) 261−276.
[25] M. Sakin, Y.C. Kiroglu, 3D printing of buildings: construction of the sustainable Houses of the future by BIM, Energy Procedia 134 (2017) 702−711.
[26] F. Craveiroa, J. Pinto Duartec, H. Bartoloa, P. Jorge Bartolod, Additive manufacturing as an enabling technology for digital construction: a perspective on Construction 4.0, Sustainable Development 4 (6) (2019).
[27] Ghaffar, S. Hamidreza, J. Corker, M. Fan, Additive manufacturing technology and its implementation in construction as an eco-innovative solution, Automation in Construction 93 (2018) 1−11.
[28] A. Kazemian, X. Yuan, E. Cochran, B. Khoshnevis, Cementitious materials for construction-scale 3D printing: laboratory testing of fresh printing mixture, Construction and Building Materials 145 (2017) 639−647.
[29] M. Kor, I. Yitmen, S. Alizadehsalehi, An investigation for integration of deep learning and digital twins towards Construction 4.0, Smart and Sustainable Built Environment (2022).
[30] S.O. Abioye, L.O. Oyedele, L. Akanbi, A. Ajayi, J.M.D. Delgado, M. Bilal, O.O. Akinade, A. Ahmed, Artificial intelligence in the construction industry: a review of present status, opportunities and future challenges, Journal of Building Engineering 44 (2021) 103299.
[31] S.K. Baduge, S. Thilakarathna, J.S. Perera, M. Arashpour, P. Sharafi, T. Bertrand, A. Shringi, P. Mendis, Artificial intelligence and smart vision for building and construction 4.0: machine and deep learning methods and applications, Automation in Construction 141 (2022) 104440.

[32] Y. Pan, L. Zhang, Roles of artificial intelligence in construction engineering and management: a critical review and future trends, Automation in Construction 122 (2021) 103517.

[33] A. Darko, A.P.C. Chan, M.A. Adabre, D.J. Edwards, M. Reza Hosseini, E.E. Ameyaw, Artificial intelligence in the AEC industry: scientometric analysis and visualization of research activities, Automation in Construction 112 (2020) 103081.

[34] M. Regona, Y. Tan, B. Xia, R. Yi, M. Li, Opportunities and adoption challenges of AI in the construction industry: a PRISMA review, Journal of Open Innovation: Technology, Market, and Complexity 8 (1) (2022) 45.

[35] K. Asadi, A.K. Suresh, A. Ender, S. Gotad, S. Maniyar, S. Anand, M. Noghabaei, K. Han, E. Lobaton, T. Wu, An integrated UGV-UAV system for construction site data collection, Automation in Construction 112 (2020) 103068.

[36] F. Elghaish, S. Matarneh, S. Talebi, M. Kagioglou, M. Reza Hosseini, S. Abrishami, Toward digitalization in the construction industry with immersive and drones technologies: a critical literature review, Smart and Sustainable Built Environment (2020).

[37] S. Alizadehsalehi, I. Yitmen, T. Celik, D. Arditi, The effectiveness of an integrated BIM/UAV model in managing safety on construction sites, International Journal of Occupational Safety and Ergonomics 26 (4) (2020) 829–844.

[38] R. Shakeri, M.A. Al-Garadi, B. Ahmed, A. Mohamed, T. Khattab, A.K. Al-Ali, K.A. Harras, M. Guizani, Design challenges of multi-UAV systems in cyber-physical applications: a comprehensive survey and future directions, IEEE Communications Surveys and Tutorials 21 (4) (2019) 3340–3385.

[39] J.M.D. Delgado, L. Oyedele, D. Peter, T. Beach, A research agenda for augmented and virtual reality in architecture, engineering and construction, Advanced Engineering Informatics 45 (2020) 101122.

[40] T. Hilfert, M. König, Low-cost virtual reality environment for engineering and construction, Visualization in Engineering 4 (1) (2016) 1–18.

[41] D. Ververidis, S. Nikolopoulos, I. Kompatsiaris, A review of collaborative virtual reality systems for the architecture, engineering, and construction industry, Architecture 2 (3) (2022) 476–496.

[42] Al-Sabbag, Z. Abbas, C.M. Yeum, S. Narasimhan, Enabling human–machine collaboration in infrastructure inspections through mixed reality, Advanced Engineering Informatics 53 (2022) 101709.

[43] E. Forcael, I. Ferrari, A. Opazo-Vega, J.A. Pulido-Arcas, Construction 4.0: a literature review, Sustainability 12 (22) (2020) 9755.

[44] O. Nagy, I. Papp, R.Z. Szabó, Construction 4.0 organisational level challenges and solutions, Sustainability 13 (21) (2021) 12321.

[45] K.-S. Schober, P. Hoff, K. Sold, Digitization in the construction industry: building Europe's road to 'Construction 4.0', Think Act (2016) 1–16.

[46] M. Ghobakhloo, Industry 4.0, digitization, and opportunities for sustainability, Journal of Cleaner Production 252 (2020) 119869.

[47] A. Gilchrist, Industry 4.0: The Industrial Internet of Things, Apress, 2016.

[48] Y. Chen, D. Huang, Z. Liu, O. Mohamed, D. Peter, Construction 4.0, industry 4.0, and building information modeling (BIM) for sustainable building development within the smart city, Sustainability 14 (16) (2022) 10028.

[49] P. Tambare, C. Meshram, C.-C. Lee, R.J. Ramteke, A.L. Imoize, Performance measurement system and quality management in data-driven Industry 4.0: a review, Sensors 22 (1) (2021) 224.

[50] M.-L. Rivera, J. Mora-Serrano, I. Valero, E. Oñate, Methodological-technological framework for construction 4.0, Archives of Computational Methods in Engineering 28 (2) (2021) 689–711.
[51] A. Sawhney, M. Riley, J. Irizarry, M. Riley, in: A. Sawhney, M. Riley, J. Irizarry (Eds.), Construction 4.0, Routledge, 2020 doi 10; 9780429398100.
[52] H. Begić, M. Galić, A systematic review of construction 4.0 in the context of the BIM 4.0 premise, Buildings 11 (8) (2021) 337.
[53] W.S. Alaloul, M.S. Liew, N.A.W.A. Zawawi, I.B. Kennedy, Industrial revolution 4.0 in the construction industry: challenges and opportunities for stakeholders, Ain Shams Engineering Journal 11 (1) (2020) 225–230.
[54] T.D. Oesterreich, T. Frank, Understanding the implications of digitisation and automation in the context of Industry 4.0: a triangulation approach and elements of a research agenda for the construction industry, Computers in Industry 83 (2016) 121–139.
[55] G. de Soto, Borja, I. Agustí-Juan, S. Joss, J. Hunhevicz, Implications of construction 4.0 to the workforce and organizational structures, International Journal of Construction Management 22 (2) (2022) 205–217.
[56] S. Demirkesen, A. Tezel, Investigating major challenges for industry 4.0 adoption among construction companies, Engineering Construction and Architectural Management 29 (3) (2021) 1470–1503.
[57] D. Hall, A. Algiers, T. Lehtinen, R.E. Levitt, C. Li, P. Padachuri, The role of integrated project delivery elements in adoption of integral innovations, in: Engineering Project Organization Conference 2014, Devil's Thumb Ranch, Colorado, July 29-31, 2014, EPOS, 2014, pp. 1–20.
[58] B. Manzoor, I. Othman, J.C. Pomares, Digital technologies in the architecture, engineering and construction (Aec) industry—a bibliometric—qualitative literature review of research activities, International Journal of Environmental Research and Public Health 18 (11) (2021) 6135.
[59] J.M.D. Delgado, L. Oyedele, A. Ajayi, L. Akanbi, O. Akinade, M. Bilal, H. Owolabi, Robotics and automated systems in construction: understanding industry-specific challenges for adoption, Journal of Building Engineering 26 (2019) 100868.
[60] F. Nex, F. Remondino, UAV for 3D mapping applications: a review, Applied Geomatics 6 (1) (2014) 1–15.
[61] M.A. Tamin, N. Darwin, Z. Majid, M.F.M. Ariff, K.M. Idris, Volume estimation of stockpile using unmanned aerial vehicle, in: 2019 9th IEEE International Conference on Control System, Computing and Engineering (ICCSCE), IEEE, 2019, pp. 49–54.
[62] S. Nanayakkara, S. Perera, S. Senaratne, G.T. Weerasuriya, H.M. Nelanga, D. Bandara, Blockchain and smart contracts: a solution for payment issues in construction supply chains, in: Informatics, vol. 8, Multidisciplinary Digital Publishing Institute, 2021, p. 36.
[63] M.-Y. Cheng, D. Kusoemo, R.A. Gosno, Text mining-based construction site accident classification using hybrid supervised machine learning, Automation in Construction 118 (2020) 103265.
[64] M. El Jazzar, H. Urban, C. Schranz, H. Nassereddine, Construction 4.0: a roadmap to shaping the future of construction, in: ISARC. Proceedings of the International Symposium on Automation and Robotics in Construction, vol. 37, IAARC Publications, 2020, pp. 1314–1321.
[65] DJI TERRA modeling framework, https://www.dji.com/dji-terra/info, [Accessed 10 2022].
[66] PIX4D modeling framework, https://www.pix4d.com/.

[67] ContextCapture modeling framework, https://docs.bentley.com/LiveContent/web/ContextCapture%20Help-v17/en/GUID994D9BBA-EC84-4A9E-ADEC-F6E75DDD1F61.html. [Accessed 10 2022].

[68] S. Kristombu Baduge, S. Thilakarathna, J.S. Perera, M. Arashpour, P. Sharafi, B. Teodosio, A. Shringi, P. Mendis, Artificial intelligence and smart vision for building and construction 4.0: machine and deep learning methods and applications, Published in Elsevier Automation in Construction Journal 141 (2022).

[69] E. Baccour, et al., Pervasive AI for IoT Applications: A Survey on Resource-Efficient Distributed Artificial Intelligence, IEEE Communications Surveys & Tutorials, 2022.

[70] H. Xu, R. Chang, M. Pan, H. Li, S. Liu, R.J. Webber, J. Zuo, N. Dong, Application of artificial neural networks in construction management: a scientometric review, MDPI Buildings (2022).

[71] I. Goodfellow, Y. Bengio, A. Courville, Y. Bengio, Deep Learning, vol. 1, MIT press Cambridge, 2016.

[72] A. Darko, A.P.C. Chan, M.A. Adabre, D.J. Edwards, M.R. Hosseini, E.E. Ameyaw, Ameyaw. Artificial intelligence in the AEC industry: scientometric analysis and visualization of research activities, Automotive Construction 112 (2020) 103081.

[73] K. Nagy, S. Gowid, S. Ghani, AI and IoT-Based Concrete Column Base Cover Localization and Degradation Detection Algorithm Using Deep Learning Techniques 11th, 14, Ain Shams Engineering Journal, 2023 102520.

[74] Portland Cement Association, Types and causes of concrete deterioration, in: PCA R&D Serial, vol. 2617, 2002, p. IS536.

[75] Y. Zhao, Y. Wong, J. Dong, Experimental study and analytical model of concrete cover spalling induced by steel corrosion, Journal of Structural Engineering 146 (6) (2020).

[76] J. Li, et al., Automatic pixel-level multiple damage detection of concrete structure using fully convolutional neural network, Computer-Aided Civil and Infrastructure Engineering 34 (2019) 616–634.

[77] H. Perez, J. Tah, A. Mosavu, Deep learning for detecting building defects using convolutional neural networks, Sensors 19 (16) (2019) 3356.

[78] D. Bhavani, A. Adhikari, D. Sumathi, Detection of building defects using convolutional neural networks, in: Proceedings of Second Doctoral Symposium on Computational Intelligence, 2021, pp. 839–855.

[79] F. Elghaish, et al., Developing a new deep learning CNN model to detect and classify highway cracks, Journal of Engineering, Design and Technology (2021), Vols. https://doi.org/10.1108/JEDT-04-2021-0192.

[80] V.a.A.I. Ostankovich, Illegal buildings detection from Satellite images using GoogLeNet and cadastral map, in: 9th IEEE International Conference on Intelligent Systems 2018 at: Madeira, Portugal, 2018, https://doi.org/10.1109/IS.2018.8710565.

Further reading

[1] MathWorks, Deep Learning Toolbox, 2021 [Online]. Available: https://www.mathworks.com/help/deeplearning/. Accessed 10 2022.

[2] F. Elghaish, M. Reza Hosseini, S. Matarneh, S. Talebi, S. Wu, I. Martek, M. Poshdar, N. Ghodrati, Blockchain and the 'Internet of Things' for the construction industry: research trends and opportunities, Automation in Construction 132 (2021) 103942.

[3] S.A. Bello, L.O. Oyedele, O.O. Akinade, M. Bilal, J.M.D. Delgado, L.A. Akanbi, A.O. Ajayi, H.A. Owolabi, Cloud computing in construction industry: use cases, benefits and challenges, Automation in Construction 122 (2021) 103441.
[4] H. Yan, N. Yang, Y. Peng, Y. Ren, Data mining in the construction industry: present status, opportunities, and future trends, Automation in Construction 119 (2020) 103331.
[5] Y. Bao, Z. Chen, S. Wei, Y. Xu, Z. Tang, H. Li, The state of the art of data science and engineering in structural health monitoring, Engineering 5 (2) (2019) 234–242.
[6] F.P. Rahimian, J.S. Goulding, S. Abrishami, S. Seyedzadeh, F. Elghaish, Industry 4.0 Solutions for Building Design and Construction: A Paradigm of New Opportunities, Routledge, 2021.
[7] T.A. Nguyen, P.T. Nguyen, S.T. Do, Application of BIM and 3D laser scanning for quantity management in construction projects, Advances in Civil Engineering (2020) 2020.

Practical prediction of ultimate axial strain and peak axial stress of FRP-confined concrete using hybrid ANFIS-PSO models

8

Denise-Penelope N. Kontoni[1,2] *and Masoud Ahmadi*[3]
[1]Department of Civil Engineering, School of Engineering, University of the Peloponnese, Patras, Greece; [2]School of Science and Technology, Hellenic Open University, Patras, Greece; [3]Department of Civil and Geomechanics Engineering, Arak University of Technology, Arak, Iran

8.1 Introduction

Due to the high cost of construction and maintenance, it is becoming more and more important to maintain structures. By carefully evaluating the behavior of concrete structures, it is evident that several factors, including design errors, improper construction, changes in occupancy, damage caused by accidental loads and impacts, corrosion of steel reinforcement, and aggressive environmental conditions, are responsible for the deterioration of concrete structures [1,2]. As a result of the change in regulations, the loading and safety coefficients have also changed, making it necessary to review the design again in order to repair or strengthen the existing structures, if necessary [3,4].

Therefore, the effort to find efficient materials and techniques for the repair and rehabilitation of existing buildings has attracted the attention of a large number of researchers and institutions [5–8], and their activities have led to the presentation of various methods such as the use of damping devices, steel sheets, reinforced concrete jacketing, steel bracing, reinforced concrete infilling, steel jacketing, composite fiber material, etc. Choosing one of these methods depends on the feasibility of implementing the retrofit method and the configuration of the structure that needs to be retrofitted. Fiber-reinforced polymer (FRP) composites have recently risen to prominence as one of the most popular techniques for strengthening existing reinforced concrete (RC) buildings.

The following are some of the reasons why composite materials are more widely used than conventional materials (concrete and steel):

- Noncorrosive
- Nonmagnetic
- Less maintenance cost
- Lightweight

Artificial Intelligence Applications for Sustainable Construction. https://doi.org/10.1016/B978-0-443-13191-2.00015-8
Copyright © 2024 Elsevier Ltd. All rights are reserved, including those for text and data mining, AI training, and similar technologies.

- High strength characteristics
- High fatigue resistance
- Low thermal expansion coefficient
- Longer service life
- Ease of installation

In contrast to this, FRP composites also have a number of disadvantages that can be mentioned, such as low fire resistance, brittle performance (elastic stress-strain curve), and relatively high fabrication costs.

FRP composite is a type of composite material made up of two parts. The first component is strong fibers, which are surrounded by the second component, the resin matrix. As a result of the fibers, the composite is strong and stiff, and the majority of the load is carried by it. In terms of fiber type, basalt, aramid, carbon, and glass are the most often utilized fibers. In FRP, the matrix connects the fibers, transfers stress to them, and protects them from environmental damage. The compatibility of the matrix with the fiber, as well as the chemical compatibility, is essential. Polymeric matrices are commonly available in thermosetting and thermoplastic forms. There are normally three distinct kinds of thermosetting resins that are utilized in the production of FRP composites. Epoxy resin, vinyl ester resin, and polyester resin are the three types listed here. It is common for epoxy and vinyl ester to be the most prevalent matrixes. Fig. 8.1 illustrates the schematic of an FRP composite.

Different forms of FRP composites can be utilized to reinforce RC structures. String, laminate, wrapping, rebar, tubes, and textiles are examples of common FRP composite materials [9–13]. FRP wrapping is commonly used to boost the confinement, ductility, and shear strength of walls, columns, and beams. This is accomplished by positioning the fibers in the direction of the hoops. A review of retrofitting structures carried out by the authors has revealed that more attention has been paid to the improvement of concrete confinement through the application of FRP materials in RC columns.

In order to fully understand how FRP wrapping affects concrete confinement in different geometric forms of reinforced concrete columns, a thorough analysis is required. A simple explanation can be made by stating that when the concrete lateral

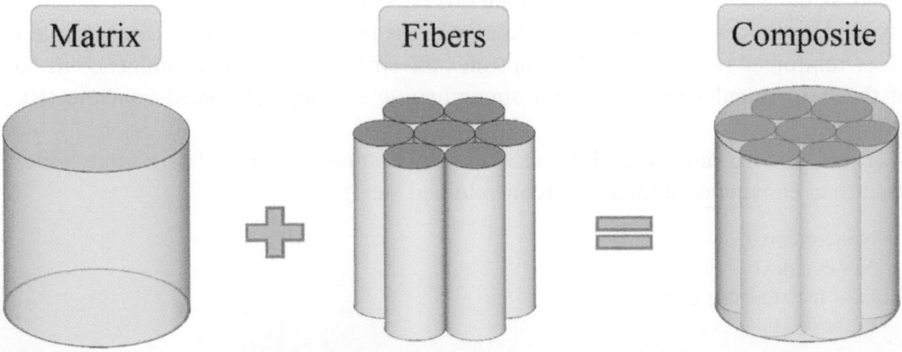

Figure 8.1 Schematic structure of the FRP composite.

strain in the column exceeds a certain limit, the cover starts to separate, and after that, the longitudinal reinforcements between the stirrups may buckle.

Consequently, if the lateral strain can be postponed until it reaches the desired value, the failure of the column will be delayed. The delay can be attributed to the full wrapping strengthening technique of the column using FRP composite material, which also improves the strength of the concrete. A significant aspect of adopting this technique is paying attention to the orientation of the composite fibers, which should be placed in the column's transverse direction. Furthermore, fiber orientation is significant since the fiber has a strong tensile capacity, and the stress induced in the FRP causes confinement by inhibiting the lateral expansion of concrete.

8.2 FRP-confined concrete

The interface between concrete and FRP, which results in concrete confinement, will be examined first in this part. The impact of confinement on the concrete strain-stress curve will then be assessed.

8.2.1 Confinement mechanism

The schematic diagram of the confinement mechanism of FRP-confined concrete (CC) in a circular cross-section is presented in Fig. 8.2. In this figure, P is the applied load; f_L and f_{frp} are the lateral restricting stress applied to the core and the uniform stress distribution in the FRP, respectively; t_{frp} is the FRP composite's thickness; L is the length of the circular sample.

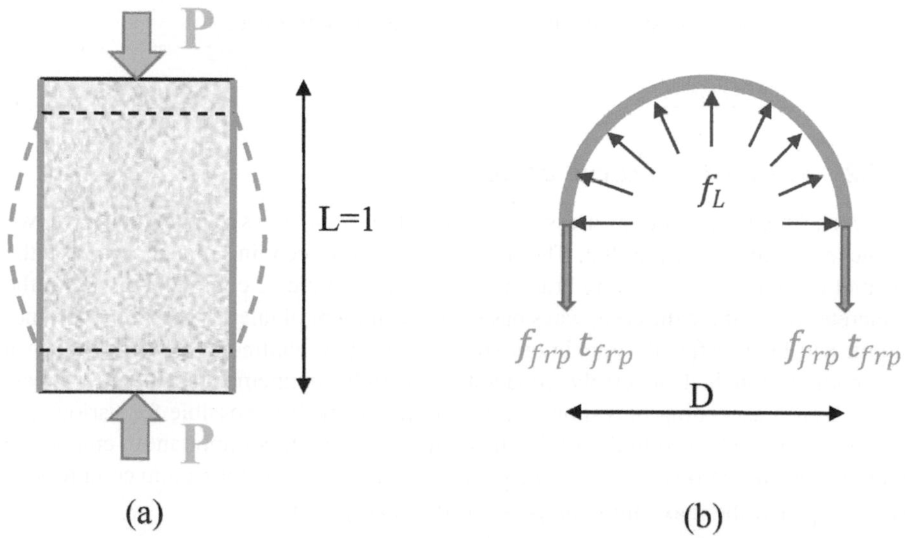

Figure 8.2 Confinement mechanism: (a) FRP wrapping, (b) confining pressure.

Eqs. (8.1) and (8.2) may be used to get the lateral confining stress, which can be determined by considering the force equilibrium.

$$f_L DL = 2f_{frp}t_{frp}L \tag{8.1}$$

$$f_L = \frac{2f_{frp}t_{frp}}{D} \tag{8.2}$$

The uniform stress distribution of the FRP layer may be calculated as follows, since the strain-stress curve in FRP composites is linear:

$$f_{frp} = E_{frp}\varepsilon_{h,frp} \tag{8.3}$$

where, E_{frp} and $\varepsilon_{h,frp}$ are Young's modulus and the hoop rupture strain of FRP composite.

As long as the fibers in the FRP hoop do not fail, the compressibility surrounding the concrete core will grow in accordance with the strains in the FRP hoop until the whole system collapses. As a result, there is a relation between the final passive stress and the ultimate strain experienced by the FRP jacket shortly before it breaks. There is a tensile fracturing of the wrap, which characterizes the final state of the wrap. FRP wraps have been observed to exhibit less tensile capacity at failure than uniaxial wraps. In order to account for the discrepancy that exists between the real strain at which the composite material ruptures and the nominal ultimate strain, the hoop rupture strain ($\varepsilon_{h,frp}$) can be considered as a percentage of the strain capacity ($\varepsilon_{t,frp}$) of FRP composite:

$$\varepsilon_{h,frp} = \beta\varepsilon_{t,frp} \tag{8.4}$$

Eqs. (8.3) and (8.4) are substituted into Eq. (8.2) to produce:

$$f_L = \frac{2E_{frp}\beta\varepsilon_{t,frp}t_{frp}}{D} \tag{8.5}$$

8.2.2 Stress-strain curve of CC

After reaching the maximum stress value in CC, the strain-stress relation may have two branches, as shown in Fig. 8.3. The first curve has a descending branch, whereas the second curve has an ascending branch. It is possible to create confinement within the concrete core in three different ways based on the amount of lateral pressure: Low, medium, and high. After the maximum stress value, low confinement will result in a descending branch. Conversely, moderate and high confinement result in a stress-strain branch ascending in response to the confinement. It is possible for various parameters to be indicated in the strain-stress curves of CC. It is important to emphasize that there are numerous challenges involved in establishing the maximum compressive stress (f'_{cc}) and the maximum compressive strain (ε_{cu}) of CC.

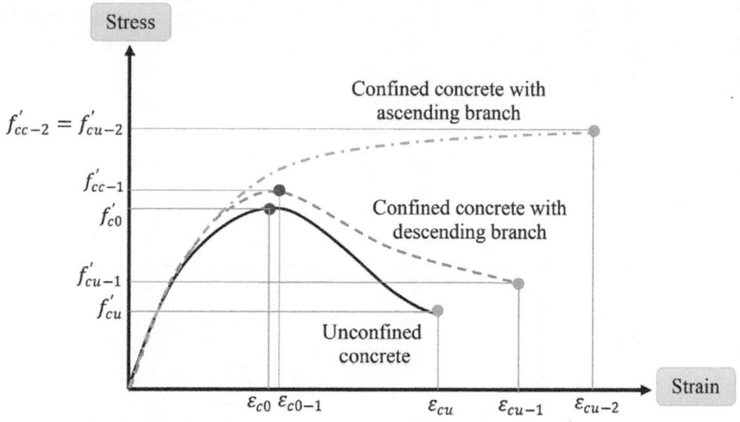

Figure 8.3 Changes in the concrete stress-strain graph due to confinement.

There have been several cases of research over the last few decades aimed at modeling and predicting the compressive behavior of CC with composite wrapping. Based on these studies, researchers have proposed relationships for the maximum compressive stress and the maximum compressive strain of CC. The common existing models for determining f'_{cc} and ε_{cu} are given in Table 8.1.

8.3 Experimental database

A database with 150 samples has been built based on experimental investigations published in the literature. The results were collected from Refs. [18,24–40].

Table 8.2 provides information on the experimental dataset. In this table, D, H, f'_{co}, ε_{c0}, t_{frp}, E_{frp}, $\varepsilon_{h,frp}$, f'_{cc}, and ε_{cu} are sample diameter, cylinder height, unconfined compression capacity, axial strain corresponding to f'_{co}, FRP thickness, Young modulus of composite, hoop rupture strain of FRP, maximum axial stress of CC, and peak axial strain of CC. Table 8.3 presents the statistical features of the experimental database used for this work. The collected database provides a wide range of mechanical and geometrical characteristics. The sample diameter varies from 100 to 153 mm, cylinder height varies from 200 to 305 mm, compressive strength of unconfined concrete (UCC) varies from 29.8 to 55.2 MPa, axial strain corresponding to f'_{co} varies from 0.188% to 0.42%, FRP thickness varies from 0.11 to 3.86 mm, Young modulus of FRP varies from 13.6 to 260 GPa, hoop rupture strain of FRP varies from 0.19% to 3.09%, ultimate axial stress of CC varies from 39 to 162.7 MPa, and peak axial strain of CC varies from 0.37% to 5.55%.

Table 8.1 The typical equations for estimating f'_{cc} and ε_{cu}.

Model	Maximum compressive stress (f'_{cc})	Maximum compressive strain (ε_{cu})	
Saadatmanesh et al. [14]	$f'_{cc} = f'_{co}\left(2.254\sqrt{1+7.94\frac{f_l}{f'_{co}}}-2\frac{f_l}{f'_{co}}-1.254\right)$	$\varepsilon_{cu} = \varepsilon_{c0}\left[1+5\left(\frac{f'_{cc}}{f'_{co}}-1\right)\right]$	–
Samaan et al. [15]	$f'_{cc} = f'_{co} + 6.0(f_l)^{0.7}$	$\varepsilon_{cu} = \frac{f'_{cc}-f_0}{E_c}$	$f_0 = 0.872f'_{co} + 0.371f_l + 6.258$ $E_c = 245.61(f'_{co})^{0.2} + 1.3456\left(\frac{E_{frp}t_{frp}}{D}\right)$
Miyauchi et al. [16]	$f'_{cc} = f'_{co} + 2.98f_l$	$\varepsilon_{cu} = \varepsilon_{c0} + \varepsilon_{c0}(15.87-0.093f'_{co})\left(\frac{f_l}{f'_{co}}\right)^{0.246+0.0064f'_{co}}$	–
Toutanji [17]	$f'_{cc} = f'_{co} + k_1f_l$	$\varepsilon_{cu} = \varepsilon_{c0}\left[1+k_2\left(\frac{f'_{cc}}{f'_{co}}-1\right)\right]$	$k_1 = 3.5\left(\frac{f_l}{f'_{co}}\right)^{-0.15}$ $k_2 = 310.57\varepsilon_l + 1.9$
Shehata et al. [18]	$\frac{f'_{cc}}{f'_{co}} = 1 + 2\frac{f_l}{f'_{co}}$	$\frac{\varepsilon_{cu}}{\varepsilon_{c0}} = 1 + 632\left(\frac{f_l}{f'_{co}} * \frac{f'_{cc}}{E_f}\right)^{0.5}$	–
Mandal et al. [1]	$\frac{f'_{cc}}{f'_{co}} = 0.0017\left(E_l\frac{f_l}{f'_{co}}\right)^2 + 0.0232\left(E_l\frac{f_l}{f'_{co}}\right) + 1$	$\frac{\varepsilon_{cu}}{\varepsilon_{c0}} = 0.0136\left(E_l\frac{f_l}{f'_{co}}\right)^2 + 0.0842\left(E_l\frac{f_l}{f'_{co}}\right) + 1$	$E_l = \frac{E_{frp}t_{frp}}{D}$
Al-Tersawy et al. [19]	$\frac{f'_{cc}}{f'_{co}} = 1 + 1.96\left(\frac{f_l}{f'_{co}}\right)^{0.81}$	$\frac{\varepsilon_{cu}}{\varepsilon_{c0}} = 1 + 8.16\left(\frac{f_l}{f'_{co}}\right)^{0.34}$	–
Youssef et al. [20]	$\frac{f'_{cc}}{f'_{co}} = 1 + 2.25\left(\frac{f_l}{f'_{co}}\right)^{\frac{5}{4}}$	$\varepsilon_{cu} = 0.003368 + 0.259\left(\frac{f_l}{f'_{co}}\right)\left(\frac{f_{frp}}{E_{frp}}\right)^{0.5}$	–

Binici [21]	$\frac{f'_{cc}}{f'_{co}} = m \quad if\left(\frac{f_l}{f'_{co}}\right) \geq 0.14$ $\frac{f'_{cc}}{f'_{co}} = n \quad if\left(\frac{f_l}{f'_{co}}\right) < 0.14$	$\frac{\varepsilon_{cu}}{\varepsilon_{c0}} = 1.75 + 12\left(\frac{f_l}{f'_{co}}\right)\left(\frac{\varepsilon_{h,frp}}{\varepsilon_{c0}}\right)^{0.45}$	$m = 1 + 2.6\left(\frac{f_l}{f'_{co}} - 0.14\right)^{0.17}$ $n = 1.8\left(\frac{f_l}{f'_{co}}\right)^{0.3}$
Benzaid et al. [22]	$\frac{f'_{cc}}{f'_{co}} = 1 + 2.2\frac{f_l}{f'_{co}}$	$\frac{\varepsilon_{cu}}{\varepsilon_{c0}} = 2 + 7.6\frac{f_l}{f'_{co}}$	
Yu and Teng [23]	$f'_{cc} = f'_{co} + 3.5E_l\left(1 - 6.5\frac{f'_{co}}{E_l}\right)\varepsilon_{h,frp}$	$\varepsilon_{cu} = 0.0033 + 0.6\left(\frac{E_l}{f'_{co}}\right)^{0.8}\left(\varepsilon_{h,frp}\right)^{1.45}$	$E_l = \frac{E_{frp}t_{frp}}{D}$

Table 8.2 Details of cylinder samples confined with FRP composites.

No.	Source (Ref.)	D (mm)	H (mm)	f'_{co} (MPa)	ε_{c0} (%)	t_{frp} (mm)	E_{frp} (GPa)	$\varepsilon_{h,frp}$ (%)	f'_{cc} (MPa)	ε_{cu} (%)
1	[24]	152	304	39.7	0.235	0.36	83	0.84	56	1.07
2	[25]	100	200	30.2	0.23	0.17	224.6	0.94	46.6	1.51
3	[25]	100	200	30.2	0.23	0.5	224.6	0.82	87.2	3.11
4	[25]	100	200	30.2	0.23	0.67	224.6	0.76	104.6	4.15
5	[25]	100	200	30.2	0.23	0.15	97.1	2.36	39	1.58
6	[25]	100	200	30.2	0.23	0.29	87.3	3.09	68.5	4.75
7	[25]	100	200	30.2	0.23	0.43	87.3	2.65	92.1	5.55
8	[26]	150	300	34.9	0.21	0.12	200	1.15	44.3	0.85
9	[26]	150	300	34.9	0.21	0.12	200	1.08	42.2	0.72
10	[27]	102	204	38	0.22	1.42	19.9	1.74	57	1.73
11	[27]	102	204	39.4	0.235	1.42	19.9	2.07	63.1	1.6
12	[27]	102	204	39.5	0.235	1.42	19.9	1.89	60.4	1.79
13	[28]	100	200	42	0.239	0.6	82.7	0.89	73.5	1.65
14	[28]	100	200	42	0.239	0.6	82.7	0.95	73.5	1.57
15	[28]	100	200	42	0.239	0.6	82.7	0.8	67.6	1.35
16	[28]	150	300	43	0.24	1.27	13.6	1.53	47.3	1.11
17	[28]	150	300	43	0.24	2.56	13.6	1.39	58.9	1.47
18	[28]	150	300	43	0.24	3.86	13.6	1.33	71	1.69
19	[29]	152	305	33.7	0.226	0.38	105	0.84	47.9	1.2
20	[29]	152	305	33.7	0.226	0.38	105	1.15	49.7	1.4
21	[29]	152	305	33.7	0.226	0.38	105	0.87	49.4	1.24
22	[29]	152	305	33.7	0.226	0.76	105	0.91	64.6	1.65
23	[29]	152	305	33.7	0.226	0.76	105	1	75.2	2.25
24	[29]	152	305	33.7	0.226	0.76	105	1	71.8	2.16
25	[29]	152	305	33.7	0.226	1.14	105	0.82	82.9	2.45
26	[29]	152	305	33.7	0.226	1.14	105	0.9	95.4	3.03
27	[29]	152	305	43.8	0.241	0.38	105	0.81	54.8	0.98

28	[29]	152	305	43.8	0.241	0.38	105	0.76	52.1	0.47
29	[29]	152	305	43.8	0.241	0.38	105	0.28	48.7	0.37
30	[29]	152	305	43.8	0.241	0.76	105	0.92	84	1.57
31	[29]	152	305	43.8	0.241	0.76	105	1	79.2	1.37
32	[29]	152	305	43.8	0.241	0.76	105	1.01	85	1.66
33	[29]	152	305	43.8	0.241	1.14	105	0.79	96.5	1.74
34	[29]	152	305	43.8	0.241	1.14	105	0.71	92.6	1.68
35	[29]	152	305	43.8	0.241	1.14	105	0.84	94	1.75
36	[29]	152	305	55.2	0.255	0.38	105	0.7	57.9	0.69
37	[29]	152	305	55.2	0.255	0.38	105	0.62	62.9	0.48
38	[29]	152	305	55.2	0.255	0.38	105	0.19	58.1	0.49
39	[29]	152	305	55.2	0.255	0.76	105	0.74	74.6	1.21
40	[29]	152	305	55.2	0.255	0.76	105	0.83	77.6	0.81
41	[29]	152	305	55.2	0.255	1.14	105	0.76	106.5	1.43
42	[29]	152	305	55.2	0.255	1.14	105	0.85	108	1.45
43	[29]	152	305	55.2	0.255	1.14	105	0.7	103.3	1.18
44	[30]	150	300	42	0.239	0.149	65	0.55	41	0.73
45	[30]	150	300	42	0.239	0.447	65	1.3	61	1.74
46	[30]	150	300	42	0.239	0.894	65	1.1	85	2.5
47	[30]	150	300	42	0.239	0.117	240	0.95	46	1.1
48	[30]	150	300	42	0.239	0.351	240	1.05	77	2.26
49	[30]	150	300	42	0.239	0.702	240	1.06	108	3.23
50	[31]	102	204	37	0.231	0.16	227	1.2	60	1.02
51	[31]	102	204	32	0.223	0.35	72	1.25	52	1.25
52	[18]	150	300	29.8	0.21	0.165	235	1.23	57	1.23
53	[18]	150	300	29.8	0.21	0.33	235	1.19	72.1	1.74
54	[32]	120	240	43	0.24	0.3	91.1	0.7	58.5	1.16
55	[32]	120	240	43	0.24	0.3	91.1	0.8	65.6	0.95
56	[32]	150	300	38	0.233	0.45	91.1	0.8	62	0.95
57	[32]	150	300	38	0.233	0.45	91.1	0.8	67.3	1.35

Continued

Table 8.2 Continued

No.	Source (Ref.)	D (mm)	H (mm)	f'_{co} (MPa)	ε_{c0} (%)	t_{frp} (mm)	E_{frp} (GPa)	$\varepsilon_{h,frp}$ (%)	f'_{cc} (MPa)	ε_{cu} (%)
58	[33]	152	305	35.9	0.203	0.165	250.5	0.969	47.2	1.106
59	[33]	152	305	35.9	0.203	0.165	250.5	0.981	53.2	1.292
60	[33]	152	305	35.9	0.203	0.165	250.5	1.147	50.4	1.273
61	[33]	152	305	35.9	0.203	0.333	250.5	0.949	71.6	1.85
62	[33]	152	305	35.9	0.203	0.333	250.5	0.988	68.7	1.683
63	[33]	152	305	35.9	0.203	0.333	250.5	1.001	69.9	1.962
64	[33]	152	305	34.3	0.188	0.495	250.5	0.799	82.6	2.046
65	[33]	152	305	34.3	0.188	0.495	250.5	0.884	90.4	2.413
66	[33]	152	305	34.3	0.188	0.495	250.5	0.968	97.3	2.516
67	[33]	152	305	38.5	0.223	1.27	21.8	1.44	51.9	1.315
68	[33]	152	305	38.5	0.223	1.27	21.8	1.89	58.3	1.459
69	[33]	152	305	38.5	0.223	2.54	21.8	1.67	77.3	2.188
70	[33]	152	305	38.5	0.223	2.54	21.8	1.76	75.7	2.457
71	[34]	152	305	41.1	0.256	0.165	250	0.81	52.6	0.9
72	[34]	152	305	41.1	0.256	0.165	250	1.08	57	1.21
73	[34]	152	305	41.1	0.256	0.165	250	1.07	55.4	1.11
74	[34]	152	305	38.9	0.25	0.33	247	1.06	76.8	1.91
75	[34]	152	305	38.9	0.25	0.33	247	1.13	79.1	2.08
76	[34]	152	305	38.9	0.25	0.33	247	0.79	65.8	1.25
77	[35]	152	305	39.6	0.263	0.17	80.1	1.869	41.5	0.825
78	[35]	152	305	39.6	0.263	0.17	80.1	1.609	40.8	0.942
79	[35]	152	305	39.6	0.263	0.34	80.1	2.04	54.6	2.13
80	[35]	152	305	39.6	0.263	0.34	80.1	2.061	56.3	1.825
81	[35]	152	305	39.6	0.263	0.51	80.1	1.955	65.7	2.558
82	[35]	152	305	39.6	0.263	0.51	80.1	1.667	60.9	1.792
83	[36]	152	305	33.1	0.309	0.17	80.1	2.08	42.4	1.303
84	[36]	152	305	33.1	0.309	0.17	80.1	1.758	41.6	1.268

85	[36]		152	305	45.9	0.243	0.17	80.1	1.523	48.4	0.813
86	[36]		152	305	45.9	0.243	0.17	80.1	1.915	46	1.063
87	[36]		152	305	45.9	0.243	0.34	80.1	1.639	52.8	1.203
88	[36]		152	305	45.9	0.243	0.34	80.1	1.799	55.2	1.254
89	[36]		152	305	45.9	0.243	0.51	80.1	1.594	64.6	1.554
90	[36]		152	305	45.9	0.243	0.51	80.1	1.94	65.9	1.904
91	[36]		152	305	38	0.217	0.68	240.7	0.977	110.1	2.551
92	[36]		152	305	38	0.217	0.68	240.7	0.965	107.4	2.613
93	[36]		152	305	38	0.217	1.02	240.7	0.892	129	2.794
94	[36]		152	305	38	0.217	1.02	240.7	0.927	135.7	3.082
95	[36]		152	305	38	0.217	1.36	240.7	0.872	161.3	3.7
96	[36]		152	305	38	0.217	1.36	240.7	0.877	158.5	3.544
97	[36]		152	305	37.7	0.275	0.11	260	0.935	48.5	0.895
98	[36]		152	305	37.7	0.275	0.11	260	1.092	50.3	0.914
99	[36]		152	305	44.2	0.26	0.11	260	0.734	48.1	0.691
100	[36]		152	305	44.2	0.26	0.11	260	0.969	51.1	0.888
101	[36]		152	305	44.2	0.26	0.22	260	1.184	65.7	1.304
102	[36]		152	305	44.2	0.26	0.22	260	0.938	62.9	1.025
103	[36]		152	305	47.6	0.279	0.33	250.5	0.902	82.7	1.304
104	[36]		152	305	47.6	0.279	0.33	250.5	1.13	85.5	1.936
105	[36]		152	305	47.6	0.279	0.33	250.5	1.064	85.5	1.821
106	[37]		152	305	48.1	0.222	1	85	1.052	80.9	1.51
107	[37]		152	305	48.1	0.222	1	85	1.124	86.6	1.53
108	[37]		152	305	48.1	0.222	2	85	0.968	109.4	2.01
109	[37]		152	305	48.1	0.222	2	85	1.221	126.7	2.66
110	[37]		152	305	48.1	0.222	3	85	1.158	162.7	3.09
111	[37]		152	305	48.1	0.222	3	85	1.035	153.6	2.89
112	[37]		152	305	48.1	0.222	1	85	1.047	84.2	1.55
113	[37]		152	305	48.1	0.222	1	85	1.216	87.9	1.69
114	[37]		152	305	48.1	0.222	2	85	1.062	123.3	2.37

Continued

Table 8.2 Continued

No.	Source (Ref.)	D (mm)	H (mm)	f'_{co} (MPa)	ε_{c0} (%)	t_{frp} (mm)	E_{frp} (GPa)	$\varepsilon_{h,frp}$ (%)	f'_{cc} (MPa)	ε_{cu} (%)
115	[37]	152	305	48.1	0.222	2	85	0.89	108.2	1.93
116	[37]	152	305	48.1	0.222	3	85	1.089	156.5	3.13
117	[37]	152	305	48.1	0.222	3	85	1.136	157	2.84
118	[37]	152	305	47.76	0.22	1.25	22	2.02	59.1	1.35
119	[37]	152	305	47.76	0.22	1.25	22	2.143	59.8	1.15
120	[37]	152	305	47.76	0.22	2.5	22	2.032	88.9	2.21
121	[37]	152	305	47.76	0.22	2.5	22	2.114	88	2.21
122	[37]	152	305	47.76	0.22	3.75	22	2.113	113.2	2.85
123	[37]	152	305	47.76	0.22	3.75	22	2.11	112.5	2.8
124	[37]	152	305	47.76	0.22	1.25	22	2.179	63.4	1.51
125	[37]	152	305	47.76	0.22	1.25	22	2.116	62.4	1.35
126	[37]	152	305	47.76	0.22	2.5	22	2.074	89.7	2.14
127	[37]	152	305	47.76	0.22	2.5	22	2.049	88.3	2.05
128	[37]	152	305	47.76	0.22	3.75	22	1.893	108	2.62
129	[38]	153	305	45	0.28	0.17	98.7	1.46	46.7	1.06
130	[38]	153	305	45	0.28	0.17	98.7	1.24	48	0.84
131	[38]	153	305	45	0.28	0.34	98.7	1.33	59.5	1.36
132	[38]	153	305	45	0.28	0.34	98.7	1.44	57.4	1.38
133	[38]	153	305	45	0.28	0.51	98.7	1.45	69.6	1.61
134	[38]	153	305	45	0.28	0.51	98.7	1.45	70	1.48
135	[39]	150	300	44.4	0.345	0.111	250	1.34	63.56	1.17
136	[39]	150	300	44.4	0.345	0.111	250	1.35	61.71	1.4
137	[39]	150	300	44.4	0.345	0.111	250	1.45	61.9	1.08
138	[39]	150	300	44.4	0.345	0.222	250	1.18	78.94	1.42
139	[39]	150	300	44.4	0.345	0.222	250	1.43	78.94	1.82
140	[39]	150	300	44.4	0.345	0.222	250	1.26	78.74	1.56
141	[39]	150	300	44.4	0.345	0.333	250	1.31	97.04	2.06

142	[39]	150	300	44.4	0.345	0.333	250	1.2	95.09	1.82
143	[39]	150	300	44.4	0.345	0.333	250	1.31	95.87	2.21
144	[40]	150	300	45	0.31	0.167	144	1.17	61.3	1.8
145	[40]	150	300	45	0.35	0.167	144	1.21	57.4	1.61
146	[40]	150	300	45	0.41	0.501	144	1.41	115.7	4.78
147	[40]	150	300	45	0.39	0.507	77.9	1.57	85.1	4.48
148	[40]	150	300	45	0.41	0.507	77.9	1.76	82	3.98
149	[40]	150	300	45	0.42	0.509	77.9	1.82	105.4	5.18
150	[40]	150	300	45	0.41	0.509	77.9	1.59	110.7	5.07

Table 8.3 Statistical characteristics of collected experimental specimens.

Statistical parameters	Minimum	Maximum	Mean	Range	Standard deviation	Coefficient of variation
D (mm)	100	153	146.413	53	15.179	0.104
H (mm)	200	305	293.427	105	30.583	0.104
f'_{co} (MPa)	29.8	55.2	42.082	25.4	6.018	0.143
ε_{c0} (%)	0.188	0.42	0.251	0.232	0.046	0.182
t_{frp} (mm)	0.11	3.86	0.813	3.75	0.858	1.055
$E_{frp}(GPa)$	13.6	260	133.387	246.4	83.775	0.628
$\varepsilon_{h,frp}(\%)$	0.19	3.09	1.254	2.9	0.482	0.384
f'_{cc} (MPa)	39	162.7	75.927	123.7	27.183	0.358
ε_{cu} (%)	0.37	5.55	1.822	5.18	0.966	0.529

8.4 ANFIS

The Takagi-Sugeno fuzzy system, a kind of artificial neural network, is utilized to build the adaptive neuro-fuzzy inference system (ANFIS). It combines the strengths of neural networks with fuzzy logic to achieve optimal performance. Fuzzy if-then expressions that may approximate nonlinear functions are used by the inference system.

The ANFIS structure is made up of five layers: Fuzzification, rule, normalization, defuzzification, and target layers.

Layer 1: This layer is composed of adaptive nodes. Nodes are defined by membership functions (MFs), such as Gaussian MF.

Layer 2: The second layer is in charge of developing the rule-based firing strengths.

Layer 3: This layer divides each value by the overall firing strength in order to normalize the computed firing strengths.

Layer 4: In this layer, adaptive nodes are present with node functions that indicate how the rules contribute to the overall result. Parameters will be referred to as subsequent parameters in this layer.

Layer 5: One node is responsible for computing the sum of all the outputs of the rules.

Choosing a model structure containing MFs, fuzzy operators, an inference type, aggregation operations, and a defuzzification mechanism is the first step in developing an ANFIS model. The system is then trained using a subset of the data known as the training class. The generated prediction model gets more accurate during the training phase by changing the system characteristics. As a feedforward network structure, ANFIS is used. Fig. 8.4 depicts the five layers of this system in schematic form for two inputs, two MFs, and one output [41].

In the first layer, the inputs are fuzzified based on the considered MFs. In this layer, A_1 and A_2 are MFs of variable x_1 and B_1 and B_2 are MFs of variable x_2. For each

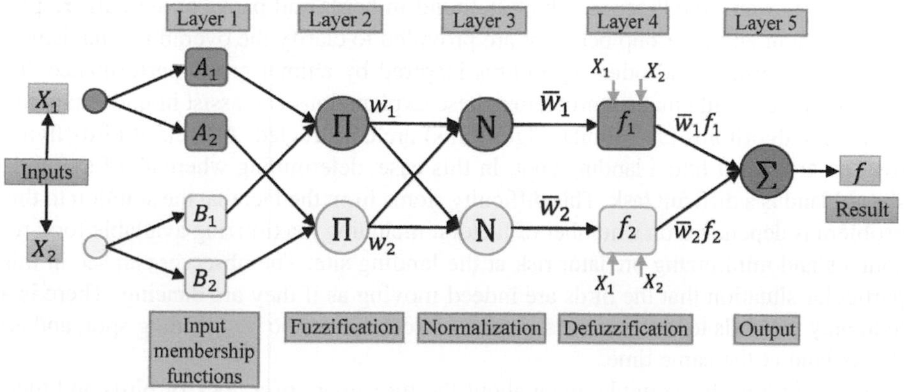

Figure 8.4 Schematic structure of ANFIS for two inputs.

variable, the type of MF (Gaussian, triangular, S-shaped, etc.) must be determined. The determined fuzzy rules are examined in the second layer, which are as follows:

$$\text{Rule 1: If } x_1 \text{ is } A_1 \text{ and } x_2 \text{ is } B_1 \text{ then } f_1 = a_1x_1 + a_2x_2 + c_1 \tag{8.6}$$

$$\text{Rule 2: If } x_1 \text{ is } A_2 \text{ and } x_2 \text{ is } B_2 \text{ then } f_2 = a_3x_1 + a_4x_2 + c_2 \tag{8.7}$$

In the above relationships, the values of a_1, a_2, a_3, a_4, c_1, and c_2 are constant coefficients, and the function f is regarded as a linear function of the inputs. The output of each rule is represented by the expression w. The values of w are obtained by multiplying the values received from the MFs if the product T-norm is employed in this phase. In the third layer, w values are normalized ($\overline{w}$). In the fourth step, the values of the normalized weights are multiplied by the corresponding output values ($\overline{w}f$). In the last step, the values obtained from the fourth step are added together. To establish ideal values corresponding to MFs, these actions are taken in each step. There are several ways to choose the right MF coefficients; in this work, the particle swarm optimization (PSO) approach is employed.

8.5 PSO

In the early 1990s, a large number of studies were undertaken on the social behavior of animal groups. In these studies, it was found that some animals, such as birds, fish, and others, are capable of sharing information between their own groups, which gives these animals a significant advantage in terms of survival. These studies served as inspiration for [42], who published an article in 1995 that debuted the PSO algorithm. Nonlinear continuous functions can be optimized using this metaheuristic algorithm.

In the article mentioned, the authors developed the PSO algorithm based on the concept of swarm intelligence, which is found in herds and packs of animals. Explanations about animal group behavior are provided to clarify the overall mechanism of the PSO algorithm and other algorithms inspired by animal group performance. To solve complex mathematical problems, these explanations can assist in understanding how PSO algorithms (and similar algorithms) are constructed. A flock of birds flying over an area must find a landing spot. In this case, determining where all of the birds should land is a difficult task. This difficulty stems from the fact that the solution to this problem is dependent on a number of factors, including maximizing available food resources and minimizing predator risk at the landing site. The observer can see in this particular situation that the birds are indeed moving as if they are dancing. There is a tendency for birds to move at the same time to determine the best landing spot, and all flocks land at the same time.

According to the example given about the movement of a flock of birds and their simultaneous landings, the members of the flock can share information with each other. Each bird in a flock will arrive at a separate location and time if they cannot communicate with one another. In accordance with bird behavior research, all birds in a flock looking for a good landing spot are able to detect when the spot has already been discovered by a member of the flock. The members of this crowd use this knowledge to balance their personal knowledge experience with their crowd knowledge experience, which is known as social knowledge. A landing place must fulfill a number of requirements before it may be considered suitable for landing, one of which is the survival circumstances that will prevail at the landing spot. This involves, as previously said, maximizing food availability and minimizing predator danger. The problem of finding the best landing point is an optimization problem. It is critical that the group choose the optimal landing spot, such as latitude and longitude, in order to optimize the odds of survival for its members.

Optimization issues seek to identify the parameter indicated by the vector $\boldsymbol{X} = [x_1, x_2, x_3, \ldots, x_n]$ and maximize or minimize it depending on the optimization formula provided by the $f(\boldsymbol{X})$ function. Alternatively, the function $f(\boldsymbol{X})$ is known as the fitness function or the objective function, and it assesses how good or terrible a position $\boldsymbol{X}$ is.

After finding a spot, the bird considers the location's suitability for landing, which is a function of the bird's flocking problem. This evaluation is made for the problem of the landing of the group of birds based on a variety of survival criteria. Consider a swarm of P particles, with each particle's position and velocity vectors defined as follows in each iteration:

$$X_i^t = (x_{i1}x_{i2}x_{i3}\ldots x_{in})^T \tag{8.8}$$

$$V_i^t = (v_{i1}v_{i2}v_{i3}\ldots v_{in})^T \tag{8.9}$$

The following equation is used to update these vectors depending on dimension j:

$$V_{ij}^{t+1} = \omega V_{ij}^t + c_1 r_1^t \left(\text{pbest}_{ij} - X_{ij}^t\right) + c_2 r_2^t \left(\text{gbest}_j - X_{ij}^t\right) \qquad i = 1, 2, 3, \ldots, P \text{ and } j = 1, 2, 3, \ldots, n \tag{8.10}$$

$$X_{ij}^{t+1} = X_{ij}^t + V_{ij}^{t+1} \quad i = 1, 2, 3, \ldots, P \text{ and } j = 1, 2, 3, \ldots, n. \tag{8.11}$$

According to Eq. (8.10), particles move as a consequence of three separate variables throughout an iteration. Eq. (8.11) updates the position of the particles. The parameter w, which is the inertia weight constant in PSO's classical form, has a constant positive value.

Choosing a suitable value for this parameter is crucial for striking a good compromise between global search (also known as discovery) and local search (when set to lower values). The first term of updated velocity in Eq. (8.10) is the inner product of the parameter w and the particle's previous velocity. r_1 is another parameter that contributes to the multiplication of the second expression. In order to maximize the possible global optimum, this random parameter plays a crucial role in preventing convergence of the parameters. The last concept is social learning. By adjusting this value, the optimal location may be broadcast to the whole swarm without regard to which individual particle found it.

8.6 Hybrid ANFIS-PSO models

In this part, two useful models for predicting the maximum compressive stress (f'_{cc}) and maximum compressive strain (ε_{cu}) of FRP-CC are described. It was necessary to create and test several hybrid models to check the best structure of the hybrid model for obtaining suitable results. At the end of the process, the best candidate was selected. Note that strength reduction factors for determining output are assumed to be one. Following the presentation of each model, the degree of accuracy in predicting the objective function is thoroughly examined.

8.6.1 Maximum compressive stress (f'_{cc})

In order to create an efficient model, choosing the input parameters of the model is a challenging problem. A careful review of the existing models, as well as the results reported in the literature, has been conducted in order to select the inputs of the suggested approach for the compressive strength of CC. The following parameters are used as input in this section:

- Diameter of the test sample, D (mm)
- Height of cylinder sample, H (mm)
- Unconfined compressive strength, f'_{co} (MPa)
- Corresponding strain to the peak stress of UCC, ε_{c0} (%)
- FRP wrap thickness, t_{frp} (mm)
- Elastic modulus of FRP, E_{frp} (GPa)
- Hoop rupture strain of FRP, $\varepsilon_{h,frp}$ (%)

There is a need to normalize the data to reduce the amount of system error in order to ensure greater homogeneity between the input vectors and target values. In this study, the data have been normalized in the range between 0.1 and 0.9. The equation used for each parameter (inputs or outputs) is as follows:

$$\theta_{i,scaled} = \frac{0.8}{[\max(\theta_i) - \min(\theta_i)]} * (\theta_i - \min(\theta_i)) + 0.1 \tag{8.12}$$

where, the values of $\theta_{i,scaled}$ and θ_i are the normalized and real values of each parameter, respectively.

8.6.1.1 Details of ANFIS-PSO hybrid model for f'_{cc}

In this section, the steps of the suggested approach are presented first, followed by the details of the best-trained system, which also performs well on test data. In order to achieve the desired structure, there are six steps to follow:

(a) The relation utilized to normalize the input parameters is represented by Eq. (8.12). A detailed description of the normalization step can be found in Table 8.4.

(b) The second step involves selecting the kind and number of MFs for each parameter and determining the optimal state of the parameters, as described in Section 8.6.1.2.

(c) At this stage, the linear functions in the fourth layer of Fig. 8.4 are calculated. Details of these functions are presented in Section 8.6.1.3.

(d) The result of the method is determined from the following equation using the hybrid ANFIS-PSO model, which is trained on train data and tested on another set.

$$\text{Output} = \frac{\sum_{j=1}^{6} W_j CL_j}{\sum_{j=1}^{6} W_j} \tag{8.13}$$

(e) Finally, the intelligent system's predicted output value, which is in the range of 0.1–0.9, should be returned to its original range, which is in the range of laboratory outputs. The following equation is used to perform this mapping.

$$f'_{cc}\ (\text{MPa}) = 154.625\left(f'_{cc\,scaled} - 0.1\right) + 39 \tag{8.14}$$

In Fig. 8.5, the optimal intelligent system based on the ANFIS-PSO algorithm is shown, which will be presented in detail later.

8.6.1.2 Details of MFs

As explained in the previous sections, each input parameter in the presented intelligent system has a certain number of MFs. In this investigation, the type of these functions is considered fixed. Based on a review of similar studies, it is recommended to determine the kind of MF and the number of clusters through trial and error, and their optimal

Table 8.4 Equations used in the data normalization step.

Type of parameter	Parameter	Normalization function
Input	D	$0.01509434 * (D - 100) + 0.1$
	H	$0.00761905 * (H - 100) + 0.1$
	f'_{co}	$0.03149606 * (f'_{co} - 29.8) + 0.1$
	ε_{c0}	$3.44827586 * (\varepsilon_{c0} - 0.188) + 0.1$
	t_{frp}	$0.21333333 * (t_{frp} - 0.11) + 0.1$
	E_{frp}	$0.00324675 * (E_{frp} - 13.6) + 0.1$
	$\varepsilon_{h,frp}$	$0.27586207 * (E_{frp} - 0.19) + 0.1$
Output	f'_{cc}	$0.00646726 * (f'_{cc} - 39) + 0.1$

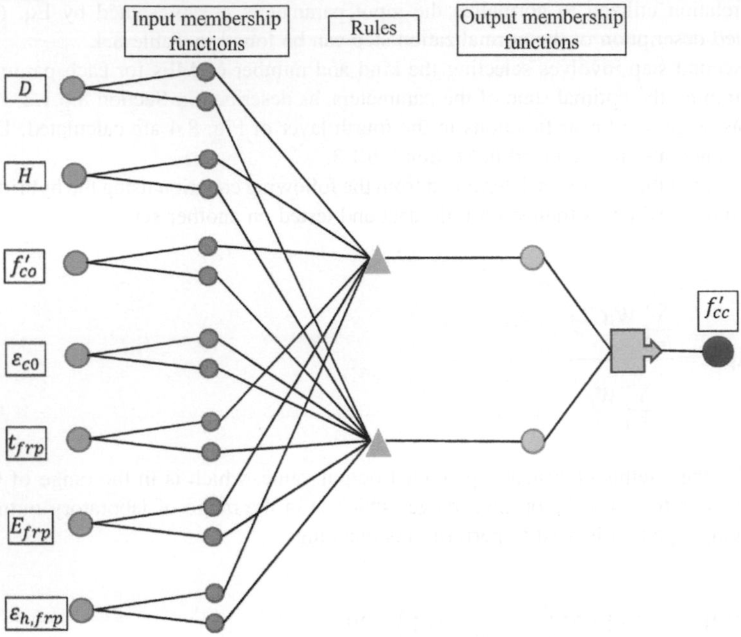

Figure 8.5 The proposed f'_{cc} model's schematic structure.

state depends on the type of problem, the number of input parameters, the number of samples, and the dispersion of parameters. In this study, various types of MFs were explored, and finally, the Gaussian MF (Eq. 8.15) was used for all input parameters. Around the average point, this function has a bell-shaped and symmetrical shape, and it tends to zero rapidly.

$$\mu(x;\sigma,c) = exp\left(-\frac{(x-c)^2}{2\sigma^2}\right) \tag{8.15}$$

where, σ and c are, respectively, the standard deviation and the mean of the Gaussian MF for variable x. The average determines the location of the peak of the MF, and the standard deviation indicates its expansion.

In the proposed model, the command genfis3 in MATLAB software is used to create the ANFIS system, and c-means classification is used to select the cluster centers. The fuzzy version of this technique is known as fuzzy c-means (FCM), and membership values are determined based on the relative distance to the cluster centers. It should be noted that, as a result, the FCM method has become susceptible to errors in data and is impacted by them. After generating the initial ANFIS, the parameters of the MFs have been optimized by the PSO algorithm. In this case, the values were obtained after several iterations, and the corresponding results are shown in Fig. 8.6. Also, the parameters of the MFs are presented in Table 8.5. It is important to mention that the consistency of the change range of the input variables is due to

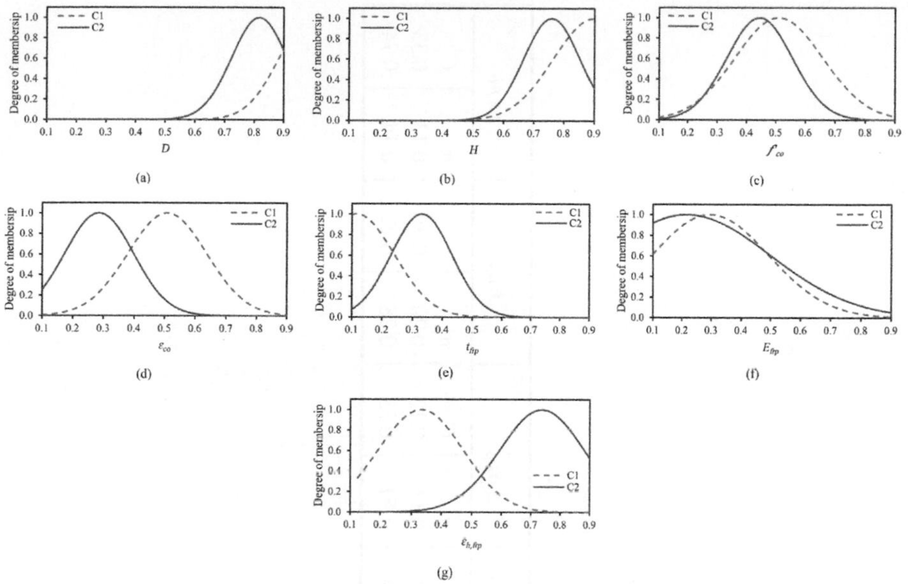

Figure 8.6 Using Gaussian MFs as input parameters: (a) D, (b) H, (c) f'_{co}, (d) ε_{c0}, (e) t_{frp}, (f) E_{frp}, and (g) $\varepsilon_{h,frp}$.

the fact that the data were normalized between 0.1 and 0.9 before training began. In Fig. 8.6, the index C_i represents the MF.

8.6.1.3 Linear functions

The linear functions created by the FCM method depend on the number of inputs and a constant number. Multiplying the constant coefficients in the inputs and adding them to a constant number achieves this goal. This equation is presented in general form in Eq. (8.16).

$$CL_j = a_1X_1 + a_2X_2 + a_3X_3 + a_4X_4 + a_5X_5 + a_6X_6 + a_7X_7 + C \quad j = 1,2 \tag{8.16}$$

In the above equation, CL_j represents a cluster. The coefficients a_1 to a_5 correspond to the number of hybrid ANFIS-PSO inputs. Additionally, the coefficient C represents the constant value of the equation. According to Table 8.6, the values relating to the parameters mentioned in Eq. (8.16) can be found.

8.6.1.4 Fuzzy rules

In order to estimate the expected target value, an adaptive fuzzy inference neural network requires a rule base. The proposed system makes use of two rules, which are listed in Table 8.7. The compressive stress of CC is represented in this table as f'_{cc}.

Table 8.5 Values of the standard deviation and mean of Gaussian MFs.

Inputs	***D***		***H***		f'_{co}		ε_{c0}		t_{frp}		E_{frp}		$\varepsilon_{h,frp}$	
Parameters	σ	C	σ	C	σ	C	σ	C	σ	C	σ	C	σ	C
C1	0.11	0.99	0.14	0.90	0.15	0.51	0.13	0.51	0.13	0.12	0.20	0.30	0.13	0.33
C2	0.09	0.82	0.09	0.76	0.11	0.44	0.11	0.29	0.10	0.33	0.28	0.22	0.14	0.74

Table 8.6 The linear function coefficients of the proposed system.

Cluster	Parameters							
	a_1	a_2	a_3	a_4	a_5	a_6	a_7	C
Cluster 1 = CL1	−5.932	5.709	0.01459	0.5052	1.49	0.1133	−0.1579	−0.05199
Cluster 2 = CL2	−6.05	5.777	0.3628	−0.7966	0.5108	1.42	−0.04826	0.02077

Table 8.7 Fuzzy rules.

Number	Rule
Rule1	If D is $C1_D$ and H is $C1_H$ and f'_{co} is $C1_{f'_{co}}$ and ε_{c0} is $C1_{\varepsilon_{c0}}$ and t_{frp} is $C1_{t_{frp}}$ and E_{frp} is $C1_{E_{frp}}$ and $\varepsilon_{h,frp}$ is $C1_{\varepsilon_{h,frp}}$ then f'_{cc} is CL_1
Rule2	If D is $C2_D$ and H is $C2_H$ and f'_{co} is $C2_{f'_{co}}$ and ε_{c0} is $C2_{\varepsilon_{c0}}$ and t_{frp} is $C2_{t_{frp}}$ and E_{frp} is $C2_{E_{frp}}$ and $\varepsilon_{h,frp}$ is $C2_{\varepsilon_{h,frp}}$ then f'_{cc} is CL_2

8.6.1.5 Stress model performance

The effectiveness of the suggested model for predicting the compressive strength of CC is assessed here. The evaluation is done in two parts. In the first part, the performance of the suggested model is assessed based on statistical parameters. The second section includes a diagram that compares the suggested model's findings to the laboratory data. The mean absolute percentage error (MAPE), correlation coefficient (R), mean square error (MSE), root mean square error (RMSE), mean absolute error (MAE), and normalized MAE were determined to measure the accuracy of the proposed relation statistically. In order to provide a clear comparison between model errors, the mentioned criteria are assessed using actual output values converted from normalized data. These statistical parameters' mathematical expressions are as follows:

$$\mathrm{MAPE} = \frac{100}{n}\sum_{i=1}^{n}\frac{\left|\left(f'_{cci,exp.} - f'_{cci,model}\right)\right|}{f'_{cci,exp.}} \tag{8.17}$$

$$R^2 = \frac{\left[\sum_{i=1}^{n}\left(f'_{cci,exp.} - \overline{f'_{ccexp.}}\right)\left(f'_{cci,model} - \overline{f'_{ccmodel}}\right)\right]^2}{\sum_{i=1}^{n}\left(f'_{cci,exp.} - \overline{f'_{ccexp.}}\right)^2 \sum_{i=1}^{n}\left(f'_{cci,model} - \overline{f'_{ccmodel}}\right)^2} \tag{8.18}$$

$$\mathrm{MSE} = \frac{\sum_{i=1}^{n}\left(f'_{cci,exp.} - f'_{cci,model}\right)^2}{n} \tag{8.19}$$

$$\mathrm{RMSE} = \sqrt{\frac{\sum_{i=1}^{n}\left(f'_{cci,exp.} - f'_{cci,model}\right)^2}{n}} \tag{8.20}$$

$$\mathrm{MAE} = \frac{1}{n}\sum_{i=1}^{n}\left|\left(f'_{cci,exp.} - f'_{cci,model}\right)\right| \tag{8.21}$$

Table 8.8 Statistical results for the f'_{cc} model.

Relation	Statistical indices				
	R	MAPE	MAE	MSE	RMSE
ANFIS-PSO	0.955	9.405	6.434	63.983	7.998
[15]	0.931	12.354	9.867	178.577	13.363
[16]	0.952	9.453	6.845	77.161	8.784
[19]	0.936	12.217	10.527	213.241	14.603

In Table 8.8, the obtained values for statistical indices are presented. A right response to the compressive strength of CC produces a higher R-value and lower MAPE, MSE, RMSE, and MAE values. As a result of the obtained findings, the suggested model has demonstrated a suitable ability to predict the compressive strength of CC.

8.6.1.6 Comparison of compressive model with existing relations

The preceding section's statistical indicators were utilized to compare the developed framework against current models. Table 8.8 summarizes the acquired findings. The results demonstrate that the suggested method fits better than the current models.

8.6.2 Maximum compressive strain (ε_{cu})

Based on the input parameters utilized in the suggested framework for maximum compressive strength, an intelligent model was created to determine the maximum compressive strain (ε_{cu}). Furthermore, the data was normalized between 0.1 and 0.9 before developing the model. In creating the proposed model for strain, the same assumptions were used as in creating the model for compressive strength. Table 8.9 contains information on the standard deviation and mean of Gaussian MFs.

Tables 8.10 and 8.11 show the linear function coefficients and fuzzy rules for the proposed ultimate strain model.

8.6.2.1 Strain model performance

To assess the findings of the suggested maximum strain model, the statistical indices presented in Section 8.6.1.5 were used. The following table shows the values associated with the proposed model and the existing relationship. The findings indicate that the recommended model is sufficiently accurate.

8.6.2.2 Comparison of strain model with existing relations

Based on the collected data, this section compares the proposed model's results to the models developed by Refs. [15,16,19]. Table 8.12 summarizes the findings. The obtained results demonstrate that the recommended model can predict the peak strain of CC with a suitable approximation.

Table 8.9 Details of standard deviation and mean of Gaussian MFs.

Inputs	D		H		f'_{co}		ε_{c0}		t_{frp}		E_{frp}		$\varepsilon_{h,frp}$	
Parameters	σ	C	σ	C	σ	C	σ	C	σ	C	σ	C	σ	C
C1	0.05	1.09	0.13	1.09	0.07	0.40	0.11	0.32	0.06	0.31	0.44	1.27	−0.05	0.34
C2	0.05	0.87	0.04	0.93	0.09	0.36	0.04	0.05	0.39	0.63	0.17	0.23	0.13	0.46
C3	2.26	0.46	0.41	0.12	0.35	0.39	0.08	0.26	0.06	0.43	0.05	0.35	0.03	0.35
C4	0.28	−0.16	0.04	0.21	0.02	0.25	−0.07	0.58	0.08	−0.69	−0.09	1.27	0.08	1.01

Table 8.10 The linear function coefficients of the ultimate strain model.

Cluster	Parameters							
	a_1	a_2	a_3	a_4	a_5	a_6	a_7	C
Cluster 1 = CL1	−4.028	3.383	−0.1057	0.3923	3.259	0.01807	0.2641	0.1479
Cluster 2 = CL2	−4.16	3.694	−0.2211	0.2268	0.3204	0.2452	0.4996	0.28
Cluster 3 = CL3	−3.791	3.767	−0.2783	0.1375	0.5661	0.319	0.2207	0.05427
Cluster 4 = CL4	−3.91	4.406	−0.293	0.3648	0.03099	0.1227	0.2183	−0.07793

Table 8.11 Fuzzy rules for the ultimate strain model.

Number	Rule
Rule1	If D is $C1_D$ and H is $C1_H$ and f'_{co} is $C1_{f'_{co}}$ and ε_{c0} is $C1_{\varepsilon_{c0}}$ and t_{frp} is $C1_{t_{frp}}$ and E_{frp} is $C1_{E_{frp}}$ and $\varepsilon_{h,frp}$ is $C1_{\varepsilon_{h,frp}}$ then f'_{cc} is CL_1
Rule2	If D is $C2_D$ and H is $C2_H$ and f'_{co} is $C2_{f'_{co}}$ and ε_{c0} is $C2_{\varepsilon_{c0}}$ and t_{frp} is $C2_{t_{frp}}$ and E_{frp} is $C2_{E_{frp}}$ and $\varepsilon_{h,frp}$ is $C2_{\varepsilon_{h,frp}}$ then f'_{cc} is CL_2
Rule3	If D is $C3_D$ and H is $C3_H$ and f'_{co} is $C3_{f'_{co}}$ and ε_{c0} is $C3_{\varepsilon_{c0}}$ and t_{frp} is $C3_{t_{frp}}$ and E_{frp} is $C3_{E_{frp}}$ and $\varepsilon_{h,frp}$ is $C3_{\varepsilon_{h,frp}}$ then f'_{cc} is CL_3
Rule4	If D is $C4_D$ and H is $C4_H$ and f'_{co} is $C4_{f'_{co}}$ and ε_{c0} is $C4_{\varepsilon_{c0}}$ and t_{frp} is $C4_{t_{frp}}$ and E_{frp} is $C4_{E_{frp}}$ and $\varepsilon_{h,frp}$ is $C4_{\varepsilon_{h,frp}}$ then f'_{cc} is CL_4

Table 8.12 Statistical findings for the strain model of CC.

	Statistical indices				
Relation	***R***	**MAPE**	**MAE**	**MSE**	**RMSE**
ANFIS-PSO	0.9072	19.9616	0.3016	0.1684	0.4104
Samaan et al. [15]	0.5313	63.6538	1.2311	0.02224	0.1497
Miyauchi et al. [16]	0.8354	20.9061	0.1639	0.00067	0.0259
Al-Tersawy et al. [19]	0.7681	26.5901	0.07	0.00011	0.0105

8.7 Conclusions

In recent years, FRP composites have been routinely employed to reinforce RC columns. To fully comprehend the behavior of these columns, it is necessary to thoroughly study the compressive behavior of CC. As part of the strain-stress curve of CC, it is important to determine two values of the maximum compressive stress and the maximum compressive strain. Based on the database compiled from the literature and the use of the hybrid ANFIS-PSO methodology, two effective models for determining these two parameters were presented in this study. The compressive strength of UCC, the corresponding strain to the peak stress of UCC, the height of the cylinder sample, the sample diameter, the FRP wrap thickness, the elastic modulus of FRP, and the hoop rupture strain of FRP are the input parameters in both models.

In order to generate efficient models, the data collected was normalized between 0.1 and 0.9. The optimal model in this study was obtained through numerous trials and errors, since reaching the final model requires the investigation of numerous parameters. Both models employ a Gaussian MF. Additionally, there are two and four input MFs in the models for maximum stress and ultimate strain, respectively.

Examining the performance of the suggested models using statistical indices used to evaluate model correctness. The proposed models' results have also been compared to the relationships developed by Refs. [15,16,19]. In this work, it was shown that the suggested models could properly estimate the maximum compressive strength and ultimate strain in FRP-CC and could be used in the preliminary evaluation of FRP-reinforced concrete columns.

References

[1] S. Mandal, A. Hoskin, A. Fam, Influence of concrete strength on confinement effectiveness of fiber-reinforced polymer circular jackets, ACI Structural Journal 102 (3) (2005) 383, https://doi.org/10.14359/14409.

[2] H. Naderpour, O. Poursaeidi, M. Ahmadi, Shear resistance prediction of concrete beams reinforced by FRP bars using artificial neural networks, Measurement 126 (2018) 299–308, https://doi.org/10.1016/j.measurement.2018.05.051.

[3] I.S. Abbood, S. Aldeen Odaa, K.F. Hasan, M.A. Jasim, Properties evaluation of fiber reinforced polymers and their constituent materials used in structures–a review, Materials Today: Proceedings 43 (2021) 1003–1008, https://doi.org/10.1016/j.matpr.2020.07.636.

[4] A.U. Al-Saadi, T. Aravinthan, W. Lokuge, Structural applications of fibre reinforced polymer (FRP) composite tubes: a review of columns members, Composite Structures 204 (2018) 513–524, https://doi.org/10.1016/j.compstruct.2018.07.109.

[5] K. Jiang, Q. Han, Y. Bai, X. Du, Data-driven ultimate conditions prediction and stress-strain model for FRP-confined concrete, Composite Structures 242 (2020) 112094, https://doi.org/10.1016/j.compstruct.2020.112094.

[6] A.F. Pour, T. Ozbakkaloglu, T. Vincent, Simplified design-oriented axial stress-strain model for FRP-confined normal-and high-strength concrete, Engineering Structures 175 (2018) 501–516, https://doi.org/10.1016/j.engstruct.2018.07.099.

[7] J. Shin, S. Park, Optimum retrofit strategy of FRP column jacketing system for non-ductile RC building frames using artificial neural network and genetic algorithm hybrid approach, Journal of Building Engineering 57 (2022) 104919, https://doi.org/10.1016/j.jobe.2022.104919.

[8] T. Vincent, T. Ozbakkaloglu, Influence of concrete strength and confinement method on axial compressive behavior of FRP confined high-and ultra high-strength concrete, Composites Part B: Engineering 50 (2013) 413–428, https://doi.org/10.1016/j.compositesb.2013.02.017.

[9] M.K. Askar, A.F. Hassan, Y.S.S. Al-Kamaki, Flexural and shear strengthening of reinforced concrete beams using FRP composites: a state of the art, Case Studies in Construction Materials (2022) e01189, https://doi.org/10.1016/j.cscm.2022.e01189.

[10] H.A. Bengar, M. Mousavi, Performance of an innovative anchorage system for strengthening RC beams in adjacency of columns with FRP laminates, Structures 28 (2020) 197–204, https://doi.org/10.1016/j.istruc.2020.08.075.

[11] W. Ge, M. Han, Z. Guan, P. Zhang, A. Ashour, W. Li, W. Lu, D. Cao, S. Yao, Tension and bonding behaviour of steel-FRP composite bars subjected to the coupling effects of chloride corrosion and load, Construction and Building Materials 296 (2021) 123641, https://doi.org/10.1016/j.conbuildmat.2021.123641.

[12] S. Reichenbach, P. Preinstorfer, M. Hammerl, B. Kromoser, A review on embedded fibre-reinforced polymer reinforcement in structural concrete in Europe, Construction and Building Materials 307 (2021) 124946, https://doi.org/10.1016/j.conbuildmat.2021.124946.

[13] Y. Zhou, X. Wang, L. Sui, F. Xing, Y. Wu, C. Chen, Flexural performance of FRP-plated RC beams using H-type end anchorage, Composite Structures 206 (2018) 11–21, https://doi.org/10.1016/j.compstruct.2018.08.015.

[14] H. Saadatmanesh, M.R. Ehsani, M.-W. Li, Strength and ductility of concrete columns externally reinforced with fiber composite straps, Structural Journal 91 (4) (1994) 434–447, https://doi.org/10.14359/4151.

[15] M. Samaan, A. Mirmiran, M. Shahawy, Model of concrete confined by fiber composites, Journal of Structural Engineering 124 (9) (1998) 1025–1031, https://doi.org/10.1061/(ASCE)0733-9445(1998)124:9(1025).

[16] K. Miyauchi, S. Inoue, T. Kuroda, A. Kobayashi, Strengthening effects with carbon fiber sheet for concrete column, Japan Concrete Institute 21 (3) (1999) 1453–1458.

[17] H. Toutanji, Stress-strain characteristics of concrete columns externally confined with advanced fiber composite sheets, Materials Journal 96 (3) (1999) 397–404, https://doi.org/10.14359/639.

[18] I.A.E.M. Shehata, L. A. v Carneiro, L.C.D. Shehata, Strength of short concrete columns confined with CFRP sheets, Materials and Structures 35 (1) (2002) 50–58, https://doi.org/10.1007/BF02482090.

[19] S.H. Al-Tersawy, O.A. Hodhod, A.A. Hefnawy, Reliability and code calibration of RC short columns confined with CFRP wraps, in: Proceedings of the 8th International Symposium on Fiber Reinforced Polymer Reinforcement for Concrete Structures, University of Patras, Patras, Greece, 2007.

[20] M.N. Youssef, M.Q. Feng, A.S. Mosallam, Stress–strain model for concrete confined by FRP composites, Composites Part B: Engineering 38 (5–6) (2007) 614–628, https://doi.org/10.1016/j.compositesb.2006.07.020.

[21] B. Binici, Design of FRPs in circular bridge column retrofits for ductility enhancement, Engineering Structures 30 (3) (2008) 766–776, https://doi.org/10.1016/j.engstruct.2007.05.012.

[22] R. Benzaid, H. Mesbah, N.E. Chikh, FRP-confined concrete cylinders: axial compression experiments and strength model, Journal of Reinforced Plastics and Composites 29 (16) (2010) 2469–2488, https://doi.org/10.1177/0731684409355199.

[23] T. Yu, J.G. Teng, Design of concrete-filled FRP tubular columns: provisions in the Chinese technical code for infrastructure application of FRP composites, Journal of Composites for Construction 15 (3) (2011) 451–461, https://doi.org/10.1061/(ASCE)CC.1943-5614.0000159.

[24] F. Picher, Confinement of concrete cylinders with CFRP, fiber composites in infrastructure, Proceedings of the First International Conference on Composites in Infrastructure (1996) 829–841.

[25] K. Watanabe, H. Nakamura, Y. Honda, M. Toyoshima, M. Iso, T. Fujimaki, M. Kaneto, N. Shirai, Confinement effect of FRP sheet on strength and ductility of concrete cylinders under uniaxial compression, in: Non-Metallic (FRP) Reinforcement for Concrete Structures. Japan Concrete Institute. Proceedings of the Third International Symposium vol 1, 1997, pp. 233–240.

[26] S. Matthys, L. Taerwe, K. Audenaert, Tests on axially loaded concrete columns confined by fiber reinforced polymer sheet wrapping, Special Publication 188 (1999) 217–228, https://doi.org/10.14359/5624.

[27] S. Kshirsagar, R.A. Lopez-Anido, R.K. Gupta, Environmental aging of fiber-reinforced polymer-wrapped concrete cylinders, Materials Journal 97 (6) (2000) 703–712, https://doi.org/10.14359/9985.

[28] P. Rochette, P. Labossiere, Axial testing of rectangular column models confined with composites, Journal of Composites for Construction 4 (3) (2000) 129–136, https://doi.org/10.1061/(ASCE)1090-0268(2000)4:3(129).
[29] Y. Xiao, H. Wu, Compressive behavior of concrete confined by carbon fiber composite jackets, Journal of Materials in Civil Engineering 12 (2) (2000) 139–146, https://doi.org/10.1061/(ASCE)0899-1561(2000)12:2(139).
[30] C. Aire, R. Gettu, J.R. Casas, Study of the compressive behavior of concrete confined by fiber reinforced composites, The International Conference on Composites in Construction 1 (2001) 239–243.
[31] F. Micelli, J.J. Myers, S. Murthy, Effect of environmental cycles on concrete cylinders confined with FRP, in: Proceedings of CCC2001 International Conference on Composites in Construction, Porto, Portugal, 2001.
[32] L. De Lorenzis, F. Micelli, A. la Tegola, Influence of specimen size and resin type on the behaviour of FRP-confined concrete cylinders, in: Advanced Polymer Composites for Structural Applications in Construction: Proceedings of the First International Conference, 2002, pp. 231–240.
[33] L. Lam, J.G. Teng, Ultimate condition of fiber reinforced polymer-confined concrete, Journal of Composites for Construction 8 (6) (2004) 539–548, https://doi.org/10.1061/(ASCE)1090-0268(2004)8:6(539).
[34] L. Lam, J.G. Teng, C.H. Cheung, Y. Xiao, FRP-confined concrete under axial cyclic compression, Cement and Concrete Composites 28 (10) (2006) 949–958, https://doi.org/10.1016/j.cemconcomp.2006.07.007.
[35] J.G. Teng, T. Yu, Y.L. Wong, S.L. Dong, Hybrid FRP–concrete–steel tubular columns: concept and behavior, Construction and Building Materials 21 (4) (2007) 846–854, https://doi.org/10.1016/j.conbuildmat.2006.06.017.
[36] T. Jiang, J.G. Teng, Analysis-oriented stress–strain models for FRP–confined concrete, Engineering Structures 29 (11) (2007) 2968–2986, https://doi.org/10.1016/j.engstruct.2007.01.010.
[37] C. Cui, S.A. Sheikh, Experimental study of normal-and high-strength concrete confined with fiber-reinforced polymers, Journal of Composites for Construction 14 (5) (2010) 553–561, https://doi.org/10.1061/(ASCE)CC.1943-5614.0000116.
[38] J.L. Zhao, T. Yu, J.G. Teng, Stress-strain behavior of FRP-confined recycled aggregate concrete, Journal of Composites for Construction 19 (3) (2015) 04014054, https://doi.org/10.1061/(ASCE)CC.1943-5614.0000513.
[39] G.M. Chen, Y.H. He, T. Jiang, C.J. Lin, Behavior of CFRP-confined recycled aggregate concrete under axial compression, Construction and Building Materials 111 (2016) 85–97, https://doi.org/10.1016/j.conbuildmat.2016.01.054.
[40] W. Wang, C. Wu, Z. Liu, H. Si, Compressive behavior of ultra-high performance fiber-reinforced concrete (UHPFRC) confined with FRP, Composite Structures 204 (2018) 419–437, https://doi.org/10.1016/j.compstruct.2018.07.102.
[41] A.F. Güneri, T. Ertay, A. Yücel, An approach based on ANFIS input selection and modeling for supplier selection problem, Expert Systems with Applications 38 (12) (2011) 14907–14917, https://doi.org/10.1016/j.eswa.2011.05.056.
[42] J. Kennedy, R. Eberhart, Particle swarm optimization, Proceedings of ICNN'95-International Conference on Neural Networks 4 (1995) 1942–1948.

9 Prediction of long-term dynamic responses of a heritage masonry building under thermal effects by automated kernel-based regression modeling

Hassan Sarmadi[1,2], *Bahareh Behkamal*[3] and *Alireza Entezami*[3]
[1]Head of Research and Development, IPESFP Company, Mashhad, Iran; [2]Department of Civil Engineering, Faculty of Engineering, Ferdowsi University of Mashhad, Mashhad, Iran; [3]Department of Civil and Environmental Engineering, Politecnico di Milano, Milan, Italy

9.1 Introduction

Heritage masonry structures reflect the identities of a nation or society. Different types of these structures from ordinary or special buildings to bridges are cultural heritage for future generations. Because the construction of heritage masonry structures date back several thousand years, governmental authorities and relevant organizations make serious attempts to conserve and preserve such structures due to their crucial importance, economic development, and prosperity for tourists. In contrast to modern structures constructed from iron, steel, and reinforced or prestressed concrete, masonry structures are often composed of heterogenous materials such as brick, stone, adobe, unreinforced concrete blocks, etc. Hence, high specific mass, low tensile and shear strengths, and limited ductility loaded out of a plane are some mechanical and structural properties of such structures [1].

The major challenges concerning masonry structures relate to their uncertain information about the mechanical properties of materials, design codes, construction techniques, and operations. An obvious fact is that the passage of time and environmental factors degrade the material characteristics and structural behavior of these structures, which often have brittle properties [2]. Similar to other civil structures, masonry buildings and bridges are prone to various external natural and man-made loads. Material deterioration, aging, and catastrophic events such as earthquakes, landslides, settlements, floods, etc. profoundly threaten their health and integrity. For these reasons, it is indispensable to assess the vulnerability and restoration of heritage masonry structures to not only conserve them but also avoid catastrophic incidents such as collapse [3]. Although regular visual inspections via trained specialists are conventional ways of monitoring of masonry structures, those have their disadvantages [1]. Recently, thanks to a recent breakthrough in sensing technologies, structural health monitoring

Artificial Intelligence Applications for Sustainable Construction. https://doi.org/10.1016/B978-0-443-13191-2.00010-9
Copyright © 2024 Elsevier Ltd. All rights are reserved, including those for text and data mining, AI training, and similar technologies.

(SHM) under non-destructive tests has become the major methodology for evaluating the stability and integrity of heritage masonry structures [2].

SHM in civil engineering is a methodology for understanding the current status of a civil structure in terms of early damage assessment [4], damage localization [5], damage type recognition [6], damage quantification [7], and damage prognosis [8]. This methodology consists of some main steps: Sensing and data acquisition (DAQ), numerical modeling and model updating, feature extraction, and feature analysis. Sensors are inevitable parts of any SHM project, thereby providing rich information about the mechanical properties and structural responses of any civil structure [9,10]. Due to its importance, SHM can be called a sensor-oriented technology. The choice of an appropriate sensor system depends on various factors such as the type of civil structure, the objective of SHM, the type of test (i.e., static or dynamic), the type of measured data, the type of sensor deployment (i.e., contact or noncontact), the geographical locations of the structure, weather conditions, the duration of monitoring (i.e., short-term or long-term), etc. For heritage masonry structures, sensing systems are intended to measure important structural responses, including displacement, compressive stress, tensile stress, strain, and acceleration, as well as some critical environmental factors such as temperature and relative humidity, and the characteristics of some obvious surface damage patterns such as cracks, spalling, etc. Some of these parameters can be measured by well-known and conventional sensors (e.g., accelerometers, stain-gauges, thermocouples, and humidity sensors) in wired or wireless communication networks [9,10] and others can be obtained by some novel sensing technologies (e.g., displacements via remote sensors, surface damage patterns by digital cameras [11], smartphones [43], smart bricks [12], etc.).

Once sensors and DAQ systems are installed in civil structures, in some cases, it is necessary to extract latent contents from raw measured data. To perform this process, feature extraction offers different techniques for discovering meaningful engineering features. For example, modal properties including natural frequencies, mode shapes, and damping ratios extracted from acceleration responses are among commonly-used dynamic features for the health monitoring of masonry structures. On the other hand, the other aspect of SHM is to develop numerical models of civil structures for further evaluations such as vulnerability assessment, rehabilitation and retrofitting, and simulations of structural behavior under various excitation loads [13]. Due to the uncertain nature and heterogenous material properties of masonry structures, there may be large discrepancies between the numerical and real models. To address this limitation, finite element model updating is prevalent in the SHM of such structures [14,15]. Using measured data or extracted features obtained from either field monitoring or numerical modeling, the final step of SHM is to analyze and infer these features for early damage assessment, localization, quantification, type recognition, and prognosis.

Ambient vibration testing (AVT) in conjunction with operational modal analysis (OMA) is a powerful and effective technique for dynamic health monitoring of masonry structures. The technique relies upon weak or strong ambient excitations caused by human activities (e.g., live loads applied to a structure), environmental (e.g., wind, temperature, earthquake, etc.), and operational (e.g., traffic on a masonry bridge) loads.

Since these excitations are unmeasurable, structural responses are the only outputs of the AVT for SHM. In this regard, the OMA presents different output-only modal identification techniques [16] to extract the key dynamic features of a civil structure; that is, natural frequencies, mode shapes, and damping ratios.

For understanding the state of a civil structure and assessing the occurrence of damage, modal frequencies are the best choices. This is because of the direct relationship between the modal frequency and structural stiffness, which is the key indicator for damage occurrence. In this case, any reduction in structural modal frequencies is representative of damage. For this reason, modal frequency-based damage assessment is an important strategy for SHM of different civil structures [17–20]. However, any change in modal frequencies cannot be interpreted as damage. In SHM, it is well accepted that structural natural frequencies are profoundly sensitive to environmental and operational variability conditions resulting from changes in temperature, humidity, wind speed and direction, excessive loadings, etc. [21]. These conditions can produce deceptive variations in modal frequencies similar to damage, while there is no damage in the structure. Notably, thermal effects can have linear or nonlinear correlations with structural modal frequencies, which are observed as downward and upward trends [22,44]. An important fact is that variability mechanisms governing thermal effects can be fundamentally different among civil structures depending upon their construction materials and structural types. In masonry structures with high structural stiffness, there exists a positive relationship between modal frequencies and temperature (i.e., an increase in natural frequencies with increasing temperature) [23]. In particular, the closure of internal microcracks in the mortar layers caused by thermal expansion is the main cause for modal frequency increases under warm temperature conditions. In contrast, this positive correlation mechanism may be different for modern civil structures such as reinforced concrete bridges or retrofitted structures [24].

More precisely, Ubertini et al. [25] assessed the influences of temperature and humidity on the modal frequencies of a masonry bell tower in a 1-year monitoring program. They reported important conclusions that temperature had further impacts on the frequency changes; the modal frequencies of bending modes made positive correlations with the temperature, while torsional modes behaved oppositely; and freezing weather led to a negative correlation between the temperature and modal frequencies. Kita et al. [23] studied the effects of seasonal temperature on the static and dynamic behavior of a historical masonry building. They demonstrated that seasonal temperature variations had negative correlations with the structure modal frequencies. Such negative correlations, which differed from other stiff masonry structures, might stem from strengthening influences and the existence of moderate damage. Zonno et al. [26] evaluated the short- and long-term effects of temperature and humidity on the structural characteristics of an adobe church building. It was demonstrated that seasonal changes of environmental parameters (i.e., temperature and humidity) in the long-term monitoring program had a significant impact (i.e., about 8%) on the dynamic responses (structural natural frequencies), while this impact decreased during the short-term monitoring scheme, especially on a daily scale. Mario Azzara et al. [27] researched into the influence of long-term ambient temperature on the modal frequencies of a masonry bell tower. Their experimental study indicated that temperature variability in a year can vary the tower natural frequencies by around 5%–6%, while no changes in the mode shapes were observed. It was also suggested that the frequencies increased when the temperature increased. An interesting result of their research was related to the linear relationship of the first and second bending modes but the nonlinear relationship of the third mode associated with the torsional mode.

Long-term SHM via AVT and OMA is a valuable and informative strategy that can provide rich information on structural responses and environmental factors. In this regard, it is necessary to use field monitoring data acquired from temporary or permanent sensor systems and extract dynamic features (i.e., modal data) for further activities. However, the implementation of any long-term monitoring program is not a trivial task. In some cases, due to the condition of a civil structure or its geographical location, this process may be unaffordable, labor-intensive, and time-consuming. Damage to sensor systems or the recording of noisy data in harsh environments are big limitations of long-term monitoring, in which case one needs to regularly inspect sensing systems and repair or replace faulty sensors. The other limitation of this process is that it may produce big data or miss some important measured variables (i.e., missing data). To deal with these important challenges, recent progress in artificial intelligence and machine learning allows civil engineers to take advantage of various deep learning and statistical learning algorithms for predicting structural responses rather than directly measuring them. Indeed, the problem of data prediction is an alternative to field monitoring, thereby forecasting upcoming or unseen structural responses via various regression models [28].

On the subject of predicting structural responses, Ubertini et al. [25] used a multivariate regression model to forecast the modal frequencies of a masonry bell tower using air temperature and relative humidity data as the main predictors. Mario Azzara et al. [27] took advantage of an autoregressive with eXogenous input model to predict the natural frequencies of a masonry bell tower by taking only the air temperature as the key environmental factor and predictor. Mu and Yuen [29] proposed a Bayesian statistical learning method for three tasks: Uncertainty quantification, sparse feature selection, and modal frequency prediction under environmental variations caused by temperature and humidity. Mu et al. [30] proposed a novel probabilistic method called Bayesian copula-based uncertainty quantification, which was able to predict structural modal frequencies with limited information. In relation to machine learning-based predictive models, Hua et al. [31] suggested a support vector regression (SVR) model and indicated that this supervised regressor with the influence of thermal inertia can achieve accurate frequency prediction of a long-span cable-stayed bridge. Laory et al. [32] proposed different supervised regression techniques developed from artificial neural networks, SVR, regression trees, and random forests to predict the long-term modal frequencies of a suspension bridge using different environmental and operational factors (i.e., temperature, wind, and traffic). They demonstrated that the SVR and random forest are the most proper choices for predicting the modal frequencies of long-span bridges subjected to temperature and traffic loadings. For other types of structural responses, Behkamal et al. [33] investigated the thermal effects on displacements of long-span bridges by using a few synthetic aperture radar images obtained from spaceborne remote sensing. In their research, the authors utilized three supervised regression techniques, including multivariate linear regression (MLR), Gaussian process regression (GPR), and SVR. It was concluded that these regressors operate well when temperature is the major environmental factor affecting the structural responses.

Despite some research studies on the prediction of dynamic responses of civil structures, how to leverage more elaborate predictive models in an automated manner with the aid of advanced machine learning algorithms is a missing piece of research activity on this topic, especially heritage masonry structures. For this reason, this chapter aims to propose an automated predictive method through kernel-based regression modeling for predicting unseen modal frequencies of a heritage masonry building. The proposed method conforms to a single-variate prediction framework, thereby forecasting the modal frequencies of one mode. On this basis, it initially makes an attempt to find the most appropriate kernel regressor between the GPR and SVR. For this purpose, an automated regressor selection algorithm with the aid of Bayesian hyperparameter optimization (BHO) is developed to choose an optimum regression model for each mode and simultaneously tune the main unknown elements (hyperparameters) of the selected kernel regressor. In the prediction process, the temperature records from some thermocouples installed in the masonry building generate an multivariate set of predictor data, while the modal frequencies of each mode make a univariate set of response data. Results show that the proposed predictive method can systematically select the most appropriate kernel regression model for each mode and it succeeds in predicting the unseen modal frequencies with high accuracy.

9.2 Kernelized regression models

A prediction problem via regression modeling is implemented into *training* and *predicting* phases. During the training phase, the underlying aim is to learn mapping from predictor (independent or input) data $\mathbf{x} = \{x_i\} \in \Re^n_{i=1}$ to response (dependent or output) data $\mathbf{y} = \{y_i\} \in \Re^n_{i=1}$, given a labeled training set of input-output pairs $\mathbb{D}_{\mathrm{Tr}} = \{x_i, y_i\} \in \Re^n_{i=1}$, where n denotes the number of training points. Once a supervised regressor is trained, the main objective in the predicting phase is to forecast m unseen response data $\hat{\mathbf{y}} = \{\hat{y}_i\} \in \Re^m_{i=1}$ by feeding m test points $\mathbb{D}_{\mathrm{Te}} = \{\widehat{x}_i\} \in \Re^m_{i=1}$, which are only related to new predictor data, into the trained regressor. With the training information, the regression model is mainly defined as:

$$\mathbf{y} = f(\mathbf{x}) + \boldsymbol{\varepsilon} \tag{9.1}$$

where $\boldsymbol{\varepsilon} \sim \mathcal{N}(0, \sigma_\varepsilon^2)$ and $\mathcal{N}$ refers to the Gaussian (normal) distribution. On this basis, most of the regression techniques aim at representing the function $f(\mathbf{x})$. Once this function is determined, it predicts the response data by applying the test data, i.e., this expression is formulated as $\hat{\mathbf{y}} = f(\widehat{\mathbf{x}})$.

To evaluate the performance of any regression model, one can exploit some tried-and-tested metrics such as R-squared (R^2) and mean squared error (MSE).

The R-squared, or coefficient of determination is a relative statistical metric varying from zero to one. This metric is mainly intended to represent the goodness-of-fit of a regression model by using the original and predicted response data. Mathematically speaking, given the original and predicted response data points, which are designated here as y and $\hat{y}$, the R-squared is derived as follows:

$$R^2 = 1 - \frac{\sum_{i=1}^{n} (y_i - \widehat{y}_i)^2}{\sum_{i=1}^{n} (y_i - \mu_y)^2} \tag{9.2}$$

where μ_y denotes the average (mean) value of the observed response data. When $R^2 = 1$, this means that the best (ideal) goodness-of-fit occurs, yielding the best prediction performance. In this case, the predicted response data exactly matches the observed response data. In contrast, when $R^2 = 0$, it makes sense of the worst prediction performance. For this condition, there are considerable discrepancies between the original and predicted response data. On the other hand, the MSE of a regressor measures the average of the squares of the regressor residuals, which is equivalent to the average squared difference between the observed and predicted response points. The MSE is expressed as follows:

$$MSE = \frac{1}{n} \sum_{i=1}^{n} (y_i - \widehat{y}_i)^2 \tag{9.3}$$

Although the MSE does not present a relative metric similar to the R-squared, it is still a useful measure for evaluating the performance of regression models, especially when comparing different regressors or automated regressor selection based on BHO. In this regard, the smaller the MSE value, the better the model's performance.

9.2.1 Introduction to kernel learning

Kernel-based machine learning involves a specific extension of linear methods to nonlinear problems. For this purpose, one can learn a kernel-based learning model by mapping the original feature set into a dimension-expanded empirical space and then applying linear techniques to a new and high-dimensional feature space [34]. Using these steps, it is possible to leverage various kernel machine learning methods for different tasks. In relation to the prediction problem, a kernel regressor mainly aims to map the original data in a low-dimensional space into a high-dimensional space and then fit a linear model to the mapped data in the high-dimensional space by minimizing regularized objective functions [34]. In cases of nonlinear behavior among features, kernel-based regression methods systematically facilitate identifying an optimum nonlinear function of the predictor (input) data to properly predict the response (output) data.

Although kernel-based regressors are often suitable for nonlinear cases, it is also feasible to exploit them for linear cases. In this regard, the utilization of a linear model

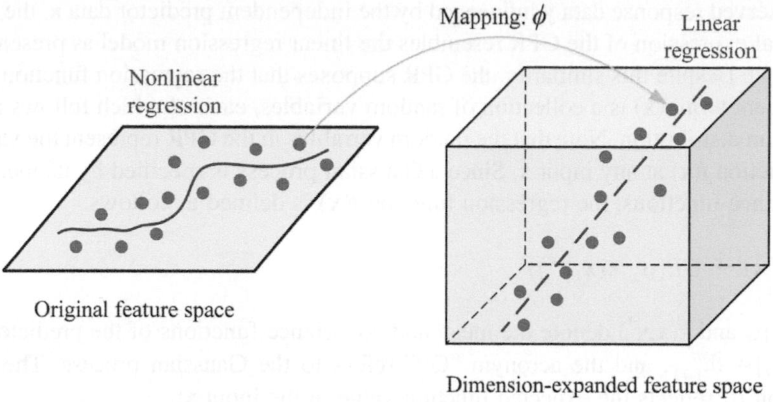

Figure 9.1 Graphical representation of the kernel trick and kernel mapping for the regression problem.

in a high-dimensional space corresponds to incorporating a nonlinear kernel function into the original data in its initial low-dimensional space. Since the representation of a nonlinear model is not trivial, the *kernel trick* is the best and most adoptable solution to kernel-based methods. Simply speaking, instead of incorporating nonlinear functions for mapping nonlinear features into the high-dimensional empirical space with linear behavior, one can exploit *dot products* of feature samples in the original space, which is the fundamental basis of the kernel trick. Accordingly, a kernel is a measure of the distance between two data samples (feature vectors). In this case, kernel methods compute the similarity between samples through inner products between mapped samples and constitute a kernel matrix. For two feature samples $\mathbf{x}$ and $\mathbf{x}^*$, the kernel trick is equivalent to:

$$\kappa(\mathbf{x}, \mathbf{x}^*) = \langle \varphi(\mathbf{x}), \varphi(\mathbf{x}^*) \rangle \tag{9.4}$$

where $\langle . \rangle$ refers to the dot product and φ is the mapping function. For the sake of convenience, Fig. 9.1 depicts the graphical representation of the kernel trick and kernel mapping for the regression problem.

9.2.2 Gaussian process regression

The GPR is a probabilistic kernel-based learning technique for predicting unseen response data. The fundamental principle of the GPR relies on the theory of the Gaussian process [35]. In machine learning, this process relates to the class of random/stochastic strategies, which aims at representing the realizations of random samples. To put it another way, the Gaussian process can make a collection of random samples, provided that any subset of these samples is jointly Gaussian [35]. The key assumption of the Gaussian process is that similar inputs can generate similar outputs, in which case one can assume a statistical model (i.e., GPR) for data prediction. Given the observed response data **y** influenced by the independent predictor data **x**, the mathematical expression of the GPR resembles the linear regression model as presented in Eq. (9.1). Despite this similarity, the GPR supposes that the regression function or independent term $f(\mathbf{x})$ is a collection of random variables, each of which follows a joint Gaussian distribution. Note that the random variables in the GPR represent the value of the function $f(\mathbf{x})$ at any input **x**. Since a Gaussian process is specified by its mean and covariance functions, the regression function $f(\mathbf{x})$ is defined as follows:

$$f(\mathbf{x}) \sim GP(\boldsymbol{\mu}_x, \kappa(\mathbf{x}, \mathbf{x}^*)) \tag{9.5}$$

where $\boldsymbol{\mu}_x$ and $\kappa(\mathbf{x},\mathbf{x}^*)$ denote the mean and covariance functions of the predictor data $\mathbf{x} = \{x_i\} \in \Re^n_{i=1}$, and the acronym "GP" refers to the Gaussian process. The mean function $\boldsymbol{\mu}_x$ reflects the expected function value at the input **x**:

$$\boldsymbol{\mu}_x = \mathbb{E}[f(\mathbf{x})] \tag{9.6}$$

which is equivalent to the average of all functions in the distribution evaluated at the input **x**. Moreover, the covariance function $\kappa(\mathbf{x},\mathbf{x}^*)$ represents the dependence between the function values at different input points (vectors) **x** and $\mathbf{x}^*$:

$$\kappa(\mathbf{x}, \mathbf{x}^*) = \mathbb{E}[(f(\mathbf{x}) - \boldsymbol{\mu}_x)(f(\mathbf{x}^*) - \boldsymbol{\mu}_{x^*})] \tag{9.7}$$

where the function κ is commonly called the *kernel* of the Gaussian process [36]. For the GPR, the exponential kernel, squared exponential kernel, and rational quadratic kernel are among the most prevalent kernel functions, which are expressed as follows:

$$\kappa(\mathbf{x}, \mathbf{x}^*) = \sigma_e^2 \exp\left(- \frac{\sqrt{(\mathbf{x} - \mathbf{x}^*)^T (\mathbf{x} - \mathbf{x}^*)}}{l_e} \right) \tag{9.8}$$

$$\kappa(\mathbf{x}, \mathbf{x}^*) = \sigma_e^2 \exp\left(- \frac{1}{2} \frac{((\mathbf{x} - \mathbf{x}^*)^T (\mathbf{x} - \mathbf{x}^*))}{l_e{}^2} \right) \tag{9.9}$$

$$\kappa(\mathbf{x}, \mathbf{x}^*) = \sigma_e^2 \left(1 + \frac{((\mathbf{x} - \mathbf{x}^*)^T (\mathbf{x} - \mathbf{x}^*))}{2\alpha l_e{}^2} \right)^{-\alpha} \tag{9.10}$$

where σ_e and l_e are the standard deviation and length scale of the kernel functions. Moreover, in Eq. (9.10), α is a positive-valued scale-mixture parameter. Since the specification of the covariance function in the GPR implies a distribution over functions (i.e., whose shape or smoothness of this distribution is defined by the kernel function), it is possible to draw samples from the distribution of functions evaluated at any number of predictor data points.

The Gaussian process for the regression problem can be an extension of Bayesian linear regression modeling with the regression function $f(\mathbf{x}) = \varphi(\mathbf{x})^T\mathbf{v}$, where **v** represents the prior information, which follows a Gaussian distribution with a zero mean and unit variance. For the GPR, the mean and covariance (kernel) functions gained by the predictor data generate the prior information. However, it is not necessarily sufficient to draw random variables from such information, in which case one should consider the whole knowledge about the training data by incorporating the response data and determining the joint posterior distributions of the random variables. Using these distributions, it is possible to predict unseen response data by sampling the posterior distributions and evaluating the mean and covariance matrices. Given the unseen predictor data $\widehat{\mathbf{x}}$, the joint distribution of the unseen response data and the function values of the test (predictor) points under the prior information can be derived as follows:

$$\begin{bmatrix} \mathbf{y} \\ \widehat{\mathbf{y}} \end{bmatrix} \sim \mathscr{N}\left(\mathbf{0}, \begin{bmatrix} \kappa(\mathbf{x},\mathbf{x}) + \sigma_\varepsilon^2 \mathbf{I} & \kappa(\mathbf{x},\widehat{\mathbf{x}}) \\ \kappa(\widehat{\mathbf{x}},\mathbf{x}) & \kappa(\widehat{\mathbf{x}},\widehat{\mathbf{x}}) \end{bmatrix}\right) \tag{9.11}$$

where $\mathbf{I}$ is an identity matrix. Therefore, the prediction of the unseen response data $\hat{\mathbf{y}}$ can be performed by deriving the following predictive equation:

$$P(\widehat{\mathbf{y}}|\mathbf{x},\mathbf{y},\widehat{\mathbf{x}}) \sim \mathscr{N}(\overline{\mathbf{y}}, \kappa(\widehat{\mathbf{y}})) \tag{9.12}$$

where

$$\overline{\mathbf{y}} \triangleq \mathbb{E}(\widehat{\mathbf{y}}|\mathbf{x},\mathbf{y},\widehat{\mathbf{x}}) = \kappa(\widehat{\mathbf{x}},\mathbf{x})\left(\kappa(\mathbf{x},\mathbf{x}) + \sigma_\varepsilon^2 \mathbf{I}\right)^{-1}\mathbf{y} \tag{9.13}$$

$$\kappa(\widehat{\mathbf{y}}) = \kappa(\widehat{\mathbf{x}},\widehat{\mathbf{x}}) - \left(\kappa(\widehat{\mathbf{x}},\mathbf{x})\left(\kappa(\mathbf{x},\mathbf{x}) + \sigma_\varepsilon^2 \mathbf{I}\right)^{-1}\kappa(\mathbf{x},\widehat{\mathbf{x}})\right) \tag{9.14}$$

9.2.3 Support vector regression

In machine learning, support vector machine (SVM) is a popular and effective supervised learning method, which is appropriate for classification and regression [37]. The basic principle of the SVM is to map the original data to a higher-dimensional space and then utilize an optimization algorithm for finding a hyperplane. This hyperplane intends to suitably separate the data in the transformed space. In the regression problem, this technique is often called SVR and aims to characterize and determine the relationship between the predictor and response data [38]. In other words, this method makes an attempt to find the best fit line (i.e., hyperplane) via the training data. This line is similar to a function that can predict a new data point within a tolerance margin or a decision limit obtained from the training data. Hence, the major goal of adjusting this margin is to minimize the prediction error.

Given the predictor data $\mathbf{x} = \{x_i\} \in \Re_{i=1}^n$ and response data $\mathbf{y} = \{y_i\} \in \Re_{i=1}^n$ from the training set $\mathbb{D}_{\mathrm{Tr}}$, the basic equation of the SVR model is written as follows:

$$\mathbf{y} = \mathbf{x}^T\mathbf{w} + \mathbf{b} \tag{9.15}$$

where **w** represents the weight vector and **b** is the bias constants. Considering this equation, the regression modeling based on the concept of SVM includes three key steps: (1) Separating the training data $\mathbb{D}_{\mathrm{Tr}}$ into support vectors; (2) mapping the support vectors into high-dimensional space via a kernel function; and (3) establishing a regressor entailing the estimated coefficients via an optimization algorithm. Based on Mercer's theorem, the second step is carried out through various kernel functions. Subsequently, one needs to minimize a convex optimization problem defined as follows [38]:

$$\min_{\mathbf{w},\mathbf{b},\mathbf{s},\widetilde{\mathbf{s}}}\left(\frac{1}{2}\mathbf{w}^{\mathrm{T}}\mathbf{w}+B\sum_{i=1}^{n}(s_i+\widetilde{s}_i)\right) \tag{9.16}$$

subject to the following constraint condition under a value e as follows:

$$\begin{cases} \widehat{y}_i-(x_iw_i+b_i)\leq e+s_i,\ \forall i=1,\dots,n \\ (x_iw_i+b_i)-\widehat{y}_i\leq e+\widetilde{s}_i,\ \forall i=1,\dots,n \\ s_i\geq 0, \widetilde{s}_i\geq 0,\ \forall i=1,\dots,n \end{cases} \tag{9.17}$$

where $\hat{y}_i$ is the ith predicted response value obtained by the SVR; the constant value B denotes the box constraint or penalty factor (i.e., a positive numeric value); s_i and $\widetilde{s}_i$ refer to nonnegative slack variables for the ith data point, which are used to deal with infeasible constraints in minimizing the objective function of Eq. (9.16). To cope with the nonlinear regression problem of the SVR, the low-dimensional data points are mapped to high-dimensional space by the mapping function $\kappa(\mathbf{x},\mathbf{x}^*)$. On this basis, the nonlinear form of the SVR is rewritten as follows [34]:

$$\mathbf{y}=\varphi(\mathbf{x})^T\mathbf{w}+\mathbf{b} \tag{9.18}$$

where $\varphi(\mathbf{x})$ is a mapped version of the predictor data $\mathbf{x}$ into a new feature space by a kernel function, and $\mathbf{w}$ is the weight vector with new variables compatible with the new feature space. For the SVR, one can exploit some kernel functions such as Gaussian kernel, linear kernel, and polynomial kernel, which are among the commonly-used kernel functions. These functions are presented in Eqs. (9.19)–(9.21):

$$\kappa(\mathbf{x},\mathbf{x}^*)=\exp\left(-\frac{\|\mathbf{x}-\mathbf{x}^*\|^2}{s}\right) \tag{9.19}$$

$$\kappa(\mathbf{x},\mathbf{x}^*)=\mathbf{x}^T\mathbf{x}^* \tag{9.20}$$

$$\kappa(\mathbf{x},\mathbf{x}^*)=(\mathbf{x}^T\mathbf{x}^*+1)^q \tag{9.21}$$

where s and q denote the Gaussian kernel parameter and the polynomial order for the Gaussian and polynomial kernel functions, respectively.

Once the kernel function (matrix) of the predictor data is generated, the SVR sets out to find new coefficients defined as $\boldsymbol{\lambda}$ and $\boldsymbol{\gamma}$ by maximizing the following Lagrange objective function [38]:

$$\max_{\boldsymbol{\lambda},\boldsymbol{\gamma}} L(\boldsymbol{\lambda},\boldsymbol{\gamma}) = -\left(\sum_{i=1}^{n} y_i(\lambda_i - \gamma_i) + e\sum_{i=1}^{n}(\lambda_i + \gamma_i) + \frac{1}{2}\sum_{i,j=1}^{n}(\lambda_i - \gamma_i)(\lambda_j - \gamma_j)\kappa(\mathbf{x},\mathbf{x}^*)\right) \quad (9.22)$$

subject to

$$\sum_{i=1}^{n}(\lambda_i - \gamma_i) = 0 \quad (9.23)$$

where $0 \le \lambda_i \le B$; $0 \le \gamma_i \le B$; and $i = 1, \ldots, n$. After determining the coefficients $\boldsymbol{\lambda}$ and $\boldsymbol{\gamma}$, the regression function regarding the SVR can be rewritten as follows [34]:

$$f(\mathbf{x}) = \sum_{i=1}^{n}(\lambda_i - \gamma_i)\kappa(\mathbf{x}_i,\mathbf{x}) \quad (9.24)$$

Using the regression function, any new (unseen) predictor data (i.e., the test point) can be applied to Eq. (9.24) to predict the unseen response data.

9.3 Proposed prediction process

The central core of the proposed predictive method is based on the fundamental principles of the BHO and kernel-based regression modeling. The main objective of the proposed method is to predict the modal frequencies of each vibrating mode by considering the thermal effects as the major reasons for the modal frequency variability. In this regard, the temperature records from temperature sensors (i.e., thermocouples) make up the multivariate set of predictor or independent data, while the modal frequencies concerning each vibrating mode generate an individual univariate response (dependent) set. Using the BHO, one initially attempts to find the best kernel regressor between the GPR and SVR by using training data $\mathbb{D}_{\mathrm{Tr}}$, which consists of the multivariate set (matrix) of the temperature and univariate set (vector) of the modal frequencies. Let n_t and n_m designate the number of temperature sensors and the number of identified vibrating modes from an OMA method, the n-dimensional training dataset is set as $\mathbb{D}_{\mathrm{Tr}} = \{\mathbf{X},\mathbf{y}_j\}$, where $\mathbf{X} = [\mathbf{x}_1, \ldots, \mathbf{x}_{n_t}]$; $\mathbf{x}_l = \{x_i\} \in \Re^n_{i=1}$; $l = 1, \ldots, n_t$; $\mathbf{y}_j = \{y_i\} \in \Re^n_{i=1}$; and $j = 1, \ldots, n_m$. Once the optimum or best kernel regressor is chosen, m unseen (new) predictor points make the test data $\mathbb{D}_{\mathrm{Te}} = \widehat{\mathbf{X}} = [\widehat{\mathbf{x}}_1, \ldots, \widehat{\mathbf{x}}_{n_t}]$, where $\widehat{\mathbf{x}}_l = \{\widehat{x}_i\} \in \Re^m_{i=1}$. This dataset is fed into the trained regressor (i.e., the regression function) of jth mode to predict the unseen modal frequencies; that is, $\hat{\mathbf{y}}_j = \{\hat{y}_i\} \in \Re^m_{i=1}$.

9.3.1 Bayesian hyperparameter optimization

Hyperparameters are unknown elements of any machine learning model that profoundly impact on its overall performance. For this reason, hyperparameter optimization is a technique for tuning such unknown elements. In contrast, the other unknown elements of machine learning are called model parameters that do not need to be tuned because those are often estimated during the learning process. In the field of machine learning, the most common hyperparameter tuning techniques are derived from manual tuning, gradient-based optimization, Bayesian optimization, and multi-fidelity optimization algorithms [39,40].

In most cases, the standard hyperparameter optimization approaches require training a model for each combination of hyperparameters. Those also need predicting validation data and calculating validation metrics. An important fact in hyperparameter optimization is that the tuned elements may vary from model to model based on the type of machine learning model and the problem at hand. For this reason, it is difficult to consider or select the best hyperparameters that are suitable for all models. Despite the popularity of manual tuning techniques such as random search and grid search, which are often suitable if the number of hyperparameters is small, these techniques optimize the unknown parameters without considering past results or values. A disadvantage of such techniques is that the optimization process is often time-consuming when the number and type of hyperparameters are large. BHO is a powerful and flexible technique for tuning any type of unknown element in machine learning models. It is mainly applicable in conditions where one cannot provide a closed-form expression for an objective function but it is feasible to obtain observations of that function at sampled values. It can also be applied to functions that are computationally expensive and non-convex. Sometimes, the determination of derivatives of some functions is difficult, in which case BHO is the best choice.

This technique is based on incorporating previously observed hyperparameter sets when determining the next set of hyperparameters to evaluate. Mathematically, the central goal of BHO is to minimize or maximize an unknown and scalar objective function by considering the problem of finding a minimizer or maximizer. Having consider a minimizer, the objective function of BHO is defined here as:

$$p^* = \underset{p \in \mathbb{S}}{\operatorname{argmin}}\, h(p) \tag{9.25}$$

where $\mathbb{S}$ refers to the search domain of the main hyperparameters, including numerical, categorical, and conditional characters in terms of values, functions, etc. Moreover, the variable p belongs to the domain search $\mathbb{S}$, which is a value in the domain; $h(p)$ represents the objective score to minimize the error rate evaluated against the validation set; and p^* is the optimum choice of p. Generally, the algorithm of BHO is comprised of three main steps of prior probabilistic modeling of the objective function, posterior model updating, and a criterion maximizing for determining the next point p for evaluation. In BHO, the fundamental prior probabilistic model for the objective function $h(p)$ is chosen from the Gaussian process with additional Gaussian noise in the observed data. In this case, one can derive the prior distribution on $h(p)$ with the mean and covariance (kernel) functions [35]. The prior mean function is often set to zero to avoid increasing posterior computations; hence, the inference is based on the covariance or kernel function. Occasionally, the squared exponential kernel function, as presented in Eq. (9.9) is the initial choice.

The BHO algorithm conforms to Bayes' theorem. Accordingly, one can state that the posterior probability of a model $\mathbb{M}$ given evidence (i.e., data or observations) $\mathbb{D}$, called here $P(\mathbb{M}|\mathbb{D})$, is proportional to the likelihood of $\mathbb{D}$ given $\mathbb{M}$, i.e., defined as $P(\mathbb{D}|\mathbb{M})$, multiplied by the prior probability of $\mathbb{M}$; that is, $P(\mathbb{M})$. Mathematically speaking, the core idea of BHO based on the aforementioned statement can be expressed as follows:

$$P(\mathbb{M}|\mathbb{D}) \propto P(\mathbb{D}|\mathbb{M})P(\mathbb{M}) \tag{9.26}$$

In other words, the principle of Bayesian optimization is to merge the prior probability of the objective function $h(p)$ with the observed data $\mathbb{D}$ (evidence) in order to obtain the posterior of the objective function. Subsequently, the posterior information is utilized to find where the function $h(p)$ is minimized based on a criterion. This criterion is represented by a utility function that is also called an *acquisition function*. This function is useful for determining the next point to evaluate. Simply speaking, once the posterior distribution of the objective function is gained, BHO exploits the acquisition function to derive the maximum of the objective function. For this purpose, one can employ some common acquisition functions such as probability of improvement, expected improvement, and Gaussian process upper confidence bound [41].

9.3.2 Automated kernel regressor selection

The selection of the best kernel regressor for each vibrating mode is of paramount importance to predicting the unseen response data. Given the multivariate predictor (temperature records) and univariate response (modal frequencies of each mode) data, the BHO-oriented regressor selection automatically chooses the best kernel regression model between the GPR and SVR along with the main hyperparameters of the best model based on a selection criterion. One of the great advantages of the proposed approach is the possibility of automated model selection and hyperparameter optimization in a simultaneous and integrated manner. On the other hand, the proposed regression selection approach uses BHO under the cross-validation algorithm by considering the only training data, $\mathbb{D}_{\mathrm{Tr}} = \{[\mathbf{x}_1, \ldots, \mathbf{x}_{n_t}], \mathbf{y}\}$. Hence, the selection criterion of the proposed approach is based on the MSE of the cross-validation algorithm by defining the following selection function:

$$C_r = \log(1 + \mathrm{MSE}_k) \tag{9.27}$$

where k denotes the number of cross-validated folds. In a simple word, the training data is divided into k new training and validation (test) datasets with a constant

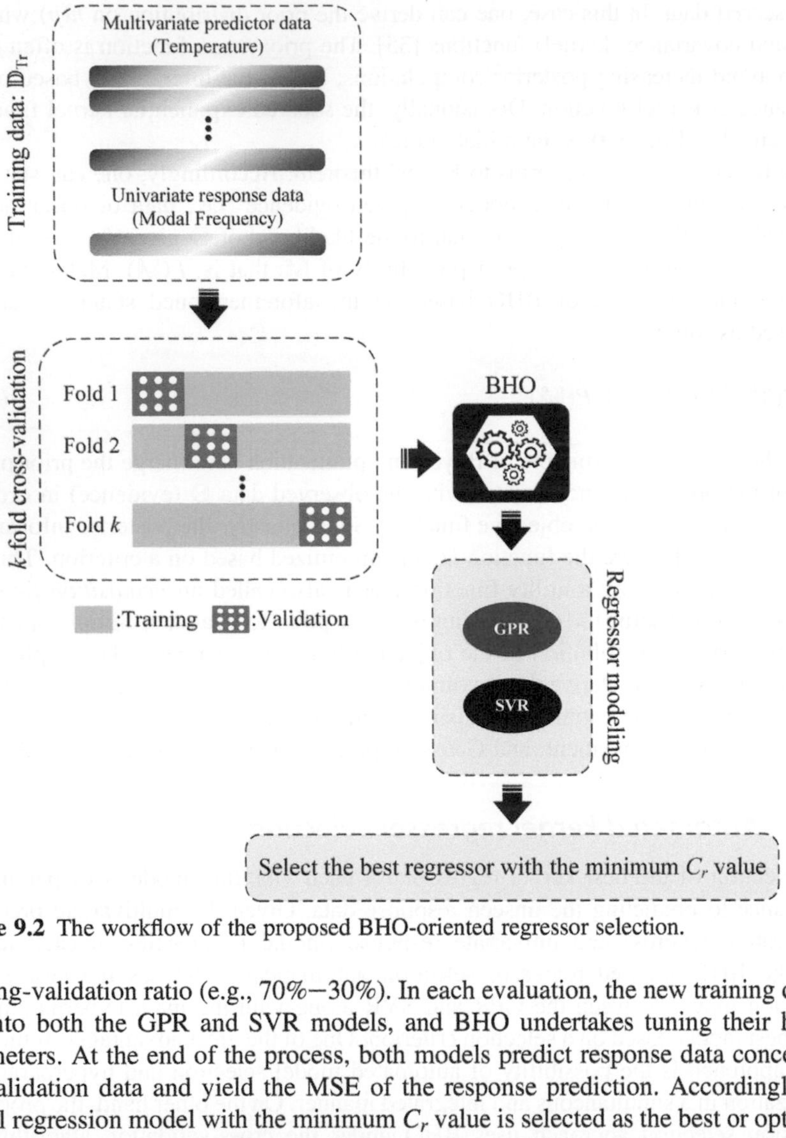

Figure 9.2 The workflow of the proposed BHO-oriented regressor selection.

training-validation ratio (e.g., 70%−30%). In each evaluation, the new training data is fed into both the GPR and SVR models, and BHO undertakes tuning their hyper-parameters. At the end of the process, both models predict response data concerning the validation data and yield the MSE of the response prediction. Accordingly, the kernel regression model with the minimum C_r value is selected as the best or optimum regressor. For simplicity, Fig. 9.2 illustrates the workflow of the proposed BHO-oriented regressor selection approach.

9.4 Application: The Consoli Palazzo

9.4.1 Description of the masonry building

The masonry structure used in this research is called the *Consoli Palazzo* (i.e., *Palazzo dei Consoli* in Italian) or Palace of Consuls, which is an important monumental building in Gubbio, Italy [42]. This building was built during 1332−1349 and used as official courts, while it has hosted the Civic Museum of Gubbio since 1901. Fig. 9.3a depicts a picture of the Consoli Palazzo. The building has a total and irregular height of 60 m in three floors with a rectangular plan of about 40 × 20 m. Fig. 9.3b shows more details of the plans for the three floors of this building. The palace has a Gothic architectural design style with calcareous stone masonry, thick masonry-bearing walls, and vaults, which comprise the structure's horizontal elements. The major façade of the building is distinguished architecturally by a fan-shaped stairway entry and an arched portal. The north façade has a height of roughly 30 m between the level of the square and the roof, whereas the south façade has a panoramic loggia and is 60 m tall from the ground level to a 13-m high bell tower rising from the roof level.

Due to locating the Consoli Palace near the Gubbio fault and observing damage to this masonry building during one of the strongest earthquakes (i.e., occurred on April 29, 1984, with a moment magnitude (Mw) of 5.6 and an epicenter located about 10 km south of Gubbio), an SHM system has been incorporated to monitor static, dynamic, and environmental parameters. This system has been considered to perform a long-term monitoring program since July 2017 and remained active until July 2020 [23]. After upgrading the sensing systems in July 2020, the building has been equipped with twelve accelerometers, four linear variable displacement transducers (LVDTs), six temperature sensors, and a DAQ system in order to conduct a continuous monitoring program from July 17, 2020 to July 18, 2021 [42]. Fig. 9.3b shows the layout of the upgraded sensing systems for continuous monitoring, which include the sensor types, numbers, and locations on the palace floors and roof. The accelerometers labeled "Acc" are high-sensitivity uniaxial piezoelectric accelerometer models with the specifications of 10 V/g, a broadband resolution of 800 μg, and a measurement range of ±0.5 g. Detecting two major cracks [23], the LVDTs with the specifications of a 50-mm measurement range and a resolution smaller than 0.3 μm have been mounted on the building. The temperature sensors (i.e., K-type thermocouples labeled as "T") have also been installed for two tasks of measuring the surface temperature of the masonry building via T1−T4 and monitoring the air temperature at the roof and the third floor by T5 and T6. All the accelerometers, two temperature sensors (T1 and T2), and two LVDTs (D1 and D2) have been connected to a DAQ device (i.e., located on the third floor) with the specifications of a dual-core atom CPU of 1.33 GHz, 4 slots, Windows Embedded Standard 7, and 16 GB of SD storage. Furthermore, the measurements of the other LVDTs (i.e., D3 and D4) and thermocouples (i.e., T3−T6) have been transferred via wireless communication to a WIFI router [42].

Ambient excitations such as traffic on the neighboring roads and wind loads have been utilized to excite the building. The ambient vibrations have been sampled at 40 Hz, while the crack amplitudes as well as the temperature data have been sampled at 0.1 Hz. A 30-min recording duration has been considered to store measurements in separated binary files. Using the measured acceleration responses, an automated time-domain OMA via covariance-based stochastic subspace identification (SSI−COV) was used by García-Macías and Ubertini [42] to identify the modal properties of the Consoli Palazzo in nine stable modes. Aiming at predicting the building modal frequencies in a 1-year continuous monitoring program, Fig. 9.4a shows the identified modal frequencies between July 17, 2020, and July 18, 2021. It should be clarified that the original modal dataset included some missing samples. To better demonstrate

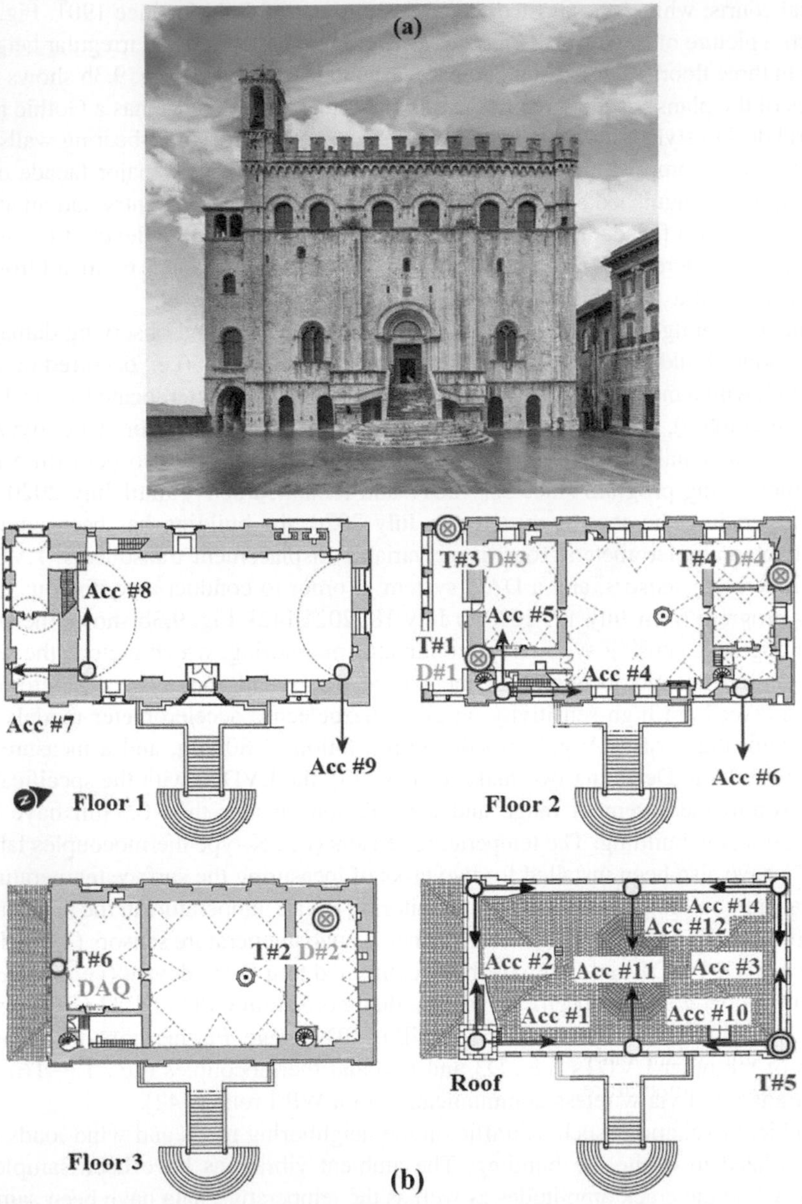

Figure 9.3 (a) A picture of the Consoli Palace, (b) the layout of the sensing system including sensor labels and locations (i.e., Acc: Accelerometer, T: Temperature, D: LVDT).

the performance of the proposed method, those samples have been eliminated by an initial data analysis strategy presented by Entezami et al. [24].

To analyze the thermal effects on the modal frequencies of the masonry building and consider them as the main predictors, Fig. 9.4b illustrates the evolution of

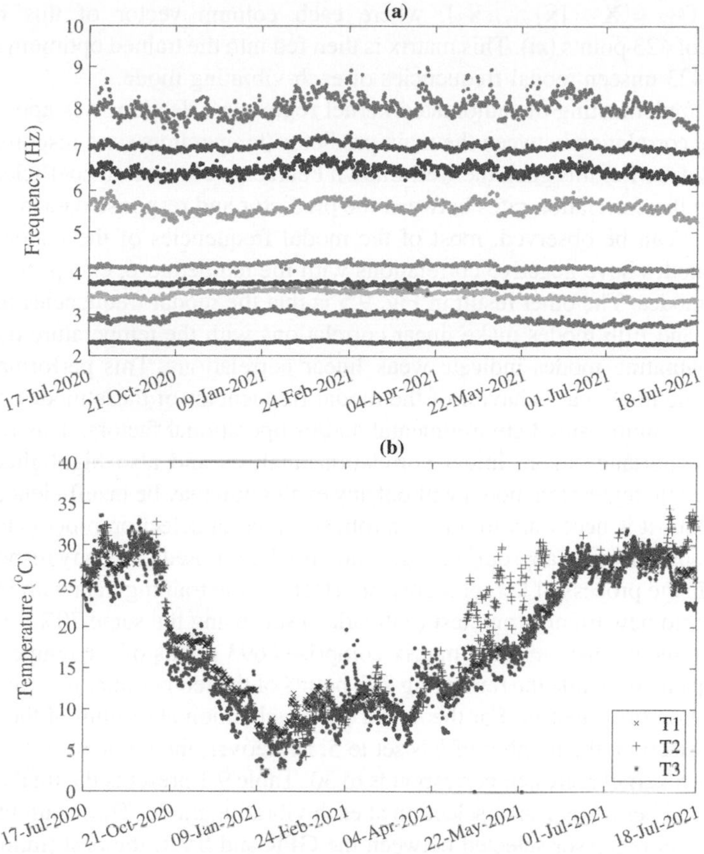

Figure 9.4 (a) Modal frequencies of the nine modes related to the continuous monitoring of the Consoli Palazzo, (b) the temperature data of the three thermocouples.

temperature records of the three thermocouples T1−T3 ($n_t = 3$). Similarly, the missing temperature samples coincided with the modal frequencies and were also removed by the same initial data analysis. In this regard, the total number of data points related to the predictor (temperature) and response (modal frequency) is identical to 1413.

9.4.2 Automated kernel regressor selection

To select and learn the optimum kernel regressor between the GPR and SVR, a ratio of 70%−30% is incorporated to divide the whole data into the training and test datasets. Accordingly, the training data (matrix) $\mathbb{D}_{\mathrm{Tr}}$ for each vibrating mode is comprised of a predictor matrix $\mathbf{X} = [\mathbf{x}_1, \ldots, \mathbf{x}_3]$ and a response vector $\mathbf{y}_j$, where $j = 1, \ldots, n_m$ and $n_m = 9$. Each column vector of the predictor matrix and the response vector include 990 data points (n), which is equivalent to 70% of all temperature and modal frequency samples. For the test data, the remaining 30% of all temperature records of T1−T3 generates $\mathbb{D}_{Te} = \widehat{\mathbf{X}} = [\widehat{\mathbf{x}}_1, \ldots, \widehat{\mathbf{x}}_3]$, where each column vector of this matrix is comprised of 423 points (m). This matrix is then fed into the trained optimum regressor to predict 423 unseen modal frequencies of each vibrating mode.

Before implementing the automated kernel regressor selection, it is appropriate to analyze the correlation between the main predictor (temperature) and response (modal frequency) data. For this goal, the well-known Pearson's correlation coefficient is used to calculate the correlation rates between the predictor and response data, as shown in Fig. 9.5. As can be observed, most of the modal frequencies of the majority of the vibrating modes have negative correlations with the temperature, except for the sixth and ninth modes. The other result in Fig. 9.5 is that the modal frequencies of the second, third, and fifth modes make linear correlations with the temperature data, while the other vibrating modes indicate weak linear correlations. This performance may stem from the nonlinear behavior of the modal frequencies of these modes or the impacts of other unmeasured environmental and/or operational factors. This result confirms the importance of an initial correlation analysis and also highlights that the use of a specific regression model without any evaluation may be insufficient and problematic. Thus, it is necessary to utilize a robust regressor selection process to systematically choose the most appropriate regression model for predicting any response data.

To begin the process of kernel regressor selection, the training data with 990 points is divided into new training and test (validation) sets using the same 70%–30% ratio. In other words, the new training matrix comprises 693 points of the temperature and modal frequencies, while the remaining 297 points of the temperature records make the new validation (test) matrix. For the k-fold cross-validation algorithm of the proposed regressor selection, the number of k is set to 5. Moreover, the number of iterations for BHO of each kernel regressor corresponds to 30. Table 9.1 presents the final outputs of the automated kernel regressor selection at each vibrating mode. These outputs contain the best kernel regressor selected between the GPR and SVR, the best (minimum) C_r value of the selected regressor, the optimum kernel function, and the kernel values.

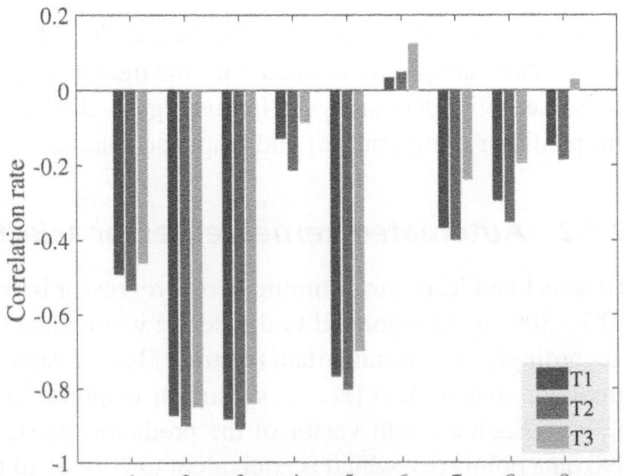

Figure 9.5 Correlation evaluation between modal frequencies and temperature by Pearson's correlation coefficient.

Table 9.1 Automated kernel regressor selection of each mode by BHO.

Mode no.	Best model	C_r ($\times 10^{-3}$)	Optimum kernel function	Kernel parameters	Kernel values
1	GPR	0.1778	Exponential	σ_e	0.1008
				l_e	0.0196
2	SVR	17.4573	Linear	–	–
3	SVR	9.1241	Linear	–	–
4	GPR	0.5388	Rational quadratic	σ_e	0.0681
				l_e	0.0072
				α	0.0131
5	SVR	8.1151	Linear	–	–
6	GPR	8.3334	Rational quadratic	σ_e	0.2306
				l_e	0.0342
				α	0.0751
7	GPR	7.4051	Exponential	σ_e	0.0889
				l_e	0.1072
8	GPR	2.7111	Rational quadratic	σ_e	0.1310
				l_e	0.2097
				α	0.0538
9	GPR	24.1574	Rational quadratic	σ_e	0.3705
				l_e	0.0315
				α	0.0555

From Table 9.1, one can realize that the proposed regressor selection chooses the SVR and the linear kernel function for the second, third, and fifth vibrating modes, for which there exist linear correlations between predictor and response data. For the other modes, the GPR with the exponential and rational quadratic kernel functions are the optimum choices. Fig. 9.6 displays the evolution of C_r under 30 iterations of the optimum kernel regressor. Furthermore, the amounts of the R-squared values between the observed and predicted response data related to the optimum kernel regressors during the training phase are shown in Fig. 9.7a. As can be observed, the selected kernel regression models yield high rates of R^2, implying reliable regression selection and modeling. To demonstrate the good performance of the kernel-based regression modeling and the proposed regressor selection, Fig. 9.7b presents the R-squared values obtained from the MLRs. Apart from the second, third, and fifth modes, it is clear that the MLRs of the other vibrating modes have poor performances. Thus, this comparison confirms the positive effect of the proposed regressor selection and kernel-based regression modeling.

9.4.3 Prediction of the building modal frequencies

Using the optimum kernel regressors selected for the nine vibrating modes via the BHO-aided regressor selection technique, the test data $\mathbb{D}_{Te}$ containing the 423 temperature samples is fed into these models to predict the unseen modal frequencies of the

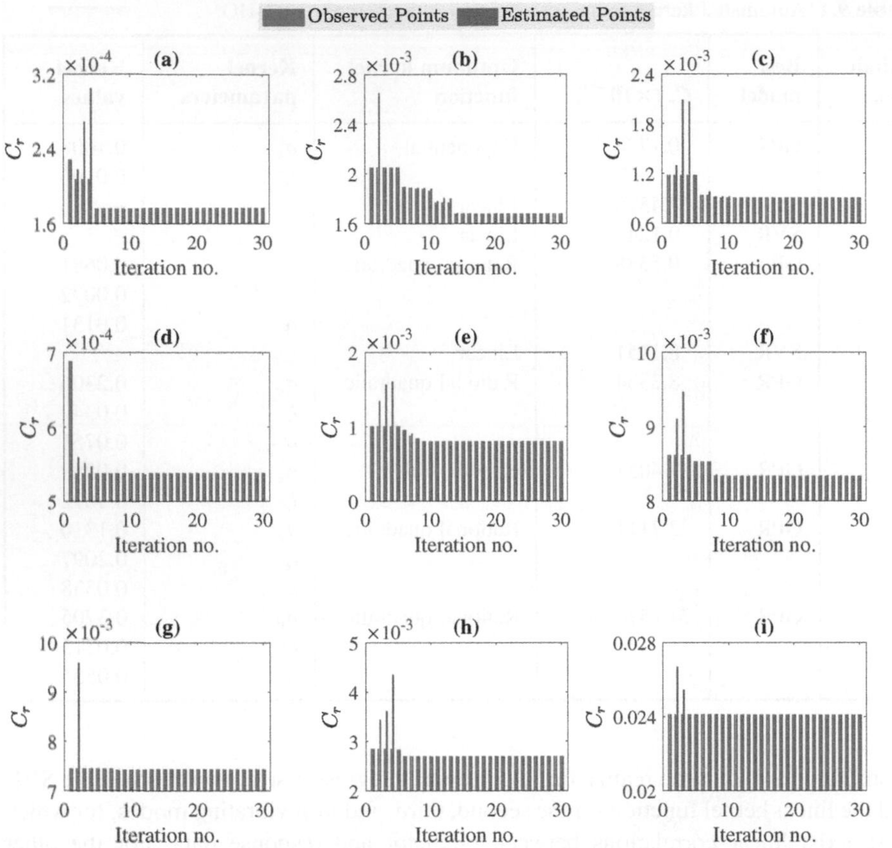

Figure 9.6 Evolution of C_r of the best kernel regression models based on the HBO in the nine modes (a)–(i).

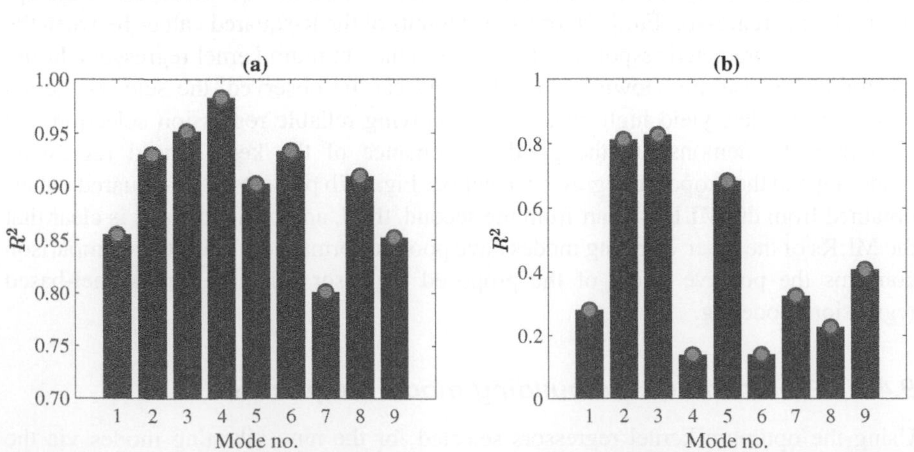

Figure 9.7 The R-squared values of the optimum kernel regressors (a) and MLRs (b) during the training phase.

Consoli Palazzo, as shown in Fig. 9.8. In this figure, there are two prediction outputs, including the predicted modal frequencies related to the training phase labeled "Predicted Training Data" and the predicted modal frequencies regarding the predicting phase labeled "Predicted Testing Data". In addition, both types of prediction outputs are compared with the entire modal frequency designated as "Original Data". From Fig. 9.8, it can be perceived that the predicted response data is in good agreement with the original response data. This reliable and reasonable performance is observed in the outputs of both the training and predicting phases.

For further evaluations, Fig. 9.9 displays the relationship between the original and predicted modal frequencies of the Consoli Palazzo, which can help us to realize

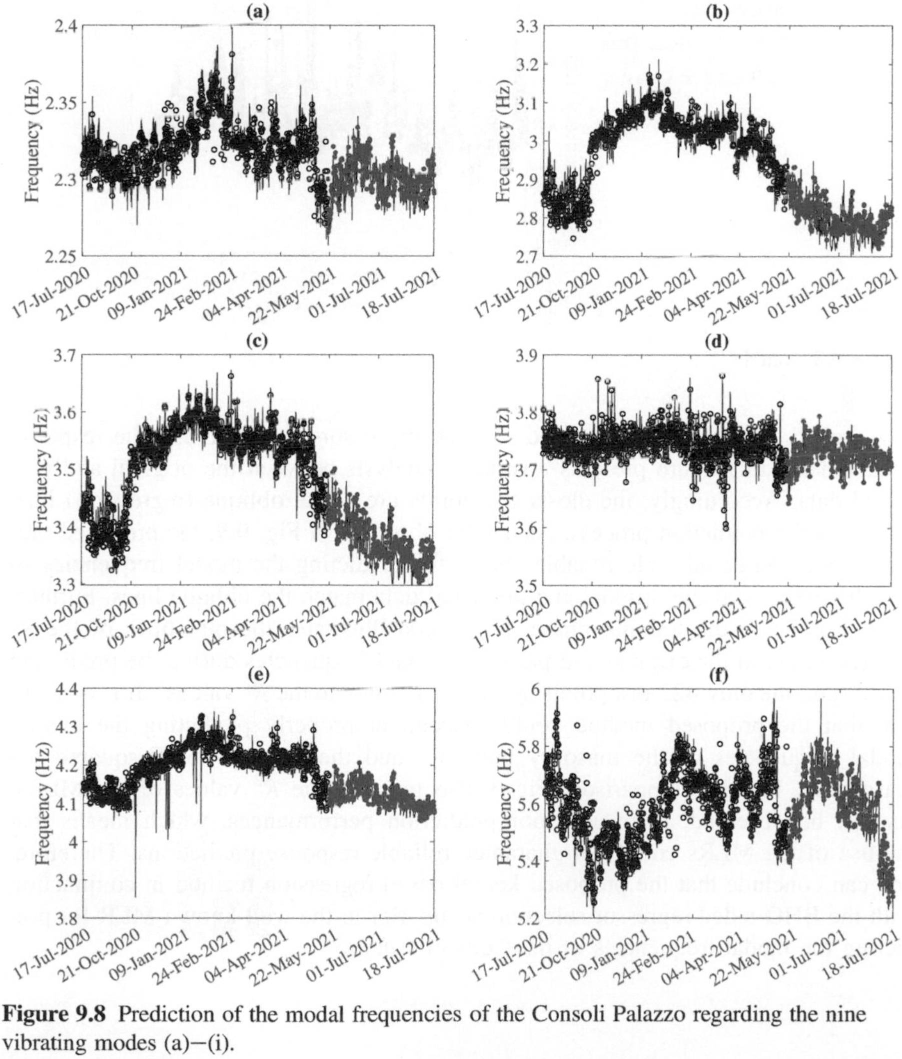

Figure 9.8 Prediction of the modal frequencies of the Consoli Palazzo regarding the nine vibrating modes (a)−(i).

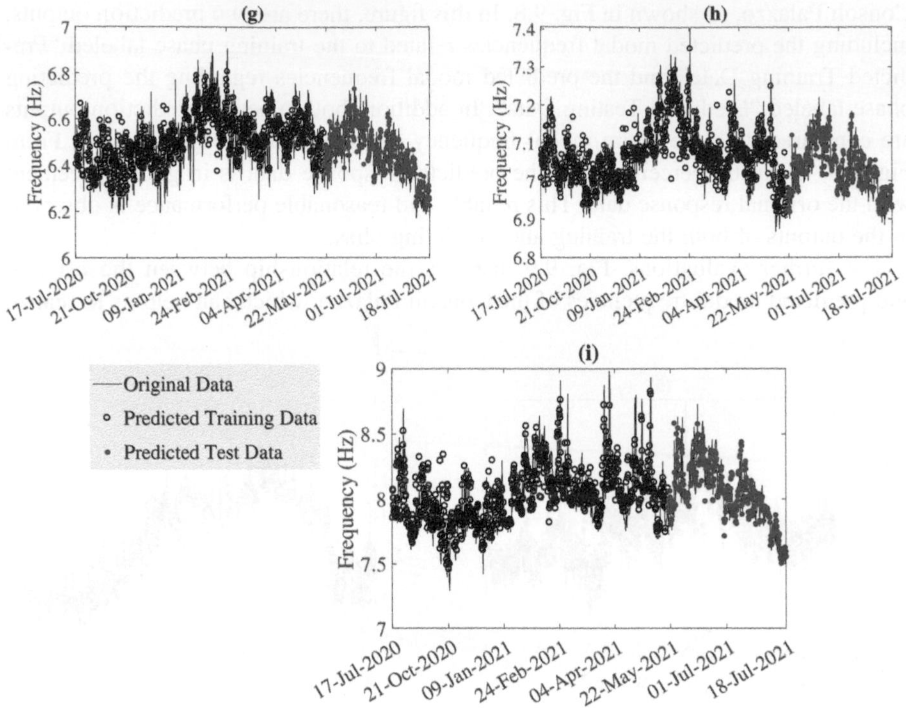

Figure 9.8 cont'd.

how well the proposed method and selected regressors could predict the response data. Indeed, this figure presents a residual analysis between the original and predicted data. Accordingly, the closer the points are to the oblique (regression) line, the better the prediction process. As can be observed in Fig. 9.9, the proposed predictive method could yield reliable results for predicting the modal frequencies so that the patterns of the original and predicted data match the oblique lines. Furthermore, similar to the training phase, Fig. 9.10a illustrates the quantities of the R-squared between the original and predicted modal frequencies during the predicting phase (i.e., the only 423 test points are used to compute the R^2 values). It is discernible that the proposed method could succeed in properly predicting the unseen modal frequencies of the masonry building, and the amounts of R-squared are reasonable. For the comparison, Fig. 9.10b presents the R^2 values of the MLRs. As can be observed, there are poor prediction performances, which means that the use of the MLRs could not guarantee reliable response predictions. Therefore, one can conclude that the proposed kernel-based regression method in conjunction with the BHO-aided regressor selection is superior to the well-known MLR for predicting the modal frequencies of the Consoli Palazzo.

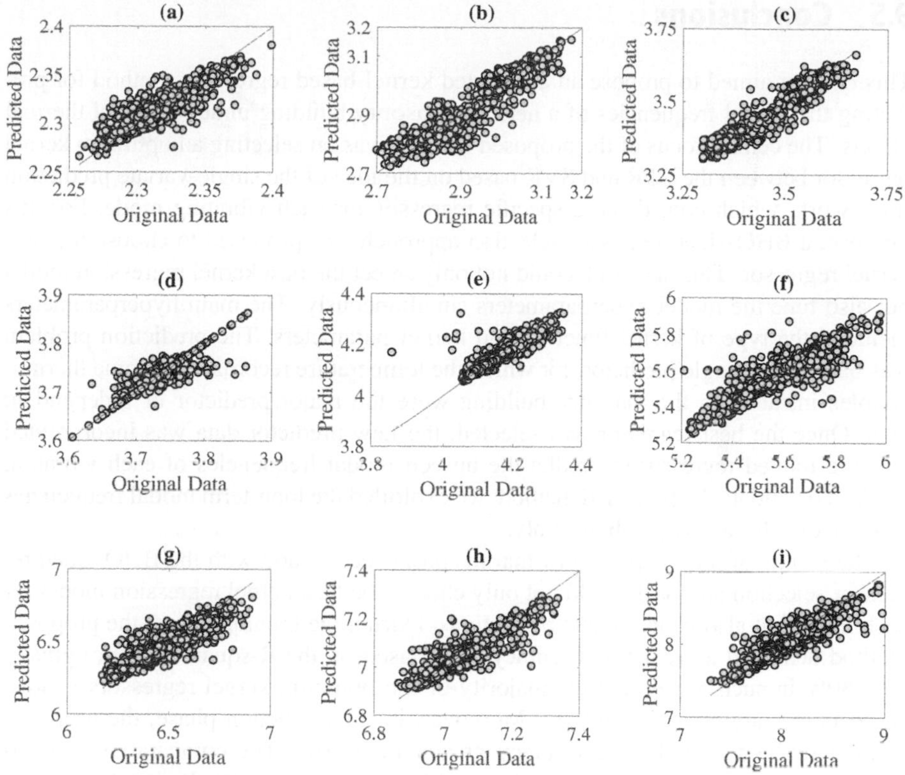

Figure 9.9 Relationship between the original and predicted modal frequencies of the Consoli Palazzo in the nine modes (a)–(i).

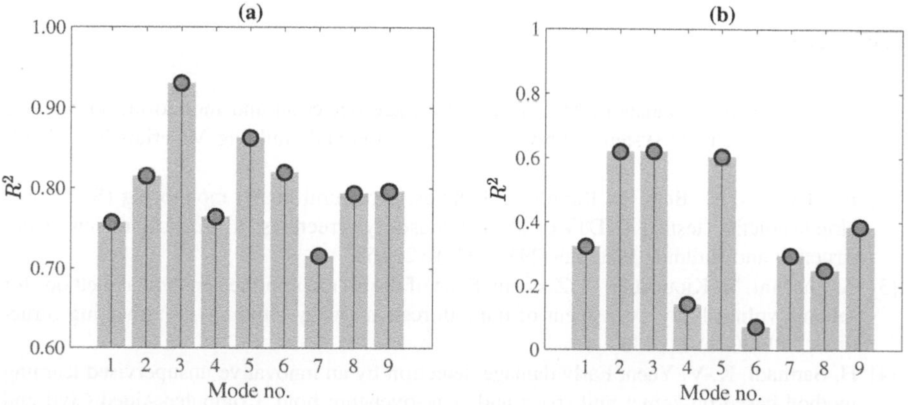

Figure 9.10 The R-squared values of the optimum kernel regressors (a) and the MLRs in predicting the modal frequencies of the Consoli Palazzo during the testing (prediction) phase.

9.5 Conclusions

This chapter aimed to propose an automated kernel-based regression method for predicting the modal frequencies of a heritage masonry building under seasonal thermal effects. The central focus of the proposed method was on selecting an optimum kernel regressor between the GPR and SVR based on the idea of the single-variate prediction framework, which considered a specific regressor for each vibrating mode. For this purpose, a BHO-aided regressor selection approach was proposed to choose the best kernel regressor. This approach could not only select the best kernel regression model but also tune the model hyperparameters simultaneously. The main hyperparameters included the type of kernel function and kernel parameters. The prediction problem was based on a single predictor, for which the temperature records from some thermocouples installed in the masonry building were the major predictor or independent data. Once the best regressor was selected, the new predictor data was incorporated into the trained regressor to predict the unseen modal frequencies of each vibrating mode. To validate the proposed method, we exploited the long-term modal frequencies of the Consoli Palazzo, Gubbio, Italy.

The results of this chapter show that the proposed method with the BHO-aided regressor selection approach could not only choose the best kernel regression model for each mode but also yield reliable predictions. During the training phase, the proposed method achieved a prediction accuracy (i.e., based on the R-squared metric) greater than 80% in such a way that the majority of the optimum kernel regressors yielded an accuracy rate greater than 90%. Moreover, in the prediction phase, the proposed method reached a prediction accuracy greater than 70%. The comparative analysis also demonstrated that the proposed kernel-based regression method outperformed the MLR.

References

[1] A. Soleymani, H. Jahangir, M.L. Nehdi, Damage detection and monitoring in heritage masonry structures: systematic review, Construction and Building Materials 397 (2023) 132402.

[2] F.J. Pallarés, M. Betti, G. Bartoli, L. Pallarés, Structural health monitoring (SHM) and Nondestructive testing (NDT) of slender masonry structures: a practical review, Construction and Building Materials 297 (2021) 123768.

[3] A. Shabani, M. Kioumarsi, M. Zucconi, State of the art of simplified analytical methods for seismic vulnerability assessment of unreinforced masonry buildings, Engineering Structures 239 (2021) 112280.

[4] H. Sarmadi, K.-V. Yuen, Early damage detection by an innovative unsupervised learning method based on kernel null space and peak-over-threshold, Computer-Aided Civil and Infrastructure Engineering 36 (2021) 1150−1167.

[5] A. Entezami, H. Sarmadi, C. De Michele, Probabilistic damage localization by empirical data analysis and symmetric information measure, Measurement 198 (2022) 111359.

[6] C. Modarres, N. Astorga, E.L. Droguett, V. Meruane, Convolutional neural networks for automated damage recognition and damage type identification, Structural Control and Health Monitoring 25 (2018) e2230.

[7] A. Entezami, H. Shariatmadar, A. Karamodin, Data-driven damage diagnosis under environmental and operational variability by novel statistical pattern recognition methods, Structural Health Monitoring 18 (2019) 1416−1443.

[8] C.R. Farrar, N.A. Lieven, Damage prognosis: the future of structural health monitoring, Philosophical Transactions of the Royal Society A: Mathematical, Physical and Engineering Sciences 365 (2007) 623−632.

[9] J.P. Lynch, H. Sohn, M.L. Wang, Sensor Technologies for Civil Infrastructures: Volume 1: Sensing Hardware and Data Collection Methods for Performance Assessment, second ed., Elsevier, Cambridge, MA, United States, 2022.

[10] J.P. Lynch, H. Sohn, M.L. Wang, Sensor Technologies for Civil Infrastructures: Volume 2: Applications in Structural Health Monitoring, second ed., Elsevier, Cambridge, MA, United States, 2022.

[11] S. Sony, S. Laventure, A. Sadhu, A literature review of next-generation smart sensing technology in structural health monitoring, Structural Control and Health Monitoring 26 (2019) e2321.

[12] A. Meoni, C. Fabiani, A. D'Alessandro, A.L. Pisello, F. Ubertini, Strain-sensing smart bricks under dynamic environmental conditions: experimental investigation and new modeling, Construction and Building Materials 336 (2022) 127375.

[13] I. Venanzi, A. Kita, N. Cavalagli, L. Ierimonti, F. Ubertini, Earthquake-induced damage localization in an historic masonry tower through long-term dynamic monitoring and FE model calibration, Bulletin of Earthquake Engineering 18 (2020) 2247−2274.

[14] S. Ereiz, I. Duvnjak, J. Fernando Jiménez-Alonso, Review of finite element model updating methods for structural applications, Structures 41 (2022) 684−723.

[15] A. Shabani, M. Feyzabadi, M. Kioumarsi, Model updating of a masonry tower based on operational modal analysis: the role of soil-structure interaction, Case Studies in Construction Materials 16 (2022) e00957.

[16] R. Brincker, C.E. Ventura, Introduction to Operational Modal Analysis, Wiley, Chichester, West Sussex, United Kingdom, 2015.

[17] A. Entezami, H. Sarmadi, B. Behkamal, A novel double-hybrid learning method for modal frequency-based damage assessment of bridge structures under different environmental variation patterns, Mechanical Systems and Signal Processing 201 (2023) 110676.

[18] A. Entezami, H. Sarmadi, B. Behkamal, C. De Michele, On continuous health monitoring of bridges under serious environmental variability by an innovative multi-task unsupervised learning method, Structure and Infrastructure Engineering (2023) 1−19. In Press.

[19] M.H. Daneshvar, H. Sarmadi, Unsupervised learning-based damage assessment of full-scale civil structures under long-term and short-term monitoring, Engineering Structures 256 (2022) 114059.

[20] S. Pereira, F. Magalhães, J.P. Gomes, Á. Cunha, J.V. Lemos, Dynamic monitoring of a concrete arch dam during the first filling of the reservoir, Engineering Structures 174 (2018) 548−560.

[21] Z. Wang, D.-H. Yang, T.-H. Yi, G.-H. Zhang, J.-G. Han, Eliminating environmental and operational effects on structural modal frequency: a comprehensive review, Structural Control and Health Monitoring 29 (2022) e3073.

[22] Q. Han, Q. Ma, J. Xu, M. Liu, Structural health monitoring research under varying temperature condition: a review, Journal of Civil Structural Health Monitoring 11 (2021) 149−173.

[23] A. Kita, N. Cavalagli, F. Ubertini, Temperature effects on static and dynamic behavior of Consoli Palace in Gubbio, Italy, Mechanical Systems and Signal Processing 120 (2019) 180−202.
[24] A. Entezami, H. Sarmadi, B. Behkamal, Long-term health monitoring of concrete and steel bridges under large and missing data by unsupervised meta learning, Engineering Structures 279 (2023) 115616.
[25] F. Ubertini, G. Comanducci, N. Cavalagli, A. Laura Pisello, A. Luigi Materazzi, F. Cotana, Environmental effects on natural frequencies of the San Pietro bell tower in Perugia, Italy, and their removal for structural performance assessment, Mechanical Systems and Signal Processing 82 (2017) 307−322.
[26] G. Zonno, R. Aguilar, R. Boroschek, P.B. Lourenço, Analysis of the long and short-term effects of temperature and humidity on the structural properties of adobe buildings using continuous monitoring, Engineering Structures 196 (2019) 109299.
[27] R. Mario Azzara, G. De Roeck, M. Girardi, C. Padovani, D. Pellegrini, E. Reynders, The influence of environmental parameters on the dynamic behaviour of the San Frediano bell tower in Lucca, Engineering Structures 156 (2018) 175−187.
[28] T. Hastie, R. Tibshirani, J. Friedman, The Elements of Statistical Learning: Data Mining, Inference, and Prediction, second ed., Springer, New York, United States, 2013.
[29] H.-Q. Mu, K.-V. Yuen, Modal frequency-environmental condition relation development using long-term structural health monitoring measurement: uncertainty quantification, sparse feature selection and multivariate prediction, Measurement 130 (2018) 384−397.
[30] H.-Q. Mu, J.-H. Shen, Z.-T. Zhao, H.-T. Liu, K.-V. Yuen, A novel generative approach for modal frequency probabilistic prediction under varying environmental condition using incomplete information, Engineering Structures 252 (2022) 113571.
[31] X.G. Hua, Y.Q. Ni, J.M. Ko, K.Y. Wong, Modeling of temperature−frequency correlation using combined principal component analysis and support vector regression technique, Journal of Computing in Civil Engineering 21 (2007) 122−135.
[32] I. Laory, T.N. Trinh, I.F.C. Smith, J.M.W. Brownjohn, Methodologies for predicting natural frequency variation of a suspension bridge, Engineering Structures 80 (2014) 211−221.
[33] B. Behkamal, A. Entezami, C. De Michele, A.N. Arslan, Investigation of temperature effects into long-span bridges via hybrid sensing and supervised regression models, Remote Sensing 15 (2023) 3503.
[34] S.Y. Kung, Kernel Methods and Machine Learning, Cambridge University Press, Cambridge, United Kingdom, 2014.
[35] C.E. Rasmussen, C.K.I. Williams, Gaussian Processes for Machine Learning, MIT Press, 2005.
[36] F. Jäkel, B. Schölkopf, F.A. Wichmann, A tutorial on kernel methods for categorization, Journal of Mathematical Psychology 51 (2007) 343−358.
[37] D.A. Pisner, D.M. Schnyer, Support vector machine, in: A. Mechelli, S. Vieira (Eds.), Machine Learning, Academic Press, 2020, pp. 101−121.
[38] A.J. Smola, B. Schölkopf, A tutorial on support vector regression, Statistics and Computing 14 (2004) 199−222.
[39] M. Feurer, F. Hutter, Hyperparameter Optimization. Automated Machine Learning, Springer, Cham, 2019, pp. 3−33.
[40] L. Yang, A. Shami, On hyperparameter optimization of machine learning algorithms: theory and practice, Neurocomputing 415 (2020) 295−316.

[41] J. Wu, X.-Y. Chen, H. Zhang, L.-D. Xiong, H. Lei, S.-H. Deng, Hyperparameter optimization for machine learning models based on Bayesian optimization, Journal of Electronic Science and Technology 17 (2019) 26–40.
[42] E. García-Macías, F. Ubertini, Least Angle Regression for early-stage identification of earthquake-induced damage in a monumental masonry palace: Palazzo dei Consoli, Engineering Structures 259 (2022) 114119.
[43] H. Sarmadi, A. Entezami, K.V. Yuen, B. Behkamal, Review on smartphone sensing technology for structural health monitoring, Measurement 223 (2023) 113716, https://doi.org/10.1016/j.measurement.2023.113716.
[44] H. Sarmadi, A. Entezami, F. Magalhães, Unsupervised data normalization for continuous dynamic monitoring by an innovative hybrid feature weighting-selection algorithm and natural nearest neighbor searching, Structural Health Monitoring 22 (6) (2023) 4005–4026, https://doi.org/10.1177/14759217231166116.

A comprehensive review on application of artificial intelligence in construction management using a science mapping approach

10

Parag Gohel[1], Rajat Dabral[1], V.H. Lad[2], K.A. Patel[1] and D.A. Patel[1]
[1]Department of Civil Engineering, Sardar Vallabhbhai National Institute of Technology (SV-NIT), Surat, Gujarat, India; [2]Department of Civil Engineering, Nirma University, Ahmedabad, Gujarat, India

10.1 Introduction

In today's society, the word "artificial intelligence" (AI) is no longer alien; rather, it is now a term that everyone uses. A branch of computer science called AI focuses on creating intelligent machines that can recognize objects and communicate with humans [1]. It has a collection of methods for addressing a variety of computer vision issues. Artificial neural networks (ANNs), fuzzy set theory, machine learning, deep learning, genetic algorithms, and others are some of these techniques [2]. AI models can deal with the fuzziness, ambiguity, and uncertainty present in both deterministic and probabilistic statistical models. AI's capacity for analyzing a vast collection of data quickly and accurately opens up various opportunities for significant productivity gains. Furthermore, after training, AI systems and techniques may quickly forecast and generalize about difficult, unpredictable practical problems [1].

The term "artificial intelligence" was coined in the 1950s by John McCarthy, who organized the famous Dartmouth Workshop in 1956, which is often considered the birth of AI as a formal field of study. Early AI researchers focused on symbolic reasoning and creating programs that could solve logical problems. This period saw the development of programs like the Logic Theorist and the General Problem Solver. Researchers shifted their focus to expert systems, which were computer programs designed to emulate the decision-making abilities of a human expert in a specific domain. Knowledge representation and inference became central themes during this period. Advances in machine learning, particularly neural networks and statistical techniques, brought about a resurgence of interest in AI. Neural networks, which simulate the structure and function of the human brain, proved effective in tasks such as image and speech recognition. The availability of massive datasets and increased

Artificial Intelligence Applications for Sustainable Construction. https://doi.org/10.1016/B978-0-443-13191-2.00006-7
Copyright © 2024 Elsevier Ltd. All rights are reserved, including those for text and data mining, AI training, and similar technologies.

computing power enabled the development of deep learning techniques, particularly deep neural networks. These techniques led to significant breakthroughs in areas like image recognition, natural language processing (NLP), and playing games like Go at a superhuman level. The current state of AI primarily involves what's known as narrow or weak AI. These AI systems excel at specific tasks but lack general intelligence. Examples include virtual assistants like Siri and Alexa, recommendation systems, and self-driving cars. The use of technology has created several new avenues for improving performance in construction-related jobs. Machine learning (ML) algorithms can be used to supplement the knowledge of construction experts with knowledge gained from observational evidence covering large numbers of hours of employee work, significantly further than the disclosure of this most experienced and largest team of specialists, while focusing on safety and injuries occurring on construction sites. This enormous volume of scientific observation can be applied to enhance safety practices throughout the phases of design, job packaging, and project execution [3].

AI technologies have also aided in the management of water infrastructure through various concepts, procedures, and models [2]. The success rate of any project is greatly influenced by the selection of the construction site's layout. The CSLP model multi-objective artificial bee colony using the Levy flights technique, which can determine the ideal worksite plan, has been the subject of some investigations to address the issue. The model's goal is to organize makeshift basic amenities at different points (phases) of a building project while achieving several goals [4]. Another example is the study's recommendation to use deep learning as well as migration for automatic detection and localization. First, a single-shot multibox detector approach is developed to identify hyperbola-containing feature points in GPR imagery [5]. In one of the papers, a method for building key metrics that classify construction sites according to the level of safety risk is described. This is like the case of safety practices on site. Then, models were trained to predict the likelihood and severity of accidents using five well-known ML algorithms [6]. Such innovations gradually penetrated the construction sector over the last few days, allowing successful project delivery, strategic planning, quality assurance, efficiency gains, pollution reduction, and even risk control [7]. According to a recent Oxford University study that examined the likelihood of automation for over 700 different job categories, architects have the lowest risk of automation, with a 2% chance, relative to just a 90% probability for steel welders as well as a 96% likelihood for measuring experts [8]. Because of its excellent learning and reasoning capabilities, NFS has evolved as a dominant technique in modeling and addressing difficult real-world issues, and it has piqued the interest of academics in a variety of business, scientific, and engineering application fields [9]. The use of enhanced graphics techniques in the construction field is currently the subject of a lot of research. Possible uses include enhancing worker safety, tracking project progress and making project management easier, determining extensive damage for damage detection and restoration, and many more [10].

This review study stands out for employing a quantitative approach to examine the conceptual underpinnings and intellectual landscape of the body of general sense about AI focusing specifically on the sector of the construction industry. The research provides for the area in several ways, including the establishment of the size and caliber of the body of existing knowledge in the construction sector, spotting gaps and weaknesses, and deciding where to focus the efforts in future research. Throughout practice, the study will assist policymakers and practitioners in planning and funding initiatives linked to AI applications within the construction sector by providing them with a useful and current reference point.

10.2 Research methodology

This study searches the Scopus indexing database for relevant papers from the last 10 years using a scientometric approach and keyword analysis. Appropriate filters were applied to narrow the search to results that were specifically related to AI applications in the construction sector. Furthermore, the final papers for this detailed study were chosen based on a citation limit. A scientometric report indicates and examines the changes in research throughout the years. This is a statistical strategy that makes use of extensive bibliographical estimating of the analysis of the research area using a variety of qualitative metrics. A thorough evaluation is used to examine the research topics and related difficulties based on scientometric results, whereas scientometric assessment has been used to realistically map the scientific knowledge domain. The study approach is made to address things like choosing the right science mapping tool, data analysis, keyword and citation analysis, and other things.

10.2.1 Selection of science mapping of tools

There are many improvement methodologies available today, each with its own capabilities and benefits. A multidisciplinary field with origins in science and technology, science mapping is fast growing. Science mapping is the creation and use of computer techniques for the presentation, evaluation, and modeling of a wide variety of technical and scientific operations. This multidisciplinary discipline has grown out of computer science's data interpretation, visualization techniques, data gathering, and knowledge discovery techniques, as well as classical library information science's scientometrics and citation analysis techniques. In this study, the advantages and disadvantages of VOSviewer, one of the several scientific mapping tools, were assessed. A piece of software called VOSviewer offers the basic features needed to build, visualize, and explore bibliometric networks. Gephi is a well-known open-source tool for exploring, visualizing, and manipulating "any sorts of graphs and networks," and it may be used to give in-depth insight into the data that is accessible from a particular graph or network [1].

10.2.2 Data collection

The Scopus index database was used for the study's literature research, leaving out "Web of Science", "ScienceDirect", and "Google Scholar". Scopus provides comprehensive scientific data and literature, as well as analytical tools, to help you stay current and ahead of the competition. Scientific research is propelled forward by discoveries.

And if the most recent research is overlooked, opportunities to rely on and improve on that research are lost (Fig. 10.1). The keywords "artificial intelligence", and "automation in construction", were used to obtain relevant publications, and thus the following query search was made:

TITLE-ABS-KEY (artificial AND intelligence AND in AND construction) AND (LIMIT-TO (PUBYEAR, 2022) OR LIMIT-TO (PUBYEAR, 2021) OR LIMIT-TO (PUBYEAR, 2020) OR LIMIT-TO (PUBYEAR, 2019) OR LIMIT-TO (PUBYEAR, 2018) OR LIMIT-TO (PUBYEAR, 2017) OR LIMIT-TO (PUBYEAR, 2016) OR LIMIT-TO (PUBYEAR, 2015) OR LIMIT-TO (PUBYEAR, 2014) OR LIMIT-TO (PUBYEAR, 2013)) AND (LIMIT-TO (SUBJAREA, "ENGI")) AND (LIMIT-TO (EXACTSRCTITLE, "Automation In Construction") OR LIMIT-TO (EXACTSRCTITLE, "IEEE Access").

The headline, description, and tags sections of publications were searched for in Scopus using the chosen search terms; there was no "date range" constraint and only the "article" document type was allowed. As of October 10, 2022, 13,927 publications were received, out of which the papers from the previous 10 years had been

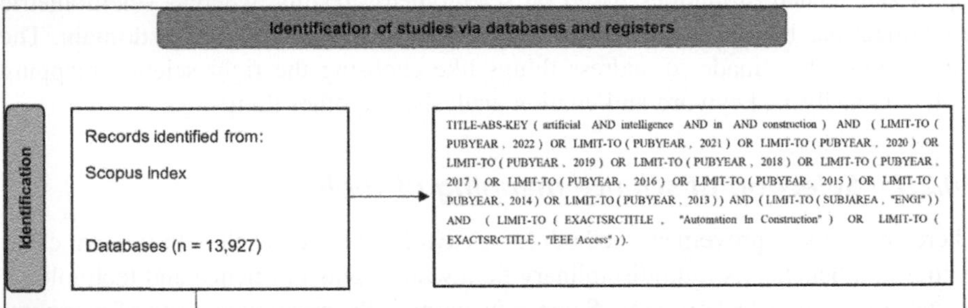

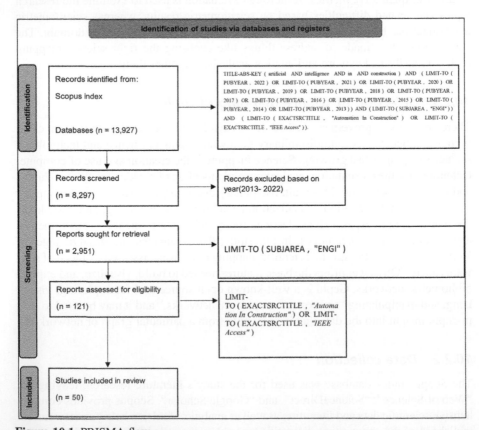

Figure 10.1 PRISMA flow.

filtered out and 8297 papers were sorted out. Further screening results in 2951 papers. In this particular study, the papers were strictly limited to two well-established journals "Automation in Construction" and "IEE Access", thus giving an output of 121 papers. Finally, out of those, 50 were narrowed down based on at least 15 citations, with 296 being the maximum citation.

10.2.3 Scientometric techniques

Beginning with "cocitation analysis", "cocountry analysis", "coauthor analysis" and "keyword search analysis" in the Scopus database using terms like "artificial intelligence" and "construction," this study's scientometric mapping is performed. By identifying patterns, trends, seasonality, and outliers, these studies will help demonstrate the intellectual, cognitive, or societal development of the subject topic.

10.3 Assessment and observations

The research is carried out utilizing all statistical techniques. The study includes a range of comprehensive summaries of the papers examined. The analysis is based on the various countries from which the sources were drawn, the keywords that were extracted, the citations, and the institutions. When conducting a qualitative study, broad themes are narrowed down and are consistently seen in every paper that is reviewed. This gives researchers insights into the specialized field in which the most recent studies are being conducted in the modern era. All of the articles include recurring topics related to "deep learning", "machine learning", "decision support systems (DSS)", "natural language processing" and the "Internet of Things". Although there are additional themes, this study confines its focus to the keywords.

10.3.1 Quantitative study

Science mapping is a fast-expanding multidisciplinary discipline with origins in computer science and engineering. The development and use of computational methods for the visualization, analysis, and modeling of a wide range of scientific and technical operations is known as science mapping. The advantages and disadvantages of VOS-viewer, one of the many science mapping tools, were assessed in the study. This software offers the basic features needed to build, visualize, and explore bibliometric networks. Thus, the scientometric analysis of the data for this study was performed using this software. The program produces mapping from various systematic analyses to give the assessment more insightful data.

10.3.1.1 Keyword analysis

As mentioned in Fig. 10.2 below, keyword analysis will allow us to know about the current research interests of the author. A network keyword can give the actual picture of the knowledge domain, provide a trend of research interest, and show how it is

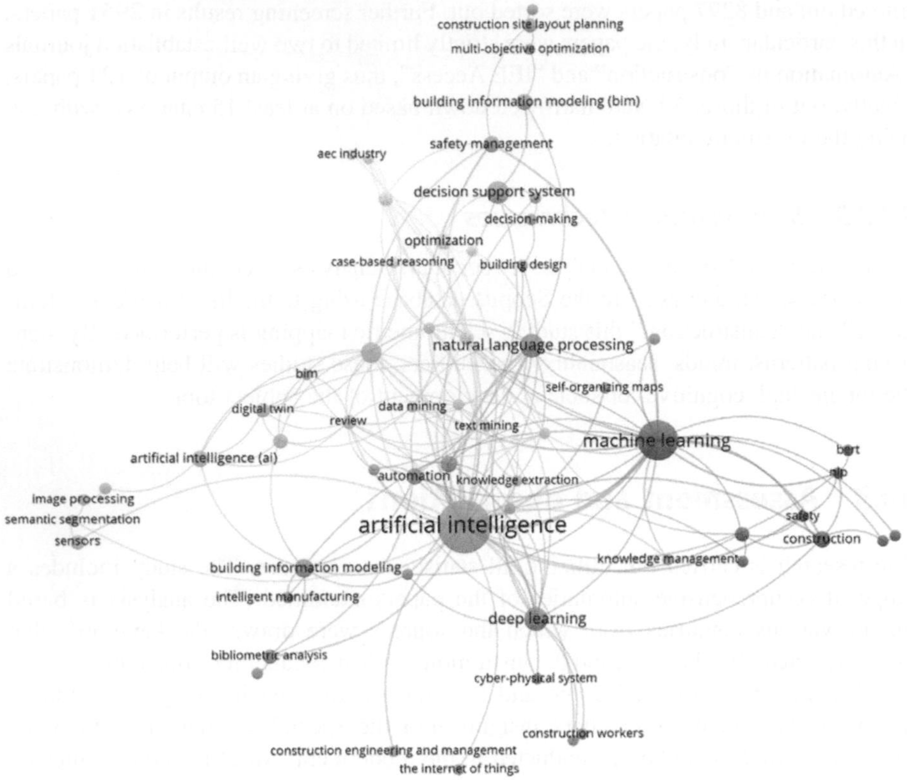

Figure 10.2 Keyword analysis.

connected to other keywords in an organized way. Thus, keyword coco-occurrence analysis was done in the VOSviewer version 1.6.18 software. The term "cooccurrence" is clearly explained as the occurrence of two or more items, in this example, a keyword.

In VOSviewer software, cooccurrence is denoted with the help of a circle and lines connected to another circle. The circle denotes the keyword, and the curvilinear line denotes the relationship among the keywords. As the network in Fig. 10.2 is a weighted network, each node not only denotes the keyword but also suggests the strength of that keyword. Bigger the circle higher the strength. In a keyword, the cooccurrence analysis strength of the keyword is computed based on how much time the keyword is used by the author in multiple publications.

To attain a visually readable output of the image, the author gave keyword-based analysis. The only limitation of this study is that it largely relies on the author's knowledge and skill of the topic for keyword selection. This limitation can be addressed by generating an image using all keywords instead of the author's keyword.

10.3.1.2 Cocountry analysis

The cocountry analysis, as mentioned in Fig. 10.3 will help the researchers understand more about which countries are working more in the relevant research area. To identify

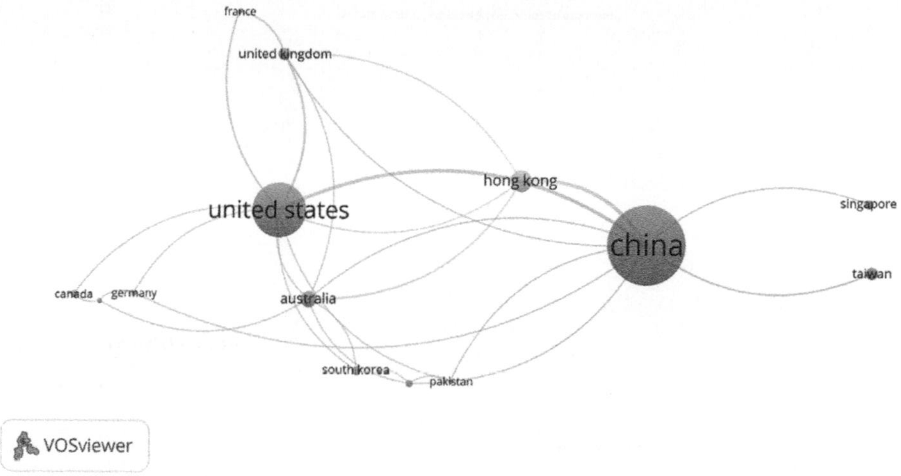

Figure 10.3 Cocountry analysis.

these countries, which are more collaborative and strengthened, a network was prepared using VOSviewer software. This study is a coauthorship analysis, and the unit of measurement is country, which is done by fractional counting. This image representation is based on the weighted average value. More the size of the circle higher the weightage is.

Based on the graph, we found the top 12 countries that are advanced in AI research and collaboration. This study showed that the United States and China are the biggest contributors to AI research in the construction sector. The strongest relationship has been seen in more developed counties and some countries are very poor in advanced research. Thus, it is proven that countries that do not have advanced research in the field of AI have to think about reforming their existing research policies.

10.3.1.3 Coinstitution analysis

For analysis purposes, VOSviewer was used to create a network. Organizations serve as the unit of measurement for this coauthorship analysis research, which employs partial numbering. As the work is fairly qualitative, the minimal number of organizational documents and the required citation count were both set at 3.8 to meet the cutoff, and the unlinked organizations were eliminated. As shown in Fig. 10.4, the University of Colorado, Ecole Polytechnique, and the University of Edinburgh have a relationship in AI research. For most of the network part, there is a lack of collaboration between major countries.

10.3.2 Qualitative study: Recent trends of AI in the construction sector

The most recent developments in AI in the building industry can reveal a lot about the applicability of the research and development work done in the publications.

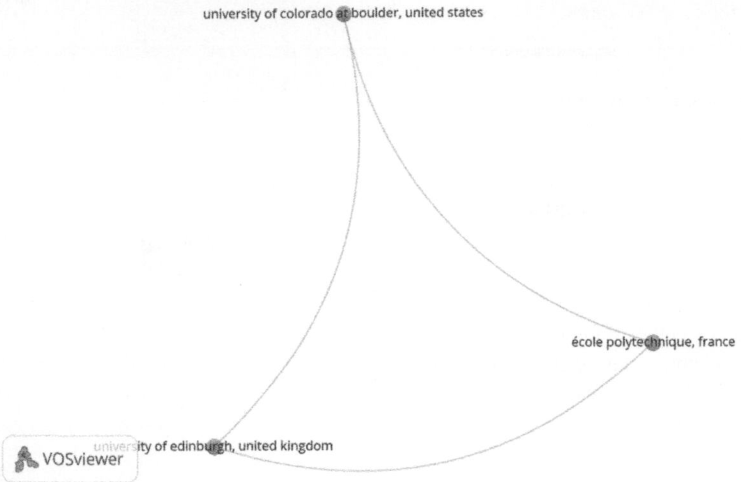

Figure 10.4 Coinstitution analysis.

Therefore, it is essential to review all these practical considerations while describing the relevant research. The current trends in the field of AI have been identified through a qualitative analysis of the selected literature from the scientometric mapping described in the previous section.

10.3.2.1 Machine learning

ML, which is ultimately a part of AI, investigates how computers can automatically learn without the assistance of programming. The model develops from AI, particularly from computational learning theory and pattern recognition. The algorithm selects the function that is best suited from the potential functions, along with a demonstration of how datasets relate to one another. Supervised, semisupervised, unsupervised, and reinforcement learning are all possible ML methods. Machines are trained using datasets and algorithms that are readily available, and they are fed a specific set of dataset features [11]. Recent applications include the creation of models that can accurately predict the type of injury, the type of energy, and the body part. The outcomes provide reliable probabilistic predictions of likely outcomes in the event of an accident, and they show significant integration potential with the help of BIM [3]. The growing trend of gathering in-service data with the help of smart sensors makes ML durable and also makes service life assessment easier to use [12].

AI as well as ML have a wide range of potential applications in the construction industry. A personal assistant that can sort through this vast amount of data is what ML is like. The system then alerts project managers to the critical things that require their attention. AI is already used in a number of applications in this manner. Its advantages range from simple spam email screening to in-depth security monitoring. ML will be used in the following sectors:

- *Removal of cost overrun*: Most large-scale projects lead to overbudgeting apart from arranging competent teams for the project. Projects employ ANNs to predict excessive amounts of costs relying on variables like the size of the project, expertise of management, and sometimes contract type. To develop realistic timetables for upcoming projects, the predictive models make use of previous data sheets, like anticipated project initiation and completion milestones. Employees can quickly advance their skills and expertise thanks to AI's ability to provide them with remote access to practical training materials. This shortens the time needed to hire new workers. The project's ending time is accelerated as a result.

- *Generative design*: Building information modeling is a method based on 3D models that helps with more effective infrastructure and building planning, design, and management. 3D models always take into consideration the MEP plans as well as the sequence of operations when planning and designing the construction of a project. Preventing conflict between the different models from the different subteams is a challenge. To find and reduce incompatibilities between the numerous models, the industry uses ML through generative design. ML algorithms can be used in software to research every potential solution and generate design options. The software produces 3D models tailored for the constraints after the user inputs their requirements and iterates for the best model.
- *Risk mitigation*: Construction projects carry a certain level of risk, appearing in the form of reliability, security, timeline, or cost issues. The risk rises with project size because multiple subcontractors perform various crafts simultaneously on work sites. Building contractors now monitor and prioritize risk on the construction site using AI-based solutions. The project team may then focus their limited attention on risks that are more serious. Automatic situational prioritization is done using AI. Construction managers can lower risk by giving contract employees a risk score and collaborating closely with high-risk teams.
- *Project planning*: Future algorithms will use "reinforcement learning," an AI strategy. Using this approach, algorithms may learn by making mistakes. Depending on previous work, it has infinite configurations and choices to consider. Because it optimizes the ideal path and gradually corrects itself, it helps with the planning stage.
- *Productivity*: Self-driving construction equipment is more efficient than its human counterparts at performing repetitive tasks like pouring concrete, laying bricks, welding, and demolishing. Trenching and site preparation tasks are carried out by fully independent or partially autonomous bulldozers. A human programmer can assist in setting up a task site to exact standards. By doing this, the project's actual completion time is halved, and the construction itself can now be carried out using human labor. The activity on the work site can be monitored by the management team in real time. To monitor employee productivity and process compliance, similar technologies such as human detection, in-person cams, and others are used.
- *Safety*: Compared to other employees, construction workers have a 5-times higher risk of being murdered at work. According to OSHA, falls were the leading cause of private sector fatalities in the construction industry, followed by being hit by an item, electric shocks, and getting stuck in the crossfire. It can look at pictures of its worksites, spot safety risks like employees who may not be wearing protective gear, and contrast the pictures with its accident rates. According to the business, it is theoretically possible to determine risk assessments for work so that safety briefings can be held when a high risk is discovered.
- *Off-site construction*: In order to create building components that are then assembled on-site by human employees, construction companies are increasingly adopting autonomous robot-staffed off-site workshops. As an illustration, autonomous robots may build walls more quickly and effectively than people on an assembly line. This may free up the human workforce to do specialty tasks like pipes, heating and cooling systems, and electronic wiring once the structure is finished.

10.3.2.2 Deep learning

Deep learning has led to the advancement of more reliable AI systems, notably in the field of image identification. Deep learning is the study of a deep NN. In this context, "deep" refers to whether the NN has more layers than a specific number. Most deep learning architectures are adaptable to a range of applications, and the architectures are occasionally combined to boost algorithm performance (Fig. 10.5).

Recent uses of DL include the automatic learning of injury precursors from unprocessed construction accident reports. The strategy involves using a dataset of unprocessed construction injury reports to automatically extract reliable accident precursors. The construction industry has implemented the NLP method for free-text data to improve document retrieval and analysis [3]. NNs come in a variety of forms, each with a unique set of computational characteristics. Most deep learning architectures can be used for a variety of tasks, and occasionally different architectures are combined to boost the efficiency of the algorithms. NNs come in a variety of forms, each with a unique set of computational characteristics. Due to its capability to grasp traceability, deep learning technology is currently regarded as one of the most hotly debated topics. Many businesses are actively researching it because it can produce impressive results in a variety of regression and classification problems, as mentioned in Fig. 10.6. Since it is considered a subset of ML in the working domain, deep learning can be thought of as an AI function that mimics how the human mind processes data.

Many studies have recommended using deep learning to increase construction site safety through safety practices to boost safety productivity and competitiveness. Two major themes emerged from a literature review: Health & security information extraction and workplace safety monitoring with the help of images, recordings, and detectors. More and more safety reports are being recorded digitally. As a result, it is now more important than ever to develop methods for utilizing this data to improve comprehension of and response to construction-related risks and hazards.

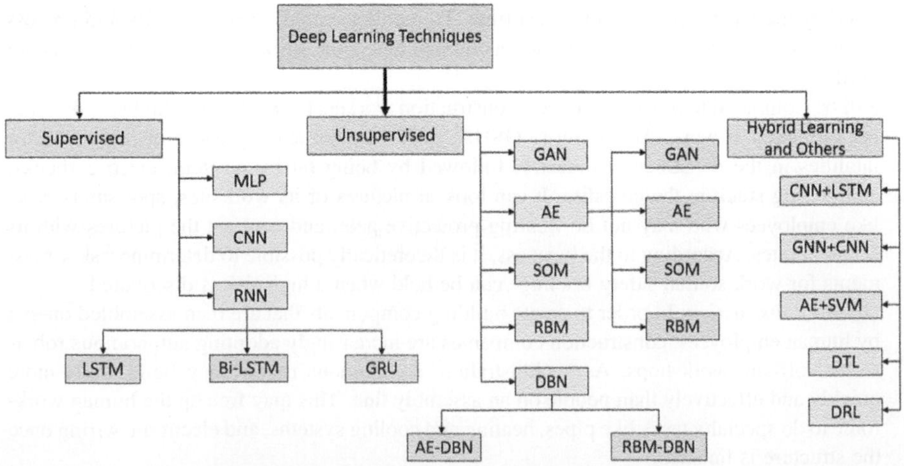

Figure 10.5 A taxonomy of deep learning techniques [12a].

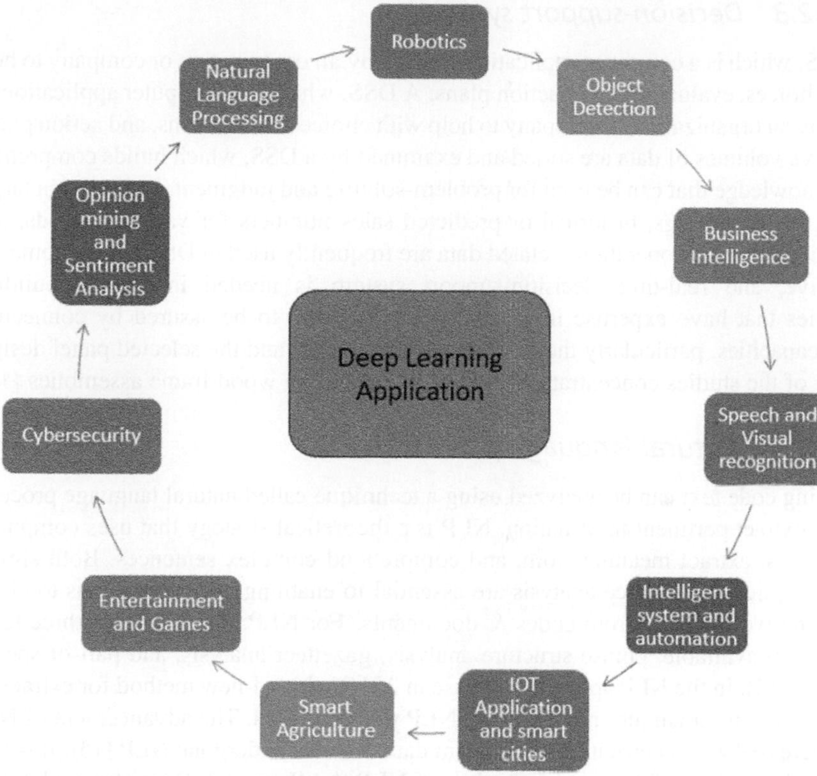

Figure 10.6 Application of deep learning in various sectors.

FastText is a state-of-the-art learning tool created by developers to find threat and security trends in text and categorize them into pertinent types of risk. Numerous studies have demonstrated the value of tracking the actions and places of construction personnel for construction safety. (ResNet-152), bounding box regression, as well as classifying out from the main image employing (R–CNN) can be used to more quickly and accurately identify construction workers. A novel deep learning algorithm can be used to track two-dimensional worker positions using color image camera footage of a worker in action. An LSTM network-based motion detection model for construction workers is created, and the usefulness of motion sensors in terms of quantity and place is investigated. Construction machine tracking is a critical step in automating construction safety surveillance. However, the present vision-based tracking systems do not attain great tracking precision. Many studies have attempted to address this problem by employing various deep-learning techniques. The leading causes of fatalities in the construction industry are clashes and brain trauma. Many worldwide health and safety institutions have mandated that contractors always uphold and keep an eye on employees' proper use of personal protective equipment. Other studies have employed deep learning techniques to monitor construction workers' compliance with wearing safety helmets.

10.3.2.3 Decision-support system

A DSS, which is a computer application, is used by an organization or company to help with choices, evaluations, and action plans. A DSS, which is a computer application, is used by an organization or company to help with choices, evaluations, and action plans. Massive volumes of data are sorted and examined by a DSS, which builds comprehensive knowledge that can be used for problem-solving and judgment. Statistics on target or expected earnings, historical or predicted sales numbers for various periods, and other inventory- or operations-related data are frequently used in DSSs. An automated, effective, and real-time decision-support system is needed in off-site building facilities that have expertise in panelized construction to be assured by connecting floor capacities, particularly the machinery capabilities, and the selected panel design. Some of the studies concentrate on its use in producing wood frame assemblies [13].

10.3.2.4 Natural language processing

Building code text can be analyzed using a technique called natural language processing to extract pertinent information. NLP is a theoretical strategy that uses computers to process, extract meaning from, and comprehend complex sentences. Both simple and complicated sentence analysis are essential to enabling NLP algorithms to effectively derive meaning from codes & documents. For NLP, the following three techniques are available: phrase structure analysis, gazetteer analysis, and part-of-speech tagging [14]. In the NLP application phase in 2016, a brand-new method for extracting attributes from raw injury reports using NLP was proposed. The advancement of NLP has increased ML's capacity to learn from data that has undergone NLP [15]. It is also helpful to predict safety outcomes using an NLP + ML model. Convolutional neural networks as well as hierarchical attention networks are used in the deep architecture used for NLP nowadays. In the area of deep learning architecture with NLP models, very little work has been done [16].

10.3.2.5 Internet of Things

Industry 4.0 is currently undergoing a revolution. All the European nations have presented their plans for the Industry 4.0 strategy since the term was first used by the German Federal Ministry of Education. Artificial intelligence, robotics, IoT, cloud computing, and neural language processing are some of the technologies driving this revolution, as mentioned in Fig. 10.7. It is entirely automated, with high levels of machine-to-machine interaction allowing for cost and time optimization in the industry [17]. The pervasive sensing capabilities of the IoT greatly facilitate data collection on a construction site. Processing complex data to extract useful information is becoming increasingly necessary. In essence, it is projected that this trend will continue as AI and Internet of Things applications proliferate. Internet of Things technologies have become increasingly popular in recent years. As a result of these technologies, a vast amount of data is generated. To be exploited effectively, this data necessitates the use of accurate data analysis tools. Deep learning, a new area of AI, has highlighted the ability for improved analysis of big data. A survey of the literature

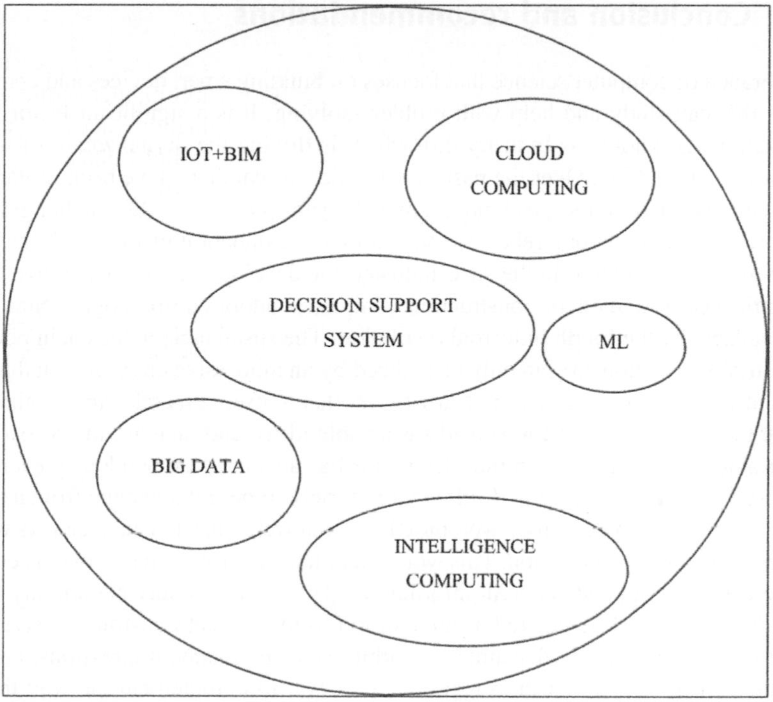

Figure 10.7 Applications of IoT.

suggests deep learning with IoT can create smart cities as well as infrastructure, with deep learning including IoT technologies and deep learning plus IoT in assessment.

- *Structures and smart cities*: By building a modern hospital assessment framework containing evaluation criteria, including subcriteria, a big issue is addressed. For the COVID-19 pandemic, this can be chosen and plotted to BIM solutions to enhance asset information management practices.
- *Construction assessment*: Deep learning techniques in end devices can be used for assessment and fault diagnosis via transfer learning as well as network cleaning, producing quick inference and little memory usage. Another study examines the technical viability of fully independent rust evaluation in concrete structures and suggests using ML and the Internet of Things to analyze the corrosion status of RC structures. Electrochemical, ultrasonic, fiber-optic, piezoelectric, wireless, fiber Bragg grating sensors, and self-sensing concrete are some of the sensors that may be utilized for accurate and real-time monitoring of agricultural building structures.

There have been some recent studies that provide data integration between IoT and BIM. One of the industry's hottest trends is the combination of IoT and cloud computing. It is necessary to work with cloud computing to process data quickly and accurately because IoT has limited data processing capabilities. The processing power of cloud computing allows for the development of high-process analysis and high-performance intelligent computing, depending on the amount of data available.

10.4 Conclusion and recommendations

AI is a branch of computer science that focuses on building smart devices and computer systems that can study and help with problem-solving. It is a significant factor in the digitalization age known as Industry 4.0, which is driving the digitalization of several industries, notably AEC. Over the past few decades, researchers have been researching how to employ AI principles, techniques, and algorithms to solve AEC challenges. This report presents the first comprehensive scientometric evaluation of the condition of AI research in the AEC sector. In the AEC industry, the development of cooperative tactics could influence how modern construction companies adopt cutting-edge technologies made available by the fourth industrial revolution. The sustainable value chain potential of Industrial Revolution 4.0 can only be realized by an innovative environment that consists of academics, businesses, and artists. In that sense, research can significantly improve the construction sector, spread sustainable ideas, and increase the accessibility of technology. The construction industry sector has advanced tremendously ever since the emergence of BIM and gained sufficient notoriety to permit a change from a stable, locked environment to a reactive, web-based one that welcomes IoT inclusion as well as a greater degree of AI application. This would facilitate the provision of smarter services, increased automation, and coherent information. The existing safety community thinks that NLP + ML could be utilized as an addition to the current construction safety approaches, which are mostly dependent on arbitrary assessments, impressions, and risk assessments. It is hoped that the suggested strategy will complement the way that the construction safety communities currently use existing techniques.

In the old construction industry, digital technology has advanced quickly in recent decades, as has the growth of big data. The adoption of AI, which aims to give machines intelligent behavior and reasoning akin to humans, has drawn a lot of attention. The construction industry has seen a substantial change thanks to the use of various AI techniques, which has led to a CEM system that is more reliable, automated, time-cutting, and economical. In dealing with complicated and dynamic situations within circumstances of high unpredictability and abundance of data, potential AI outperforms traditional computing methods and expert assessments, increasing the possibility that answers will be precise and compelling for calculated decisions [18].

Based on the study conducted, five domains have been found to actively participate in the application of AI. These are deep learning, ML, decision support systems, NLP, and the Internet of Things. Future research will focus on many topics, including administration of the construction site, surveillance of the site's safety, and assessment of the construction resources' productivity [11].

References

[1] A. Darko, A.P.C. Chan, M.A. Adabre, D.J. Edwards, M.R. Hosseini, E.E. Ameyaw, Artificial intelligence in the AEC industry: scientometric analysis and visualization of research activities, Automation in Construction 112 (2020), https://doi.org/10.1016/j.autcon.2020.103081.

[2] T. Dawood, E. Elwakil, H.M. Novoa, J.F.G. Delgado, Artificial intelligence for the modelling of water pipes deterioration mechanisms, Automation in Construction 120 (2020), https://doi.org/10.1016/j.autcon.2020.103398.

[3] A.J.-P. Tixier, M.R. Hallowell, B. Rajagopalan, D. Bowman, Application of machine learning to construction injury prediction, Automation in Construction 69 (2016) 102−114, https://doi.org/10.1016/j.autcon.2016.05.016.

[4] M. Yahya, M.P. Saka, Construction site layout planning using multi-objective artificial bee colony algorithm with Levy flights, Automation in Construction 38 (2014) 14−29, https://doi.org/10.1016/j.autcon.2013.11.001.

[5] H. Liu, C. Lin, J. Cui, L. Fan, X. Xie, B.F. Spencer, Detection and localization of rebar in concrete by deep learning using ground penetrating radar, Automation in Construction 118 (2020), https://doi.org/10.1016/j.autcon.2020.103279.

[6] C.Q.X. Poh, C.U. Ubeynarayana, Y.M. Goh, Safety leading indicators for construction sites: a machine learning approach, Automation in Construction 93 (2018) 375−386, https://doi.org/10.1016/j.autcon.2018.03.022.

[7] Z. You, L. Feng, Integration of industry 4.0 related technologies in construction industry: a framework of cyber-physical system, IEEE Access 8 (2020) 122908−122922, https://doi.org/10.1109/ACCESS.2020.3007206.

[8] E. Karan, S. Asadi, Intelligent designer: a computational approach to automating design of windows in buildings, Automation in Construction 102 (2019) 160−169, https://doi.org/10.1016/j.autcon.2019.02.019.

[9] G.G. Tiruneh, A.R. Fayek, V. Sumati, Neuro-fuzzy systems in construction engineering and management research, Automation in Construction 119 (July) (2020) 103348, https://doi.org/10.1016/j.autcon.2020.103348. Elsevier.

[10] K. Mostafa, T. Hegazy, Review of image-based analysis and applications in construction, Automation in Construction 122 (2021), https://doi.org/10.1016/j.autcon.2020.103516.

[11] R. Khallaf, M. Khallaf, Classification and analysis of deep learning applications in construction: a systematic literature review, Automation in Construction 129 (June) (2021) 103760, https://doi.org/10.1016/j.autcon.2021.103760. Elsevier B.V.

[12] W.Z. Taffese, E. Sistonen, Machine learning for durability and service-life assessment of reinforced concrete structures: recent advances and future directions, Automation in Construction 77 (2017) 1−14, https://doi.org/10.1016/j.autcon.2017.01.016.

[12a] I.H. Sarker, Deep learning: a comprehensive overview on techniques, taxonomy, applications and research directions, SN Computer Science 2 (2021) 420, https://doi.org/10.1007/s42979-021-00815-1.

[13] S. An, P. Martinez, M. Al-Hussein, R. Ahmad, BIM-based decision support system for automated manufacturability check of wood frame assemblies, Automation in Construction 111 (June 2019) (2020) 103065, https://doi.org/10.1016/j.autcon.2019.103065.

[14] J. Zhang, N.M. El-Gohary, Integrating semantic NLP and logic reasoning into a unified system for fully-automated code checking, Automation in Construction 73 (2017) 45−57, https://doi.org/10.1016/j.autcon.2016.08.027.

[15] M.Y. Cheng, D. Kusoemo, R.A. Gosno, Text mining-based construction site accident classification using hybrid supervised machine learning, Automation in Construction 118 (May) (2020) 103265, https://doi.org/10.1016/j.autcon.2020.103265. Elsevier.

[16] H. Baker, M.R. Hallowell, A.J.-P. Tixier, AI-based prediction of independent construction safety outcomes from universal attributes, Automation in Construction 118 (2020), https://doi.org/10.1016/j.autcon.2020.103146.

[17] F. Bianconi, M. Filippucci, A. Buffi, Automated design and modelling for mass-customized housing. A web-based design space catalog for timber structures, Automation in Construction 103 (2019) 13–25, https://doi.org/10.1016/j.autcon.2019.03.002.

[18] Y. Pan, L. Zhang, Roles of artificial intelligence in construction engineering and management: a critical review and future trends, Automation in Construction 122 (2021), https://doi.org/10.1016/j.autcon.2020.103517.

Textile-reinforced mortar-masonry bond strength calibration using machine learning methods

11

Atefeh Soleymani[1], *Danial Rezazadeh Eidgahee*[2] *and Hashem Jahangir*[3]
[1]Structural Engineering, Shahid Bahonar University of Kerman, Kerman, Iran; [2]Department of Civil Engineering, Ferdowsi University of Mashhad, Mashhad, Iran; [3]Department of Civil Engineering, University of Birjand, Birjand, Iran

11.1 Introduction

Most of the old structures all around the world, specifically in Europe, were built with masonry materials [1]. These masonry structures can experience very high levels of damage, which should be reduced using strengthening techniques [2]. Among the strengthening techniques, fiber-reinforced polymer (FRP) composites attract the researcher's attention with their advantages, such as their high strength, low weight, and ease of application on different geometrical structural elements [3]. Oppositely, FRP composites have some disadvantages because of their use of polymer-based epoxy, such as not bonding well to the uneven surface of structural elements, not working well in a humid environment, and being highly sensitive to temperature [4]. Therefore, the researchers introduced textile-reinforced mortar (TRM) composites as a proper alternative to FRP composites [5].

In TRM composites, the mineral mortars replaced the polymer-based epoxy in FRP composites, and as a result, the TRM composites converted the disadvantages of FRP composites into advantages. The TRM composites are not sensitive to temperature and can be applied to uneven surfaces in humid environments [6]. To evaluate the performance of TRM composites, their bond strength can be investigated experimentally by conducting direct shear tests. On the other hand, some analytical studies such as Wu et al. [7], De Lorenzis et al. [8], Yuan et al. [9], Izumo [10], Iso [10], Sato [10], Dai et al. [11], Lu et al. [12], Van Gemert [13], Tanaka [14], Hiroyuki and Wu [15], Maeda et al. [16], Khalifa et al. [17], Adhikary and Mutsuyoshi [18], and international codes such as fib Bulletin 14 [19], CNR-DT 200/2004 [20], ACI 440.2R [21], and HB 305 [22], proposed some models to estimate the bond strength of TRM composites.

Artificial Intelligence Applications for Sustainable Construction. https://doi.org/10.1016/B978-0-443-13191-2.00001-8
Copyright © 2024 Elsevier Ltd. All rights are reserved, including those for text and data mining, AI training, and similar technologies.

11.2 Research significance

Investigating the existing models for estimating the bond strength of TRM composites in empirical studies and international codes showed their performance needs to be improved, as most of them were proposed for FRP composites or concrete substrates. In this chapter, by gathering an experimental database from a previous study conducted by Jahangir and Esfahani [23], the performance of existing models was evaluated, and the best one was selected. Then, by utilizing the genetic algorithm (GA) technique, it is tried to enhance the performance of the selected model to reach more accurate results in predicting the TRM-masonry bond strength.

11.3 Existing TRM-masonry bond strength models

In this chapter, as presented in Table 11.1, eighteen existing models for calculating the TRM-masonry bond strength proposed in previous empirical studies and codes [7–22] were considered to be evaluated. As illustrated in Fig. 11.1, in the considered models, the bond strength (P_u) can be obtained from input parameters, including the width of the masonry prism (b_m), compressive strength of the masonry prism (f_m), elastic modulus and thickness of the fibers (E_f and t_f, respectively), and the bond length and width of the TRM composite (L_b and b_f, respectively). More details regarding these models can be found in Ref. [24].

11.4 Experimental direct shear test specimens

To investigate the performance of the considered existing models provided in Table 11.1, in this chapter, TRM-masonry direct shear tests were compiled from a previous experimental study conducted by Jahangir and Esfahani [23], which are reported in Table 11.2. Fig. 11.2 showcases the average P_u in different experimental specimen series as well as their corresponding range coefficient of variation (CoV).

11.5 Performance of existing models

To evaluate the accuracy and performance of the existing models, the influential parameters of the experimental specimens, reported in Table 11.2, were introduced to their presented equations in Table 11.1. Fig. 11.3 shows the absolute percentage error (APE, presented in Eq. (11.20)) and the corresponding CoV of all eighteen models for each specimen series. Moreover, the standard deviation (SD), correlation coefficient (R, presented in Eq. (11.19)), and mean absolute percentage error (MAPE, presented in Eq. (11.21)) for each model as well as their rank based on the best performance (higher R and lower MAPE and difference between the experimental and estimated SD values) are presented in Table 11.3. Furthermore, the Taylor diagram of the existing models with respect to experimental results is depicted in Fig. 11.4.

Table 11.1 Existing bond strength models.

References	Model	Equation
Wu et al. [7] – (M1)	$P_u = \beta_w b_f \sqrt{2\left(1 + \frac{\lambda}{\Sigma}\right) E_f t_f G_{cc}}$; $\beta_w = \sqrt{\frac{2 - b_f/b_m}{1 + b_f/b_m}}; \lambda = t_d/t_f$; $t_d = 3.5$ mm; $\Sigma = E_f/E_m$; $G_{cc} = 0.17$ N/mm;	(11.1)
fib Bulletin 14 [19] – (M2)	$P_u = \begin{cases} \alpha c_1 k_c k_b b_f \sqrt{E_f t_f f_{mt}} & \text{if} L_b \geq L_e \\ \alpha c_1 k_c k_b b_f \sqrt{E_f t_f f_{mt}} \frac{L_b}{L_e}\left(2 - \frac{L_b}{L_e}\right) & \text{if} L_b < L_e \end{cases} K_b = 1.06\sqrt{\frac{2 - b_f/b_m}{1 + b_f/400}} \geq 1; L_e = \sqrt{\frac{E_f t_f}{c_2 f_{mt}}}; \alpha = 0.9; k_c = 1; c_1 = 0.64; c_2 = 2$	(11.2)
CNR-DT 200/ 2004 [20] – (M3)	$P_u = \begin{cases} b_f \sqrt{2E_f t_f \Gamma_f} & \text{if} L_b \geq L_e \\ b_f \sqrt{2E_f t_f \Gamma_f} \frac{L_b}{L_e}\left(2 - \frac{L_b}{L_e}\right) & \text{if} L_b < L_e \end{cases} \Gamma_f = 0.03 k_b \sqrt{f_m f_{mt}}; k_b = \sqrt{\frac{2 - b_f/b_m}{1 + b_f/400}} \geq 1$; $L_e = \sqrt{\frac{E_f t_f}{2 f_{mt}}}$;	(11.3)
Van Gemert [13] (M4)	$P_u = 0.5 f_{mt} b_f L_b$	(11.4)
Tanaka [14] – (M5)	$P_u = (6.13 - \ln L_b) b_f L_b$	(11.5)
Hiroyuki and Wu [15] – (M6)	$P_u = b_f L_b \left(5.88 L_b^{-0.669}\right)$	(11.6)
Maeda et al. [16] – (M7)	$P_u = 110.2 \times 10^{-6} \cdot E_f t_f b_f L_e L_e = e^{6.13 - 0.580 \ln(E_f t_f)}; E_f(\text{GPa}), t_f(\text{mm})$	(11.7)

Continued

Table 11.1 Continued

References	Model	Equation
Khalifa et al. [17] – (M8)	$P_u = k\left(\frac{f_m}{42}\right)^{2/3} E_f t_f b_f L_e$; $k = 110.2\times10^{-6}$; $L_e = e^{6.13-0.580\ln\frac{E_f t_f}{1000}} \leq L_b$;	(11.8)
Adhikary and Mutsuyoshi [18] – (M9)	$P_u = b_f L_b\left(0.25 f_m^{2/3}\right)$	(11.9)
Chen and Teng (ACI 440.2R) [21] – (M10)	$P_u = \alpha\beta_w\beta_l\sqrt{f_m} b_f L_e$; $\beta_l = \begin{cases} 1 & \text{if} L_b \geq L_e \\ \sin\frac{\pi L_b}{2L_e} & \text{if} L_b < L_e \end{cases}$ $\beta_w = \sqrt{\frac{2 - b_f/b_m}{1 + b_f/b_m}}$; $L_e = \sqrt{\frac{E_f t_f}{\sqrt{f_m}}}$; $\alpha = 0.427$	(11.10)
De Lorenzis et al. [8]– (M11)	$P_u = b_f\sqrt{2E_f t_f G_f}$; $G_f = 1.06\ \text{N}\cdot\text{mm/mm}^2$	(11.11)
Yuan et al. [9] – (M12)	$P_u = \eta b_f\sqrt{2E_f t_f G_f}$; $\eta = 1.212$; $G_f = \gamma\beta_w^2 f_{mt}^d$; $\gamma = 0.420$; $d = 0.695$; $\beta_w = \sqrt{\frac{2.25 - b_f/b_m}{1.25 + b_f/b_m}}$	(11.12)

Izumo [10] (M13)	$P_u = \left(3.8f_m^{2/3} + 15.2\right) L_b b_f E_f t_f \times 10^{-3}$	(11.13)
Iso [10] (M14)	$P_u = 0.93 f_m^{0.44} b_f L_e$; $L_e = 0.125\left(E_f t_f\right)^{0.57} \leq L_b$;	(11.14)
Sato [10] (M15)	$P_u = 2.68\left(b_f + 2\Delta b_f\right) L_e f_m^{0.2} E_f t_f \times 10^{-5}$; $L_e = 1.89\left(E_f t_f\right)^{0.4}$; $\Delta b_f = 3.7$ mm	(11.15)
Dai et al. [11] – (M16)	$P_u = \left(b_f + 2\Delta b_f\right)\sqrt{2E_f t_f G_f}$; $G_f = 0.514(f_m)^{0.236}$; $\Delta b_f = 3.7$ mm;	(11.16)
Lu et al. [12] – (M17)	$P_u = \beta_l b_f \sqrt{2E_f t_f G_f}$ $\beta_l = \begin{cases} 1 & \text{if} L_b \geq L_e \\ \dfrac{L_b}{L_e}\left(2 - \dfrac{L_b}{L_e}\right) & \text{if} L_b < L_e \end{cases}$ $L_e = a + \dfrac{1}{2\lambda_1}\ln\left[\dfrac{\lambda_1 + \tan(\lambda_2 a)}{\lambda_1 - \tan(\lambda_2 a)}\right]$; $G_f = 0.308\beta_w^2\sqrt{f_{mt}}$; $\lambda_1 = \sqrt{\dfrac{\tau_{\max}}{S_0 E_f t_f}}$; $\tau_{\max} = 1.5\beta_w f_{mt}$; $\lambda_2 = \sqrt{\dfrac{\tau_{\max}}{(S_f - S_0)E_f t_f}}$; $S_0 = 0.0195\beta_w f_{mt}$; $S_f = \dfrac{2G_f}{\tau_{\max}}$; $a = \dfrac{1}{\lambda_2}\arcsin\left[0.99\sqrt{\dfrac{(S_f - S_0)}{S_f}}\right]$ $\beta_w = \sqrt{\dfrac{2.5 - b_f/b_m}{1.25 + b_f/b_m}}$	(11.17)

Continued

Table 11.1 Continued

References	Model	Equation
Seracino et al. (HB 305) [22] – (M18)	$P_u = 0.853\beta_l\left(\frac{d_f}{b_f}\right)^{0.25}(f_m)^{0.33}\sqrt{L_{per}E_fA_f}$; $\beta_l = \begin{cases} 1 & \text{if} L_b \geq L_e \\ \frac{L_b}{L_e} & \text{if} L_b < L_e \end{cases} L_e = \frac{\pi}{2\sqrt{(\tau_f L_{per}/\delta_f E_f A_f)}}; L_{per} = 2d_f + b_f$ $\tau_f = \left(0.802 + 0.078\frac{d_f}{b_f}\right)(f_m)^{0.6}; d_f = 1\text{ mm}$ $\delta_f = \frac{0.73}{\tau_f}\left(\frac{d_f}{b_f}\right)^{0.5}(f_m)^{0.67}; A_f = b_f t_f$	(11.18)

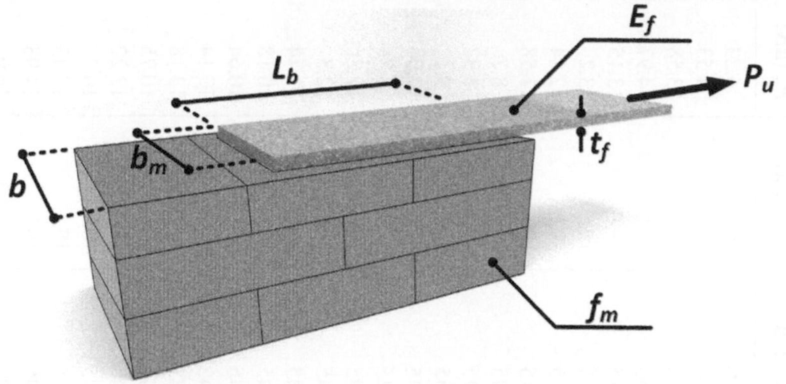

Figure 11.1 The influential parameters on TRM-masonry bond strength.

$$R = \frac{\sum_{i=1}^{n}(P_{ui} - \overline{P}_u)\left(\widetilde{P}_{ui} - \overline{\widetilde{P}}_u\right)}{\sqrt{\sum_{i=1}^{n}(P_{ui} - \overline{P}_u)^2 \sum_{i=1}^{n}\left(\widetilde{P}_{ui} - \overline{\widetilde{P}}_u\right)^2}} \tag{11.19}$$

$$\text{APE} = \left(\frac{\sum_{i=1}^{n}|P_{ui} - \widetilde{P}_{ui}|}{\sum_{i=1}^{n}|P_{ui}|}\right)\times 100 \tag{11.20}$$

$$\text{MAPE} = \frac{1}{n}\left(\frac{\sum_{i=1}^{n}|P_{ui} - \widetilde{P}_{ui}|}{\sum_{i=1}^{n}|P_{ui}|}\right)\times 100 \tag{11.21}$$

In Eqs. (11.19) to (21), the P_{ui} and $\overline{P}_u$ are respectively the individual and average experimental TRM-masonry bond strength for direct shear specimens, $\widetilde{P}_{ui}$ and $\overline{\widetilde{P}}_u$ are respectively the individual and average estimated TRM-masonry bond strength for direct shear specimens, and n is the number of specimens.

As can be inferred from Table 11.3 and Fig. 11.4, the best existing models, which were placed in the first and second ranks, are Hiroyuki and Wu [15]. (M6) with R and MAPE values of 0.83% and 25.64%, and CNR-DT 200/2004 [20] (M3) with R and MPAE values of 0.71% and 31.48%, respectively.

11.6 Calibration of the best existing models

In this section, the best existing models, including Hiroyuki and Wu [15] and CNR-DT 200/2004 [20] models, were selected to be calibrated by a GA with a defined target of obtaining maximum R and minimum MAPE values. To achieve this goal, some of the

Table 11.2 Experimental TRM-masonry direct shear tests [23].

Series	Specimen name	b_m (mm)	f_m (MPa)	t_f (mm)	b_f (mm)	L_b (mm)	E_f (GPa)	P_u (kN)
S_150_50	S_150_50_C_R1_1	120	50	0.084	50	150	192	6.23
	S_150_50_C_R1_2	120	50	0.084	50	150	192	4.53
	S_150_50_C_R1_3	120	50	0.084	50	150	192	3.58
	S_150_50_C_R1_4	120	50	0.084	50	150	192	4.97
S_215_50	S_215_50_C_R1_1	120	50	0.084	50	215	192	13.15
	S_215_50_C_R1_2	120	50	0.084	50	215	192	13.2
	S_215_50_C_R1_3	120	50	0.084	50	215	192	12.84
	S_215_50_C_R1_4	120	50	0.084	50	215	192	11.58
S_345_40	S_345_40_C_R1_1	120	50	0.084	40	345	192	8.7
	S_345_40_C_R1_2	120	50	0.084	40	345	192	8.32
	S_345_40_C_R10_3	120	50	0.084	40	345	192	9.1
	S_345_40_C_R10_4	120	50	0.084	40	345	192	7.93
	S_345_40_C_R15_5	120	50	0.084	40	345	192	8.88
S_345_50	S_345_50_C_R1_1	120	50	0.084	50	345	192	11.26
	S_345_50_C_R1_2	120	50	0.084	50	345	192	12.3
	S_345_50_C_R1_3	120	50	0.084	50	345	192	11.74
	S_345_50_C_R1_4	120	50	0.084	50	345	192	10.18
	S_345_50_R_R1_1	120	50	0.084	50	345	192	10.64
	S_345_50_R_R1_2	120	50	0.084	50	345	192	13.14
	S_345_50_R_R1_3	120	50	0.084	50	345	192	13.18
	S_345_50_R_R1_4	120	50	0.084	50	345	192	10.75
	S_345_50_R_R1_5	120	50	0.084	50	345	192	12.22
	S_345_50_R_R1_6	120	50	0.084	50	345	192	10.3
S_345_75	S_345_75_C_R1_1	120	50	0.084	75	345	192	17.43
	S_345_75_C_R1_2	120	50	0.084	75	345	192	14.93
	S_345_75_C_R1_3	120	50	0.084	75	345	192	16.9
	S_345_75_C_R1_4	120	50	0.084	75	345	192	16.47

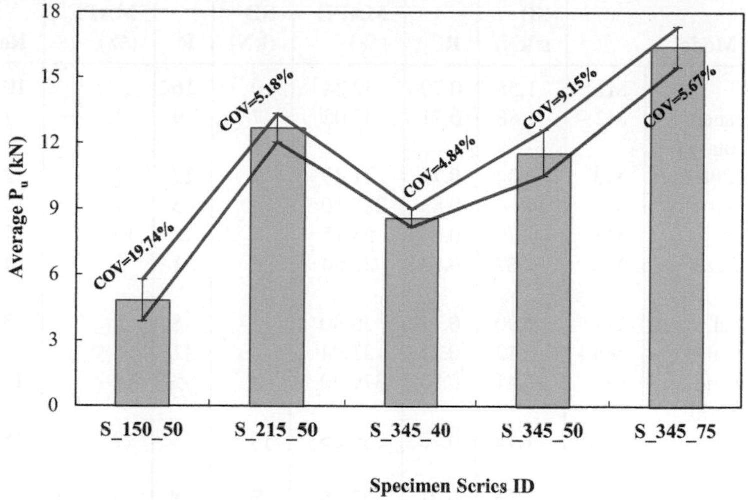

Figure 11.2 The average P_u in the experimental specimen series.

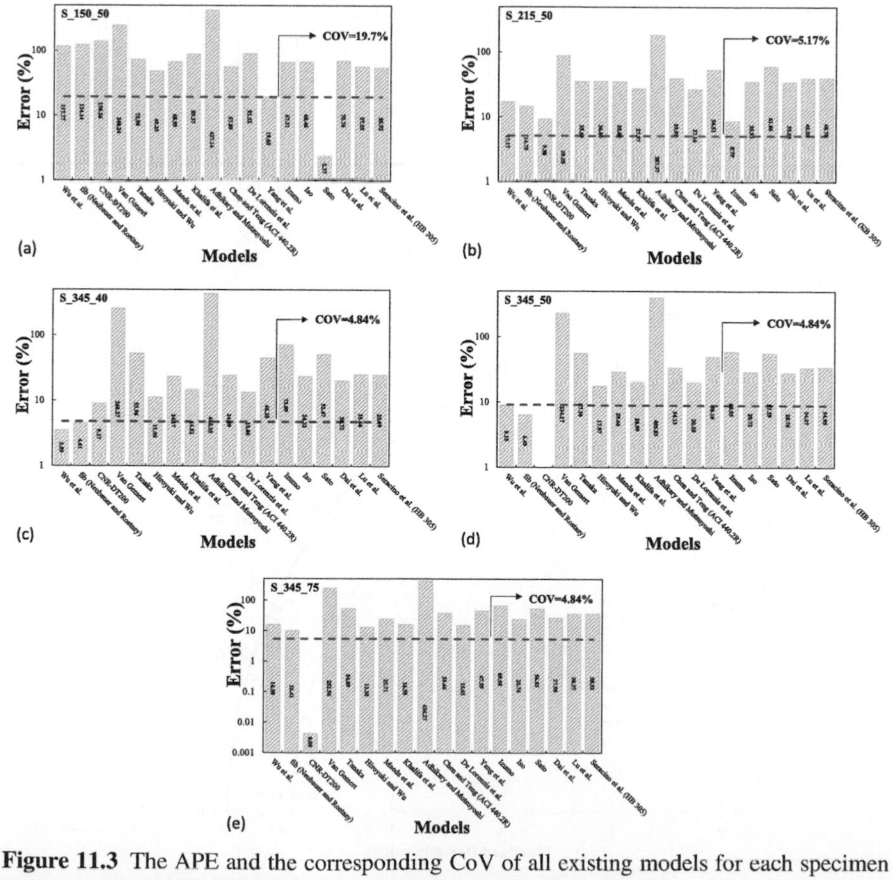

Figure 11.3 The APE and the corresponding CoV of all existing models for each specimen series are: (a) S_150_50; (b) S_215_50; (c) S_345_40; (d) S_345_50; and (e) S_345_75.

Table 11.3 Performance and error evaluation parameters of existing models.

Model		SD s(kN)	R	MAPE (%)	Rank: SD (kN)	Rank: R	Rank: MAPE (%)	Rank: Rerank
Wu et al.	M1	1.58	0.70	32.84	10	16	4	10
fib (Neubauer and Rostasy)	M2	1.88	0.71	32.08	7	14	3	7
CNR-DT200	**M3**	**2.34**	**0.71**	**31.48**	**2**	**12**	**2**	**2**
Van Gemert	M4	14.14	0.83	217.10	17	3	17	13
Tanaka	M5	1.79	0.04	55.15	8	18	15	17
Hiroyuki and Wu	**M6**	**2.57**	**0.83**	**25.64**	**1**	**1**	**1**	**1**
Maeda et al.	M7	1.90	0.71	36.80	5	8	8	5
Khalifa et al.	M8	2.13	0.71	33.94	4	11	6	5
Adhikary and Mutsuyoshi	M9	21.41	0.83	379.99	18	3	18	15
Chen and Teng (ACI 440.2R)	M10	1.14	0.70	39.26	14	17	12	18
De Lorenzis et al.	M11	2.16	0.71	33.65	3	8	5	2
Yang et al.	M12	1.35	0.71	43.54	11	5	13	9
Izumo	M13	6.80	0.83	55.86	15	2	16	11
Iso	M14	1.90	0.71	36.82	6	5	9	4
Sato	M15	1.00	0.71	46.00	16	7	14	13
Dai et al.	M16	1.67	0.71	36.65	9	8	7	7
Lu et al.	M17	1.23	0.70	39.19	13	15	11	15
Seracino et al. (HB 305)	M18	1.24	0.71	39.07	12	13	10	12

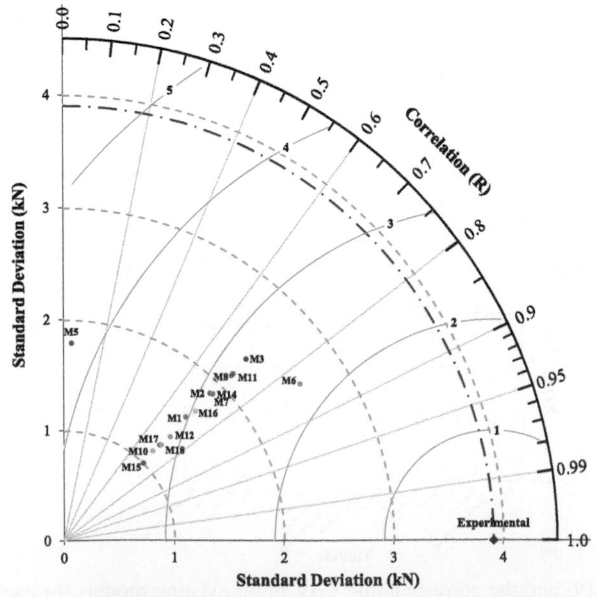

Figure 11.4 Taylor diagram of existing models.

constants in these models were replaced by two parameters named C and α, as presented in Eqs. (11.22) and (11.23) with red colors ($P_{u\_H\&M}$ and $P_{u\_CNR\text{-}DT}$), to be calibrated based on experimental data with the GA technique. The results of the calibration process are reported in Table 11.4. Moreover, Table 11.5 revealed the R and MAPE values for calibrated models as well as the considered ranges for C and α in each model. Fig. 11.5 shows the predicted versus experimental values of P_u in both calibrated models.

$$P_{u\_H\&W} = b_f L_b \left(C \cdot L_b^{-\alpha} \right)$$

$$P_{u\_CNR-DT} = \begin{cases} b_f \sqrt{C \cdot E_f t_f \Gamma_f} & \text{if } L_b \geq L_e \\ b_f \sqrt{C \cdot E_f t_f \Gamma_f} \dfrac{L_b}{L_e}(2 - \dfrac{L_b}{L_e}) & \text{if } L_b < L_e \end{cases}$$

$$\Gamma_f = 0.03 k_b \left(f_m f_{mt} \right)^{\alpha};$$

$$k_b = \sqrt{\frac{2 - b_f / b_m}{1 + b_f / 400}} \geq 1;$$

$$L_e = \sqrt{\frac{E_f t_f}{2 f_{mt}}};$$

The presented results in Table 11.5 and Fig. 11.5 showed that the calibration process could increase the R values and reduce the MAPE values of selected models. As illustrated in Fig. 11.6, the R values of the Hiroyuki and Wu models improved from

Table 11.4 The obtained results of the calibration process.

Calibrated method	***C***		***α***	
$P_{u\_H\&M}$	Considered range	0.1–50	Considered range	0.01–10
	Obtained value	**1.0487**	**Obtained value**	**0.1416**
$P_{u\_CNR\text{-}DT}$	Considered range	0.1–10	Considered range	−10 to 10
	Obtained value	**1.9767**	**Obtained value**	**0.4763**

Table 11.5 The obtained results of the calibration process.

Calibrated method	***R***	***R*²**	**MAPE (%)**
$P_{u\_H\&M}$	0.84	0.70	12.18
$P_{u\_CNR\text{-}DT}$	0.71	0.50	30.79

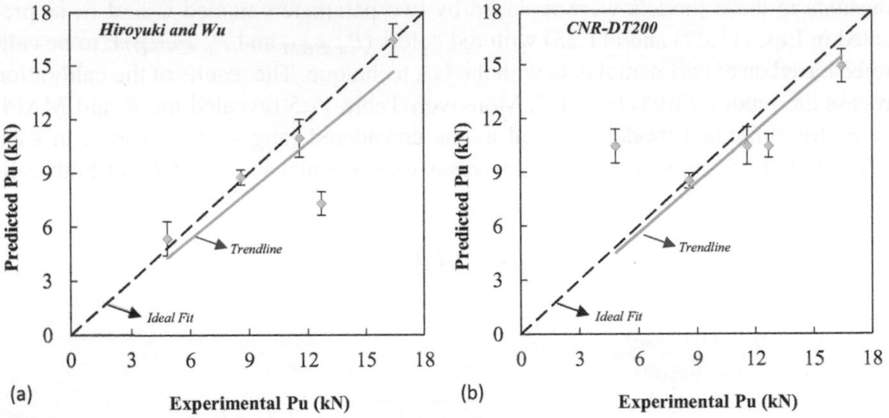

Figure 11.5 Predicted versus experimental values of P_u in calibrated models: (a) Hiroyuki and Wu, and (b) CNR-DT 200.

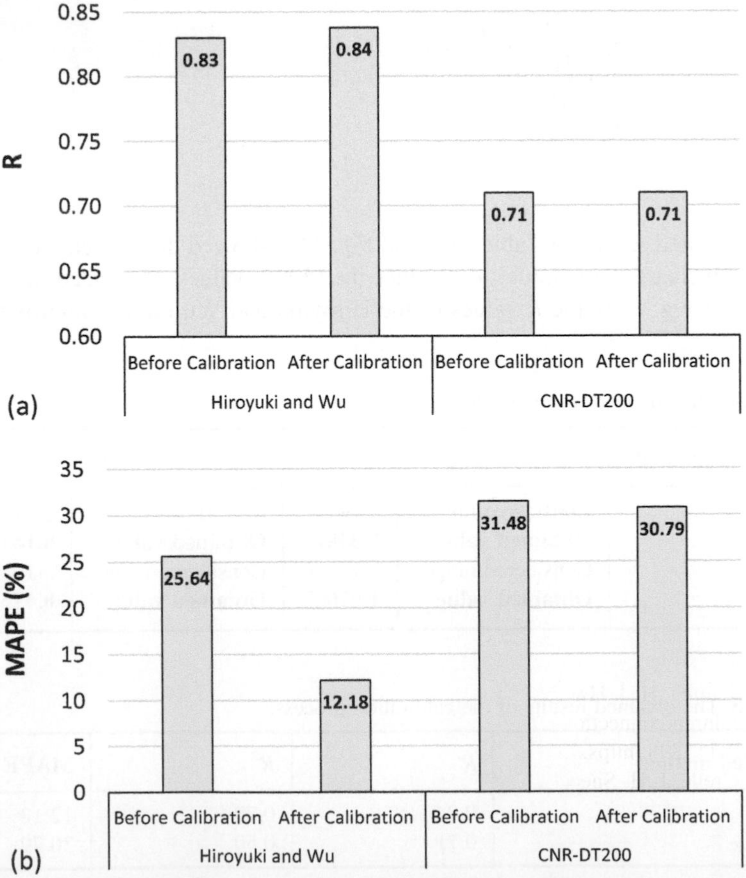

Figure 11.6 The performance of Hiroyuki and Wu and the CNR-DT200 models before and after the calibration process: (a) *R* and (b) MAPE (%).

0.83 to 0.84, while the R values of the CNR-DT 200 model remained constant and equal to 0.71 after the calibration process. On the other hand, the MAPE values of the Hiroyuki and Wu models decreased significantly from 25.464% to 12.18%, and the MAPE values of the CNR-DT 200 model decreased slightly from 31.48% to 30.79% after the calibration process.

11.7 Conclusions

In this chapter, eighteen existing TRM-masonry bond strength models were compiled, and their performance was evaluated by experimental direct shear test specimens gathered from previous research work, including 27 experimental direct shear tests. At this stage, the Hiroyuki and Wu models and the CNR-DT 200 models were selected as the best two existing models. Using the GA technique and considering two unknown coefficients in these two models, the coefficients were optimized with the target of obtaining higher R and lower MAPE values. The outcomes can be summarized below:

- The first and second best existing models among the other eighteen ones were Hiroyuki and Wu, with R and MAPE values of 0.83% and 25.64%, and CNR-DT 200 with R and MPAE values of 0.71% and 31.48%, respectively.
- The results showed that the calibration process could increase the R values of selected models. The R values of the Hiroyuki and Wu models improved from 0.83 to 0.84, while the R values of the CNR-DT 200 model remained constant and equal to 0.71 after the calibration process.
- The suggested calibration process could decrease the error values. The MAPE values of the Hiroyuki and Wu models decreased significantly from 25.464% to 12.18%, and the MAPE values of the CNR-DT 200 model decreased slightly from 31.48% to 30.79% after the calibration process.

References

[1] G. Livitsanos, N. Shetty, E. Verstrynge, M. Wevers, D. Van Hemelrijck, D.G. Aggelis, Acoustic emission health monitoring of historical masonry to evaluate structural integrity under incremental cyclic loading, Proceedings 2 (2018), https://doi.org/10.3390/ICEM18-05417.

[2] F. Ubertini, A. D'Alessandro, A. Downey, E. García-Macías, S. Laflamme, R. Castro-Triguero, Recent advances on SHM of reinforced concrete and masonry structures enabled by self-sensing structural materials, Proceedings 2 (2018), https://doi.org/10.3390/ecsa-4-04889.

[3] G.T. Truong, H.-J. Hwang, C.-S. Kim, Assessment of punching shear strength of FRP-RC slab-column connections using machine learning algorithms, Engineering Structures 255 (2022) 113898, https://doi.org/10.1016/j.engstruct.2022.113898.

[4] S.V. Grelle, L.H. Sneed, Review of anchorage systems for externally bonded FRP laminates, International Journal of Concrete Structures and Materials 7 (2013), https://doi.org/10.1007/s40069-013-0029-0.

[5] F. Zhou, H. Liu, Y. Du, L. Liu, D. Zhu, W. Pan, Uniaxial tensile behavior of carbon textile reinforced mortar, Materials 12 (2019) 374, https://doi.org/10.3390/ma12030374.
[6] G. Cerniauskas, Z. Tetta, D.A. Bournas, L.A. Bisby, Concrete confinement with TRM versus FRP jackets at elevated temperatures, Materials and Structures/Materiaux et Constructions 53 (2020) 1−14, https://doi.org/10.1617/s11527-020-01492-x.
[7] Y. Wu, Z. Zhou, Q. Yang, W. Chen, On shear bond strength of FRP-concrete structures, Engineering Structures 32 (2010) 897−905, https://doi.org/10.1016/j.engstruct.2009.12.017.
[8] L. De Lorenzis, B. Miller, A. Nanni, Bond of FRP laminates to concrete, ACI Materials Journal 98 (2001) 256−264.
[9] C. Yuan, W. Chen, T.M. Pham, H. Hao, Effect of aggregate size on bond behaviour between basalt fibre reinforced polymer sheets and concrete, Composites Part B: Engineering 158 (2019) 459−474, https://doi.org/10.1016/j.compositesb.2018.09.089.
[10] Japan Concrete Institute (JCI), Technical report of technical committee on retrofit technology, in: Proceedings of the International Symposium on the Latest Achievement of Technology and Research on Retrofitting Concrete Structures, 2003.
[11] J. Dai, T. Ueda, Y. Sato, Development of the nonlinear bond stress−slip model of fiber reinforced plastics sheet−concrete interfaces with a simple method, Journal of Composites for Construction 9 (2005) 52−62, https://doi.org/10.1061/(ASCE)1090-0268(2005)9:1(52).
[12] X.Z. Lu, J.G. Teng, L.P. Ye, J.J. Jiang, Bond−slip models for FRP sheets/plates bonded to concrete, Engineering Structures 27 (2005) 920−937, https://doi.org/10.1016/j.engstruct.2005.01.014.
[13] D. Van Gemert, Force transfer in epoxy bonded steel/concrete joints, International Journal of Adhesion and Adhesives 1 (1980) 67−72, https://doi.org/10.1016/0143-7496(80)90060-3.
[14] T. Tanaka, Shear Resisting Mechanism of Reinforced Concrete Beams with CFS as Shear Reinforcement, Hokkaido University, Japan, 1996.
[15] Y. Hiroyuki, Z. Wu, Analysis of debonding fracture properties of CFS strengthened member subject to tension, in: Non-Metallic (FRP) Reinforcement for Concrete Structures, Proceedings of the 3rd International Symposium, Japan Concrete Institute, Tokyo, Japan, 1997, pp. 287−294.
[16] T. Maeda, Y. Asano, Y. Sato, T. Ueda, Y. Kakuta, A study on bond mechanism of carbon fiber sheet, in: Non-Metallic (FRP) Reinforcement for Concrete Structures, Proceedings of the 3rd International Symposium, Japan Concrete Institute, Tokyo, Japan, 1997, pp. 279−285.
[17] A. Khalifa, W.J. Gold, A. Nanni, M.I. Abdel Aziz, Contribution of externally bonded FRP to shear capacity of RC flexural members, Journal of Composites for Construction 2 (1998) 195−202, https://doi.org/10.1061/(ASCE)1090-0268(1998)2:4(195).
[18] B. Adhikary, H. Mutsuyoshi, Study on the bond between concrete and Externally bonded CFRP sheets, in: Proceeding of the 5th International Symposium on Fiber Reinforcement Structures, Cambridge, UK, 2001, pp. 371−378.
[19] T. Triantafillou, S. Matthys, K. Audenaert, G. Balázs, M. Blaschko, H. Blontrock, et al., Externally bonded FRP reinforcement for RC structures, in: Bulletin FIB vol 14, International Federation for Structural Concrete (fib), Lausanne, Switzerland, 2001.
[20] CNR-DT200, Guide for the Design and Construction of Externally Bonded FRP Systems for Strengthening Existing Structures, 2004.
[21] American Concrete Institute, ACI 440.2R-08. Guide for the Design and Construction of Externally Bonded FRP Systems for Strengthening Existing Structures, Farmington Hills, MI, 2008.
[22] HB 305-2008, Design Handbook for RC Structures Retrofitted with FRP and Metal Plates: Beams and Slabs, Standards Australia, Sydney, NSW, 2008.

[23] H. Jahangir, M.R. Esfahani, Bond behavior investigation between steel reinforced grout composites and masonry substrate, Iranian Journal of Science and Technology, Transactions of Civil Engineering (2022), https://doi.org/10.1007/s40996-022-00826-9.
[24] H. Jahangir, D. Rezazadeh Eidgahee, A new and robust hybrid artificial bee colony algorithm – ANN model for FRP-concrete bond strength evaluation, Composite Structures 257 (2021) 113160, https://doi.org/10.1016/j.compstruct.2020.113160.

Forecasting the compressive strength of FRCM-strengthened RC columns with machine learning algorithms

12

Prashant Kumar[1,2,3], *Harish Chandra Arora*[1,2] *and Aman Kumar*[1,2]
[1]Academy of Scientific and Innovative Research (AcSIR), Ghaziabad, Uttar Pradesh, India; [2]Department of Structural Engineering, CSIR—Central Building Research Institute, Roorkee, Uttarakhand, India; [3]Department of Civil Engineering, COER University, Roorkee, Uttarakhand, India

12.1 Introduction

Reinforced concrete (RC) structures are deteriorating day by day nowadays. There are numerous reasons of degrading RC elements, such as the aging of structures, inadequate transverse reinforcement, rebar corrosion, inappropriate concrete mix design, increase in loads, poor workmanship, etc. The major cause of the collapse of structures is found to be the corrosion of steel rebars. Corrosion can be caused by the carbonation phenomenon or by the ingress of chloride ions. In the case of carbonation-induced corrosion, carbon dioxide from the atmosphere penetrates the alkaline concrete, lowering the pH value of the concrete and breaking the passiveness of steel rebar, causing rust in reinforced steel. On the other hand, chloride ions penetrate the concrete from a saline environment, which breaks the barrier of the passive layer, resulting in corrosion [1]. There has been an emergent need to strengthen or restore existing structures rather than demolish or replace them due to economic or heritage concerns [2]. During the past 3 decades, researchers from the scientific community have worked enormously to find solutions to restore the strength of structural elements and enhance the service life of structures [3]. Since 1980s, the need for composite materials like fiber-reinforced polymer (FRP) has emerged as a promising retrofitting technique to strengthen existing structures [4].

FRP has gradually gained the attention of researchers because it possesses high strength, a lesser weight, easy applications in laying on concrete, lower tensile strength, is noncorrodible, has a higher seismic capability, and is stiffer [5]. FRP consists of fibers embedded in resins, which are utilized to affirm the bond behavior between the substrate and FRP. But at variable temperatures, epoxy resins show incompatible behavior with the concrete substrate, which results in the delamination of FRP with the substrate [6—13]. However, recent findings show that FRP displays poor performance at higher temperatures, resulting in poor bond compatibility with

Artificial Intelligence Applications for Sustainable Construction. https://doi.org/10.1016/B978-0-443-13191-2.00005-5
Copyright © 2024 Elsevier Ltd. All rights are reserved, including those for text and data mining, AI training, and similar technologies.

substrate, which has been the major drawback of FRP. When FRP is exposed to elevated temperatures (300–500°C), the matrix of the polymer disintegrates and heat is released, and when treated to low temperatures (100–200°C), the matrix gets softened [14]. The thermal compatibility of FRP is inferior, and its application becomes challenging in humid conditions and low temperatures. These constraints hinder the utilization of FRP in modern-day construction projects.

In modern construction practices, fabric-reinforced cementitious matrix (FRCM) systems, alternatively known as fiber-reinforced mortar, textile-reinforced mortar (TRM), or textile-RC (TRC), have emerged as a recent alternative to address the constraints associated with FRP [15]. FRCM is advantageous due to the inherent temperature-compatible, vapor-permeable, and long-term durability characteristics [15]. Earlier, textile-based material was named TRM or TRC in Europe, and later, it was named FRCM in the United States. FRCM consists of a number of layers of inorganic cement-based matrix embedded with textiles such as carbon, glass, etc. in the form of mesh or grid. Researchers found that organic matrix usually epoxy resins can be replaced by organic mortar matrix, but due to the large grain size, penetration of fibers in the mortar has proven difficult in these conditions as compared to resins [16]. To improve penetration and saturation levels between fibers and matrices, fibers are replaced with nonmetallic textiles [17]. The different varieties of textiles used are carbon [18–22], basalt [19–22], alkali-resistant [20–23], polyphenylene benzobisoxazole (PBO) [24,25], and natural fibers [26] which are generally applied orthogonally onto the surface. The application of FRCM composites on concrete surfaces is more or less similar to that of FRP in the process of repair and rehabilitation of concrete and masonry structures. It entails spreading mortar over the substrate, then laying a layer of fabric mesh, followed by another coating of mortar. The process is carried out until the desired number of layers is obtained.

The axial strength of a column is preferred as a crucial parameter in defining the strength and safety of the structures, and numerous techniques have been discovered to define the compressed strength of RC members [27,28]. Due to the novelty of FRCM systems, scarce literature is available in published studies in comparison to FRP composites. In the past times, researchers have developed analytical models and performed many experiments on the strengthening of columns using FRCM subjected to varying conditions to evaluate the axial strength of columns [29–31]. The results of these investigations show that the strength of retrofitted FRCM columns is higher than that of specimens reinforced with FRP.

Machine learning (ML) algorithms such as random forest (RF), Gaussian process regression (GPR), and support vector machine (SVM) are employed in this study to forecast the compressive strength (CS) of FRCM-reinforced RC columns. This chapter is bifurcated into five segments. The causes of the deterioration of structures, the advantages and disadvantages of FRP, and the evolution of FRCM are discussed in Part 1. Section 12.2 describes the past work performed by renowned researchers on FRCM composites and the significance of the present work. Section 12.3 talks about the methodology of the present research, database organization, data processing, application of performance indices, analytical models, and ML models on FRCM-confined columns. Results and discussion are described in the last section of the chapter.

12.2 Literature review

The need for artificial intelligence (AI) is emerging at a faster rate in the domain of civil engineering. To the best of the author's knowledge, only four published literature are available concerning the application of AI to FRCM-strengthened concrete columns.

Le-Nguyen et al. [32] developed a deep neural network (NN) algorithm to forecast the FRCM-confined CS of columns. The authors used a dual- and triple-hidden-layer artificial NN (ANN) model. A total of 330 specimens were amassed considering input parameters including geometrical specifications of columns, strength properties of concrete, and properties of fiber and mortar. The output of the ANN model was contrasted with that of the SVM, GPR, decision tree, and XGBoost models. The ANN model with two hidden layers performs best, whereas the decision tree model, among others, exhibits lower accuracy.

Irandegani et al. [33] explored the ANN method and Monte Carlo simulation to assess the load-bearing strength of FRCM-wrapped compression members. A total of 10 columns were simulated using the finite element modeling (FEM) method, and 43 datasets from published literature under eccentric loading conditions have been considered. The dataset contains eight input parameters, including concrete CS, reinforcement detailing, and FRCM properties. The findings reveal that a sharp decline in the exceedance probability was noticed with the surge in compressed capacity. The probability of not more than 4.51% was expected with a load beyond 930 KN.

Irandegani et al. [34] also evaluated the CS of columns wrapped with FRCM by employing ML techniques. For this purpose, a large database was trained with the ANN technique to examine the confined CS of columns. Afterward, using a genetic algorithm, an explicit relationship between input variables and target variables was established. In addition, to develop a reliable model, exceedance probability-based results were presented by utilizing Monte Carlo simulation. It was discovered that probability exceedance was sharply reduced to 2% when confined CS was increased above 68 MPa. Also, most of the diameter and height of the column fall in the range of 25 and 35 MPa CS. Moreover, effective fiber percentages lie in the range of 45 and 49 MPa CS.

Some literature based on FRCM-strengthening on concrete columns excluding applications of ML techniques is also shown below:

Bournas and Triantafillou [35] investigated the bar buckling behavior of compressed members exposed to earthquake loading strengthened with FRP and TRM composites. The onset of buckling of bars at the location of plastic hinges is carried out through measurements by strain gauges. It was noticed that bar buckling was not fully removed; in fact, a significant delay was observed in three to seven cycles in comparison to unreinforced specimens. Contrary to FRP, TRM specimens deformed outwards without early fiber rupture by developing stress concentration at the corners without failure. In addition, TRM jackets show an increase in postbuckling reserve capacity. Triantafillou and Papanicolaou [36] explored experimental investigations of TRM and resin-based FRP to define the shear and axial behavior of confined columns. The authors noticed that there was a substantial gain in CS when members were wrapped with TRM. In addition, the effectiveness of TRM jackets in terms of strength was reduced as compared to their FRP counterparts for the specific mortar used. It was also discussed that, with modifications to the properties of mortar, the effectiveness of TRM jackets can be increased. The failure of TRM-based specimens was less abrupt in contrast to resin-based samples. The TRM confinement on rectangular columns also shows an effective response with reference to strength and deformability. Also, the shear capacity of TRM-confined RC columns was found to be substantially increased.

Del Zoppo et al. [37] investigated the seismic resistance of FRCM-retrofitted RC columns. It was observed that without anchorage, the FRCM thin jacket was not able to enhance the strength of columns in cases of flexure. Although premature failure was prevented, ductility was found to be increased by 76%−132%. In the shear critical column case, brittle failure was prevented, and the ductility ratio was achieved between 2 and 2.8. Talo et al. [38] performed FEM of PBO-FRCM-reinforced RC columns to examine the temperature effect. The quantity of FRCM coatings was used to investigate the insulation effect by plotting the temperature-time curve. The authors' conclusion states that augmenting the quantity of FRCM coatings results in enhanced thermal resistance. Moreover, each additional layer of FRCM corresponds to a 6% reduction in the temperature at the concrete substrate.

Tello et al. [39] examined the PBO-FRCM confinement effect on circular and square columns. It was observed that circular columns demonstrated higher capacity as compared to square-shaped reinforced structures with two to four layers, while one layer of wrapping depicts lower capacity. Therefore, circular columns showed a pronounced confinement effect of FRCM layers, and higher strength was noticed in the circular columns than in the latter ones. Colajanni et al. [40] performed experimental research on medium-strength concrete columns of varied cross-sections and explored the effect of carbon-FRCM (C-FRCM) on columns. The confinement effect, cross-section shape, and corner radius were investigated on 30 RC columns under static and lateral loading. C-FRCM confinement produces a considerable increase in CS, deformability, and energy absorption. The authors also observed that circular columns showed effective jacketing confinement. The corner radius has a smaller nfluence on prismatic cross-sections compared to its FRP counterpart. Further, C-FRCM-confined specimens showed less abrupt behavior in comparison to FRP.

Ombres [41] investigated the effect of PBO-based FRCM composites on concrete cylinders to estimate the confinement impact and to develop the stress-strain mathematical relationship. It has been noted that by applying confinement, there is a notable enhancement in both strength and axial strain. Furthermore, an increase in the number of PBO-FRCM layers and aligning the fibers along the specimen's axis were found to enhance the ductility of the specimens. Faleschini et al. [42] discovered the wrapping effect of C-FRCM on 12 full-scale RC columns under monotonic loading. Geometry of cross-section, longitudinal and transverse rebar details, and fiber percentage were the input variables. The CS of columns increased with FRCM layers, and the cylindrical cross-section exhibits superior strength among other specimens. The square cross-section specimen exhibits premature breakdown due to the formation of stresses at the edges and poor ductility. In prismatic specimens, localized stresses persist at the corner edges even when a corner radius of 20 mm is provided. In cylindrical columns, the presence of longitudinal rebars induces local stresses that initiate the formation of cracks as well.

CS is one of the practicable properties of concrete, which includes laborious manual work along with costly setups. There is an emerging need in the field of concrete engineering to predict CS with better accuracy and reliable techniques without using skilled labor and high-cost equipment. ML algorithms not only generate reliable and precise outcomes but also use less time. This chapter focuses on the application of FRCM composites on concrete columns, considering ML algorithms. The primary goal of this work is to generate an optimal ML method to forecast the FRCM-confined CS of columns.

12.3 Methodology

12.3.1 Data collection and description

The dataset of 229 column specimens on FRCM confinement has been accumulated from a rigorous literature survey of published studies. The input parameters considered in this research are the concrete grade (f_c'), unconfined CS of concrete (f_{co}'), height (h) and diameter (D) of the specimen, cross-sectional shape of the specimen (C_s), corner radius (r_c), fabric type (F_{ty}), number of FRCM layers (n), FRCM thickness (t_f), FRCM Young's modulus (E_f), FRCM tensile strength (f_f), and confinement volumetric ratio (ρ). The target output parameter taken in the study is confined CS of column (f_{cc}').

The value of f_c' ranges between 14.28 and 50 MPa with a mean strength of 26.88 MPa. h and D are considered from 200 to 700 mm with an average value of 365.62 mm and 84.85–447.21 mm with an average of 207.19 mm, respectively. C_s of specimen used in the study are of square, circular, and rectangular shapes. For this work, 85 square columns, 126 circular columns, and 18 rectangular column samples are chosen for the evaluation of the effective confinement of FRCM composites. The maximum value of r_c used in the study is 62.5 mm for square specimens. F_{ty} chosen in the work are carbon, glass, hybrid (carbon & glass), PBO, and basalt types, whereas n ranges from 1 to 6. The value of t_f ranges from 0.045 to 0.1035 mm. The number of specimens wrapped with carbon, glass, hybrid, PBO, and basalt FRCM fibers is 126, 18, 24, 35, and 26, respectively. E_f and f_f of fabric varies from 30.7 to 270 GPa and 48.6–5800 MPa, respectively. The value of ρ ranges from 0.094% to 1.08%, and the output variable f_{cc}' limits from 14.2 to 63.77 MPa. The statistical indicators of all input parameters chosen from an extensive literature review [40,41,43–50] are listed in Table 12.1.

12.3.1.1 Data processing

To achieve higher accuracy of ML models, standardization is necessary in the field of AI. The dataset can be normalized in different ranges, such as 0–1, 0–9, and −1 to +1. In this work, standardization is carried out between −1 and +1 to obtain a

Table 12.1 Statistical indicators of input parameters.

Parameters	Symbol	Unit	Min.	Max.	Mean	Std.	Kurtosis	Skewness	Type
CS of concrete	f_c'	MPa	14.28	50	26.88	7.7894	0.0039	0.627	Input
Unconfined CS	f_{co}'	MPa	12	43.1	21.95	6.99	0.2968	0.8022	Input
Height	h	mm	200	700	365.62	128.6	−0.2013	1.0488	Input
Diameter	D	mm	84.85	447.21	207.19	96.43	1.1219	1.3526	Input
Cross-sectional shape	C_s	NA	1	3	1.68	0.591	−0.6151	0.223	Input
Corner radius	r_c	mm	0	62.5	11.38	16.8283	1.08555	1.38392	Input
Fabric type	F_{ty}	NA	1	5	2.15	1.47407	−1	0.786	Input
No. of layers	n	NA	1	6	2.22	0.95581	−0.6323	0.87724	Input
Thickness of fiber	t_f	mm	0.045	0.125	0.0704	0.03216	−0.6974	1.0158	Input
Elastic modulus of fabric	E_f	GPa	30.7	270	164.8	92.9806	−0.8541	0.9541	Input
Tensile strength of fabric	f_f	MPa	48.6	5800	2966	1762.2	−1.2423	0.433	Input
Confinement volumetric ratio	ρ	%	0.094	1.08	0.403	0.27	0.2351	1.2321	Input
Confined CS	f_{cc}'	MPa	14.2	63.77	31.98	11.9276	−0.5905	0.58909	Output

higher efficient and reliable model. The equation used to normalize the data is shown below:

$$Z_{normalized} = \left[2 \times \left(\frac{z - z_{min}}{z_{max} - z_{min}}\right)\right] - 1 \quad (12.1)$$

where, $Z_{normalized}$ is the normalized value, z is the nonnormalized original value, z_{min} and z_{max} are the minimum and maximum of the nonnormalized original value, respectively.

After normalizing the database, the dataset was split into a machine training and a testing stage in proportions of 70% and 30%, respectively.

12.3.1.2 Performance indices

The outcome of ML models to compute FRCM-confined CS of columns is assessed through statistical indicators such as MAE (mean absolute error), R (Pearson correlation coefficient), RMSE (root mean square error), MAPE (mean absolute percentage error), a20-index, and NS (Nash–Sutcliffe). Positive results from the models are shown by the least error values of MAE, MAPE, and RMSE and the highest values of R, NS, and a20-index. The details of performance indicators were taken from Refs. [51–56].

$$R = \frac{\sum_{i=1}^{N}(E_i - \overline{E})(P_i - \overline{P})}{\sqrt{\sum_{i=1}^{N}(E_i - \overline{E})^2 (P_i - \overline{P})^2}} \quad (12.2)$$

$$\text{MAE} = \frac{1}{N}\sum_{i=1}^{N}|E_i - P_i| \quad (12.3)$$

$$\text{MAPE} = \frac{1}{N}\sum_{i=1}^{N}\left|\frac{E_i - P_i}{E_i}\right| \times 100 \quad (12.4)$$

$$\text{RMSE} = \sqrt{\frac{\sum_{i=1}^{N}(E_i - P_i)^2}{N}} \quad (12.5)$$

$$NS = 1 - \frac{\sum_{i=1}^{N}(E_i - P_i)^2}{\sum_{i=1}^{N}(E_i - \overline{P_i})^2} \quad (12.6)$$

$$a20 - \text{index} = \frac{m20}{N} \quad (12.7)$$

where N is the number of data points; E_i and $\overline{E}$ refer to actual and average actual values; P_i and $\overline{P}$ are the anticipated and average anticipated values, respectively; and $m20$ is the number of specimens with actual and anticipated values ranging from 0.8 to 1.2.

12.3.2 Mathematical models

In this chapter, two analytical models, i.e., Triantafillou's model [43] and De Caso's model [57] are selected from the literature available on the web. The details of both models, including formulations, are shown below.

12.3.2.1 Triantafillou's model

Triantafillou et al. [43] investigated the efficacy of TRM on the confinement of concrete cylinders. The findings demonstrate that increasing the number of TRM layers led to a significant rise in CS. The analytical model of TRM-confined specimens presented by the authors is shown below:

$$\frac{f'_{cc}}{f'_{co}} = 1 + 1.9\left(\frac{f_{lu}}{f'_{co}}\right)^{1.27} \tag{12.8}$$

$$f_{lu} = k_e \frac{b+h}{bh} t_f E_f \varepsilon_{fu} \tag{12.9}$$

$$k_e = 1 - \frac{b_n^2 + h_n^2}{3A_c} \tag{12.10}$$

where, f'_{cc} is CS of FRCM-confined concrete specimens, f'_{co} is unconfined CS of concrete specimens, f_{lu} is lateral confining stress, k_e is effectiveness coefficient, t_f is the thickness of jacket, E_f is the elastic modulus of fiber, ε_{fu} is the lateral strain in jacket, b and h are the dimensions of the cross-section, $b_n = b - 2r_c$, $h_n = h - 2r_c$ and A_c = total cross-sectional area.

12.3.2.2 De Caso's model

De Caso's et al. [57] presented a semiempirical relation to determine the FRCM-confined CS of specimens. The model is expressed below:

$$\frac{f'_{cc}}{f'_{co}} = 1 + 2.87\left(\frac{f_{lu}}{f'_{co}}\right)^{0.775} \tag{12.11}$$

$$f_{lu} = \frac{2nt_f E_f \varepsilon_{fu}}{D} \tag{12.12}$$

where, n is the number of layers and D is the specimen's diameter

12.3.3 Model development

This section outlines the application and utilization of ML models to compute the FRCM-confined CS of columns. GPR, SVM, and RF are the ML algorithms implemented in this study, which are briefly described below.

12.3.3.1 Gaussian process regression (GPR)

The Gaussian process (GP) is a statistical model that relies on the normal distribution, named after Carl Friedrich Gauss [58]. GPs are widely used in ML algorithms and are a collection of random variables that offer several benefits, including their ability to handle small datasets and provide uncertainty estimates for predictions [59]. A GP can be represented by a subset of joint Gaussian distributions [60] and is composed of a mean function and a covariance function. The mean function is typically assumed to be zero in GPR.

The mean function

$$\mu(x) = E[y(x)] \tag{12.13}$$

and covariance function

$$C\left(x, x'\right) = E\left[(y(x) - \mu(x))\left(y\left(x'\right) - \mu\left(x'\right)\right)\right] \tag{12.14}$$

considering $\mu(x) = 0$

All the above formulations are taken from [61].

12.3.3.2 Support vector machine (SVM)

The SVM was initially proposed by Vapnik and team [62]. The SVM uses hypothetical linear functions. The simplest SVM model is linear SVM (LSVM). It operates effectively in cases where the data is linearly distributed within the original space. Nevertheless, real-world problems often exhibit limitations and necessitate the application of nonlinear problem analysis techniques. In the 1990s, nonlinear functions, also called kernel functions were introduced based on the theory of LSVM. The model related to nonlinear functions is known as nonlinear SVM (NSVM). Let S be taken as a sample for training with N sets of vectors.

$X_i \in R^n$ with $i = 1, \ldots\ldots, N$.

$$S = ((x_1, y_1), \ldots\ldots, (x_N, y_N)) \tag{12.15}$$

Each vector x_i belongs to $y_i \in \{-1, 1\}$

$$\left(w^T x_i\right) + b = 0 \tag{12.16}$$

The pair of (w, b) defines a hyperplane and is called a separating hyperplane. The determination of the optimal separating hyperplane involves solving an optimization problem defined by

minimize $d(w) = \frac{1}{2}(w^T w)$

$$\text{subject to } y_i\left(\left(w^T x_i\right) + b\right) \geq 1,\ i = 1, 2, \ldots..N \tag{12.17}$$

The resulting SVM is often referred to as a hard margin SVM because it does not permit any errors.

All the above formulations are extracted from [63].

12.3.3.3 Random forest (RF) regression

Breiman proposed RF regression for the first time in 2001 [64]; it is regarded as a significant classification technique for regression. It exhibits characteristics in developing relationships between input and output variables with high speed, accuracy, and flexibility. Moreover, RF constitutes a large data handling capacity with high precision and accuracy. RF has various applications in distinguished domains such as predicting the response of customers in banking services [65], forecasting stock market activities [66], medicine or pharmaceutical industry [67], e-commerce [68], etc. RF is based on the bagging algorithm, an ensemble learning (EL) technique that functions by creating a lower-level model and predicting a higher-level model.

The following is an example of a typical EL approach with n ML models.

$$F^{EL}(x) = \sum_{i=1}^{n} F_i^{ML}(x) \tag{12.18}$$

where x is the input parameter, $F^{EL}(x)$ is the final output mode, and $\sum_{i}^{ML} x$ is the ML algorithm applied, also known as the basis model.

The fundamental concept behind this technique is to create multiple data subsets from the original training sample by random selection. Each subset of the data is used to train the decision tree model as presented in Fig. 12.1. A fresh training dataset, comprised of bootstrap samples, is generated by randomly selecting and replacing data points from the original training set. The excluded training samples are assembled into a distinct set referred to as "out-of-bag samples" and are employed to validate the model until the desired level of accuracy is attained.

12.4 Results and discussion

This section contains individual outcomes of analytical models (Triantafillou's model and De Caso's model) and implemented ML models (GPR, SVM, and RF regression models). Furthermore, the performances of ML models are compared with each other.

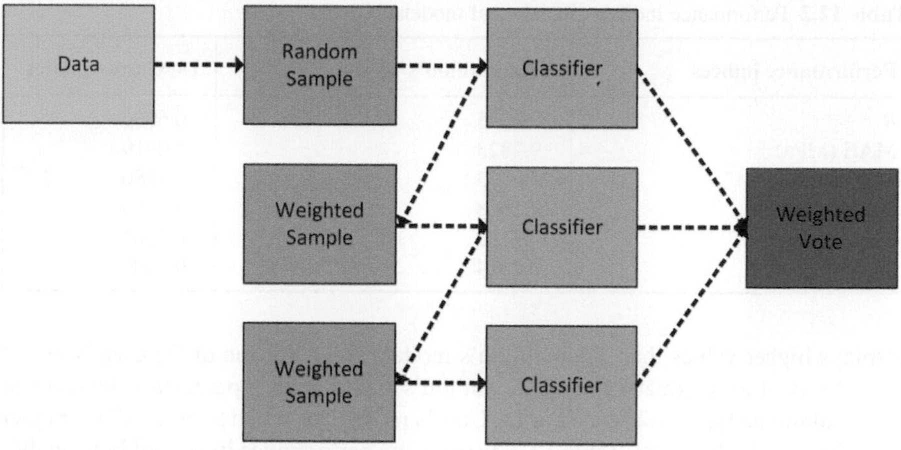

Figure 12.1 Random forest regression.

In addition, a comparison between the best-performing analytical model (De Caso's model) and ML models is demonstrated in subsequent sections. The results and comparisons are also demonstrated in the form of scatter diagrams, errors, and Taylor's plots.

12.4.1 Results of analytical models

In this chapter, the details of two mathematical models, i.e., Triantafillou's model and De Caso's model have been considered. The details of the mentioned models have also been discussed in the prior Section 12.3.2. The chosen input variables in the study have undergone processing using these models. The performance of these models has been measured by considering the statistical indices mentioned in Section 12.3.1.2.

The *R*-values of Triantafillou's model and De Caso's model were measured as 0.8295 and 0.6980, respectively. The error values of Triantafillou's model, such as MAE, MAPE, and RMSE, were found to be 9.7823 MPa, 27.6573%, and 12.1628 MPa, sequentially, whereas for De Caso's model, they were found to be a little lower than Triantafillou's model, i.e., 7.0919 MPa, 23.7680%, and 8.8103 MPa, sequentially. In contrast, De Caso's model represents higher *NS* (0.4581) and *a*20-index (0.5109) values than those of Triantafillou's model.

Based on the ranking criteria of both models, De Caso's model was considered a better model than Triantafillou's model. Although the *R*-value of Triantafillou's model was measured to be 18.83% higher than De Caso's model, in terms of error parameters, De Caso's model shows better performance than its counterpart Triantafillou's model. The MAE value of De Caso's model was found to be lower than Triantafillou's model with a percentage reduction of 27.50%. Similarly, the MAPE value of De Caso's model was observed to be 14.06% lower than Triantafillou's model. Moreover, the RMSE value of De Caso's model was found to be lower than Triantafillou's model by 27.56%. Furthermore, in terms of *NS* and *a*20-index values, De Caso's model

Table 12.2 Performance indices of analytical models.

Performance indices	Triantafillou's model	De Caso's model
R	0.8295	0.6980
MAE (MPa)	9.7823	7.0919
MAPE (%)	27.6573	23.7680
RMSE (MPa)	12.1628	8.8103
NS	0.3755	0.4581
a20-index	0.2314	0.5109

displays higher values than Triantafillou's model. The NS value of De Caso's model was observed to be greater than Triantafillou's model with a percentage increase of 22%. Additionally, the a20-index of De Caso's model was noted to be 120.78% higher than Triantafillou's model. Table 12.2 displays the performance indices of both analytical models.

In addition, the performance of analytical models was also shown by the scatter plots (Fig. 12.2). The scatter plot was drawn between the experimental confined CS of columns and the predicted confined CS of columns, illustrating a correlation between the two values. In the case of Triantafillou's model, approximately all the data values were concentrated off the diagonal line (45° line), representing an inadequate fit, whereas in De Caso's model, some data points were lying on the diagonal line and some were dispersed away from the line. It was noticeable from the scatter plot that De Caso's model was better than Triantafillou's model. However, both models represent inadequate performance in terms of precision.

12.4.2 Results of GPR model

Table 12.3 demonstrates the outcomes of the GPR model. The GPR model exhibits an R-value of 0.9830, 0.5507, and 0.8845 for training, testing, and all datasets

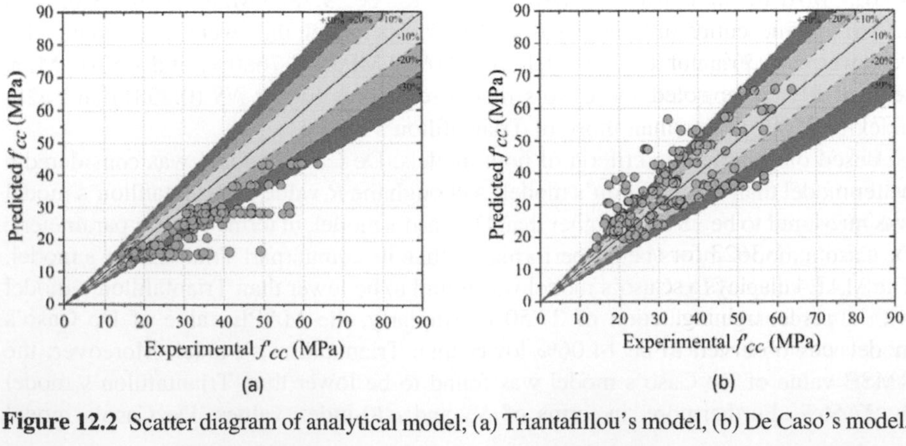

Figure 12.2 Scatter diagram of analytical model; (a) Triantafillou's model, (b) De Caso's model.

Table 12.3 Performance of ML models.

Model		*R*	MAPE (%)	MAE (MPa)	RMSE (MPa)	*a*20-index	*NS*
	Training	0.9830	3.5971	1.0980	2.1230	0.9812	0.9655
GPR	Testing	0.5507	27.4095	8.7136	10.4627	0.3478	0.4750
	All	0.8845	10.7721	3.3927	1.5859	0.7903	0.7709
SVM	Training	0.9778	5.9355	1.7735	2.7415	0.9750	0.9425
	Testing	0.6202	33.0936	11.0601	13.5116	0.3623	0.1445
	All	0.8033	14.1185	4.5717	2.0652	0.7903	0.6201
RF	Training	0.9871	3.5528	1.0381	1.8981	0.9812	0.9724
regression	Testing	0.9738	6.8272	1.8937	2.9671	0.9420	0.9453
	All	0.9828	4.5394	1.2959	2.2737	0.9694	0.9635

sequentially. The RMSE values for the GPR model were observed as 2.1230 MPa, 10.4627 MPa, and 1.5959 MPa for the training, testing, and entire data stages, respectively. Similarly, MAPE values for all three sets were 3.5971%, 27.4095%, and 10.7721%, respectively. In addition, the MAE values for training, testing, and the whole data set were 1.0980 MPa, 8.7136 MPa, and 3.3927 MPa, respectively. Moreover, *NS* values for training, testing, and the entire data set were demonstrated as 0.9655, 0.4750, and 0.7709, respectively. Further, the *a*20-index was found to be 0.9812, 0.3478, and 0.7903 for training, testing, and the whole set, sequentially.

Fig. 12.3 displays scatter plots for the GPR model. For the training set, the majority of data values are closely concentrated along the best-fit line, showing better accuracy, whereas for the testing phase, the data points are randomly distributed and disseminated nonuniformly around the best-fit line, representing poor accuracy.

12.4.3 Results of SVM model

The SVM model demonstrates *R* values of 9.778 for training, 0.6202 for testing, and 0.8033 for the entire dataset. The MAPE values were determined as 5.9355% for training, 33.0936% for testing, and 14.1185% for the entire dataset. Likewise, the MAE values were noted as 1.7735 MPa for training, 11.0601 MPa for testing, and 4.5717 MPa for the entire dataset. The RMSE values followed a sequence of 2.7415 MPa for training, 13.5116 MPa for testing, and 2.0652 MPa for the whole dataset. Furthermore, *NS* values were determined as 0.9425 for training, 0.1445 for testing, and 0.6201 for the entire dataset. Conversely, the *a*20-index exhibited values of 0.9750 for training, 0.3623 for testing, and 0.7903 for the complete dataset.

Fig. 12.4 displays a scatter diagram between experimental and forecasted confined CS of columns for training, testing, and all datasets, respectively. In the training set, most of the data points cluster closely around the best-fit line, indicating good agreement between experimental and predicted values, whereas for the testing set, they are not positioned along the 45° line, with many of them lying horizontally, indicating poor accuracy.

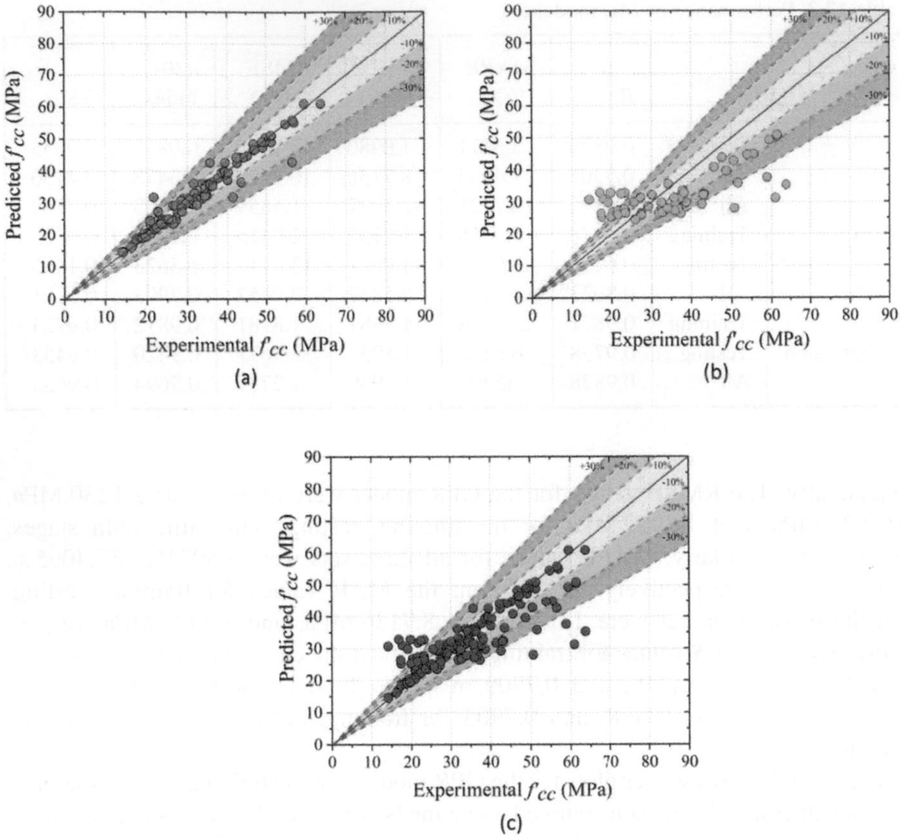

Figure 12.3 Scatter plot of GPR model; (a) training stage, (b) testing stage, and (c) whole data set.

12.4.4 Results of RF model

The RF model exhibits higher accuracy and precision in comparison with the GPR and SVM models. The training, testing, and all data sets have *R*-values of 0.9871, 0.9738, and 0.9828, sequentially. The MAPE values were found to be 3.5528% for training, 6.8272% for testing, and 4.5394% for the entire dataset. Similarly, MAE values were identified as 1.0381 MPa for training, 1.8937 MPa for testing, and 1.2959 MPa for all datasets, respectively. Similarly, lesser values of RMSE were also observed, i.e., 1.8981 MPa, 2.9671 MPa, and 2.2737 MPa for all three stages, respectively. The *NS* values for training, testing, and all datasets are sequentially represented as 0.9812, 0.9420, and 0.9694. In addition, *a*20-index values were also the highest, indicating 0.9724, 0.9453, and 0.9635 for all three sets, correspondingly. The scatter plots (Fig. 12.5) indicate that a significant portion of the data points align closely with the 45° line. A smaller number of data values shows a deviation from the diagonal line.

Fig. 12.6 displays the error distribution in the form of a raincloud plot. Among the analytical models, Triantafillou's model demonstrates an error range between −0.87

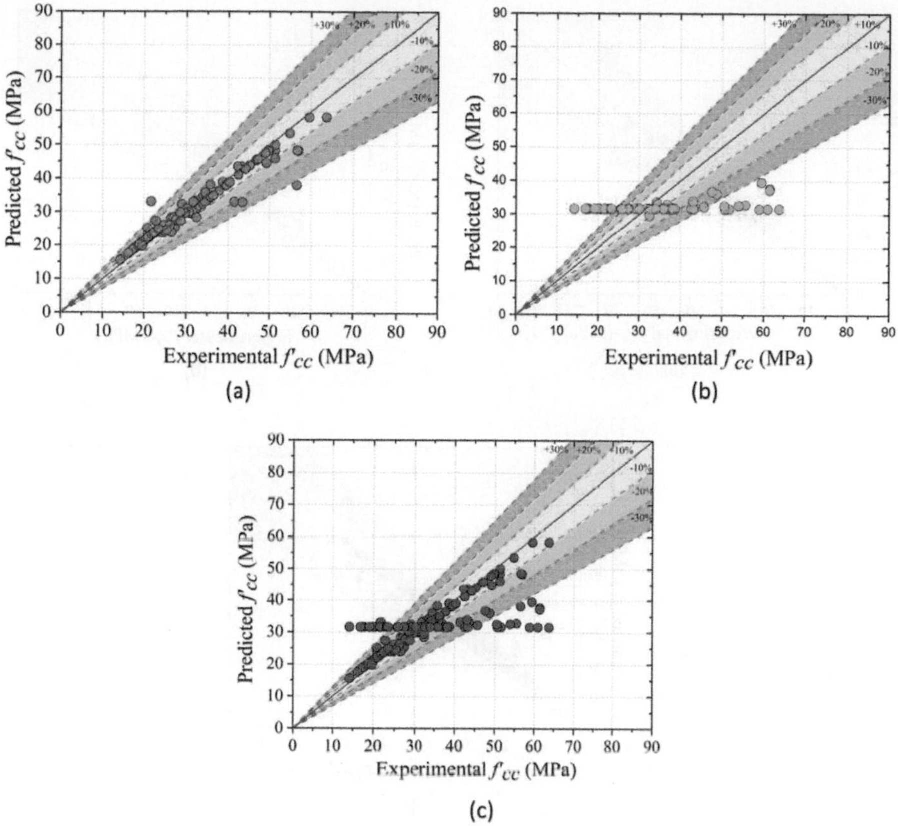

Figure 12.4 Scatter plot of SVM model; (a) training, (b) testing, and (c) whole datasets.

and 32.73 MPa, whereas De Caso's model shows error values between −24.26 and 0.066 MPa. In the case of ML-based models, the RF model achieves superior accuracy and covers a small range of errors. It displays errors ranging between −10.89 and 9.10 MPa. However, the SVM model illustrates a large range of errors between −17.40 and 32.16 MPa. Also, the GPR model demonstrates errors between −16.51 and 29.74 MPa.

12.4.5 Comparison of implemented ML models

The comparison between ML techniques is presented in Table 12.3. Based on the outcomes yielded by the applied ML models, it is evident that the RF model stands out as the most accurate method in comparison to SVM and GPR. The *R*-value of the RF regression model exceeded the GPR model with a percentage increase of 11.11%, whereas its MAPE value was observed to be lower than the GPR model with a percentage reduction of 57.86%. Likewise, the RF model exhibited lower MAE and RMSE values compared to the GPR model, resulting in percentage reductions of 61.80%

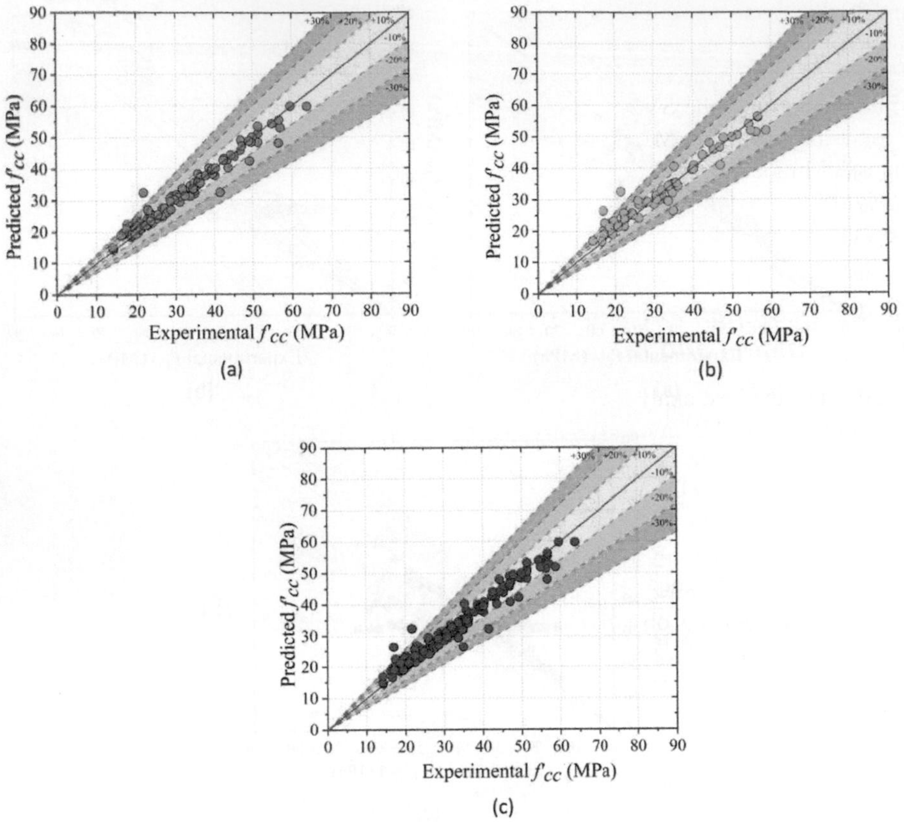

Figure 12.5 Scatter plot of RF model; (a) training, (b) testing, and (c) whole datasets.

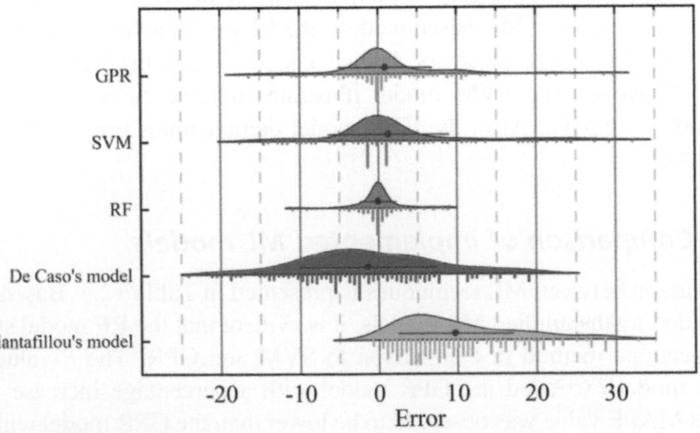

Figure 12.6 Error distribution with raincloud plot.

and 30.25%, respectively. However, $a20$-index and NS values are higher than the GPR model, with percentage increases of 22.68% and 24.98%, respectively.

Further, in comparison to the SVM model, the R-value of the RF model is observed to be greater than the SVM model. with a percentage increase of 22.34%. Similarly, error indices such as MAPE, RMSE, and MAE values of the RF model are lesser than SVM model, with a percentage decrease of 67.84%, 9.17%, and 71.65%, respectively. Moreover, the $a20$-index and NS values of the RF regression model are greater than the SVM model by 22.66% and 55.37%, respectively.

12.4.6 Comparison between performances of best analytical model and ML models

In this part, the best analytical model, i.e., De Caso's model has been compared with the implemented ML models individually. Table 12.4 shows that the accuracy (R-value) of the GPR model exceeded De Caso's model with a percentage increase of 26.71%. Similarly, the error indices, such as MAE and RMSE values of the GPR model were lower than those of De Caso's model, with percentage reductions of 85.72% and 81.99%, respectively. However, the MAPE value of the GPR model was found to be 51.89% higher than De Caso's model. The value of the $a20$-index of the GPR model was observed to be 72.51% greater than De Caso's model. In addition, the NS value of the GPR model surpassed De Caso's model with a percentage increase of 50.89%.

In a comparison of De Caso's and SVM models, the SVM model's R-value was observed to be 15.08% higher than De Caso's model, whereas the MAPE value exceeded De Caso's model with a percentage increase of 99.07%. Moreover, MAE and RMSE values of the SVM model decrease with percentage reductions of 80.76% and 76.55%, respectively, with respect to De Caso's model. Furthermore, the values of $a20$-index and NS of the SVM model were found to be 72.51% and 21.37% greater than De Caso's model, sequentially.

At last, De Caso's model has been compared with the RF model, which has been observed to be the best ML model. The R-value of the RF model was noticed to be 40.80% higher than De Caso's model, representing the highest accuracy in comparison with the other two ML models. Additionally, the values of RMSE, MAE, and MAPE

Table 12.4 Comparison of best predicted analytical model with ML models.

Model	R	MAPE (%)	MAE (MPa)	RMSE (MPa)	$a20$-index	NS
De Caso's model	0.6980	7.0919	23.7680	8.8103	0.4581	0.5109
GPR	0.8845	10.7721	3.3927	1.5859	0.7903	0.7709
SVM	0.8033	14.1185	4.5717	2.0652	0.7903	0.6201
RF regression	0.9828	4.5394	1.2959	2.2737	0.9694	0.9635

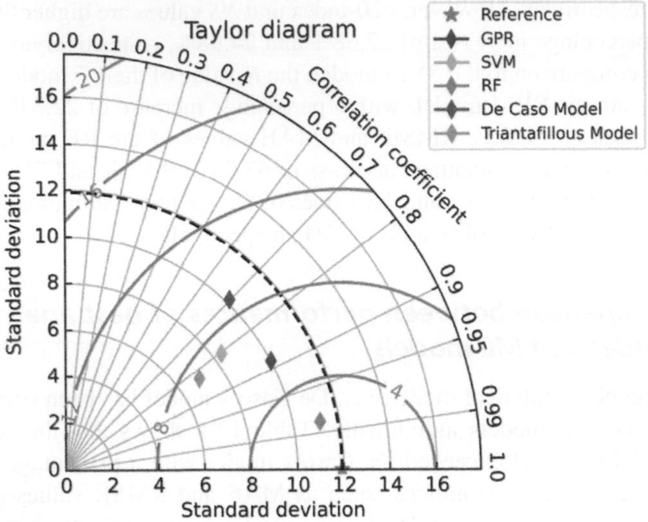

Figure 12.7 Taylor's diagram for estimated FRCM-confined CS.

of the RF model were observed to be lower than De Caso's model, with a percentage reduction of 74.19%, 94.54%, and 35.99%, sequentially.

Taylor diagrams are used to evaluate the accuracy of models by simply visualizing the values of the standard deviation, R, and RMSE. The azimuth angle indicates the standard deviation values of different models, whereas radial distances indicate R-values. Fig. 12.7 represents Taylor's diagram, which demonstrates the accuracy of applied ML and analytical models. The blue line indicates the reference line, which displays the standard deviation of the experimental value. The graph shows that analytical models are inferior in comparison to ML models. The closest model to the reference line is the RF model, which exhibits higher accuracy and reliability. The GPR model is also positioned under the reference line, whereas the SVM model is located far away from the reference line. The model with the orange diamond shape is observed as the most accurate best-fitted model, having a greater R-value and a lower RMSE value.

12.5 Conclusions and future recommendations

In this chapter, a database of 229 column specimens was collected to predict the CS of FRCM-reinforced columns. A total of 12 input parameters, such as $f_c^{'}$, $f_{co}^{'}$, h, D, C_s, r_c, F_{ty}, n, t_f, E_f, f_f, and ρ has been selected to predict the CS ($f_{cc}^{'}$). The dataset has been normalized within the -1 to $+1$ range with a split ratio of 70% for training and 30% for testing. The ML models (GPR, SVM, and RF regression) were implemented in this research and evaluated using performance measures. The following conclusions were taken from the findings:

1. The RF model was observed as the top-performing model compared to other ML models. It achieved an accuracy of 0.9871 for the training set and 0.9738 for the testing set.
2. The error parameters of the RF model, such as MAE, MAPE, and RMSE, were 1.2959 MPa, 4.5394%, and 2.2737 MPa, respectively, which were less than those of the GPR and SVM models.
3. The SVM model was considered the worst model based on its accuracy and error values. The *R*-value of the SVM model was 0.8033, which was the minimum with respect to other models, whereas MAPE, MAE, and RMSE values were 14.1185%, 4.5717 MPa and 2.0652 MPa sequentially, respectively, illustrating greater errors than other ML models.
4. De Caso's model was the best analytical model compared to Triantafillou's model. The *R*-value of De Caso's model was 0.6980, and MAPE, MAE, and RMSE values were 7.0919%, 23.7680 MPa, and 8.8103 MPa, respectively.
5. Following the evaluation criteria, there is a sequential decline in accuracy across the ML models, starting with the RF regression model, followed by the GPR and SVM models.

The suggested results will help scientists and practitioners forecast the CS of FRCM-confined columns. Both nature-inspired algorithms and advanced ML techniques can be employed to improve the model's accuracy. Moreover, a large dataset with varying input parameters is required to obtain reliable prediction models.

Nomenclature

ACI	American Concrete Institute
A_c	Total cross-sectional area
AI	Artificial intelligence
ANN	Artificial neural network
b_n, h_n	Free edge length between fillet radii of rectangular columns
b	Width of column
C-FRCM	Carbon-FRCM
CS	Compressive strength
C_s	Cross-sectional shape
D	Diameter of column
E_f	Elastic modulus of fabric
EL	Ensemble learning
f_c'	CS of concrete
f_{cc}'	Confined CS
f_{co}'	Unconfined CS
FEM	Finite element modeling
f_f	Tensile strength of fabric
f_{lu}	Confining stress
FRCM	Fiber-reinforced cementitious matrix
FRM	Fiber-reinforced mortar
FRP	Fiber-reinforced polymer
F_{ty}	Fabric type
GP	Gaussian process
GPR	Gaussian process regression
h	Height of column

k_e	Effectiveness coefficient
LSVM	Linear SVM
MAE	Mean absolute error
MAPE	Mean absolute percentage error
ML	Machine learning
NN	Neural network
n	No. of fabric layers
NS	Nash–Sutcliffe
NSVM	Nonlinear SVM
OSH	Optimal separating hyperplane
PBO	Polyphenylene benzobisoxazole
R	Pearson correlation coefficient
r_c	Corner radius
RF	Random forest
RMSE	Root mean square error
SVM	Support vector machine
t_f	Thickness of fiber
TRC	Textile-reinforced concrete
TRM	Textile-reinforced mortar
ρ	Confinement volumetric ratio
ε_{fu}	Strain in jacket

References

[1] J. Broomfield, Corrosion of Steel in Concrete: Understanding, Investigation and Repair, first ed., CRC Press, 1996.

[2] M.J. Masia, T.N. Gale, N.G. Shrive, Size effects in axially loaded square-section concrete prisms strengthened using carbon fibre reinforced polymer wrapping, Canadian Journal of Civil Engineering 31 (1) (2004) 1–13.

[3] S.M. Homam, S.A. Sheikh, P. Collins, G. Pernica, J. Daoud, Durability of fiber reinforced polymers used in concrete structures, in: International Conference on Advanced Materials in Bridges and Structures, vol. 3, August 2000, pp. 751–758.

[4] O. Awani, T. El-Maaddawy, N. Ismail, Fabric-reinforced cementitious matrix: a promising strengthening technique for concrete structures, Construction and Building Materials 132 (2017) 94–111.

[5] L.S. Lee, R. Jain, The role of FRP composites in a sustainable world, Clean Technologies and Environmental Policy 11 (2009) 247–249.

[6] J.G. Teng, J.F. Chen, S.T. Smith, Debonding failures in FRP-strengthened RC beams: failure modes, existing research and future challenges, in: Composites in Construction: A Reality, 2001, pp. 139–148.

[7] T. Blanksvärd, B. Täljsten, A. Carolin, Shear strengthening of concrete structures with the use of mineral-based composites, Journal of Composites for Construction 13 (1) (2009) 25–34.

[8] O. Buyukozturk, O. Gunes, E. Karaca, Progress on understanding debonding problems in reinforced concrete and steel members strengthened using FRP composites, Construction and Building Materials 18 (1) (2004) 9–19.

[9] T. El Maaddawy, K. Soudki, Strengthening of reinforced concrete slabs with mechanically-anchored unbonded FRP system, Construction and Building Materials 22 (4) (2008) 444−455.
[10] T. El Maaddawy, S. Sherif, FRP composites for shear strengthening of reinforced concrete deep beams with openings, Composite Structures 89 (1) (2009) 60−69.
[11] T. El-Maaddawy, Y. Chekfeh, Retrofitting of severely shear-damaged concrete t-beams using externally bonded composites and mechanical end anchorage, Journal of Composites for Construction 16 (6) (2012) 693−704.
[12] T. El-Maaddawy, Y. Chekfeh, Shear strengthening of T-beams with corroded stirrups using composites, ACI Structural Journal 110 (5) (2013) 779.
[13] T. El-Maaddawy, B. El-Ariss, Behavior of concrete beams with short shear span and web opening strengthened in shear with CFRP composites, Journal of Composites for Construction 16 (1) (2012) 47−59.
[14] A.P. Mouritz, A.G. Gibson, Fire Properties of Polymer Composite Materials, vol 143, Springer Science & Business Media, 2007.
[15] ACI Committee 549, ACI 549.6 R-20 Guide to Design and Construction of Externally Bonded Fabric-Reinforced Cementitious Matrix (FRCM) and Steel-Reinforced Grout (SRG) Systems for Repair and Strengthening Masonry Structures, American Concrete Institute, Farmington Hills, MI, USA, 2020.
[16] L.N. Koutas, Z. Tetta, D.A. Bournas, T.C. Triantafillou, Strengthening of concrete structures with textile reinforced mortars: state-of-the-art review, Journal of Composites for Construction 23 (1) (2019) 03118001.
[17] X. Wang, C.C. Lam, V.P. Iu, Bond behaviour of steel-TRM composites for strengthening masonry elements: experimental testing and numerical modelling, Construction and Building Materials 253 (2020) 119157.
[18] F.G. Carozzi, A. Bellini, T. D'Antino, G. de Felice, F. Focacci, Ł. Hojdys, L. Laghi, E. Lanoye, F. Micelli, M. Panizza, C. Poggi, Experimental investigation of tensile and bond properties of Carbon-FRCM composites for strengthening masonry elements, Composites Part B: Engineering 128 (2017) 100−119.
[19] X. Wang, C.C. Lam, V.P. Iu, Comparison of different types of TRM composites for strengthening masonry panels, Construction and Building Materials 219 (2019) 184−194.
[20] F.G. Carozzi, C. Poggi, E. Bertolesi, G. Milani, Ancient masonry arches and vaults strengthened with TRM, SRG and FRP composites: experimental evaluation, Composite Structures 187 (2018) 466−480.
[21] C. Papanicolaou, T. Triantafillou, M. Lekka, Externally bonded grids as strengthening and seismic retrofitting materials of masonry panels, Construction and Building Materials 25 (2) (2011) 504−514.
[22] F.A. Kariou, S.P. Triantafyllou, D.A. Bournas, L.N. Koutas, Out-of-plane response of masonry walls strengthened using textile-mortar system, Construction and Building Materials 165 (2018) 769−781.
[23] F.G. Carozzi, G. Milani, C. Poggi, Mechanical properties and numerical modeling of fabric reinforced cementitious matrix (FRCM) systems for strengthening of masonry structures, Composite Structures 107 (2014) 711−725.
[24] T. D'Antino, L.H. Sneed, C. Carloni, C. Pellegrino, Influence of the substrate characteristics on the bond behavior of PBO FRCM-concrete joints, Construction and Building Materials 101 (2015) 838−850.
[25] C. Caggegi, F.G. Carozzi, S. De Santis, F. Fabbrocino, F. Focacci, Ł. Hojdys, E. Lanoye, L. Zuccarino, Experimental analysis on tensile and bond properties of PBO and aramid

fabric reinforced cementitious matrix for strengthening masonry structures, Composites Part B: Engineering 127 (2017) 175–195.

[26] C.B. de Carvalho Bello, I. Boem, A. Cecchi, N. Gattesco, D.V. Oliveira, Experimental tests for the characterization of sisal fiber reinforced cementitious matrix for strengthening masonry structures, Construction and Building Materials 219 (2019) 44–55.

[27] M. Ahmadi, H. Naderpour, A. Kheyroddin, Utilization of artificial neural networks to prediction of the capacity of CCFT short columns subject to short term axial load, Archives of Civil and Mechanical Engineering 14 (3) (2014) 510–517.

[28] P. Sarir, S.L. Shen, Z.F. Wang, J. Chen, S. Horpibulsuk, B.T. Pham, Optimum model for bearing capacity of concrete-steel columns with AI technology via incorporating the algorithms of IWO and ABC, Engineering with Computers 37 (2021) 797–807.

[29] X. Li, J. Wang, Y. Bao, G. Chen, Cyclic behavior of damaged reinforced concrete columns repaired with high-performance fiber-reinforced cementitious composite, Engineering Structures 136 (2017) 26–35.

[30] M.M. Kadhim, M.J. Altaee, A.H. Adheem, A.R. Jawdhari, A robust 3D finite element model for concrete columns confined by FRCM system, in: MATEC Web of Conferences, vol 281, EDP Sciences, 2019, p. 01006.

[31] M. Kyaure, F. Abed, Finite element parametric analysis of RC columns strengthened with FRCM, Composite Structures 275 (2021) 114498.

[32] K. Le-Nguyen, Q.C. Minh, A. Ahmad, L.S. Ho, Development of deep neural network model to predict the compressive strength of FRCM confined columns, Frontiers of Structural and Civil Engineering (2022) 1–20.

[33] M.A. Irandegani, D. Zhang, M. Shadabfar, Compressive strength of concrete cylindrical columns confined with fabric-reinforced cementitious matrix composites under monotonic loading: application of machine learning techniques, in: Structures, vol 42, Elsevier, August 2022, pp. 205–220.

[34] M.A. Irandegani, D. Zhang, M. Shadabfar, Probabilistic assessment of axial load-carrying capacity of FRCM-strengthened concrete columns using artificial neural network and Monte Carlo simulation, Case Studies in Construction Materials 17 (2022) e01248.

[35] D.A. Bournas, T.C. Triantafillou, Bar buckling in RC columns confined with composite materials, Journal of Composites for Construction 15 (3) (2011) 393–403.

[36] T.C. Triantafillou, C.G. Papanicolaou, Textile reinforced mortars (TRM) versus fiber reinforced polymers (FRP) as strengthening materials of concrete structures, Special Publication 230 (2005) 99–118.

[37] M. Del Zoppo, C. Menna, M. Di Ludovico, D. Asprone, A. Prota, Opportunities of light jacketing with fibre reinforced cementitious composites for seismic retrofitting of existing RC columns, Composite Structures 263 (2021) 113717.

[38] R. Talo, S. Khalaf, M. Kyaure, F. Abed, FEA of strengthened RC columns with PBO FRCM exposed to fire, in: 2022 Advances in Science and Engineering Technology International Conferences (ASET), IEEE, February 2022, pp. 1–4.

[39] N. Tello, Y. Alhoubi, F. Abed, A. El Refai, T. El-Maaddawy, Circular and square columns strengthened with FRCM under concentric load, Composite Structures 255 (2021) 113000.

[40] P. Colajanni, M. Fossetti, G. Macaluso, Effects of confinement level, cross-section shape and corner radius on the cyclic behavior of CFRCM confined concrete columns, Construction and Building Materials 55 (2014) 379–389.

[41] L. Ombres, Concrete confinement with a cement based high strength composite material, Composite Structures 109 (2014) 294–304.

[42] F. Faleschini, M.A. Zanini, L. Hofer, C. Pellegrino, Experimental behavior of reinforced concrete columns confined with carbon-FRCM composites, Construction and Building Materials 243 (2020) 118296.
[43] T.C. Triantafillou, C.G. Papanicolaou, P. Zissimopoulos, T. Laourdekis, Concrete confinement with textile-reinforced mortar jackets, ACI Structural Journal 103 (1) (2006) 28.
[44] P. Colajanni, F. De Domenico, A. Recupero, N. Spinella, Concrete columns confined with fibre reinforced cementitious mortars: Experimentation and modelling, Construction and Building Materials 52 (2014) 375−384.
[45] P. Colajanni, F. Di Trapani, G. Macaluso, M. Fossetti, M. Papia, Cyclic axial testing of columns confined with fiber reinforced cementitiuos matrix, in: Proceedings of the 6th International Conference on FRP Composites in Civil Engineering (CICE '12), June 2012.
[46] J. Gonzalez-Libreros, M.A. Zanini, F. Faleschini, C. Pellegrino, Confinement of low-strength concrete with fiber reinforced cementitious matrix (FRCM) composites, Composites Part B: Engineering 177 (2019) 107407.
[47] H.Y. Yin, Y. Xun, C.B. Ji, S. Sun, Experimental investigation of concrete confinement with textile reinforced concrete, in: Applied Mechanics and Materials, vol 752, Trans Tech Publications Ltd, 2015, pp. 702−710.
[48] M. Di Ludovico, A. Prota, G. Manfredi, Structural upgrade using basalt fibers for concrete confinement, Journal of Composites for Construction 14 (5) (2010) 541−552.
[49] D. Messerer, K. Holschemacher, Confinement of RC columns with CFRCM, Proceedings of International Structural Engineering and Construction 7 (2020) 2.
[50] M. Di Ludovico, A. Prota, G. Manfredi, Concrete confinement with BRM systems: experimental investigation, in: Proceedings of the 4th International Conference on FRP Composites in Civil Engineering—CICE, 2008, pp. 22−24.
[51] P. Kumar, H.C. Arora, A. Bahrami, A. Kumar, K. Kumar, Development of a reliable machine learning model to predict compressive strength of FRP-confined concrete cylinders, Buildings 13 (4) (2023) 931.
[52] A. Kumar, H.C. Arora, K. Kumar, H. Garg, Performance prognosis of FRCM-to-concrete bond strength using ANFIS-based fuzzy algorithm, Expert Systems with Applications 216 (2023) 119497.
[53] A. Kumar, H.C. Arora, N.R. Kapoor, K. Kumar, M. Hadzima-Nyarko, D. Radu, Machine learning intelligence to assess the shear capacity of corroded reinforced concrete beams, Scientific Reports 13 (1) (2023) 2857.
[54] A. Kumar, H.C. Arora, M.A. Mohammed, K. Kumar, J. Nedoma, An optimized neuro-bee algorithm approach to predict the FRP-concrete bond strength of RC beams, IEEE Access 10 (2021) 3790−3806.
[55] R. Singh, H.C. Arora, A. Bahrami, A. Kumar, N.R. Kapoor, K. Kumar, H.S. Rai, Enhancing sustainability of corroded RC structures: estimating steel-to-concrete bond strength with ANN and SVM algorithms, Materials 15 (23) (2022) 8295.
[56] A. Kumar, H.C. Arora, N.R. Kapoor, K. Kumar, Prognosis of compressive strength of fly-ash-based geopolymer-modified sustainable concrete with ML algorithms, Structural Concrete (2022).
[57] F.J.D.C. y Basalo, F. Matta, A. Nanni, Fiber reinforced cement-based composite system for concrete confinement, Construction and Building Materials 32 (2012) 55−65.
[58] A. Kumar, H.C. Arora, K. Kumar, M.A. Mohammed, A. Majumdar, A. Khamaksorn, O. Thinnukool, Prediction of FRCM−Concrete bond strength with machine learning approach, Sustainability 14 (2) (2022) 845.

[59] N.R. Kapoor, A. Kumar, A. Kumar, A. Kumar, M.A. Mohammed, K. Kumar, S. Kadry, S. Lim, Machine learning-based CO2 prediction for office room: a pilot study, Wireless Communications and Mobile Computing (2022).
[60] D. Fang, X. Zhang, Q. Yu, T.C. Jin, L. Tian, A novel method for carbon dioxide emission forecasting based on improved Gaussian processes regression, Journal of Cleaner Production 173 (2018) 143–150.
[61] C. Williams, C. Rasmussen, Gaussian processes for regression, Advances in Neural Information Processing Systems 8 (1995).
[62] V. Vapnik, S. Golowich, A. Smola, Support vector method for function approximation, regression estimation and signal processing, Advances in Neural Information Processing Systems 9 (1996).
[63] A. Mita, H. Hagiwara, Quantitative damage diagnosis of shear structures using support vector machine, KSCE Journal of Civil Engineering (1776).
[64] L. Breiman, Random forests, Machine Learning 45 (2001) 5–32.
[65] V. Svetnik, A. Liaw, C. Tong, J.C. Culberson, R.P. Sheridan, B.P. Feuston, Random forest: a classification and regression tool for compound classification and QSAR modeling, Journal of Chemical Information and Computer Sciences 43 (6) (2003) 1947–1958.
[66] J. Patel, S. Shah, P. Thakkar, K. Kotecha, Predicting stock market index using fusion of machine learning techniques, Expert Systems with Applications 42 (4) (2015) 2162–2172.
[67] H. Jiang, Y. Deng, H.S. Chen, L. Tao, Q. Sha, J. Chen, C.J. Tsai, S. Zhang, Joint analysis of two microarray gene-expression data sets to select lung adenocarcinoma marker genes, BMC Bioinformatics 5 (1) (2004) 1–12.
[68] A.M. Prasad, L.R. Iverson, A. Liaw, Newer classification and regression tree techniques: bagging and random forests for ecological prediction, Ecosystems 9 (2006) 181–199.

Assessment of shear capacity of a FRP-reinforced concrete beam without stirrup: A machine learning approach

13

Prashant Kumar [1,2,3], *Harish Chandra Arora* [1,2] *and Aman Kumar* [1,2]
[1]Academy of Scientific and Innovative Research (AcSIR), Ghaziabad, Uttar Pradesh, India; [2]Department of Structural Engineering, CSIR—Central Building Research Institute, Roorkee, Uttarakhand, India; [3]Department of Civil Engineering, COER University, Roorkee, Uttarakhand, India

13.1 Introduction

Shear failure is characterized as a sudden and brittle collapse of structural elements. The design of beams to withstand both bending and shear is crucial, as failure in either mode can lead to an overall structural breakdown. Excessive shear forces can result in the abrupt fracturing of beams. In order to enhance the structural integrity of concrete elements, civil engineering construction includes the incorporation of steel bars [1]. However, as time passes, the strength of reinforced concrete (RC) structures degrades, creating safety concerns. If the RC beam already has weak shear and if its shear capacity is lower than its bending strength, measures for enhancing shear capacity must be taken into account [2].

A growing number of existing buildings are becoming structurally or functionally unsound for a variety of reasons. The main issues with the deteriorating structures that need to be rectified to support larger loads are aging and environmental deterioration, poor construction quality, lack of maintenance, and natural disasters [3]. Consequently, a wide range of constructions require reinforcement. It is important to emphasize that evaluating the shear capacity of reinforced beams is crucial, as the optimal method for reinforcing RC structures is frequently chosen based on the success of shear reinforcement. Presently, a diverse range of methods are employed to enhance shear capacity of various types of RC structures [4].

In this context, composites like fiber-reinforced polymer (FRP) sheets, bars, and laminates are growing quickly in modern times. It has been discovered that adding FRP to concrete enhances both the tensile and compressive strengths of the material. FRP is an excellent material in relation to strength-weight ratio, corrosive resistance, flexibility, durability, fatigue resistance, and simplicity of maintenance. FRP-strengthened RC structures indicate different shear behaviors due to their anisotropic linear-elastic property and their comparatively lower elastic modulus [5].

Artificial Intelligence Applications for Sustainable Construction. https://doi.org/10.1016/B978-0-443-13191-2.00016-X
Copyright © 2024 Elsevier Ltd. All rights are reserved, including those for text and data mining, AI training, and similar technologies.

A composite material cosists of a polymer matrix that is strengthened by filaments. Commonly employed are four categories of FRP, i.e., carbon (CFRP), glass (GFRP), aramid (AFRP), and basalt (BFRP), which give composites a high level of strength and stiffness [6]. FRP bars can be employed as alternatives to rebar due to their exceptional ability to provide resistance to corrosion in building components [7]. Hence, FRP sheets are extensively employed globally to effectively repair or retrofit structural elements [8−11]. Many research investigations have been carried out regarding the performance of RC elements enhanced with FRP confinement. A preliminary study on the shear strength of concrete beams wrapped with FRP sheets was conducted by Berset in 1992 [12]. The study provided an analytical approach to forecast the shear behavior of beam elements confined with FRP using certain experimental data. Additionally, the shear strength of CFRP-strengthened beams can also be estimated using the model developed by Uji in 1992 [13]. Similar to this, Dolan et al. [14] researched the effects of AFRP, CFRP, and GFRP in order to increase the shear strength of beams. Gamino et al. [15] also used a number of CFRP sheets to enhance the nonbrittleness and load-bearing capacity of concrete elements.

However, assessing the shear strength of RC beams with FRP reinforcement can be challenging due to the need for costly equipment and the time-consuming nature of the procedure. To tackle these challenges, the beams shear capacity can be anticipated through the application of analytical models and machine learning (ML) techniques. For forecasting beam shear capacity, ML models like Gaussian process regression (GPR), linear regression (LR), artificial neural network (ANN), support vector machines (SVM), etc., can be utilized [16].

13.2 Literature review

Jumaa and Yousif [17] introduced three novel predictive models in their work, including ANN, nonlinear regression analysis (NLR), and gene expression programming (GEP) for shear capacity prediction. Their study involved compiling a dataset comprising 269 shear test results from FRP-wrapped RC beams found in published literature. The effectiveness of these proposed models was assessed in comparison to available standards and empirical equations. Remarkably, the ANN model exhibited remarkable performance, notably demonstrating superior accuracy when compared to alternative existing models. According to Kara [18], a GEP model was utilized to evaluate the shear strength of a beam with FRP reinforcement. This study focused on cases where the aspect ratio of the beam exceeded 2.5 and where no stirrups were incorporated. The results of the proposed model were found to be superior to previously derived shear equations. Alam et al. [19] develop a hybrid intelligence model in combination with the Bayesian optimization algorithm (BOA) for the prediction of the shear behavior of RC beams. The database contains 216 specimens without stirrups in total. Beam depth (d), strength of concrete ($f_c^{'}$), shear span to depth ratio (a/d), and effective reinforcement ratio (ρ_{eff}) were used as input parameters. It was concluded that predicted outcomes using the suggested method was superior to other alternative methods.

Naderpour et al. [3] utilized the ANN algorithm to forecast the beam shear capacity (V_u) that was reinforced with FRP bars. For this purpose, 110 datasets were compiled. The study utilizes a set of input parameters, including f_c', FRP reinforcement ratio (p), FRP's elastic modulus (E_f), a/d, web width (b_w), and d. The results reveal that the proposed model presented higher values compared to earlier works on this subject. Yang and Liu [16] reviewed the accuracy of codified existing provisions and four suggested ML methods. The database of 219 beam samples without stirrups was used in the study. The study incorporated various input parameters, including beam geometry aspects such as d and b_w, as well as testing setup factors like a/d. Additionally, concrete properties like f_c' and flexural bar characteristics, including tensile strength (f_u), E_f, p, and total shear force $(V_{u,exp})$ were considered. The predictive accuracy of the proposed ML techniques demonstrated a noteworthy improvement in estimating the ultimate capacity. Nikoo et al. [7] evaluated the shear strength of FRP-RC beams by considering 140 experimental results from the published literature. In the study, the ANN algorithm was optimized using the bat algorithm. The results revealed that the ANN-BAT optimized algorithm produced better results in comparison to other ML and analytical models used in the study.

13.3 Research significance

In order to upgrade existing structures, using FRP to increase the strength of concrete beams has shown to be a highly successful strategy. Still, it is very difficult to determine the shear strength of RC beams. The complexity gets worse when FRP bars are employed to boost the shear capacity of RC beams [20]. Numerous mathematical models have been established in order to calculate the shear strength of RC beams when FRP bars are used. However, mathematical models are based on a limited number of assumptions and a limited dataset, which restricts their utilization in intricate problems. In addition, shear strength estimation of RC beams is a time-consuming process, and highly skilled labor with high-cost equipment is required. This proves that the use of computational intelligence methods is required to anticipate the shear strength of RC beams. This chapter utilizes ML algorithms such as LR, SVM, GPR, and ANN.

13.4 Artificial intelligence (AI)

Artificial intelligence (AI), a branch of computer science, strives to create machines with the ability to carry out tasks that traditionally require human intelligence. The field involves designing models, algorithms, and systems with the ability to think, learn, and make decisions, enabling them to tackle problems and achieve defined goals.

By improving efficiency, reliability, and decision-making processes, AI has enormous promise in the civil engineering domain. There are some areas, like structural design, where AI can predict how members will behave under various loading conditions and support planning and scheduling construction projects more effectively by considering resource allocation, project sequencing, and risk assessment. Along with routine maintenance and cost-cutting measures, it can be used to continuously monitor the condition of infrastructure, including buildings, bridges, and dams.

AI includes a subfield designated as ML, which revolves around developing algorithms and models that allow computers to learn, predict, and make decisions separately without having to be explicitly programmed. Computers can detect patterns, calculate insightful conclusions, and make reliable predictions thanks to ML algorithms, which learn from data and continuously improve their performance. ML algorithms come in diverse forms, including supervised, unsupervised, and reinforcement learning.

Different algorithms produce functions that establish a connection between inputs and their corresponding desired outputs. A common representation of supervised learning involves the classification challenge. In this scenario, the learning process involves approximating a function that takes a vector as input and assigns it to a specific class from a set of options. This approximation is achieved by analyzing multiple instances where inputs and outputs are provided. Typical supervised learning includes neural networks, SVM, decision trees, random forests, and LR. Conversely, unsupervised learning algorithms lack predefined classifications. The primary objective of unsupervised learning is to autonomously generate classification labels. These algorithms examine similarities between data fragments to see whether they may be grouped and categorized. These resulting groups are referred to as clusters, which in the field of ML refers to a broad category of clustering approaches. In the case of reinforcement learning, the algorithm develops an action plan based on an observation of the outside world. Every action has an effect on the environment in some way, and the environment gives the learning algorithm feedback. It has achieved success in many areas, including robotics, gaming, and driving autonomous vehicles [21] (Table 13.1).

13.4.1 Proposed ML models to predict shear capacity

13.4.1.1 LR

LR is a supervised learning approach that involves analyzing the correlation between a target variable and one or more predictor variables. This analysis, known as regression analysis, is commonly employed to investigate how changes in the independent variables relate to the variations in the dependent variable. The objective is to create a model capable of forecasting the forthcoming values of the dependent variable based on the independent variables. Regression analysis with an individual predictor variable is referred to as simple regression analysis. Conversely, when employing multiple predictor variables, it is termed multiple regression analysis. LR models are defined by linear equations, while NLR models are characterized by equations that do not adhere to linearity [26].

13.4.1.2 SVM

SVM is an ML method employed in regression as well as classification analysis. It is a kind of supervised learning algorithm that transforms the input data using a nonlinear mapping function. Following their predicted values, SVM creates a hyperplane that

Table 13.1 Summary of previous ML models.

S. no.	References	Input variables	AI algorithm employed	Shear strength	*R*-value	No. of samples
1.	[1]	a/d, w_f, E_f, t_f, a/d, f_c'	Multigene genetic programming (MGP) machine	–	0.994 (MGP)	103
2.	[2]	d, a/d, f_c, ρ_l, E_f	Generalized regression neural network (GRNN)–	10–250	0.911 (GRNN)	196
3.	[3]	f_c', ρ_l, E_f, b, d, a/d	ANN–	8.8–190	0.96	110
4.	[5]	f_c', ρ_l, E_f, b, d, a/d	ANN–	8.9–174	0.98	106
5.	[7]	a/d, b, d, E_f, f_c'	Particle swarm optimization (PSO), ANN, and genetic algorithm (GA)–	10–220	0.995	140
6.	[16]	b, d, a/d, f_c', E_f, ρ_l	Linear regression, extreme gradient boosting, decision tree, random forest–	9.1–396.3	0.96 (XG boost)	219
7.	[18]	f_c', b_w, b, d, ρ_l, E_f	Genetic programming	8.8–190	–	104
8.	[19]	b, d, a/d, f_c', ρ_{eff}, V_{exp}	BOA and ANN–	8.76–264.80	0.970	216
9.	[22]	E_f, a/d	Adaptive neuro-fuzzy inference system (ANFIS)–	0–160	0.988	84
10.	[23]	b, d, f_c', a/d, A_f, f_y, ρ_l	Strut-and-tie models and GA–	10–800	0.950	24
11.	[24]	f_c', b, d, a/d	ANN	17.58-49.78	–	78
12.	[25]	a/d, b, ρ_l	Traditional truss model and regression method	–	0.568	200

maximally divides the input data into two classes. The objective is to identify the hyperplane that maximizes the margin, which is the space between the hyperplane and the closest data points, while minimizing the prediction error on the training data. The SVM helps it to robustly handle outliers and noise in the data by implying a greater emphasis on maximizing the margin between the predicted values and the hyperplane.

When dealing with high-dimensional data and nonlinear correlation between the input and target variables, SVM is an accurate method for performing regression analysis. However, it has some limits because it needs to have certain parameters, like the regularization parameter and kernel function, carefully tuned. In general, SVM is a successful method for performing regression analysis, particularly when working with complex and high-dimensional data. It has been successfully used in several civil engineering disciplines [27].

13.4.1.3 GPR

"Gaussian processes (GPs) derive their name from Carl Friedrich Gauss, as they are founded upon the Gaussian distribution, also known as the normal distribution." A GP can be described as an infinite set of random data points with a defined joint Gaussian range over all of its bounded ranges. GPR presents several benefits, including the ability to handle small datasets and offer predictive uncertainty assessments. A GP is represented by the mean and covariance functions [28,29]. The mean function is frequently considered to be zero because GPs are constructed by linearly arranging random variables that follow a normal distribution.

$$f(x) \sim GP(\mu(x), k(x, x')) \tag{13.1}$$

where, $\mu(x)$ = mean function and $k(x, x') = cov(f(x), f(x'))$

$k(x, x')$ can be best expressed as:

$$k(x, x') = \sigma_f^2 \, exp\left\{-\frac{1}{2}\left(\frac{x_i - x_j}{l^2}\right)\right\} \tag{13.2}$$

where, σ_f^2 = maximum permissible variance and l = length of scale.

13.4.1.4 ANN

ANNs are an AI methodology that has been influenced by the neural system found in the human brain [30]. ANN is a computational technique that employs analytical computation to model intricate correlations and functions, exhibiting improved accuracy compared to conventional approaches [31]. Artificial neurons are the interconnected building blocks that construct ANN. These neurons build a network by connecting to one another. A neuron's basic duty is to gather input information from neighboring neurons, analyze that data through the activation function, and then send the processed output to neighboring neurons in the subsequent layer. These

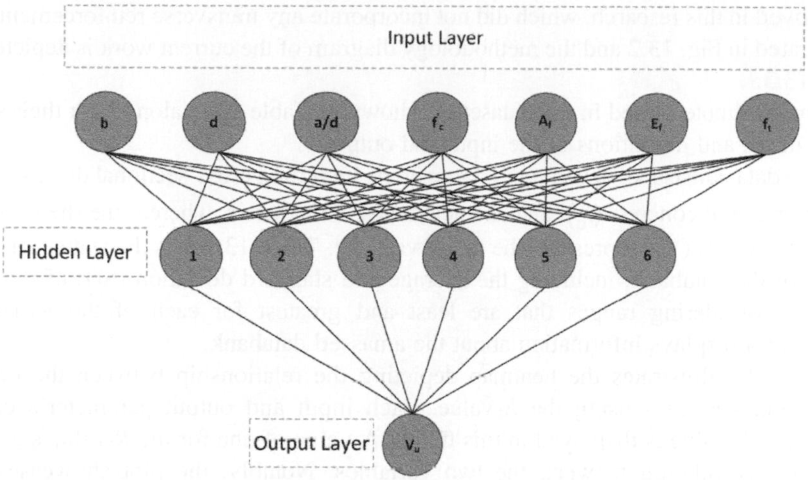

Figure 13.1 Architecture of artificial neural network.

layers include input, hidden, and output layers. Within an ANN structure, the input layer holds the independent variables, and the output layer generates the model's final output. Hidden layers act as connectors, linking the neurons between the input layer and output layer. The number of hidden layers as well as the number of neurons within these hidden layers significantly impact both the complexity and precision of the model.

Fig. 13.1 displays the employed ANN architecture. There are different types of learning algorithms that play a significant part in the ANN framework. The most commonly utilized is the backpropagation algorithm [32]. This algorithm involves the transmission of information from the input layer to neurons located within hidden layers and then forwarding it to the resulting layer. The algorithm computes the error that arises during this operation by comparing the real values with the obtained values. This error is then utilized to adjust and reset the weights of the interconnected neurons. The entire procedure is iteratively repetitive with the aim of minimizing error [32]. The error generated by the model in each iteration is evaluated using a metric known as the loss function. This loss function serves as a measurement criterion to guide the optimization process toward identifying the optimal weights within the network, aiming to achieve the lowest possible error.

13.5 Methodology

13.5.1 Experimental database

This chapter compiles 215 sets of experimental data from previous studies. The beam data utilized in these studies lacked stirrup reinforcement. Each specimen in the studies underwent a shear test and failed. The structural characteristics of the beam specimens employed in this research, which did not incorporate any transverse reinforcement, are illustrated in Fig. 13.2 and the methodology diagram of the current work is depicted in Fig. 13.3.

The parameters used in the dataset are shown in Table 13.2, along with their symbols, units, and definitions of the input and output.

The data consists of seven input parameters, including cross-sectional details of the beam, grade of concrete (f_c'), and characteristics of the fabric, whereas the shear capacity of the beam (V_u) represents the target variable. Table 13.3 lists the statistical variables of the databank, including the average and standard deviation (SD) of the data points, considering ranges that are least and greatest for each of the variables. Table 13.4 displays information about the amassed databank.

Fig. 13.4 illustrates the heatmap depicting the relationship between the output and input variables using the R-value. Each input and output parameter's corresponding R-value is displayed in this figure. A value of one for the R-value signifies a robust correlation between the two variables. Notably, the plot showcases the strongest correlation with a value of 0.95, which is observed between the correlation of E_f and f_t

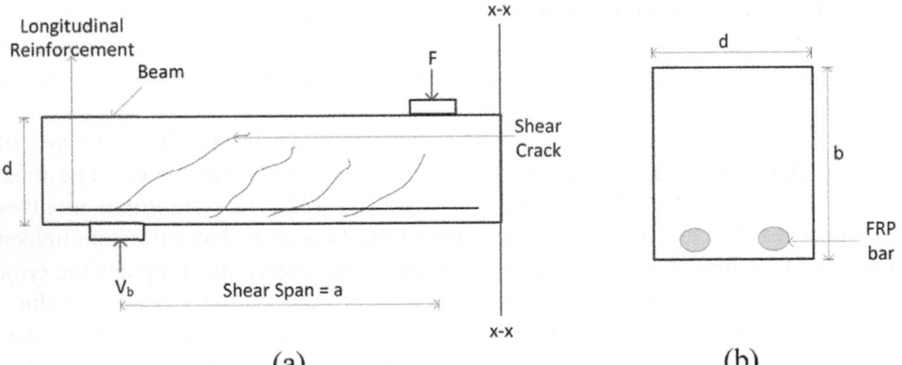

Figure 13.2 (a) Longitudinal cross-section of specimen, (b) cross section of specimen.

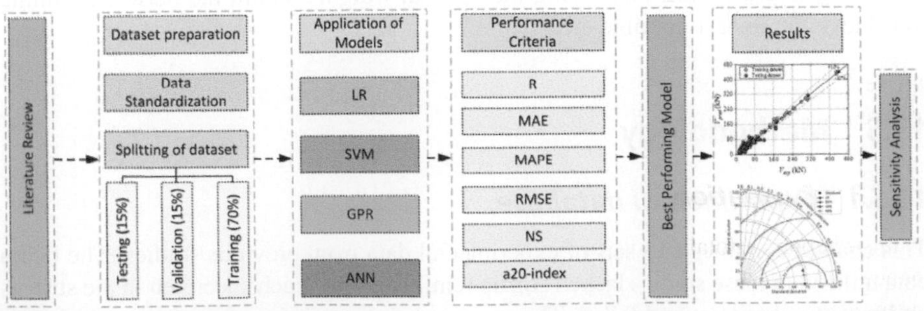

Figure 13.3 Methodology chart.

Table 13.2 Details of used parameters.

S. no.	Parameter	Symbol used	Unit	Input/ output
1.	Shear span to depth ratio	a/d	—	Input
2.	Area of fabric	A_f	mm^2	Input
3.	Beam width	b	mm	Input
4.	Effective depth of beam	d	mm	Input
5.	Modulus of FRP	E_f	GPa	Input
6.	Concrete characteristic compressive strength	f_c'	MPa	Input
7.	Tensile strength of FRP	f_t	MPa	Input
8.	Shear capacity of beam	V_u	kN	Output

Table 13.3 Statistical parameters of the database.

S. no.	Parameter	b	d	a/d	f_c'	A_f	E_f	f_t	V_u
	Units	mm	mm	—	MPa	mm^2	GPa	MPa	kN
1.	Minimum	89.0	73.0	1.0	20.0	60.0	29.0	397.0	9.8
2.	Maximum	1000.0	937.0	16.2	93.0	8608.0	148.0	2640.0	526.0
3.	Mean	250.9	262.4	3.7	40.1	721.7	73.9	1157.0	86.1
4.	SD	163.4	145.1	1.8	15.1	1049.9	43.9	578.8	94.8

13.5.2 Data standardization

Making the data unitless using standardization allows ML algorithms to easily understand it. Using the standardization technique, all values are distributed between a range of two numbers. In this chapter, normalization of the dataset is done between 0 and 1. Without normalization, large values have a much greater impact on training than small values, which could cause the model to train incorrectly and produce inaccurate outputs.

$$A_{normalized} = \left[2 \times \frac{g - g_{min}}{g_{max} - g_{min}}\right] - 1 \tag{13.3}$$

where $A_{normalized}$ stands for normalized value, g stands for original value, g_{max} and g_{min} referred to greatest and lowest value of g, correspondingly

13.5.3 Evaluation metrics

In this chapter, a total of six measuring indices are utilized to determine the performance of employed models, including the R-value, mean absolute error (MAE), mean absolute percentage error (MAPE), root mean square error (RMSE), a20-

Table 13.4 Details of the databank.

S. no.	References	Specimens	b	d	a/d	f_c'	A_f	E_f	f_t	V_u
1.	[33]	6	150	167.5–267.5	2.49–3.98	34–59–	113.062–461.437	32–38	650–705–	25–60
2.	[34]	7	200	225	1.82–4.5	40.5–49–	112.5–396	145	2250	27.87-42.33
3.	[35]	12	130–160	310–346	2.75–3.71	34.1–43.2–	284–800	42–120	600–1551–	42.7–63.7
4.	[36]	18	178–279	224–225	4	36.3–	567.05–1134.11	40.366	689.5–	56.2–102
5.	[37]	9	200	230	1–1.52	43–65–	423.2–846.4	51	1050–	233.1–474.7
6.	[38]	53	150–200	213.5–215.5	1.5–4.5	30–	71.3–126.7	48.2–147.9	940.6–2130–	15.9–85.1
7.	[39]	167	89–1000	73–937	1.12–16.22	20–93–	60–8608	29–148	397–2640–	9.8–526

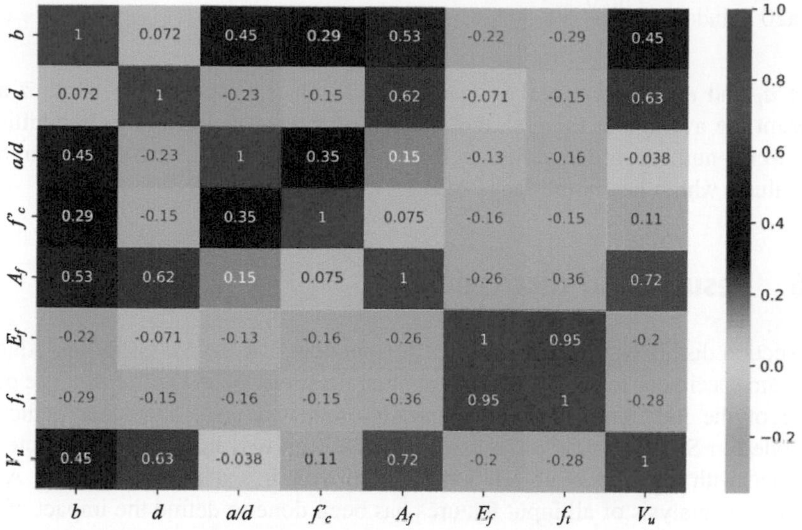

Figure 13.4 Correlation coefficient plot between input and output parameters.

index, and Nash-Sutcliffe efficiency index (NS). MAPE, MAE, and RMSE were employed to quantify errors, and a lower error value indicates a higher level of model accuracy, whereas *R*, a20-index, and NS values closer to one also indicate higher precision [40].

$$R = \frac{\sum_{i=1}^{N}(a_i - \overline{a})(c_i - \overline{c})}{\sqrt{\sum_{i=1}^{N}(a_i - \overline{a})^2 \sum_{i=1}^{N}(c_i - c)}^2} \tag{13.4}$$

$$\text{MAE} = \frac{1}{N}\sum_{i=1}^{N}|a_i - c_i| \tag{13.5}$$

$$\text{RMSE} = \sqrt{\frac{1}{N}\sum_{i=1}^{N}(a_i - c_i)^2} \tag{13.6}$$

$$\text{MAPE} = \frac{1}{N}\sum_{i=1}^{N}\left|\frac{a_i - c_i}{a_i}\right| \times 100 \tag{13.7}$$

$$\text{NS} = 1 - \frac{\sum_{i=1}^{N}(a_i - c_i)^2}{\sum_{i=1}^{N}(a_i - \overline{c})^2} \tag{13.8}$$

$$\text{a20} - \text{index} = \frac{\text{m20}}{N} \tag{13.9}$$

where a_i and c_i represent real and anticipated values, respectively, while $\overline{a}$ and $\overline{c}$ represent the average of the real and anticipated values, sequentially. In addition, N refers to the number of data points in the databank, and m20 indicates the number of data values, which lies in the range of 0.8–1.2.

13.6 Results and discussion

This section discusses the application of ML models (LR, SVM, GPR, and ANN) on 215 beam specimens to predict their shear strength strengthened with FRP. The performance of the mentioned ML algorithms is evaluated using statistical indices, as mentioned in Section 13.5.3. The results of the proposed methods are depicted and explained with the help of correlation, error distribution, and Taylor's plot. At last, a sensitivity analysis of all input features has been done to define the impact of input parameters on the target variable.

13.6.1 Performance of established models

To forecast the shear strength of RC beams, ML models (LR, SVM, GPR, and ANN) have been established. The performance of established models has been evaluated on the basis of six performance indicators, as discussed in Section 13.5.3.

As depicted in Table 13.5, for the established LR model, the R value is revealed to be 0.9932 for the training stage, 0.9336 for the testing stage, and 0.9768 for the entire databank, sequentially. In addition, the error measurements, i.e., MAPE, RMSE, and MAE, are observed at 16.77%, 8.52 KN, and 6.70 KN for the training phase; 33.83%, 25.25 KN, and the 15.28 KN for testing phase; and 21.86%, 15.53 KN, and 9.27 KN for the entire dataset, respectively. Also, NS values observed for the training phase are 0.9922, the testing phase is 0.8684, and the complete databank is 0.9539, respectively. The scatter plot of the LR model is shown in Fig. 13.5a. It has been observed that in the training set, the data values are lying within the error range of between -10% and 10%, whereas the testing dataset exhibits an uneven spread of data along the 45° line.

It is clear from the SVM model that the precision achieved during the training stage was 99.34%, which was higher than that during the testing phase. Moreover, the accuracy for testing and entire datasets is 93.44% for both, sequentially. Furthermore, it's noteworthy that the error values are minimal when considering the training set, indicating a strong fit and a robust relationship between the training set values and the model. The SVM model yields an MAPE of 16.76%, an RMSE of 8.45 KN, and an MAE of 6.71 KN for the training set; these values are 33.62%, 25.04 KN, and 15.25 KN for the testing set; and they are 21.8%, 15.50 KN, and 9.26 KN for the all-data set, sequentially. Fig. 13.5b displays the scatter plot for the SVM method, which illustrates that a significant number of data points cluster closely along the best-fit line, revealing a good correlation with the model within the training set.

Table 13.5 Outcomes of employed models.

S. no.	Model		Values						SD
			R	a20-index	NS	MAPE	RMSE	MAE	
1.	LR	Training	0.9932	0.7097	0.9922	16.7720	8.5196	6.7064	71.8427
		Testing	0.9336	0.4394	0.8684	33.8242	25.2523	15.2898	
		Validation	0.9768	0.6290	0.9539	21.8645	15.5353	9.2697	
2.	SVM	Training	0.9934	0.7226	0.9867	16.7678	8.4574	6.7177	71.5235
		Testing	0.9344	0.3939	0.8704	33.6204	25.0446	15.2549	
		All	0.9771	0.6244	0.9546	21.8007	15.4105	9.2673	
3.	GPR	Training	0.9935	0.7226	0.9870	16.4934	8.3753	6.6831	71.8689
		Testing	0.9335	0.4242	0.8680	32.6992	25.2860	15.1922	
		All	0.9769	0.6335	0.9541	21.3331	15.4966	9.2243	
4.	ANN	Training	0.9930	0.7032	0.9861	16.5595	8.7721	6.9054	72.0792
		Testing	0.9411	0.6061	0.8765	24.0146	13.6531	9.0388	
		All	0.9872	0.3939	0.9739	26.2434	14.1089	11.0493	
		Validation	0.9893	0.6425	0.9787	19.1187	10.5607	7.8427	

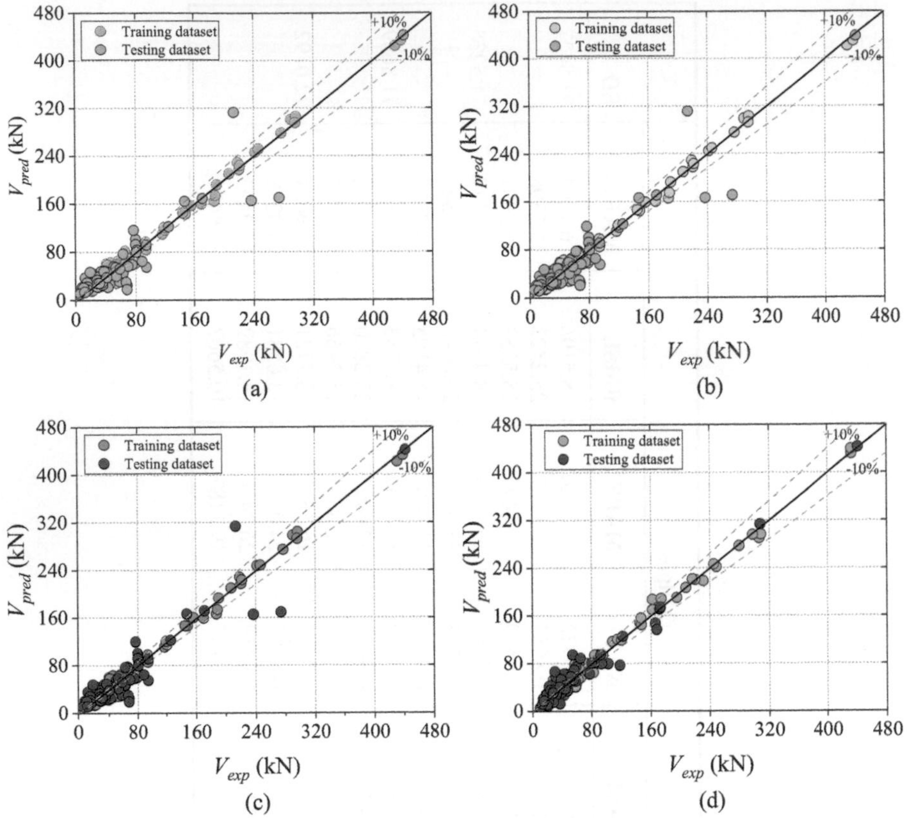

Figure 13.5 Scatter plot between experimental and predicted shear capacity: (a) LR, (b) SVM, (c) GPR, and (d) ANN.

The GPR model achieves relatively more precision in comparison to other models. The accuracy rate of the GPR model is observed at 99.35%, 93.35%, and 97.69% for training, testing, and the whole dataset, respectively, which is relatively more in comparison to other models. Furthermore, for the training set, the MAPE, RMSE, and MAE are 16.49%, 8.37 KN, and 6.68 KN, sequentially. Fig. 13.5c displays a scatter plot for the GPR model, which illustrates unequal data distribution in the testing phase covering a large number of errors in comparison to the training phase.

Likewise, in the case of the established ANN model, the *R* values are observed as 0.9930 for the training phase, 0.9411 for the testing phase, and 0.9872 for the validation phase. Furthermore, RMSE, MAPE, and MAE are determined to be 8.77 KN, 16.55%, and 6.90 KN for the training set; 13.65 KN, 24.01%, and 9.03 KN for the testing phase; 14.10 KN, 26.24%, and 11.04 KN for the validation phase; and 10.56 KN, 19.11%, and 7.84 KN for the complete databank, correspondingly. Additionally, it is shown that the NS values are measured at 0.9861 for the training, 0.8765 for the testing, 0.9739 for the validation, and 0.9787 for the combined dataset, sequentially. However, the a 20-index values of all four data sets are not found to be close to their ideal values. The correlation diagram of the ANN method is displayed in Fig. 13.5d. It has been observed that in the training phase, the data values are concentrated along the best-fit line, whereas in the testing phase, the data values are not lying along the diagonal line. Moreover, for the training set, the majority of data values range between −10% and 10%, while for the testing set, data points display a high range of errors.

13.6.2 Discussion

To determine the accuracy and dependability of the proposed models, the evaluation of proposed ML models has been done in this part. From the results presented in Table 13.5, it can be concluded that the GPR model exhibits the highest level of precision and reliability. The ANN model demonstrates an R value that is 1.27% higher than that of the LR model while simultaneously exhibiting MAPE, RMSE, and MAE values that are 12.58%, 32%, and 15.42% lower than those observed in the LR model, respectively. In comparison to the SVM model, the accuracy rate of the ANN model is 1.25% higher than that of the SVM model. In addition, the error values of the ANN model, i.e., MAPE, RMSE, and MAE, are 12.33%, 31.47%, and 15.42% lower than those of the SVM model, respectively. Similarly, in comparison to the GPR model, the precision of the ANN model is 1.26% higher than that of GPR. Moreover, MAPE, RMSE, and MAE values of the ANN model are 10.40%, 31.87%, and 15% lower than those of the GPR method.

The violin plot presented in Fig. 13.6 visually depicts the error distribution, showcasing the range of errors across the suggested ML models. It can be seen that all proposed models illustrate similar error ranges. However, the ANN model shows less error in comparison to other models.

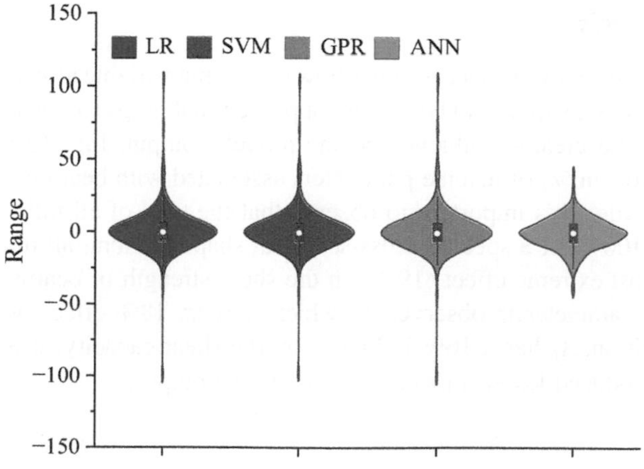

Figure 13.6 Error plot.

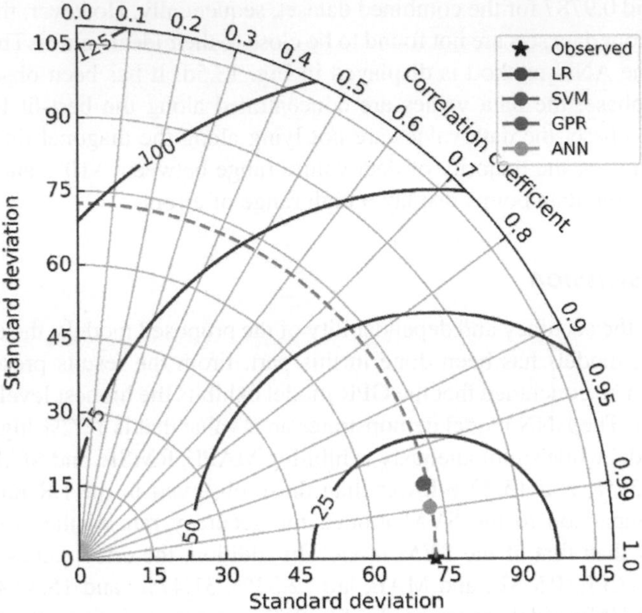

Figure 13.7 Taylor plot.

Fig. 13.7 demonstrates the relationship between the SD, RMSE values, and R values of all proposed ML models with the help of Taylor's diagram. The green line is referred to as the reference line, depicting the SD of the experimental dataset. The Taylor's plot represents that all proposed models are lying on the reference line and performed with good accuracy. However, the ANN model represented by the gray dot is lying close to the maximum correlation coefficient value.

13.6.3 Sensitivity analysis

This section presents a sensitivity analysis that examines how each input variable affects the shear strength of the beam. Sensitivity analysis is an approach that helps to determine the variables that have the greatest influence on the model's output. Fig. 13.8 shows the influencing indicator (in %) of diverse parameters associated with beam geometry and mechanical properties. It's important to observe that the total of all influencing indicators equates to 100% for a specific cross-sectional shape. Among all the features, a/d ratio has the most extreme effect (19%) on the shear strength of beams. The second most influential parameter is observed f_t, whicht has an 18% effect on the outcome variable. In addition, A_f has a 16% influence on the shear capacity. It is also noted that b, d, and E_f produced lesser impact (12%) on target output.

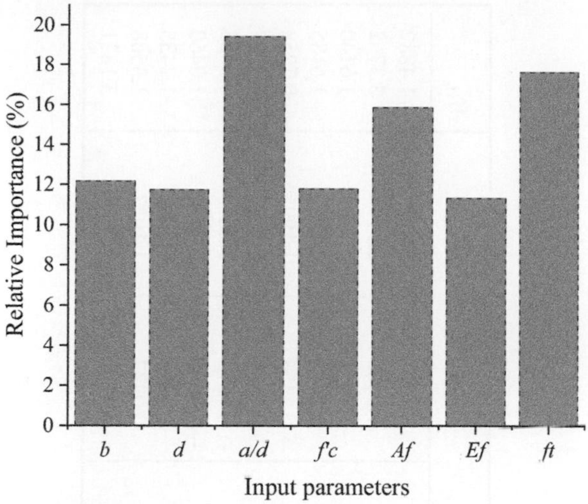

Figure 13.8 Sensitivity analysis.

13.6.4 ANN-based formulation

$$\mathbf{V}_u = f_{(H-O)} \sum_{i=1}^{N} \mathbf{W}_{i(O)} N_i + \mathbf{B}_{(O)} \tag{13.10}$$

where $W_{i(O)}$ is the weights of the output layer, $B_{(O)}$ is the output bias as shown in Eq. (13.10), and N_i can be calculated from Eq. (13.12).

$$\begin{aligned}\mathbf{V}_u = \mathbf{0.4102A_1} + \mathbf{0.9675A_2} + \mathbf{0.9653A_3} - \mathbf{1.3185A_4} + \mathbf{0.6940A_5}\\ - \mathbf{1.3314A_6} - \mathbf{0.9770A_7} + \mathbf{1.7911A_8} - \mathbf{0.7719A_9}\\ - \mathbf{0.3199A_{10}} - \mathbf{2.5508A_{11}} - \mathbf{0.6899}\end{aligned} \tag{13.11}$$

$$N_i = f_{(I-H)} \sum_{i=1}^{N} \mathbf{W}_{i(H)} X_i + \mathbf{B}_{(H)} \tag{13.12}$$

where $W_{i(H)}$ is the weight of the hidden layer and $B_{(H)}$ is the hidden layer bias (Table 13.6).

Table 13.6 Weight and biases of the hidden layer.

Neurons	W_H							B_H
1.	−0.2753	0.2129	0.9464	1.9038	−0.6264	1.3305	0.8098	−1.4882
2.	0.8242	0.9036	−1.8467	1.1736	1.6821	1.0505	2.5948	3.3762
3.	−0.1137	1.1554	1.5291	1.0068	0.1712	−2.5494	1.1124	1.6140
4.	−0.8980	−0.6574	−1.6159	−1.9987	−0.2367	0.0082	2.2261	−1.6185
5.	−1.1534	0.8186	−0.3476	−1.5483	−0.3059	−0.2453	−1.1995	0.7258
6.	−1.5632	−0.6524	−0.2176	−0.0481	2.8892	−0.5157	1.1087	−1.0097
7.	−0.4389	0.5632	−0.5470	0.1912	0.1672	0.1875	1.2690	0.5063
8.	1.0831	1.2792	1.9472	1.3548	−0.5385	−1.8882	−0.2972	1.0460
9.	−0.7182	0.9972	−0.7724	0.3092	0.8499	−0.4253	−0.8361	−1.2435
10.	0.7538	−0.2770	0.2631	1.5666	0.3882	0.0237	−1.1927	1.3508
11.	0.4790	−0.6819	3.0977	0.0594	−0.8502	0.6392	−0.5549	3.1821

13.7 Conclusions

The study focused on developing an ML model to forecast the shear strength of FRP-strengthened RC beam specimens with the best accuracy rate. A collected database of 215 specimens was utilized for developing four models (ANN, SVM, GPR, and LR) with seven input parameters $\left(b, d, a/d, f_c', E_f, f_t \text{ and } A_f\right)$ to forecast shear capacity (V_u). The statistical parameters (R, a20-index, NS, MAPE, RMSE, and MAE) of developed ML models were computed and compared to evaluate their relative performance. Following are the conclusions obtained from the study:

1. The ANN model demonstrates higher accuracy among all the ML models, with R values of 0.9930 for the training dataset and 0.9411 for the testing dataset.
2. The ANN model yielded the lowest error values across all cases. Specifically, the MAE, MAPE, and RMSE values for the ANN model were noted at 6.9054 KN, 16.55%, and 8.77 KN, respectively.
3. The LR model demonstrates lower performance in comparison to other models, as depicted in the Taylor diagram and violin plot.
4. The sensitivity analysis indicated that shear span to depth ratio (a/d) has the maximum impact on the shear strength of the beam (V_u) with a relative importance of 19%, followed by the tensile strength of the fabric (f_t) with a relative importance of 18%.

Appendix A

$$\begin{aligned}\mathbf{V}_u = &\ \mathbf{0.4102A_1 + 0.9675A_2 + 0.9653A_3 - 1.3185A_4 + 0.6940A_5} \\ &\mathbf{- 1.3314A_6 - 0.9770A_7 - 1.7911A_8 - 0.7719A_9 - 0.3199A_{10}} \\ &\mathbf{- 2.5508A_{11} - 0.6899}\end{aligned} \tag{13.13}$$

Considered input parameters are: (normalized values)

$$b = -0.7563, d = -0.6481, a/d = -0.7806, f_c' = -0.2055, A_f = -0.9614,$$
$$E_f = 0.9496, \text{and } f_t = 0.6523$$

$$\begin{bmatrix} A_1 \\ A_2 \\ A_3 \\ A_4 \\ A_5 \\ A_6 \\ A_7 \\ A_8 \\ A_9 \\ A_{10} \\ A_{11} \end{bmatrix} = \text{tansig} \begin{bmatrix} -0.2753 & 0.2129 & 0.9464 & 1.9038 & -0.6264 & 1.3305 & 0.8098 \\ 0.8242 & 0.9036 & -1.8467 & 1.1736 & 1.6821 & 1.0505 & 2.5948 \\ -0.1137 & 1.1554 & 1.5291 & 1.0068 & 0.1712 & -2.5494 & 1.1124 \\ -0.8980 & -0.6574 & -1.6159 & -1.9987 & -0.2367 & 0.0082 & 2.2261 \\ -1.1534 & 0.8186 & -0.3476 & -1.5483 & -0.3059 & -0.2453 & -1.1995 \\ -1.5632 & -0.6524 & -0.2176 & -0.0481 & 2.8892 & -0.5157 & 1.1087 \\ -0.4389 & 0.5632 & -0.5470 & 0.1912 & 0.1672 & 0.1875 & 1.2690 \\ 1.0831 & 1.2792 & 1.9472 & 1.3548 & -0.5385 & -1.8882 & -0.2972 \\ -0.7182 & 0.9972 & -0.7724 & 0.3092 & 0.8499 & -0.4253 & -0.8361 \\ 0.7538 & -0.2770 & 0.2631 & 1.5666 & 0.3882 & 0.0237 & -1.1927 \\ 0.4790 & -0.6819 & 3.0977 & 0.0594 & -0.8502 & 0.6392 & -0.5549 \end{bmatrix} \times \begin{bmatrix} -0.7563 \\ -0.6481 \\ -0.7806 \\ -0.2055 \\ -0.9614 \\ 0.9496 \\ 0.6523 \end{bmatrix} + \begin{bmatrix} -1.4882 \\ 3.3762 \\ 1.6140 \\ -1.6185 \\ 0.7258 \\ -1.0097 \\ 0.5063 \\ 1.0460 \\ -1.2435 \\ 1.3508 \\ 3.1821 \end{bmatrix}$$

The final values of A_1 to A_{11} neurons are shown in Table 13.7.

$$\begin{aligned} V_{u,normalized} = {} & (0.4102 \times (-0.1529)) + (0.9675 \times (0.9997)) \\ & + (0.9653 \times (-0.9805)) - (1.3185 \times (0.9933)) \\ & + (0.6940 \times (0.7333)) - (1.331 \times (-0.9435)) \\ & - (0.9770 \times (0.9361)) - (1.7911 \times (-0.9991)) \\ & - (0.7719 \times (-0.9884)) - (0.3199 \times (-0.6016)) \\ & - (2.5508 \times (0.9557)) - 0.6899 \end{aligned} \tag{13.14}$$

Table 13.7 Values of A_1 to A_{11}.

A_1	A_2	A_3	A_4	A_5	A_6	A_7	A_8	A_9	A_{10}	A_{11}
−0.1529	0.9997	−0.9805	0.9933	0.7333	−0.9435	0.9361	−0.9991	−0.9884	−0.6016	0.9557

$$V_{u,normalized} = -0.8843$$

$$V_{u,unnormalized} = \left[\frac{-0.8843 + 1}{2}\right] \times 431.6 + 9.8$$

$V_{u,unnormalized} = 34.7680$ KN, which is 1% less than the $V_{u,exp} = 35.12$ KN.

Nomenclatures

a/d	Shear span to depth ratio
A_f	Area of fiber reinforcement
AFRP	Aramid fiber-reinforced polymer
AI	Artificial intelligence
ANFIS	Adaptive Neuro-Fuzzy Inference System
ANN	Artificial neural network
b	Width of beam
BFRP	Basalt fiber-reinforced polymer
BOA	Bayesian optimization algorithm
b_w	Web width
CFRP	Carbon fiber-reinforced polymer
d	Effective depth of cross-section
E_f	Elastic modulus of FRP
f_c'	Concrete compressive strength
FRP	Fiber-reinforced polymer
f_t/f_u	FRP tensile strength
GA	Genetic algorithm
GEP	Gene expression programming
GFRP	Glass fiber-reinforced polymer
GP	Gaussian processes
GPR	Gaussian process regression
GRNN	Generalized regression neural network
LR	Linear regression
MAE	Mean absolute error
MAPE	Mean absolute percentage error
MGP	Multigene genetic programming
ML	Machine learning
NLR	Nonlinear regression analysis
NS	Nash-Sutcliffe
PSO	Particle swarm optimization
R	Correlation coefficient
RC	Reinforced concrete
RMSE	Root mean square error
SD	Standard deviation
SVM	Support vector regression
t_f	FRP thickness
V_{exp}	Experimental shear strength

$V_{u,exp}/V_u$	Shear capacity of beam
w_f	Width of FRP
ρ_{eff}	Effective reinforcement ratio
ρ_l	Reinforcement ratio
ρ	Ratio of FRP reinforcement

References

[1] R. Kamgar, M.H. Bagherinejad, H. Heidarzadeh, A new formulation for prediction of the shear capacity of FRP in strengthened reinforced concrete beams, Soft Computing 24 (2020) 6871–6887, https://doi.org/10.1007/s00500-019-04325-4.

[2] M.S. Alam, U. Gazder, Shear strength prediction of FRP reinforced concrete members using generalized regression neural network, Neural Computing & Applications 32 (2020) 6151–6158, https://doi.org/10.1007/s00521-019-04107-x

[3] H. Naderpour, M. Haji, M. Mirrashid, Shear capacity estimation of FRP-reinforced concrete beams using computational intelligence, in: Structures, vol 28, Elsevier, December 2020, pp. 321–328, https://doi.org/10.1016/j.istruc.2020.08.076.

[4] M.Z. Naser, Machine learning assessment of fiber-reinforced polymer-strengthened and reinforced concrete members, ACI Structural Journal 117 (6) (2020) 237–251, https://doi.org/10.14359/51728073.

[5] S. Lee, C. Lee, Prediction of shear strength of FRP-reinforced concrete flexural members without stirrups using artificial neural networks, Engineering Structures 61 (2014) 99–112, https://doi.org/10.1016/j.engstruct.2014.01.001.

[6] R.A. Hawileh, J.A. Abdalla, M.Z. Naser, Modeling the shear strength of concrete beams reinforced with CFRP bars under unsymmetrical loading, Mechanics of Advanced Materials and Structures 26 (15) (2019) 1290–1297, https://doi.org/10.1080/15376494.2018.1432803.

[7] M. Nikoo, B. Aminnejad, A. Lork, Predicting shear strength in FRP-reinforced concrete beams using Bat algorithm-based artificial neural network, Advances in Materials Science and Engineering 2021 (2021) 1–13, https://doi.org/10.1155/2021/5899356.

[8] F. Colomb, H. Tobbi, E. Ferrier, P. Hamelin, Seismic retrofit of reinforced concrete short columns by CFRP materials, Composite Structures 82 (4) (2008) 475–487, https://doi.org/10.1016/j.compstruct.2007.01.028.

[9] A. Niroomandi, A. Maheri, M.R. Maheri, S.S. Mahini, Seismic performance of ordinary RC frames retrofitted at joints by FRP sheets, Engineering Structures 32 (8) (2010) 2326–2336, https://doi.org/10.1016/j.engstruct.2010.04.008.

[10] O. Ozcan, B. Binici, G. Ozcebe, Improving seismic performance of deficient reinforced concrete columns using carbon fiber-reinforced polymers, Engineering Structures 30 (6) (2008) 1632–1646, https://doi.org/10.1016/j.engstruct.2007.10.013.

[11] G. Promis, E. Ferrier, P. Hamelin, Effect of external FRP retrofitting on reinforced concrete short columns for seismic strengthening, Composite Structures 88 (3) (2009) 367–379, https://doi.org/10.1016/j.compstruct.2008.04.019.

[12] J.D. Berset, Strengthening of Reinforced Concrete Beams for Shear Using FRP Composites, Doctoral dissertation, Massachusetts Institute of Technology, 1992.

[13] K. Uji, Improving shear capacity of existing reinforced concrete members by applying carbon fiber sheets, Transactions of the Japan Concrete Institute 14 (1992).

[14] C.W. Dolan, W. Rider, M.J. Chajes, M. DeAscanis, Prestressed concrete beams using nonmetallic tendons and external shear reinforcement, Special Publication 138 (1993) 475–496.

[15] A.L. Gamino, J.L.A.O. Sousa, O.L. Manzoli, T.N. Bittencourt, R/C structures strengthened with CFRP part II: analysis of shear models, Revista IBRACON de Estruturas e Materiais 3 (2010) 24–49, https://doi.org/10.1590/S1983-41952010000100003.

[16] Y. Yang, G. Liu, Data-driven shear strength prediction of FRP-reinforced concrete beams without stirrups based on machine learning methods, Buildings 13 (2) (2023) 313, https://doi.org/10.3390/buildings13020313.

[17] G.B. Jumaa, A.R. Yousif, Predicting shear capacity of FRP-reinforced concrete beams without stirrups by artificial neural networks, gene expression programming, and regression analysis, Advances in Civil Engineering 2018 (2018) 1–16, https://doi.org/10.1155/2018/5157824.

[18] I.F. Kara, Prediction of shear strength of FRP-reinforced concrete beams without stirrups based on genetic programming, Advances in Engineering Software 42 (6) (2011) 295–304, https://doi.org/10.1016/j.advengsoft.2011.02.002.

[19] M.S. Alam, N. Sultana, S.Z. Hossain, M.S. Islam, Hybrid intelligence modeling for estimating shear strength of FRP reinforced concrete members, Neural Computing & Applications 34 (9) (2022) 7069–7079, https://doi.org/10.1007/s00521-021-06791-0.

[20] G. Sas, B. Täljsten, J. Barros, J. Lima, A. Carolin, Are available models reliable for predicting the FRP contribution to the shear resistance of RC beams? Journal of Composites for Construction 13 (6) (2009) 514–534, https://doi.org/10.1061/(ASCE)CC.1943-5614.0000045.

[21] R. Singh, H.C. Arora, A. Bahrami, A. Kumar, N.R. Kapoor, K. Kumar, H.S. Rai, Enhancing sustainability of corroded RC structures: estimating steel-to-concrete bond strength with ANN and SVM algorithms, Materials 15 (23) (2022) 8295, https://doi.org/10.3390/ma15238295.

[22] H. Naderpour, S.A. Alavi, A proposed model to estimate shear contribution of FRP in strengthened RC beams in terms of Adaptive Neuro-Fuzzy Inference System, Composite Structures 170 (2017) 215–227, https://doi.org/10.1016/j.compstruct.2017.03.028.

[23] R. Perera, J. Vique, A. Arteaga, A. De Diego, Shear capacity of reinforced concrete members strengthened in shear with FRP by using strut-and-tie models and genetic algorithms, Composites Part B: Engineering 40 (8) (2009) 714–726, https://doi.org/10.1016/j.compositesb.2009.06.008.

[24] H.M. Tanarslan, M. Secer, A. Kumanlioglu, An approach for estimating the capacity of RC beams strengthened in shear with FRP reinforcements using artificial neural networks, Construction and Building Materials 30 (2012) 556–568, https://doi.org/10.1016/j.conbuildmat.2011.12.008.

[25] C. Ji, W. Li, C. Hu, F. Xing, Data analysis on fiber-reinforced polymer shear contribution of reinforced concrete beam shear strengthened with U-jacketing fiber-reinforced polymer composites, Journal of Reinforced Plastics and Composites 36 (2) (2017) 98–120, https://doi.org/10.1177/0731684416671423.

[26] G. Tripepi, K.J. Jager, F.W. Dekker, C. Zoccali, Linear and logistic regression analysis, Kidney International 73 (7) (2008) 806–810.

[27] M. Ahmadi, A. Kheyroddin, A. Dalvand, M. Kioumarsi, New empirical approach for determining nominal shear capacity of steel fiber reinforced concrete beams, Construction and Building Materials 234 (2020) 117293, https://doi.org/10.1016/j.conbuildmat.2019.117293.

[28] R.R. Richardson, M.A. Osborne, D.A. Howey, Gaussian process regression for forecasting battery state of health, Journal of Power Sources 357 (2017) 209−219, https://doi.org/10.1016/j.jpowsour.2017.05.004.

[29] L. Cheng, S. Ramchandran, T. Vatanen, N. Lietzén, R. Lahesmaa, A. Vehtari, H. Lähdesmäki, An additive Gaussian process regression model for interpretable non-parametric analysis of longitudinal data, Nature Communications 10 (1) (2019) 1798, https://doi.org/10.1038/s41467-019-09785-8.

[30] A. Burkov, The Hundred-Page Machine Learning Book, vol 1, Andriy Burkov, Quebec City, QC, Canada, 2019, p. 32.

[31] H. Ceylan, M.B. Bayrak, K. Gopalakrishnan, Neural networks applications in pavement engineering: a recent survey, International Journal of Pavement Research & Technology 7 (6) (2014).

[32] M. Ling, X. Luo, S. Hu, F. Gu, R.L. Lytton, Numerical modeling and artificial neural network for predicting J-integral of top-down cracking in asphalt pavement, Transportation Research Record 2631 (1) (2017) 83−95, https://doi.org/10.3141/2631-10.

[33] J.R. Yost, S.P. Gross, D.W. Dinehart, Shear strength of normal strength concrete beams reinforced with deformed GFRP bars, Journal of Composites for Construction 5 (4) (2001) 268−275, https://doi.org/10.1061/(ASCE)1090-0268(2001)5:4(268).

[34] A.F. Ashour, Flexural and shear capacities of concrete beams reinforced with GFRP bars, Construction and Building Materials 20 (10) (2006) 1005−1015, https://doi.org/10.1016/j.conbuildmat.2005.06.023.

[35] A.G. Razaqpur, B.O. Isgor, S. Greenaway, A. Selley, Concrete contribution to the shear resistance of fiber reinforced polymer reinforced concrete members, Journal of Composites for Construction 8 (5) (2004) 452−460, https://doi.org/10.1061/(ASCE)1090-0268(2004)8:5(452).

[36] M. Tariq, J.P. Newhook, Shear testing of FRP reinforced concrete without transverse reinforcement, in: Proceedings, Annual Conference of the Canadian Society for Civil Engineering, June 2003, pp. 1330−1339.

[37] F. Abed, H. El-Chabib, M. AlHamaydeh, Shear characteristics of GFRP-reinforced concrete deep beams without web reinforcement, Journal of Reinforced Plastics and Composites 31 (16) (2012) 1063−1073.

[38] C.H. Kim, H.S. Jang, Concrete shear strength of normal and lightweight concrete beams reinforced with FRP bars, Journal of Composites for Construction 18 (2) (2014) 04013038.

[39] A.G. Razaqpur, S. Spadea, Shear strength of FRP reinforced concrete members with stirrups, Journal of Composites for Construction 19 (1) (2015) 04014025, https://doi.org/10.1061/(ASCE)CC.1943-5614.0000483.

[40] P. Kumar, H.C. Arora, A. Bahrami, A. Kumar, K. Kumar, Development of a reliable machine learning model to predict compressive strength of FRP-confined concrete cylinders, Buildings 13 (4) (2023) 931.

Estimating the load carrying capacity of reinforced concrete beam-column joints via soft computing techniques

14

Danial Rezazadeh Eidgahee [1], Atefeh Soleymani [2], Hashem Jahangir [3], Mohaddeseh Nikpay [3], Harish Chandra Arora [4] and Aman Kumar [4,5]
[1]Department of Civil Engineering, Ferdowsi University of Mashhad, Mashhad, Iran; [2]Structural Engineering, Shahid Bahonar University of Kerman, Kerman, Iran; [3]Department of Civil Engineering, University of Birjand, Birjand, Iran; [4]Department of Structural Engineering, CSIR—Central Building Research Institute, Roorkee, Uttarakhand, India; [5]Academy of Scientific and Innovative Research (AcSIR), Ghaziabad, Uttar Pradesh, India

14.1 Introduction

In most reinforced concrete (RC) structures, the beams are directly connected to columns and based on the strong column-weak beam principle, most of the damages in RC structures would occur in beam-column joints. Consequently, determining the RC beam-column joints' load capacity is an essential issue among researchers and engineers. Some researchers conducted experimental tests to obtain the load capacity and shear strength of the RC beam-column joints [1−8]. In some international codes and other studies, empirical-based formulas were suggested to approximate the shear capacity of RC beam-column joints. Investigating the performance of empirical-based formulas for quantifying the RC beam-column joints' shear capacity demonstrated that the proposed equations were not successful in approximating the exact shear capacity values [10,11]. Therefore, to find more reliable equations for determining the shear capacities of the RC beam-column joints, the researchers seek more powerful and accurate techniques. Among the novel powerful analytical techniques, soft computing techniques and machine learning (ML) methods can reveal the most accurate results [12−14]. Some of the ML techniques such as the ANN method can suggest high-performance low-level estimation but with more complicated analytical equations [14]. On the other hand, methods such as the group method of data handling (GMDH) and genetic expression programming (GEP) propose closed-form equations with probably lower performance and greater error levels compared to the ANN approach.

Artificial Intelligence Applications for Sustainable Construction. https://doi.org/10.1016/B978-0-443-13191-2.00014-6
Copyright © 2024 Elsevier Ltd. All rights are reserved, including those for text and data mining, AI training, and similar technologies.

14.2 Research significance

In this chapter, multi-linear regression equipped with a genetic algorithm (MLR-GA) and ANN techniques were utilized to quantify the shear capacity of RC beam-column joints. An experimental database including 149 RC beam-column specimens gathered from previous research works. Moreover, a suggested empirical-based formula was selected to be compared with the proposed MLR-GA and ANN models in this study. Some known performance criteria and error values were utilized to conduct the performance evaluation. Eventually, to determine the factors that have the greatest impact on the RC beam-column joints' shear capacity, a sensitivity analysis was carried out.

14.3 Experimental database

In this chapter, 149 experimental RC beam-column joints were compiled from previous literature reviews [1–32], which were also reported in Kotsovou et al. [33] research work. In the considered database, the influential input parameters, which were introduced in Table 14.1, affected the shear capacity of the RC beam-column joint as an output parameter. The statistical properties of the inputs and output parameters in the compiled database, including minimum, maximum, range, average, standard deviation, coefficient of variation, kurtosis, and skewness, are presented in Table 14.2. The distribution of each input parameter with respect to the output

Table 14.1 Introduction of input parameters.

Input parameter	Unit	Definition
f_c	MPa	Cylindrical compressive strength of concrete
h_c	mm	Height of column's cross section
b_c	mm	Width of column's cross section
ρ_{col}	%	Longitudinal reinforcement ratio of column
M_{Rc}	kN.m	Flexural capacity of column
h_b	mm	Height of beam's cross section
b_b	mm	Width of beam's cross section
ρ_t	%	Tensile longitudinal reinforcement ratio of beam
ρ_c	%	Compressive longitudinal reinforcement ratio of beam
M_{Rb}	kN.m	Flexural capacity of beam
$\rho_s f_{ys}$	MPa	Yield stress × reinforcement ratio of the joint stirrups
$\rho_d f_{yd}$	MPa	Yield stress × reinforcement ratio of the horizontal component of the diagonal reinforcement of the joints
ρ_{sv}	%	Ratio of vertical reinforcement
H_c	mm	Overall height of column
L_b	mm	Distance of applied load from the beam-joint interface

Table 14.2 Statistical properties of the database.

Property	Inputs															Output
	f_c (Mpa)	h_c (mm)	b_c (mm)	ρ_{col} (%)	M_{Rc} (kNm)	h_b (mm)	b_b (mm)	ρ_t (%)	ρ_c (%)	M_{Rb} (kNm)	$\rho_{sf_{ys}}$ (Mpa)	$\rho_{df_{yd}}$ (Mpa)	ρ_{sv} (Mpa)	H_c (mm)	L_b (mm)	Experimental
Minimum	18.10	200.00	100.00	0.58	13.00	200.00	100.00	0.29	0.22	17.00	0.00	0.00	0.00	800.00	600.00	65.00
Maximum	93.80	500.00	500.00	5.99	599.00	610.00	406.00	2.87	2.87	488.00	6.53	3.30	1.20	3581.00	3048.00	1316.40
Range	75.70	300.00	400.00	5.41	586.00	410.00	306.00	2.58	2.65	471.00	6.53	3.30	1.20	2781.00	2448.00	1251.40
Average	37.79	287.84	260.28	2.65	147.84	350.00	225.87	1.22	1.17	151.49	1.92	0.30	0.35	2111.17	1201.65	452.79
SD	15.45	85.59	99.83	1.24	150.97	112.92	77.53	0.56	0.60	131.76	1.85	0.86	0.34	752.78	494.53	335.96
CoV (%)	40.90	29.74	38.36	46.65	102.11	32.26	34.33	45.66	50.99	86.97	96.05	286.46	97.17	35.66	41.15	74.20
Kurtosis	1.08	−0.77	−0.62	0.01	0.84	−1.11	−0.77	0.81	0.45	−0.78	−0.40	6.50	−1.04	−1.14	3.32	−0.05
Skewness	1.28	0.63	0.35	0.67	1.32	0.17	0.17	0.93	0.69	0.77	0.81	2.79	0.42	0.46	1.77	0.95

was depicted in Fig. 14.1. Moreover, the correlation matrix of the input parameters and the output is depicted in Fig. 14.2.

As it can be seen in Figs. 14.1 and 14.2, the considered ranges of each input covered the most possible data between the minimum and maximum values, as well as a good correlation exists among the inputs and outputs. With such a vast distribution, the estimation of the shear capacity of the RC beam-column joints via machine learning techniques can lead to acceptable results.

14.4 Machine learning methods

In this chapter, two ML techniques, including multilinear regression optimized by a genetic algorithm (MLR-GA) and artificial neural networks (ANNs), were utilized to estimate the shear capacity of the RC beam-column joints.

14.4.1 MLR-GA model

To estimate the shear capacity of the RC beam-column joints via the MLR-GA method, all the 149 experimental data was divided into training (75%), testing (25%), and, as presented in Eq. (14.1), the linear relationship between the output and the multiple input parameters was optimized via genetic algorithm (GA):

$$V_{c\text{-MLR-GA}} = \alpha + \beta \cdot f_c + \chi \cdot h_c + \delta \cdot b_c + \gamma \cdot \rho_{col} + \eta \cdot M_{Rc} + \kappa \cdot h_b + \lambda \cdot b_b + \mu \cdot \rho_t + \nu \cdot \rho_c + \tau \cdot M_{Rb} + \upsilon \cdot \rho_s f_{ys} + \omega \cdot \rho_d f_{yd} + \psi \cdot \rho_{sv} + \zeta \cdot H_c + \varphi \cdot Lb \tag{14.1}$$

By conducting a trial-and-error process, the unknown coefficients in Eq. (14.1) were determined by the GA method, and the optimal MLR-GA method for estimating the shear capacity of the RC beam-column joints is reported in Eq. (14.2):

$$V_{c\text{-MLR-GA}} = 0.0109 - 0.0687 f_c + 0.0842 h_c - 0.3308 b_c - 0.0350 \rho_{col} + 0.2137 M_{Rc} - 0.2351 h_b + 0.4320 b_b + 0.2538 \rho_t + 0.0203 \rho_c + 0.7872 M_{Rb} - 0.0012 \rho_s f_{ys} - 0.0131 \rho_d f_{yd} - 0.0182 \rho_{sv} + 0.0873 H_c - 0.2284 Lb \tag{14.2}$$

To evaluate the behavior of the predicted shear capacity of RC beam-column joints with respect to individual inputs, a parametric analysis was conducted in which each individual input parameter ranged between its minimum and maximum values, while the other input parameters were fixed on their average values. Fig. 14.3 shows the results of parametric analysis for the MLR-GA technique. As it can be inferred from Fig. 14.3, the descending and ascending orders of shear capacity of the RC beam-column joints with respect to each individual input are compatible with their rational expectation manner.

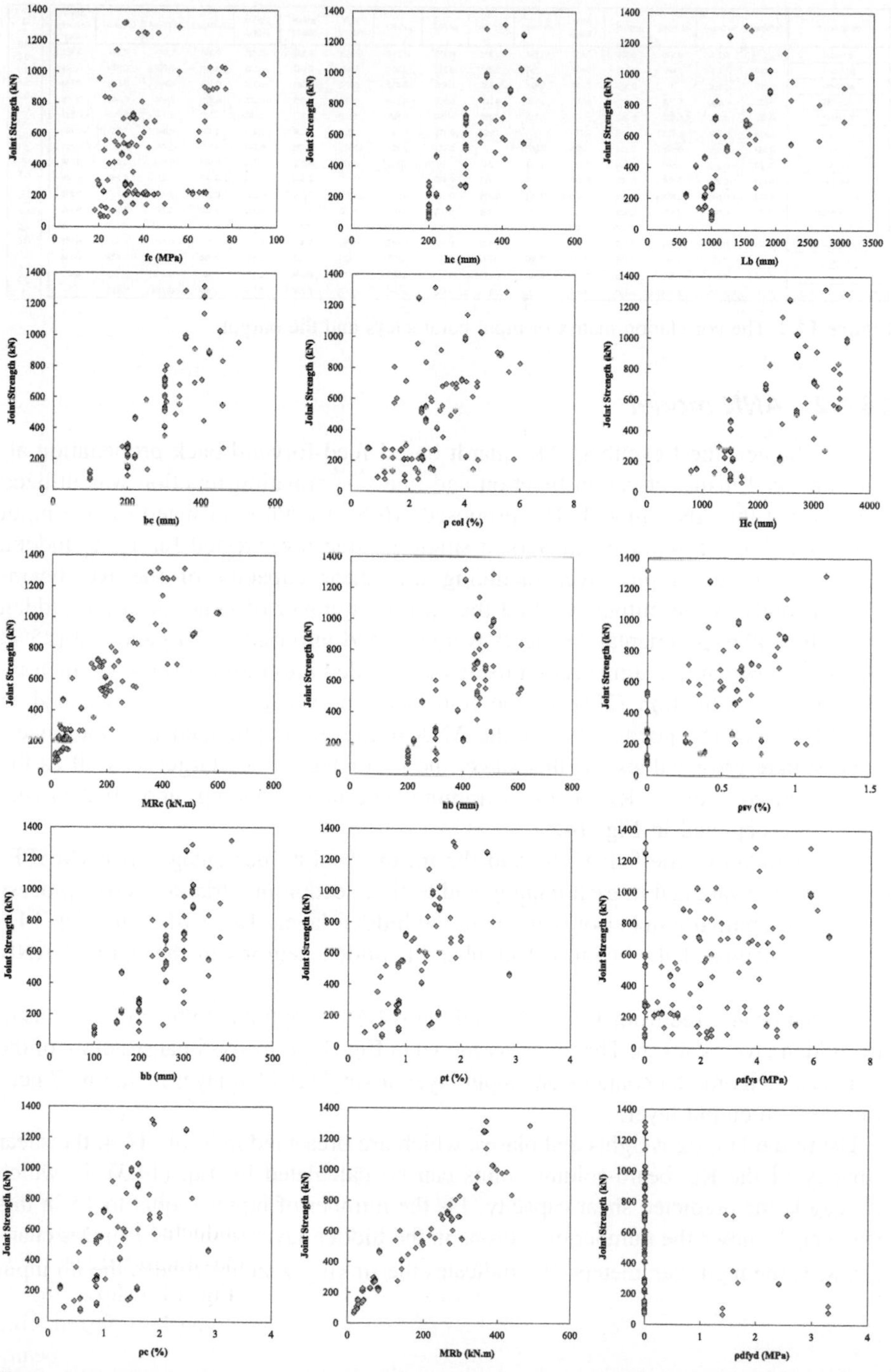

Figure 14.1 The distribution of each input data with respect to output (shear strength of the joint).

Parameter	fc (Mpa)	hc (mm)	bc (mm)	pcol (%)	MRc (kNm)	hb (mm)	bb (mm)	pt (%)	pc (%)	Mfb (kNm)	psFyh (Mpa)	pdFyd (Mpa)	psv (Mpa)	Hc (mm)	Lb (mm)	Shear Capacity (kN)
fc (Mpa)	1	0.260	0.390	0.506	0.455	0.354	0.327	0.259	0.164	0.421	0.155	-0.235	0.381	0.425	0.167	0.382
hc (mm)	0.260	1	0.896	0.330	0.877	0.754	0.836	0.332	0.365	0.845	0.082	-0.135	0.424	0.707	0.740	0.825
bc (mm)	0.390	0.896	1	0.508	0.881	0.825	0.926	0.409	0.409	0.878	0.143	-0.314	0.554	0.745	0.678	0.849
pcol (%)	0.506	0.330	0.508	1	0.593	0.516	0.509	0.328	0.286	0.614	0.425	-0.115	0.810	0.498	0.340	0.569
MRc (kNm)	0.455	0.877	0.881	0.593	1	0.715	0.828	0.393	0.384	0.899	0.157	-0.181	0.580	0.714	0.747	0.885
hb (mm)	0.354	0.754	0.825	0.516	0.715	1	0.887	0.113	0.065	0.851	0.234	-0.238	0.471	0.866	0.628	0.732
bb (mm)	0.327	0.836	0.926	0.509	0.828	0.887	1	0.231	0.213	0.879	0.198	-0.274	0.519	0.776	0.661	0.841
pt (%)	0.259	0.332	0.409	0.328	0.393	0.113	0.231	1	0.966	0.420	0.011	-0.268	0.444	0.173	0.254	0.558
pc (%)	0.164	0.365	0.409	0.286	0.384	0.065	0.213	0.966	1	0.407	0.042	-0.224	0.432	0.127	0.267	0.546
Mfb (kNm)	0.421	0.845	0.878	0.614	0.899	0.851	0.879	0.420	0.407	1	0.299	-0.210	0.602	0.837	0.700	0.960
psFyh (Mpa)	0.155	0.082	0.143	0.425	0.157	0.234	0.198	0.011	0.042	0.299	1	-0.102	0.460	0.257	0.244	0.220
pdFyd (Mpa)	-0.235	-0.135	-0.314	-0.115	-0.181	-0.238	-0.274	-0.268	-0.224	-0.210	-0.102	1	-0.233	-0.167	-0.138	-0.220
psv (Mpa)	0.381	0.424	0.554	0.810	0.580	0.471	0.519	0.444	0.432	0.602	0.460	-0.233	1	0.456	0.501	0.569
Hc (mm)	0.425	0.707	0.745	0.498	0.714	0.866	0.776	0.173	0.127	0.837	0.257	-0.167	0.456	1	0.633	0.748
Lb (mm)	0.167	0.740	0.678	0.340	0.747	0.628	0.661	0.254	0.267	0.700	0.244	-0.138	0.501	0.633	1	0.616
Shear Capacity (kN)	0.382	0.825	0.849	0.569	0.885	0.732	0.841	0.558	0.546	0.960	0.220	-0.220	0.569	0.748	0.616	1

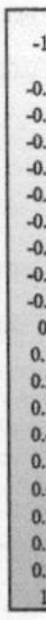

Figure 14.2 The correlation matrix of input parameters and the output.

14.4.2 ANN model

In this chapter, the Levenberg Marquardt based feed-forward back propagation algorithm via Tansig activation function and linear summation function was utilized to construct the ANN model. The proposed ANN structure consisted of one input layer including 15 input parameters, a single hidden layer tested for 1−20 hidden neurons, and an output layer including the shear capacity of the RC beam-column joints as the output. To find the optimal number of neurons in the hidden layer, the 149 experimental databases were divided into train (75%) and test (25%) sections, in which, the train section itself was divided into training (80% of the train section) and validation (20% of the train section) subsections. As presented in Fig. 14.4, after 18 epochs (cycles), the ANN model was optimized with the lowest mean square error values. At this stage, the predicted versus target as well as the correlation coefficient (R) values in training, validation, testing, and all data sections were depicted in Fig. 14.5.

The correlation coefficient (R) and the mean absolute percentage error (MAPE) criteria were evaluated in each training and testing section in a trial-and-error process to find the optimal number of neurons in the hidden layer. Table 14.3 and Fig. 14.6 show the results of the suggested number of hidden neurons in the optimal ANN structure.

As can be inferred from Table 14.3 and Fig. 14.6, the optimal number of neurons in the hidden layer is seven. Therefore, as shown in Fig. 14.7, the optimal structure of the proposed ANN model contains one input layer, a single hidden layer including 7 neurons, and an output layer.

Using the linking weights and biases, which are presented in Table 14.4, the shear capacity of the RC beam-column joints can be calculated by Eq. (14.3), in which V_{c_ANN} is the predicted shear capacity, I is the number of inputs (equal to 15 in this chapter), K shows the number of neurons in the hidden layer (equal to 7 in this chapter), X_i is the input parameters, W^1_{ki} indicates the linking weight between the ith input

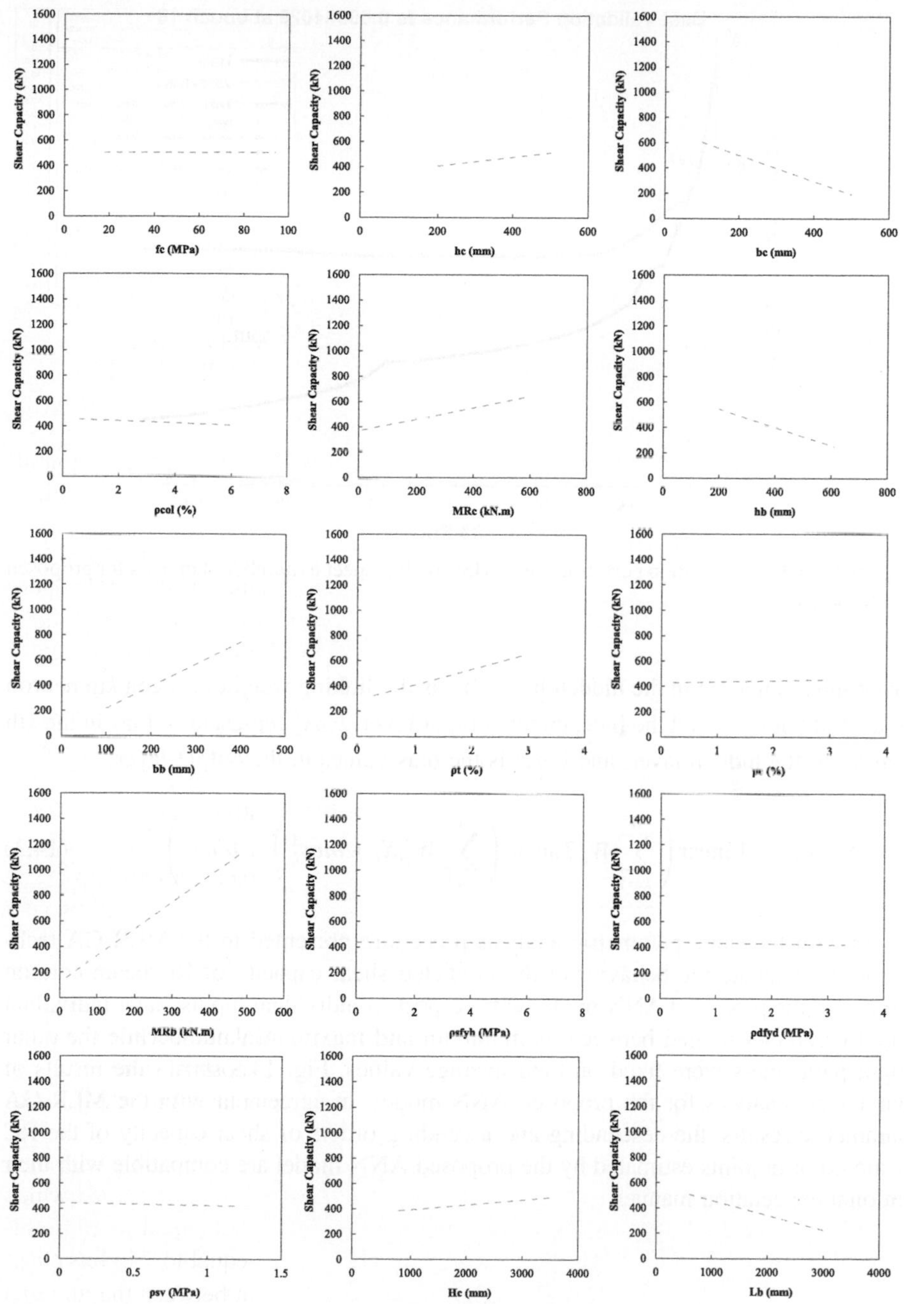

Figure 14.3 The parametric analysis in MLR-GA technique.

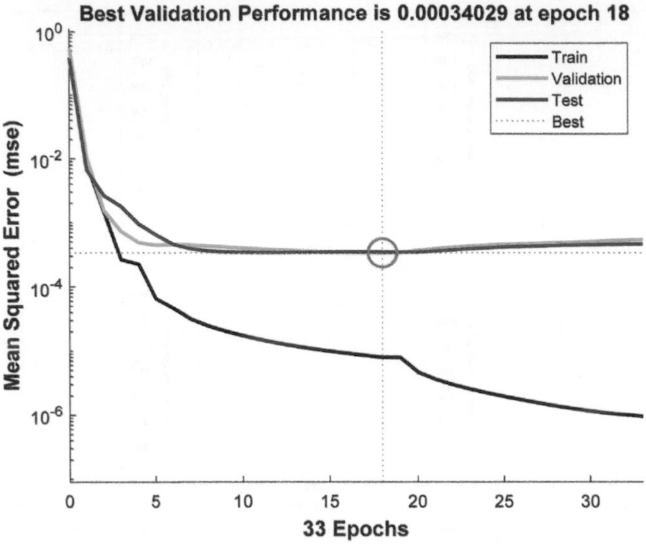

Figure 14.4 Best validation performance of MSE with respect to number of epochs for proposed ANN model.

layer and kth neuron in the hidden layer, W_k^2 is the linking weight between kth neuron in the hidden layer and the independent output layer, $bias_k^1$ represents a bias in the kth neuron of the hidden layer, and $bias^2$ is the bias values in the output layer:

$$V_{c\text{-ANN}} = \text{Linear}\left(\sum_{k=1}^{K} W_k^2 \text{Tansig}\left(\sum_{i=1}^{I} W_{ki}^1 X_i + bias_k^1\right) + bias^2\right) \tag{14.3}$$

As with the same parametric analysis procedure presented in the MLR-GA technique, to evaluate the behavior of the predicted shear capacity of RC beam-column joints via the proposed ANN model with respect to individual inputs, each individual input parameter ranged between its minimum and maximum values, while the other input parameters were fixed on their average values. Fig. 14.8 shows the results of parametric analysis for the proposed ANN model. In agreement with the MLR-GA parametric results, the descending and ascending orders of shear capacity of the RC beam-column joints estimated by the proposed ANN model are compatible with their rational expectation manner.

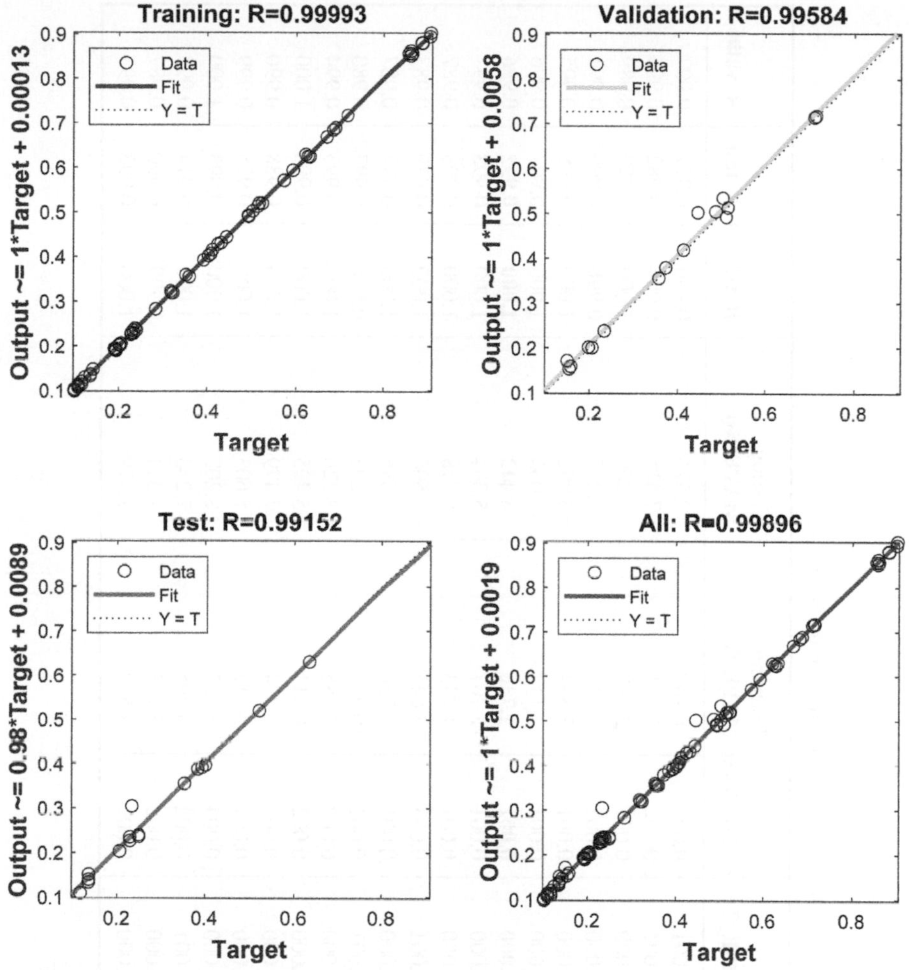

Figure 14.5 The predicted versus target values in the proposed ANN model.

14.5 Performance evaluation of the proposed and existing models

14.5.1 Existing model

One of the most regular existing models for estimating the shear capacity of RC beam-column joints presented in Kotsovou and Mouzakis [33] research work, named $V_{c\_K\&M}$, is as below:

$$V_{c\text{-}K\&M} = \frac{z_c}{3} w_j f_c \sin a \tag{14.4}$$

Table 14.3 The trial and error process to find the optimal number of neurons in the hidden layer.

Neuron number	Training R	Testing R	MSE_Tr	MSE_Ts	Training MAPE(%)	Testing MAPE(%)	R_train	R_test	R_valid
1	0.993	0.985	0.001	0.001	6.469	7.576	0.996	0.977	0.997
2	0.993	0.979	0.001	0.002	5.111	7.415	0.996	0.982	0.996
3	0.998	0.988	0.000	0.001	3.098	6.200	1.000	0.999	0.989
4	0.996	0.975	0.000	0.002	5.446	9.267	0.998	0.990	0.991
5	0.998	0.985	0.000	0.001	3.334	4.890	1.000	0.998	0.995
6	0.999	0.987	0.000	0.001	2.114	6.063	1.000	0.997	0.996
7	**0.999**	**0.993**	**0.000**	**0.001**	**1.736**	**4.442**	**1.000**	**0.992**	**0.996**
8	0.999	0.990	0.000	0.001	2.707	5.304	1.000	0.998	0.997
9	0.998	0.985	0.000	0.001	1.743	5.718	1.000	0.998	0.987
10	0.995	0.962	0.001	0.003	3.645	6.508	1.000	0.979	0.983
11	0.998	0.980	0.000	0.001	1.756	5.586	1.000	0.997	0.987
12	0.994	0.975	0.001	0.002	10.252	10.293	0.996	0.997	0.980
13	0.998	0.978	0.000	0.002	3.888	7.526	1.000	0.993	0.994
14	0.998	0.977	0.000	0.002	0.911	5.405	1.000	0.990	1.000
15	0.996	0.967	0.000	0.003	2.324	7.479	1.000	0.988	0.990
16	0.997	0.987	0.000	0.001	3.212	5.603	1.000	0.975	0.999
17	0.999	0.952	0.000	0.004	1.122	8.203	1.000	0.993	1.000
18	0.990	0.967	0.001	0.002	3.295	7.263	1.000	0.914	0.993
19	0.996	0.979	0.000	0.002	5.103	7.314	0.999	0.990	0.987
20	0.998	0.969	0.000	0.002	4.558	7.479	1.000	0.993	0.997

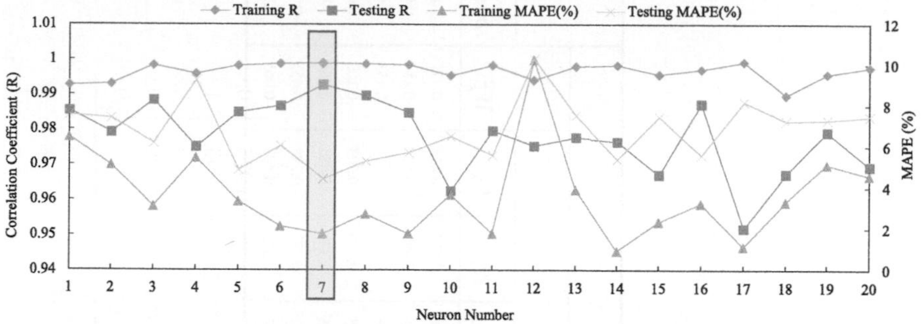

Figure 14.6 The optimal number of neurons in the hidden layer.

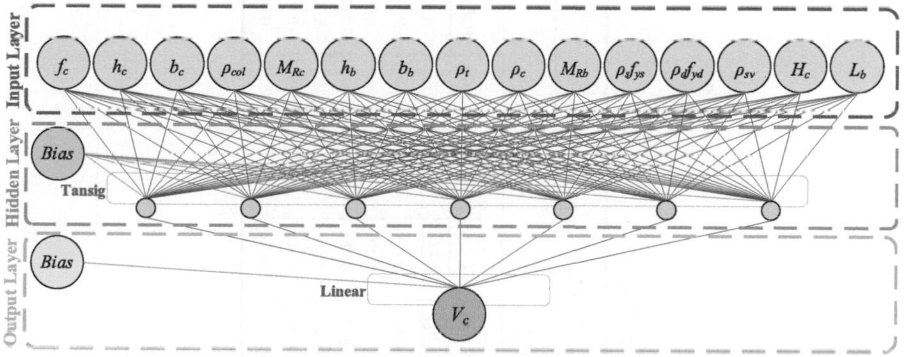

Figure 14.7 The optimal structure of the proposed ANN model.

In Eq. (14.4), z_c is the difference between h_c (height of the column's cross-section) and x_c (the depth of the column's compressive section), w_j is the width of the RC beam-column joint, and a is the inclination angle of the RC beam-column joint diagonal.

14.5.2 Performance evaluation

As presented in Table 14.5, considering the V_{c_i} and $\widetilde{V}_{c_i}$ as the experimental shear capacity of the RC beam-column joints and the corresponding estimated values, respectively, n as the number of input data, $\overline{V}_c$ as the average of the experimental V_{c_i} values and $\overline{\widetilde{V}}_c$ as the mean value of the corresponding estimated values, the most known performance and error evaluation parameters [34] were utilized in this study to compare the results of the proposed models (ANN and MLR-GA) and the existing model (Kotsovou and Mouzakis [33]). Table 14.6 shows the obtained results of performance and error evaluation parameters for the training, testing, and all data sections in the proposed models and the existing model.

Table 14.4 The linking weights and biases in the proposed optimal ANN model.

Neuron number	Weight																Bias	
	W_{ki}^1																	
	f_c	h_c	b_c	ρ_{col}	M_{Rc}	h_b	b_b	ρ_t	ρ_c	M_{Rb}	$\rho_s f_{ys}$	$\rho_d f_{yd}$	ρ_{sv}	H_c	L_b	W_k^2	$bias_k^1$	$bias^2$
1	−0.106	0.302	−0.169	0.108	0.317	−0.001	0.060	−0.258	0.234	1.261	0.053	−0.104	−0.181	−0.173	0.425	0.771	1.639	−0.207
2	0.328	−0.284	0.615	0.383	−0.175	0.818	−0.383	0.260	−0.264	−0.769	−0.284	−0.373	0.261	−0.093	0.325	−0.154	−0.993	
3	−0.529	0.318	−0.639	−0.281	−0.260	0.719	0.474	0.032	−0.090	−0.158	−0.244	0.624	0.606	0.563	0.016	0.153	0.626	
4	−0.665	−0.147	0.136	−0.287	0.221	0.446	−0.179	0.093	0.170	0.476	−0.581	−0.420	−0.468	0.730	0.503	−0.100	0.039	
5	−0.362	0.285	0.344	−0.025	0.978	−1.257	0.218	0.061	0.274	0.855	0.180	0.139	−0.183	0.421	−0.085	0.563	−0.614	
6	0.465	−0.908	−0.093	−0.259	−0.450	−0.207	0.387	0.852	−0.342	0.831	−0.325	−0.168	0.121	0.073	−0.298	0.453	−0.709	
7	0.782	0.552	0.313	−0.455	0.159	−0.375	0.197	−0.267	−0.057	−0.180	−0.276	0.642	−0.567	−0.315	0.408	−0.065	1.506	

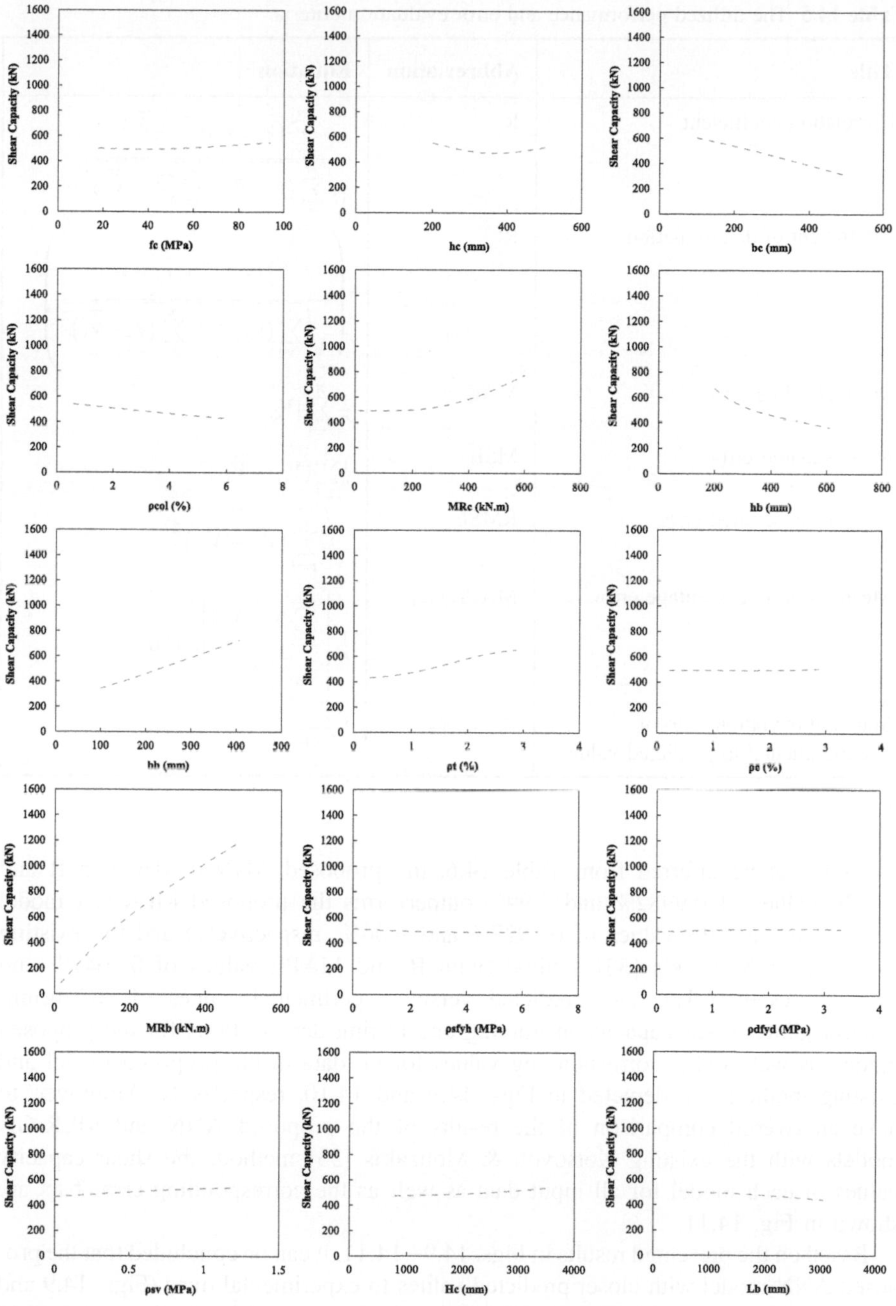

Figure 14.8 The parametric analysis of the proposed ANN model.

Table 14.5 The utilized performance and error evaluation criteria.

Title	Abbreviation	Equation
Correlation coefficient	R	$\frac{\sum_{i=1}^{n}\left(V_{c_i}-\overline{V}_c\right)\left(\widetilde{V}_{c_i}-\overline{\widetilde{V}}_c\right)}{\sqrt{\sum_{i=1}^{n}\left(V_{c_i}-\overline{V}_c\right)^2\sum_{i=1}^{n}\left(\widetilde{V}_{c_i}-\overline{\widetilde{V}}_c\right)^2}}$
Coefficient of determination	R^2	$\left(\frac{\sum_{i=1}^{n}\left(V_{c_i}-\overline{V}_c\right)\left(\widetilde{V}_{c_i}-\overline{\widetilde{V}}_c\right)}{\sqrt{\sum_{i=1}^{n}\left(V_{c_i}-\overline{V}_c\right)^2\sum_{i=1}^{n}\left(\widetilde{V}_{c_i}-\overline{\widetilde{V}}_c\right)^2}}\right)^2$
Mean absolute error	MAE	$\frac{1}{n}\sum_{i=1}^{n}\left\|V_{c_i}-\widetilde{V}_{c_i}\right\|$
Mean squared error	MSE	$\frac{1}{n}\sum_{i=1}^{n}\left(V_{c_i}-\widetilde{V}_{c_i}\right)^2$
Root mean squared error	RMSE	$\sqrt{\frac{1}{n}\sum_{i=1}^{n}\left(V_{c_i}-\widetilde{V}_{c_i}\right)^2}$
Mean absolute percentage error	MAPE (%)	$\frac{1}{n}\left[\frac{\sum_{i=1}^{n}\left\|V_{c_i}-\widetilde{V}_{c_i}\right\|}{\sum_{i=1}^{n}\left\|V_{c_i}\right\|}\right]\times 100$
Standard deviation ratio of experimental to predicted values	$\sigma_{Exp}/\sigma_{Pre}$	$\frac{Exp_{STD}}{Pre_{STD}}$

As it can be inferred from Table 14.6, the proposed ANN model with R and MAPE values of 0.9982% and 2.99% outperforms the proposed MLR-GA model (with R and MAPE values of 0.9927% and 9.06%, respectively) and the existing Kotsovou & Mouzakis [33] method (with R and MAPE values of 0.6647% and 47.61%, respectively). The predicted versus experimental values of RC beam-column joints' shear capacity in training and testing data sections for the proposed model, as well as the corresponding values for all data in the proposed model and existing method, are depicted in Figs. 14.9 and 14.10, respectively. Moreover, to have an overall comparison of the results of the proposed ANN and MLR-GA models with the existing Kotsovou & Mouzakis [33] method, the shear capacity values of each model for all input data as well as the corresponding error bars are shown in Fig. 14.11.

Based on the presented results in Figs. 14.9–14.11, it can be concluded that the proposed ANN model with closer predicted values to experimental ones (Figs. 14.9 and 14.10) and lower error bars (Fig. 14.11) can be announced as the best method in

Table 14.6 The obtained performance and error evaluation parameters.

		Performance and error evaluation parameters						
Method	**Data section**	**R**	R^2	**MAE (kN)**	**MSE**	**RMSE**	**MAPE (%)**	$\sigma_{Exp}/\sigma_{Pre}$
ANN	Training	0.9994	0.9987	6.41	158.93	12.61	2.34	1.00
	Testing	0.9935	0.9870	20.44	1114.29	33.38	4.89	1.02
	All data	**0.9982**	**0.9964**	**9.99**	**402.58**	**20.06**	**2.99**	**1.00**
MLR-GA	Training	0.9935	0.9870	27.38	1592.84	39.91	9.22	1.00
	Testing	0.9897	0.9796	27.56	1767.92	42.05	8.58	1.03
	All data	0.9927	0.9854	27.42	1637.49	40.47	9.06	1.01
Kotsovou & Mouzakis [33]	All data	0.6647	0.4419	215.31	129,008.31	359.18	47.61	0.70

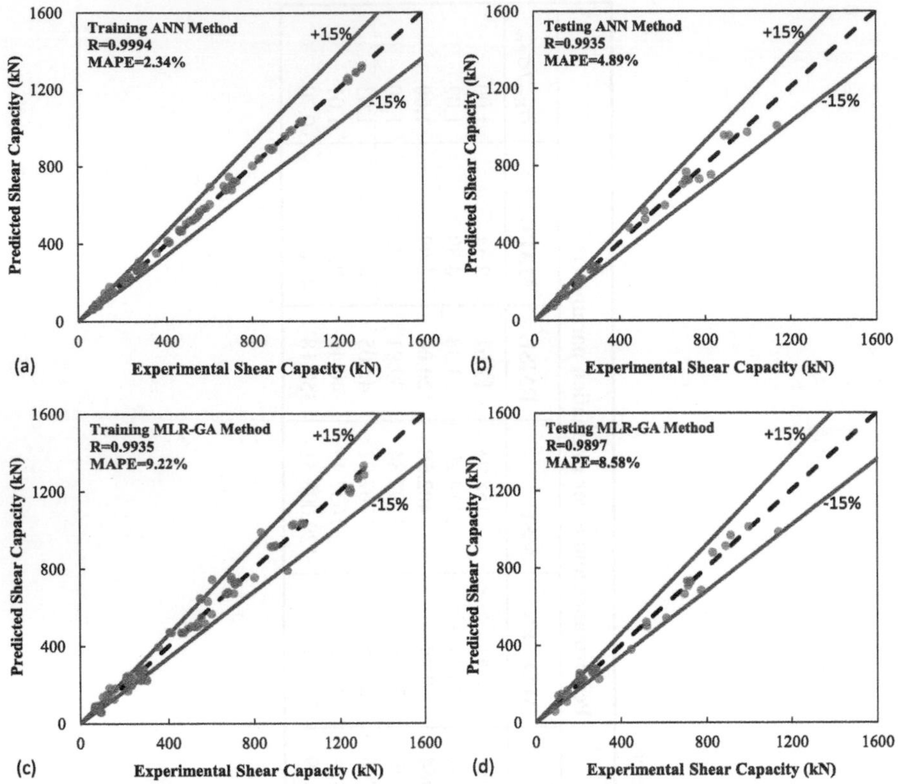

Figure 14.9 The predicted versus experimental shear capacity values for training and testing data sections in the proposed models are: (a and b) ANN; (c and d) MLR-GA.

comparison to the proposed MLR-GA and the existing Kotsovou & Mouzakis [33] method. Moreover, the proposed MLR-GA model outperforms the existing Kotsovou & Mouzakis [33] method with lower errors and higher R values.

14.6 Sensitivity analysis

To find out the influence percentage of each input parameter on the output (shear capacity of RC beam-column joints), the sensitivity analysis was done by utilizing the Milne method [35] and by considering the proposed ANN model as the best method in this study. In the Milne sensitivity analysis method, using the linking

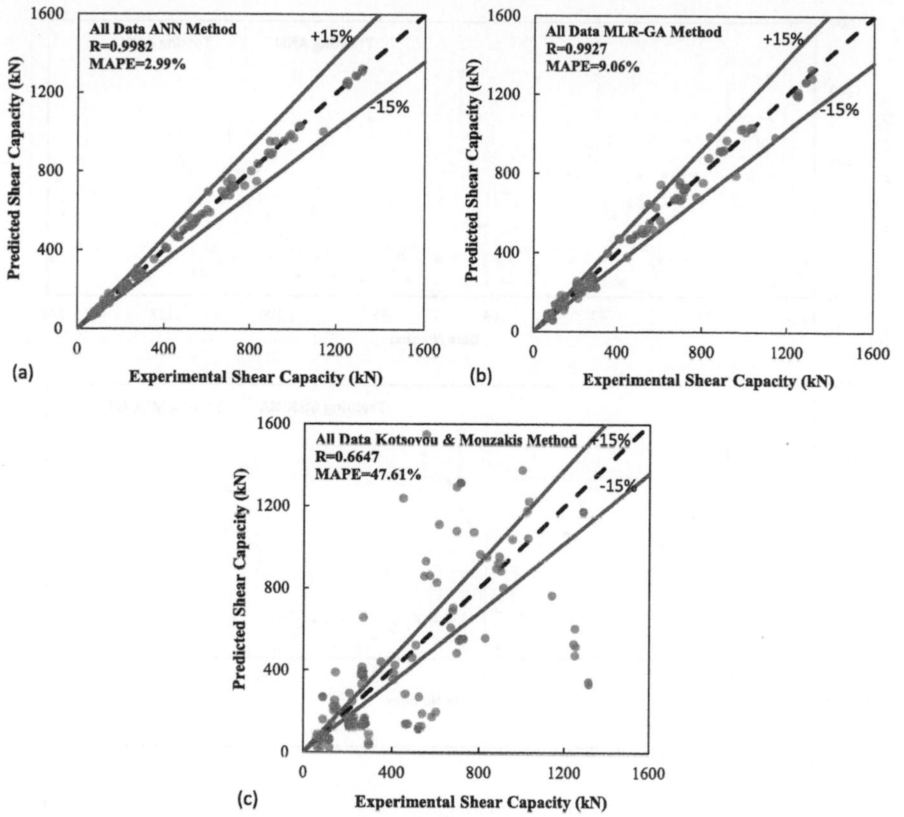

Figure 14.10 The predicted versus experimental shear capacity values for all data sections in the proposed and existing models: (a) ANN; (b) MLR-GA; and (c) Kotsovou & Mouzakis.

weights of the ANN model, the relative importance of each input is presented in percentages such that the summation of the importance of all input data will be 100%. More details regarding the Milne sensitivity analysis method can be found in Ref. [34]. Fig. 14.12 showcases the results of the sensitivity analysis in this study.

The presented results in Fig. 14.12 show that the most influential input parameters on the shear capacity of RC beam-column joints are M_{Rb} (flexural capacity of beam) with a 13.07% relative importance value, and the least effective input parameters are ρ_c (compressive longitudinal reinforcement ratio of beam) with a 3.92% relative importance value.

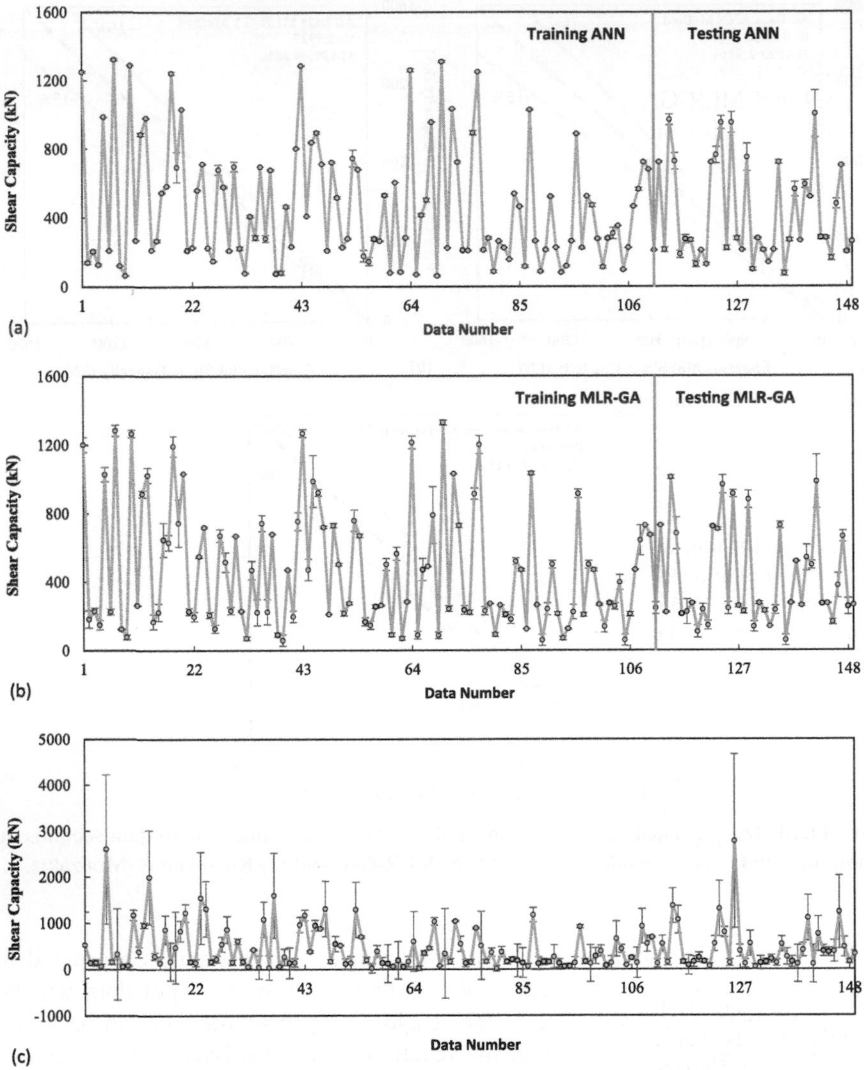

Figure 14.11 The shear capacity values for all input data as well as the corresponding error bars in models: (a) ANN; (b) MLR-GA; and (c) Kotsovou & Mouzakis.

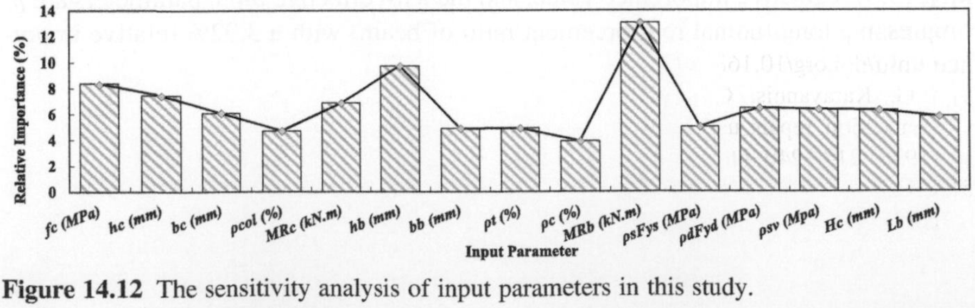

Figure 14.12 The sensitivity analysis of input parameters in this study.

14.7 Conclusions

In this chapter, the shear capacity of RC beam-column joints was estimated by proposing ANN and MLR-GA models. To achieve this goal, 149 experimental data were compiled from previous publications, and the performance of one of the existing models (Kotsovou & Mouzakis) to predict the shear capacity was compared with the obtained results of the proposed ANN and MLR-GA models. At the end, the sensitivity analysis was done to find out the effect of each input parameter on the shear capacity of the RC beam-column joints as the output. The following outcomes can be summarized in this study:

- The good correlation between the input parameters and the output showed the 149 experimental databases were selected properly and could lead to a suitable estimation of the shear capacity of RC beam-column joints.
- The results of parametric analysis for the MLR-GA and ANN models showed that the descending and ascending orders of shear capacity of the RC beam-column joints with respect to each individual input are compatible with their rational expectations.
- The optimal structure of the proposed ANN model with an input layer, a single hidden layer containing 1–20 neurons, and an output layer was selecting 7 neurons in the hidden layer with R values of 0.999 and 0.993 and MAPE values of 1.736% and 4.442% in the training and testing data sections.
- The investigations showed that the proposed ANN model with R and MAPE values of 0.9982% and 2.99% outperforms the proposed MLR-GA model (with R and MAPE values of 0.9927% and 9.06%, respectively) and the existing Kotsovou & Mouzakis method (with R and MAPE values of 0.6647% and 47.61%, respectively).
- The sensitivity analysis results illustrated that the most influential input parameters on the shear capacity of RC beam-column joints are M_{Rb} (flexural capacity of beam) with a 13.07% relative importance value, and the least effective input parameters are ρ_c (compressive longitudinal reinforcement ratio of beam) with a 3.92% relative importance value.

References

[1] L.M. Megget, Cyclic behaviour of exterior reinforced concrete beam-column joints, Bulletin of the New Zealand Society for Earthquake Engineering 7 (1974) 27–47, https://doi.org/10.5459/bnzsee.7.1.27-47.

[2] S.M. Uzumeri, Strength and ductility of cast-in-place beam-column joints, in: The American Concrete Institute Annual Convention, Symposium on Reinforced Concrete Structures in Seismic Zones, San Francisco, USA, 1974.

[3] G.-J. Ha, J.-K. Kim, L. Chung, Response of reinforced high-strength concrete beam–column joints under load reversals, Magazine of Concrete Research 44 (1992) 175–184, https://doi.org/10.1680/macr.1992.44.160.175.

[4] C.G. Karayannis, C.E. Chalioris, K.K. Sideris, Effectiveness of RC beam-column connection repair using epoxy resin injections, Journal of Earthquake Engineering 2 (1998) 217–240, https://doi.org/10.1142/S1363246998000101.

[5] C.C. Chen, G.K. Chen, Cyclic behavior of reinforced concrete eccentric beam-column corner joints connecting spread-ended beams, Structural Journal 96 (1999) 443–449.

[6] S. Hakuto, R. Park, H. Tanaka, Seismic load tests on interior and exterior beam-column joints with substandard reinforcing details, Structural Journal 97 (2000) 11−25.

[7] C.P. Pantelides, C. Clyde, L.D. Reaveley, Performance-based evaluation of reinforced concrete building exterior joints for seismic excitation, Earthquake Spectra 18 (2002) 449−480, https://doi.org/10.1193/1.1510447.

[8] M. Gebman, Application of Steel Fiber Reinforced Concrete in Seismic Beam-Column Joints, San Diego State University, 2001.

[9] T. El-Amoury, A. Ghobarah, Seismic rehabilitation of beam−column joint using GFRP sheets, Engineering Structures 24 (2002) 1397−1407, https://doi.org/10.1016/S0141-0296(02)00081-0.

[10] C.P. Pantelides, J. Hansen, J.U.S.T.I.N. Nadauld, L.D. Reaveley, Assessment of Reinforced Concrete Building Exterior Joints with Substandard Details, 2002.

[11] C.P. Antonopoulos, T.C. Triantafillou, Experimental investigation of FRP-strengthened RC beam-column joints, Journal of Composites for Construction 7 (2003) 39−49, https://doi.org/10.1061/(ASCE)1090-0268(2003)7:1(39).

[12] A.E.-N. Atta, S.E.-D.F. Taher, A.H. Khalil, S.E.-D. El-Metwally, Behaviour of reinforced high-strength concrete beam—column joint. Part 1: experimental investigation, Structural Concrete 4 (2003) 175−183.

[13] T. Paulay, A. Scarpas, The behaviour of exterior beam-column joints, Bulletin of the New Zealand Society for Earthquake Engineering 14 (1981) 131−144, https://doi.org/10.5459/bnzsee.14.3.131-144.

[14] N. Chutarat, R.S. Aboutaha, Cyclic response of exterior reinforced concrete beam-column joints reinforced with headed bars—experimental investigation, Structural Journal 100 (2003) 259−264.

[15] S.J. Hwang, L. Hung-Jen, T.F. Liao, W. Kuo-Chou, H.H. Tsai, Role of hoops on shear strength of reinforced concrete beam-column joints, ACI Structural Journal 102 (2005) 445−459.

[16] H. Shiohara, F. Kusuhara, Benchmark test for validation of mathematical models for nonlinear and cyclic behavior of R, C Beam-Column Joints (2006).

[17] A.G. Tsonos, Cyclic load behaviour of reinforced concrete beam-column subassemblages of modern structures, WIT Transactions on The Built Environment 81 (2005) 1−11, https://doi.org/10.2495/ERES050421.

[18] C.G. Karayannis, C.E. Chaliori, G.M. Sirkelis, Exterior RC beam-column joints with diagonal reinforcement, in: Proceedings of the 15th Concrete Congress, Hellenic Technical Chamber, Alexandroupolis, Greece, 2006, pp. 368−377.

[19] C.G. Karayannis, C.E. Chalioris, G.M. Sirkelis, Local retrofit of exterior RC beam−column joints using thin RC jackets—an experimental study, Earthquake Engineering & Structural Dynamics 37 (2008) 727−746, https://doi.org/10.1002/eqe.783.

[20] C.E. Chalioris, M.J. Favvata, C.G. Karayannis, Reinforced concrete beam−column joints with crossed inclined bars under cyclic deformations, Earthquake Engineering & Structural Dynamics 37 (2008) 881−897, https://doi.org/10.1002/eqe.793.

[21] H.F. Wong, J.S. Kuang, Effects of beam—column depth ratio on joint seismic behaviour, Proceedings of the Institution of Civil Engineers—Structures and Buildings 161 (2008) 91−101, https://doi.org/10.1680/stbu.2008.161.2.91.

[22] G.-J. Ha, C.-G. Cho, Strengthening of reinforced high-strength concrete beam−column joints using advanced reinforcement details, Magazine of Concrete Research 60 (2008) 487−497, https://doi.org/10.1680/macr.2008.60.7.487.

[23] G. Kotsovou, H. Mouzakis, Seismic behaviour of RC external joints, Magazine of Concrete Research 63 (2011) 247−264, https://doi.org/10.1680/macr.9.00194.

[24] R. Park, J.R. Milburn, Comparison of recent New Zealand and United States seismic design provisions for reinforced concrete beam-column joints and test results from four units designed according to the New Zealand code, Bulletin of the New Zealand Society for Earthquake Engineering 16 (1983) 3–24, https://doi.org/10.5459/bnzsee.16.1.3-24.
[25] G. Kotsovou, H. Mouzakis, Seismic design of RC external beam-column joints, Bulletin of Earthquake Engineering 10 (2012) 645–677, https://doi.org/10.1007/s10518-011-9303-1.
[26] G. Kotsovou, H. Mouzakis, Exterior RC beam–column joints: new design approach, Engineering Structures 41 (2012) 307–319, https://doi.org/10.1016/j.engstruct.2012.03.049.
[27] M.R. Ehsani, J.K. Wight, Exterior reinforced concrete beam-to-column connections subjected to earthquake-type loading, Journal Proceedings 82 (1985) 492–499.
[28] M.R. Ehsani, A.E. Moussa, C.R. Valenilla, Comparison of inelastic behavior of reinforced ordinary-and high-strength concrete frames, Structural Journal 84 (1987) 161–169.
[29] T. Kaku, H. Asakusa, Ductility estimation of exterior beam-column subassemblages in reinforced concrete frames, Special Publication 123 (1991) 167–186.
[30] S. Fujii, S. Morita, Comparison between interior and exterior R/C beam-column joint behavior, Special Publication 123 (1991) 145–166.
[31] F. Alameddine, M.R. Ehsani, High-strength RC connections subjected to inelastic cyclic loading, Journal of Structural Engineering 117 (1991) 829–850, https://doi.org/10.1061/(ASCE)0733-9445(1991)117:3(829).
[32] Y. Kurose, Recent Studies on Reinforced Concrete Beam-Column Joints in Japan, University of Texas at Austin., 1987.
[33] G.M. Kotsovou, D.M. Cotsovos, N.D. Lagaros, Assessment of RC exterior beam-column Joints based on artificial neural networks and other methods, Engineering Structures 144 (2017) 1–18, https://doi.org/10.1016/j.engstruct.2017.04.048.
[34] D. Rezazadeh Eidgahee, H. Jahangir, N. Solatifar, P. Fakharian, M. Rezaeemanesh, Data-driven estimation models of asphalt mixtures dynamic modulus using ANN, GP and combinatorial GMDH approaches, Neural Computing and Applications (2022), https://doi.org/10.1007/s00521-022-07382-3.
[35] L. Milne, Feature selection using neural networks with contribution measures, in: AI-Conference, World Scientific Publishing, 1995, p. 571.

Global seismic damage assessment of RC framed buildings using machine learning techniques

15

V. Vasugi [1], M. Helen Santhi [1] and G. Malathi [2]
[1]School of Civil Engineering, Vellore Institute of Technology, Chennai, Tamil Nadu, India; [2]School of Computer Science and Engineering, Vellore Institute of Technology, Chennai, Tamil Nadu, India

15.1 Introduction

Reinforced concrete (RC) multistory building structures are becoming popular in this modern era and are designed and constructed based on the requirements of the owners/clients. Most of these structures are deficient in terms of strength and ductility, and many are under seismic threat [1]. The seismic damage assessment of existing RC multistory building structures is helpful in finding the probable damage levels and upgrading their performance during future earthquakes [2]. The seismic damage assessment of structures can be carried out by determining local and measures of overall damage [3,4]. Based on observational methods for evaluating building damage, the damage states divide structural damage into categories ranging from undamaged to completely damaged (collapsed) states. Damage states help to comprehend the building's postearthquake status. The same assessment methods could be adopted for the existing structures as well. For the assessment of earthquake damage to buildings [5], utilized deformation-based and nonaccruing criteria, whereas [6] used deformation-based cumulative parameters. Exponential and hyperbolic functions of energy terms were used by Refs. [7−11]. The damage indices were created utilizing structural response parameters that were determined by an analytical analysis of structural reactions. Overall damage to the structures from structural analysis was documented by Refs. [12−23].

Artificial intelligence (AI) has been widely used nowadays for the rapid method of seismic damage assessment of large stocks of existing structures and damage levels after postearthquake events [24−27]. Also, when evaluating engineering design parameters, AI-based solutions are viable alternatives to testing that significantly reduce human efforts. Machine learning (ML) algorithms in the previous research ensured that they are very effective in seismic risk assessment of buildings using large datasets obtained from either experimental or numerical analysis of buildings [28−30].

Artificial Intelligence Applications for Sustainable Construction. https://doi.org/10.1016/B978-0-443-13191-2.00011-0
Copyright © 2024 Elsevier Ltd. All rights are reserved, including those for text and data mining, AI training, and similar technologies.

This paper explains the method of estimating the damage levels of a typical RC structure with the help of dynamic responses such as base shear and drift. Also, simple ML algorithms are developed for the same. The goal of the investigation is to identify the seismic risk of the RC building considered under study from its capacity and demand characteristics and the requirements for a safe and economical design to resist seismic load using ML algorithms.

15.2 Methodology

The methodology followed in the current study is shown in Fig. 15.1. A typical G+9-story RC building is modeled and analyzed under earthquake loading using ETABS software. For the time history analysis, El Centro earthquake data is considered and scaled for 0.1−1.0 g PGA using Seismosoft software. Based on base shear and drift responses, the seismic damage level of the RC structure is evaluated.

15.3 Modeling and analysis of a building

In this study, a G+9-story 3-bay by 2-bay RC building is considered and modeled in ETABS software. The beams and columns are defined as frame elements, and the slab is defined as a shell element. The load of the brick masonry wall is assigned to the beams. The base of the RC building is fixed, and the diaphragm action is defined at

Figure 15.1 Methodology.

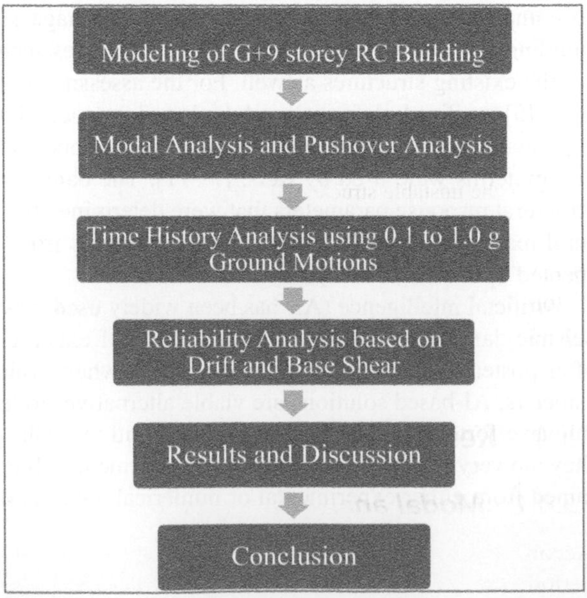

Table 15.1 Structure parameters.

Sl.No.	Description	Values
1	Materials	M25 concrete, Fe415 steel
2	No. of storeys	G+9
3	Plan size	12 × 8 m
4	Beam size	300 × 600 mm
5	Column size	700 × 700 mm
6	Height of floor	3.0 m
7	Thickness of slab	150 mm
8	Types of analysis	Pushover analysis, time history analysis
9	Z	0.36
10	R	5
11	I	1.2
12	Soil	Type II Soil

each floor level. The beams and columns are rigidly connected. Beams and columns are given the proper hinges. The system parameters are given in Table 15.1 and the plan and 3-D view of the RC building are shown in Fig. 15.2a and b. The live load considered is 1.5 and 3 kN/m^2 for the terrace and other floors, respectively. Modal analysis is performed on the building, and dynamic characteristics such as time period and mode shapes are found.

The pushover information on study provides the different performance levels of buildings. Building performance levels that are commonly used are shown below:

i. Operational: This performance level gives an idea that the building can be used after the earthquake event since the deformations in the buildings are minor.
ii. Immediate occupancy: There is very limited structural damage, and damage to life is negligible. The components retain almost all preearthquake characteristics.
iii. Life safety: This level ensures the life safety of the occupants from structural or nonstructural building component damage.
iv. Structural stability: This stage shows the partial or total collapse of the building, which reveals the unstable structural elements.

El-Centro earthquake data with a maximum PGA of 0.3172 g (Fig. 15.3) is scaled to have ground motions from 0.1 to 1.0 g, as shown in Fig. 15.4a–j. From the time history analysis, the drift and base shear values of the RC building are observed under 0.1–1.0 g ground motions.

15.4 Results and discussion

15.4.1 Modal analysis results

According to the results of the modal analysis, the RC structure's fundamental time period is 0.7158 s, and Fig. 15.5 displays the first three mode shapes.

Figure 15.2 Details of G+9-story RC building.

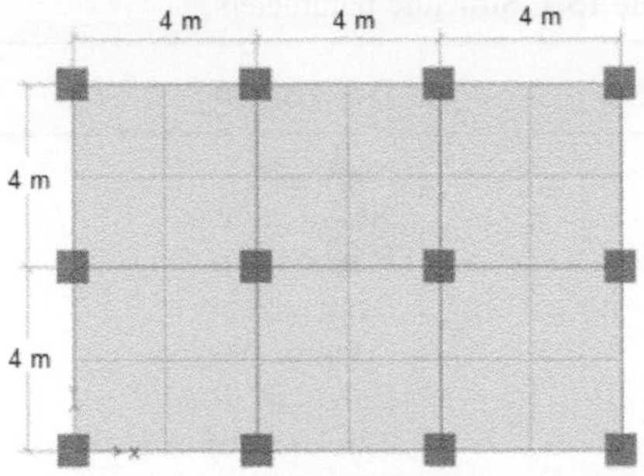

(a) Plan

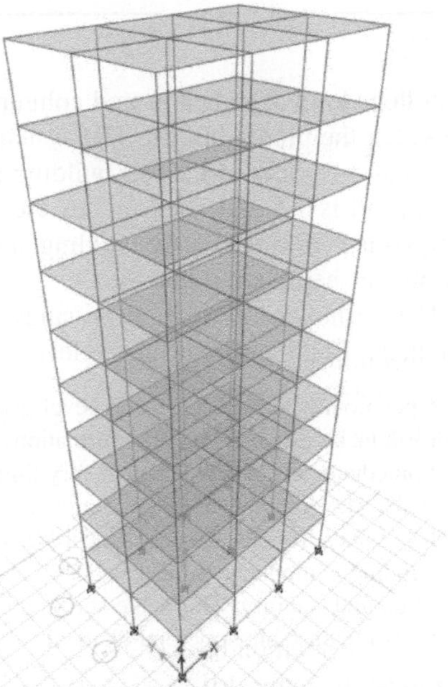

(b) 3-D View

15.4.2 Pushover analysis results

The pushover curve is obtained from the pushover analysis, as is the number of hinges formed at different performance levels. From Table 15.2 it is observed that at the final step, around 75% of the total hinges are formed in the elastic range, 24% in the immediate-to-life safety range, and 0.5% in the beyond-collapse prevention range. There are three critical hinges beyond the CP performance state that are not desirable for seismic resistance (Fig. 15.6). It is suggested to strengthen the frames on the ground

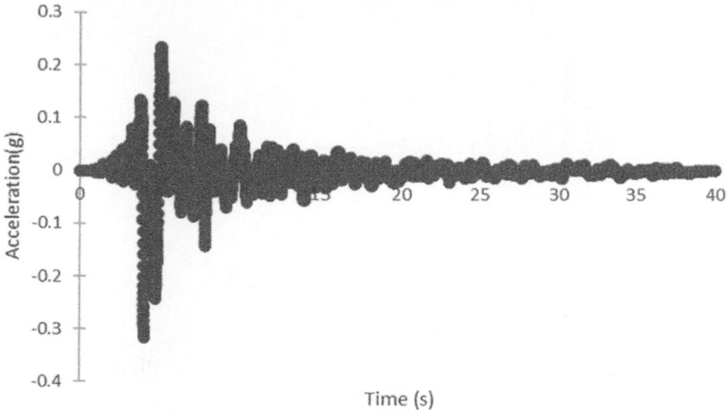

Figure 15.3 El-Centro earthquake.

floor and the first floor to resist the higher seismic load. The capacity curve and bilinear curve of the RC building are shown in Fig. 15.7. Early yielding of the frame is observed, with a displacement ductility of almost 8. The capacity versus seismic demand is given in Fig. 15.8, which gives the desired spectral acceleration and spectral displacement for the RC building under study.

15.5 Analysis of seismic reliability

The recommendations for a building's safety against earthquakes are provided by the seismic reliability study of buildings. In this paper, the Weibull reliability approach is adopted to assess the performance levels, such as IO, LS, and CP, at 90%, 50%, and 20%, respectively. Based on the reliability analysis, the base shear of the RC building at IO, LS, and CP is 1500, 8800, and 19,500 kN, respectively, as shown in Fig. 15.9. If the RC building under study is to be designed for the performance level of IO, then the base shear of 1500 kN is to be considered. However, this consideration is not economical. For the economical and safe design of the RC building, the next performance level, i.e., LS with a base shear of 8800 kN, could be considered.

The maximum drift limitations listed below in Table 15.3 are taken from performance-based seismic codes, and they may be of interest.

Fig. 15.10 shows the reliability index versus drift variation of the RC building for the ground motions of 0.1–1.0 g PGA. The drift limit of 0.002 may be considered equivalent to the IO performance state. According to the Indian code IS 1893:2016, the maximum drift allowed is 0.004, and the corresponding performance state is considered LS. Therefore, up to the drift limit of 0.004 is taken as the operational condition.

For the seismic design of RC structures both base shear and drift are important parameters to be considered. Based on the base shear, drift levels, and performance states, it is concluded that the RC building under study can sustain an earthquake of PGA 0.4 g.

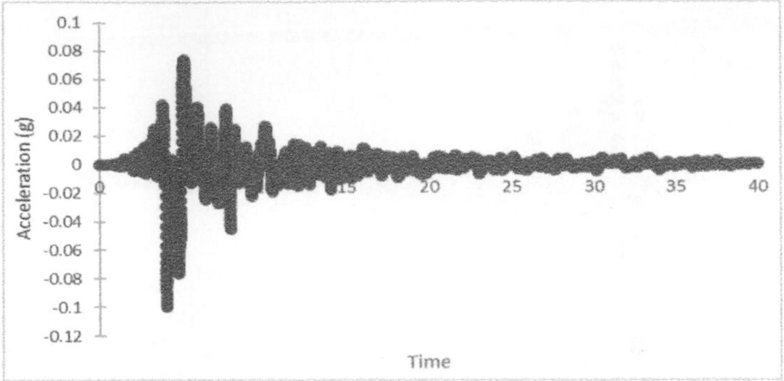

(a) 0.1g

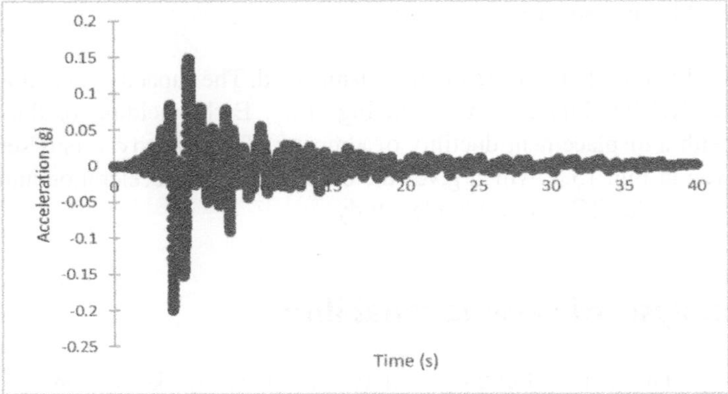

(b) 0.2g

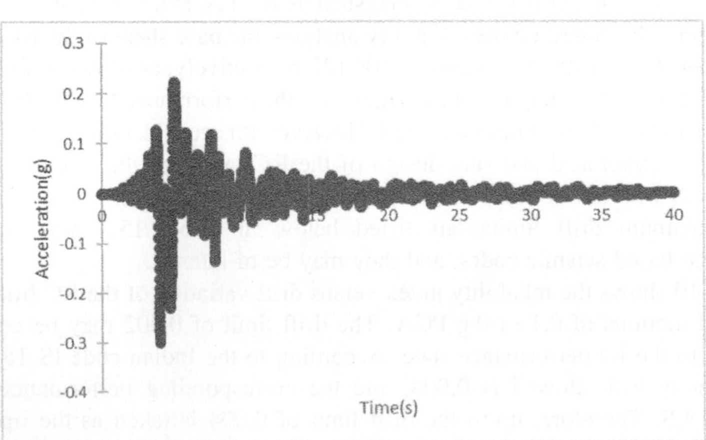

© 0.3g

Figure 15.4 Scaled ground motions.

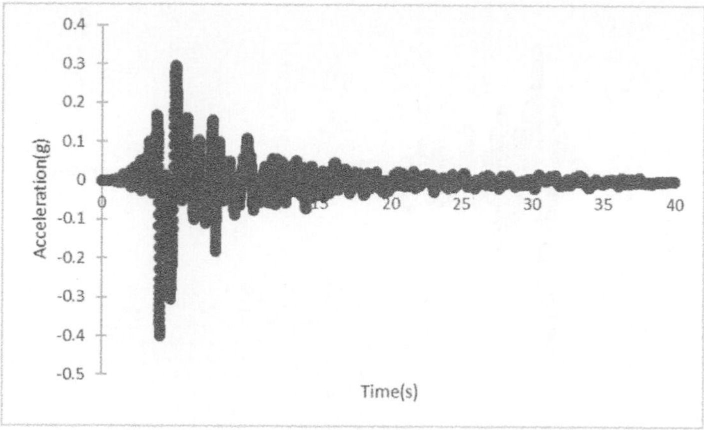

(d) 0.4g

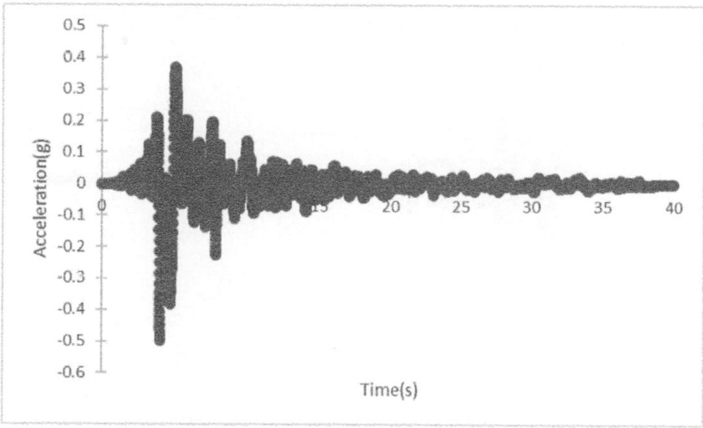

(e) 0.5g

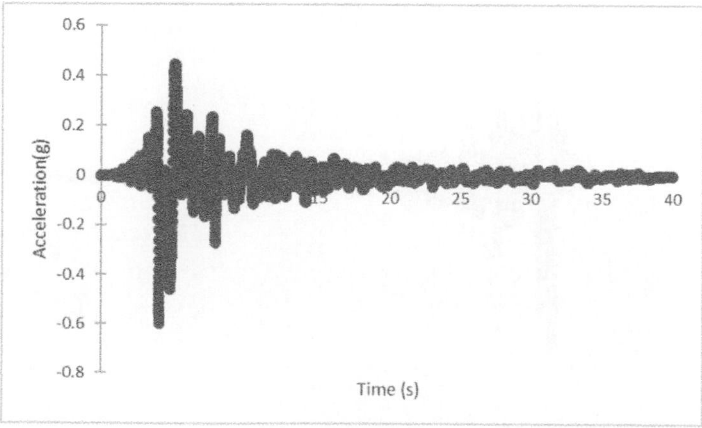

(f) 0.6g

Figure 15.4 cont'd.

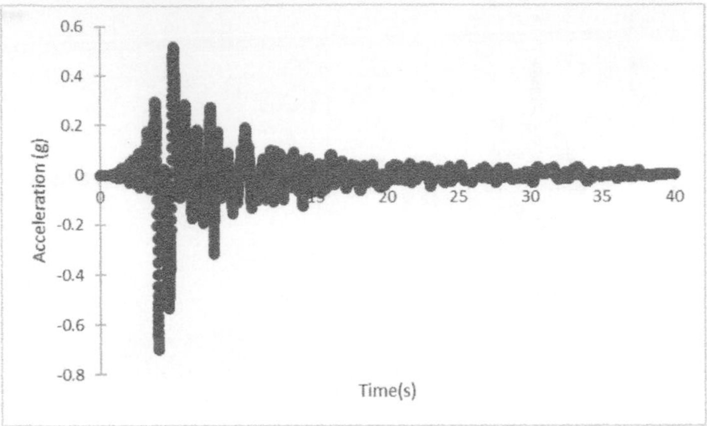

(g) 0.7g

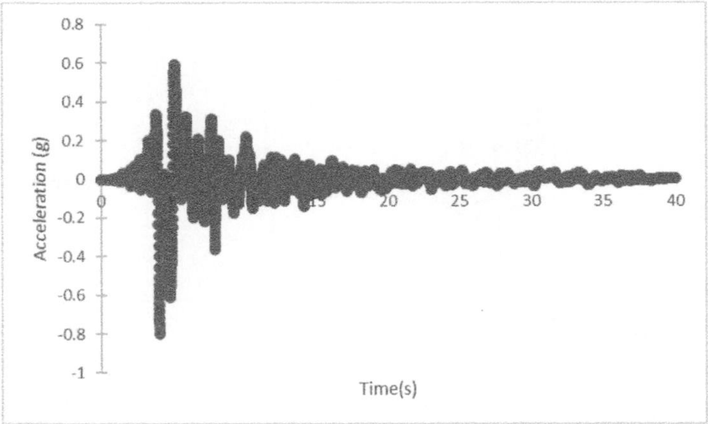

(h) 0.8g

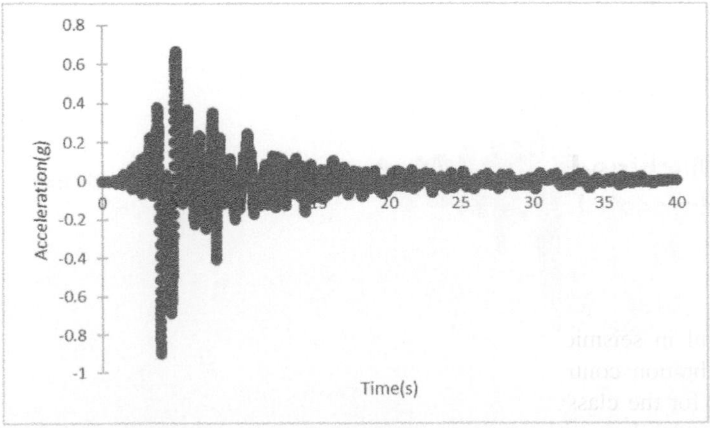

(i) 0.9g

Figure 15.4 cont'd.

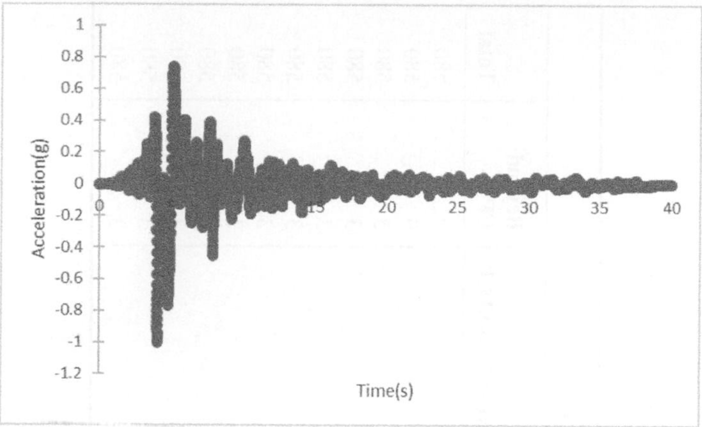

(j) 1.0g

Figure 15.4 cont'd.

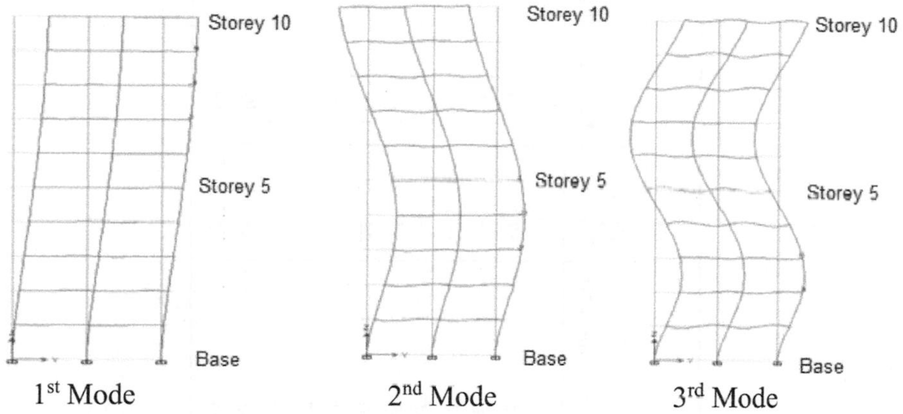

Figure 15.5 Mode shapes of the RC building.

15.6 Machine learning algorithms for classification of damage level

ML techniques have been used in the earthquake engineering field in the recent past because of the difficulty in handling a large amount of seismic data sets manually. It is helpful in seismic damage detection, classification of seismic damage levels, seismic vibration control, and mitigation. In this investigation, ML algorithms are developed for the classification of seismic damage states of the typical RC building based on reliability index and damage assessment of the building based on drift, base shear, and reliability index. The same algorithm could be extended to other types of buildings as well.

Table 15.2 Results of pushover analysis.

Stage	Top displacement (mm)	Shear at bottom (kN)	No. of hinges				
			Upto IO	Between IO and LS	Between LS and CP	Beyond CP	Total
0	0	0	580	0	0	0	580
1	14.867	1108.874	580	0	0	0	580
2	135.019	4759.446	453	124	0	3	580
3	149.162	5042.274	444	133	0	3	580
4	149.205	5042.227	444	133	0	3	580
5	152.025	5096.005	440	137	0	3	580
6	152.110	5096.956	440	137	0	3	580
7	153.478	5122.923	438	139	0	3	580
8	153.520	5123.277	438	139	0	3	580
9	153.691	5126.913	438	139	0	3	580
10	153.777	5127.543	438	139	0	3	580
11	154.119	5133.755	438	139	0	3	580

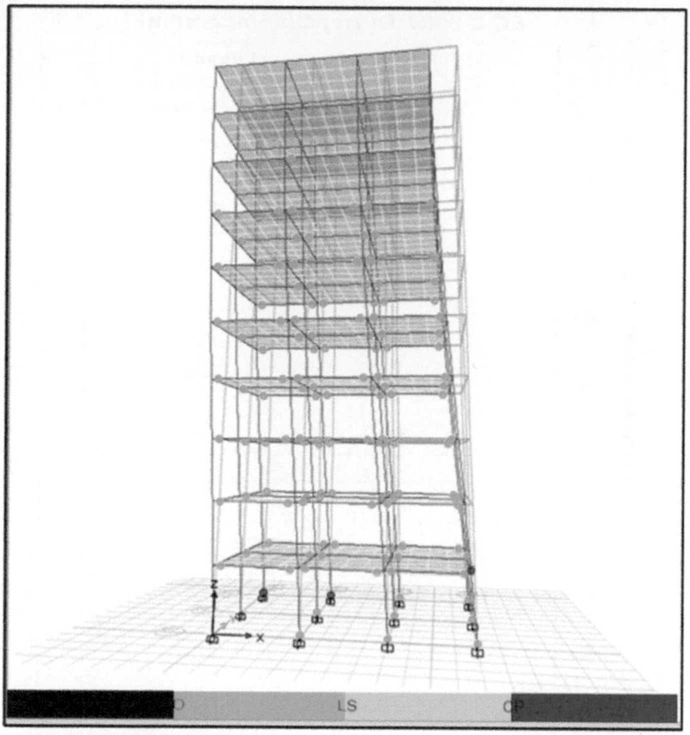

Figure 15.6 Hinges at the final step.

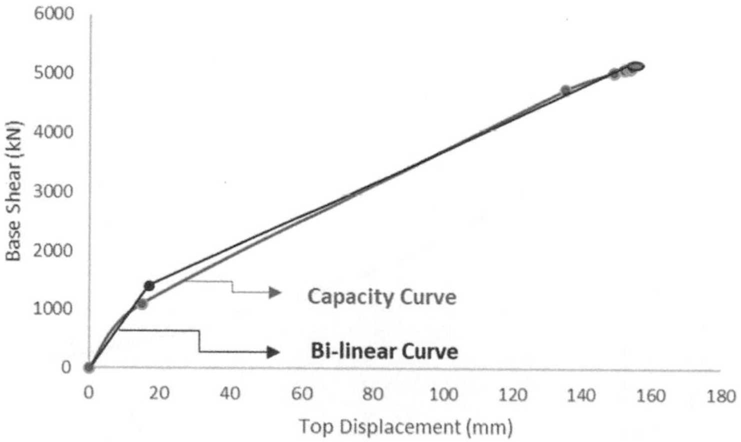

Figure 15.7 Capacity curve and bilinear curve.

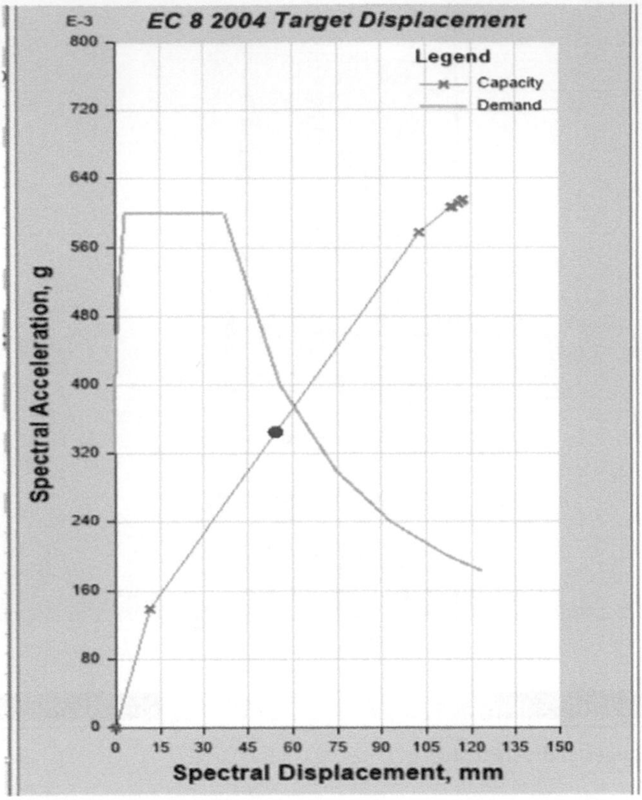

Figure 15.8 Typical capacity versus demand behavior of the RC building.

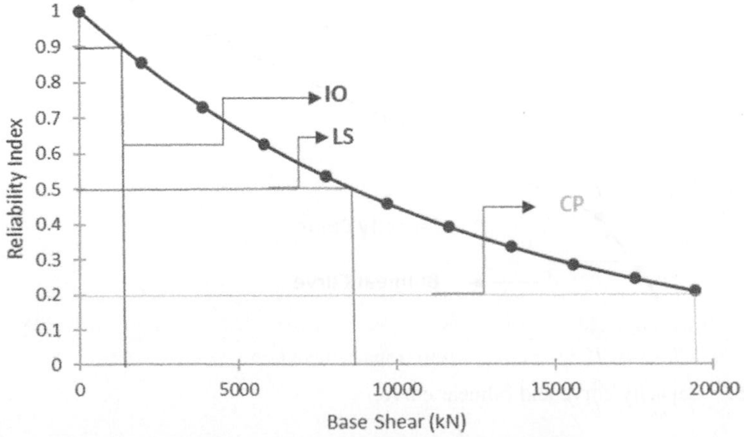

Figure 15.9 Reliability index versus base shear of the RC building.

Table 15.3 Performance state based on drift.

Sl.No.	Drift	Performance state
1	0.002	Fully operational
2	0.005	Operational
3	0.015	Life safety
4	0.025	Near collapse

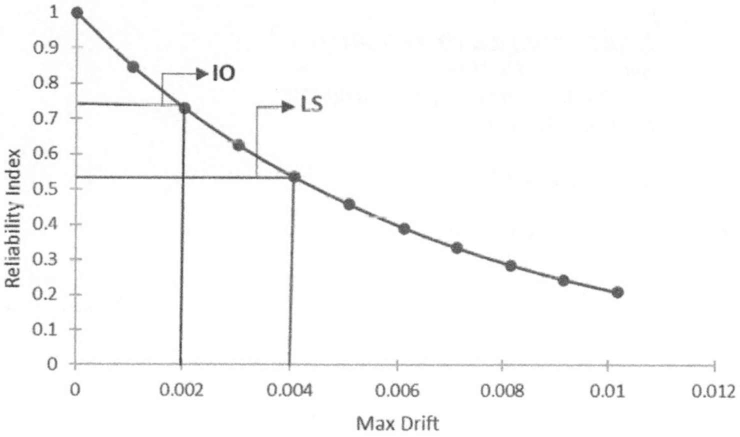

Figure 15.10 Reliability index versus drift of the RC building.

Algorithm 15.1. Classification of damage level of building based on reliability index

Input Variables:
RI as Reliability Index.

Output Variables.
IO, LS, CP, C refer to the classification of damage level referred as Class

Steps:
If (RI ==0.1)
Set Class as 'C'
If ((RI ≥ 0.2) && (RI < 0.5))
Set Class as 'CP'
If ((RI ≥ 0.5) && (RI <0.6))
Set Class as 'LS'
If ((RI ≥ 0.6) && (RI <0.8))
Set Class as 'IO to LS'
If ((RI ≥ 0.8) && (RI ≤1))
Set Class as 'IO'
End

Algorithm 15.2. Damage assessment of building based on drift, base shear and reliability index

Input Variables:
RI as Reliability Index;D as Drift; BS as Base Shear in kN

Output Variables.
IO, LS, CP, C refer to the classification of damage level referred as Class

Steps:
```
If ((RI ≥ 0.5) && (RI ≤ 1))
        {
        If ((BS ≤ 8000) && (D == 0.0041))
        Set Class as 'LS'1888
        Else If ((BS ≤ 8000) && (D < 0.0041))
        Set Class as 'IO to LS'
        Else
        Set Class as 'CP'
        }
ElseIf ((RI < 0.5) && (RI ≥ 0.2)
        Set Class as 'CP'
Else
Set Class as 'C'
End
```

15.7 Conclusions

The seismic damage states of a G+9-story RC building are assessed using a numerical study, and reliability analysis is performed to obtain the desired base shear for the design.

- The results from the pushover analysis underestimate the shear capacity and displacement of the RC building; therefore, the design based on this approach may not be economical.
- The RC building's time history study under various PGAs (0.1–1.0 g) provides adequate estimates of base shear and drift, and it is determined that the RC building is secure up to PGAs of 0.4 g.
- For the safe and cost-effective design of the buildings, the design engineers would find the seismic reliability analysis of RC buildings based on the base shear and drift under 0.1–1.0 g PGAs useful.
- ML algorithms have proven to be effective in assessing the seismic damage states of buildings under low to high earthquake ground motions.

The current study focuses on the seismic damage assessment of a typical RC multi-story building using numerical and ML algorithms. The same methodology can be extended for evaluating the damage levels due to earthquakes in large stocks of existing buildings, thereby taking remedial measures to mitigate earthquake effects.

References

[1] Y. Lu, Comparative study of seismic behaviour of multi-storey reinforced concrete framed structures, Journal of Structural Engineering, ASCE 128 (2) (2001) 169−178.

[2] R. Sinha, A. Goyal, A National Policy of Seismic Vulnerability Assessment of Buildings and Procedure for Rapid Visual Screening of Buildings for Potential Seismic Vulnerability, Department of Civil Engineering, IIT Bombay, 2004.

[3] E. DiPasquale, J.-W. Ju, A. Askar, A.S. Cakmak, Relation between global damage indices and local stiffness degradation, Journal of Structural Engineering 116 (5) (1990) 1440−1456.

[4] T.-H. Kim, K.-M. Lee, Y.-S. Chung, H.M. Shin, Seismic damage assessment of reinforced concrete bridge columns, Engineering Structures 27 (2005) 576−592.

[5] G.H. Powell, R. Allahabadi, Seismic damage prediction by deterministic methods: concept and procedure, Earthquake Engineering & Structural Dynamics 16 (5) (1988) 719−734.

[6] S. Mehanny, G. Deierlein, Seismic damage and collapse assessment of composite moment frames, Journal of Structural Engineering, ASCE 127 (9) (2001) 1045−1053.

[7] A. Colombo, P. Negro, A damage index of generalised applicability, Engineering Structures 27 (8) (2005) 1164−1174.

[8] E. DiPasquale, A.S. Cakmak, On the Relation between Local and Global Damage Indices, Technical Report NCEER-89-0034, State University of New York at Buffalo, 1989.

[9] A. Massumi, E. Moshtagh, A new damage index for RC buildings based on variations of nonlinear fundamental period, The Structural Design of Tall and Special Buildings 21 (1) (2010) 50−61.

[10] Y.J. Park, A.H.S. Ang, Y.K. Wen, Seismic damage analysis of reinforced concrete buildings, Journal of Structural Engineering, ASCE 111 (4) (1985) 740−757.

[11] M.E. Rodriguez, D. Padilla, A damage index for the seismic analysis of reinforced concrete members, Journal of Earthquake Engineering 13 (3) (2009) 364−383.

[12] V.V. Cao, H. Ronagh, M. Ashraf, H. Baji, A new damage index for reinforced concrete structures, Earthquakes and Structures 6 (6) (2014) 581−609.

[13] E. Cosenza, G. Manfredi, Damage indices and damage measures, Progress in Structural Engineering and Materials 2 (1) (2000) 50−59.

[14] M.A. Erberik, Fragility-based assessment of typical mid-rise and low-rise RC buildings in Turkey, Engineering Structures 30 (5) (2008) 1360−1374.

[15] A. Ghobarah, H. Abou-Elfath, A. Biddah, Response based damage assessment of structures, Earthquake Engineering & Structural Dynamics 28 (1) (1999) 29−104.

[16] J. Hancock, J.J. Bommer, P.J. Sttaford, Numbers of scaled and matched accelerograms required for inelastic dynamic analyses, Earthquake Engineering & Structural Dynamics 37 (14) (2008) 1585−1607.

[17] P. Lamego, P.B. Lourenço, M.L. Sousa, R. Marques, Seismic vulnerability and risk analysis of the old building stock at urban scale: application to a neighbourhood in Lisbon, Bulletin of Earthquake Engineering 15 (2017) 2901−2937.

[18] J. Ortega, G. Vasconcelos, H. Rodrigues, M. Correia, Assessment of the influence of horizontal diaphragms on the seismic performance of vernacular buildings, Bulletin of Earthquake Engineering 16 (2018) 3871−3904.

[19] M.J.N. Priestley, G.M. Calvi, M.J. Kowalsky, Displacement Based Seismic Design of Structures, IUSS Press, Pavia, 2007.

[20] M.B. Sørensen, D.H. Lang, Incorporating simulated ground motion in seismic risk assessment-application to the Lower Indian Himalayas, Earthquake Spectra 31 (1) (2014) 71−95.

[21] D. Vamvatsikos, C.A. Cornell, The incremental dynamic analysis, Earthquake Engineering & Structural Dynamics 31 (3) (2002) 491−514.
[22] M.S. Williams, R.G. Sexsmith, Seismic damage indices for concrete structures: a State-of-the-Art Review, Earthquake Spectra 11 (2) (1995) 740−757.
[23] Y.F. Vargas, L.G. Pujades, A.H. Barbat, J.E. Hurtado, Probabilistic seismic damage assessment of RC buildings based on nonlinear dynamic analysis, The Open Civil Engineering Journal 9 (Suppl. 1, M 12) (2015) 344−350.
[24] J. Bialas, T. Oommen, U. Rebbapragada, E. Levin, Object-based classification of earthquake damage from high-resolution optical imagery using machine learning, Journal of Applied Remote Sensing 10 (3) (2016) 1−16.
[25] T. Kim, J. Song, O.S. Kwon, Pre- and post-earthquake regional loss assessment using deep learning, Earthquake Engineering & Structural Dynamics 49 (2020) 657−678.
[26] S. Mangalathu, H. Sun, C.C. Nweke, Z. Yi, H.V. Burton, Classifying earthquake damage to buildings using machine learning, Earthquake Spectra 36 (1) (2020) 183−208.
[27] Y. Zhang, H.V. Burton, H. Sun, M. Shokrabadi, A machine-learning framework for assessing post-earthquake structural safety, Structural Safety 72 (2018) 1−16.
[28] N.W. Chi, J.P. Wang, J.H. Liao, et al., Machine learning-based seismic capability evaluation for school buildings, Automation in Construction 118 (2020) 103274.
[29] E. Harirchian, V. Kumari, K. Jadhav, S. Rasulzade, T. Lahmer, R. Raj Das, A synthesized study based on machine learning approaches for rapid classifying earthquake damage grades to RC buildings, Applied Sciences 11 (16) (2021) 7540.
[30] J. Hegde, B. Rokseth, Applications of machine learning methods for engineering risk assessment—a review, Safety Science 122 (2020) 104492.

Further reading

[1] ETABS® Version 18.1.1 © 2020 Computers and Structures, Inc.
[2] H. Salehi, R. Burgueño, Emerging artificial intelligence methods in structural engineering, Engineering Structures 171 (2018) 170−189.
[3] Seismosoft, Seismosoft Earthquake Engineering Software Solutions, 2002.

Index

Note: 'Page numbers followed by *f* indicate figures and *t* indicate tables.'

Printed and bound by CPI Group (UK) Ltd, Croydon, CR0 4YY
03/10/2024
01040847-0009